MW00700719

SAXON MATH™

6/5

Student Edition

SAXON MATH™ 6/5

Stephen Hake
John Saxon

SAXON™
PUBLISHERS

Saxon Publishers gratefully acknowledges the contributions of the following individuals in the completion of this project:

Authors: Stephen Hake, John Saxon

Consultants: Diane Blank, Shirley McQuade Davis

Editorial: Chris Braun, Brian E. Rice, Mary Burleson, Sherri Little, Rodney Clint Keele, Bo Björn Johnson, Brian Smith, Brooke Butner, Andrew Kershen, Dana Nixon, Sean G. Douglas

Editorial Support Services: Christopher Davey, Jay Allman, Jean Van Vleck, Shelley Turner, Darlene C. Terry

Production: Adriana Maxwell, Karen Hammond, Brenda Lopez, Debra Sullivan, Diane Readnour, Donna Jarrel, Ryan LaCroix, Alicia Britt, Nancy Rimassa, Cristi D. Whiddon

Project Management: Angela Johnson, Becky Cavnar

Printed in the United States of America

ISBN: 1-56577-505-8

Manufacturing Code: 03S0504

CONTENTS

LETTER FROM AUTHOR STEPHEN HAKE

Dear Student,

We study mathematics because of its importance to our lives. Our school schedule, our trip to the store, the preparation of our meals, and many of the games we play involve mathematics. You will find that the word problems in this book are often drawn from everyday experiences.

As you grow into adulthood, mathematics will become even more important. In fact, your future in the adult world may depend on the mathematics you have learned. This book was written to help you learn mathematics and to learn it well. For this to happen, you must use the book properly. As you work through the pages, you will see that similar problems are presented over and over again. ***Solving each problem day after day is the secret to success.***

Your book is made up of daily lessons and investigations. Each lesson has four parts. The first part is a Warm-Up that includes practice of basic facts and mental math. These exercises improve your speed, accuracy, and ability to do math "in your head." The Warm-Up also includes a problem-solving exercise to familiarize you with strategies for solving complicated problems. The second part of the lesson is the New Concept. This section introduces a new mathematical concept and presents examples that use the concept. In the next section, the Lesson Practice, you have a chance to solve problems involving the new concept. The problems are lettered a, b, c, and so on. The final part of the lesson is the Mixed Practice. This problem set reviews previously taught concepts and prepares you for concepts that will be taught in later lessons. Solving these problems helps you remember skills and concepts for a long time.

Investigations are variations of the daily lesson. The investigations in this book often involve activities that fill an entire class period. Investigations contain their own set of questions instead of a problem set.

Remember, solve every problem in every practice set, every problem set, and every investigation. Do not skip problems. With honest effort, you will experience success and true learning that will stay with you and serve you well in the future.

Stephen Hake
Temple City, California

ACKNOWLEDGMENTS

I wish to acknowledge the following contributors to the revision of *Saxon Math 5/4–8/7:*

Barbara Place, who conceived the project.

Dr. Gerald Beer, who provided lesson content and problems on topics of pattern recognition, functions, scale drawings and models, probability, data analysis, and problem solving.

Shirley McQuade Davis, for her ideas on teaching story-problem thinking patterns.

Brian Rice, for his conceptual work on story-problem lessons and for his editorial work on the *Assessments and Classroom Masters.*

Dan Shippey, who designed the Mercury Freedom 7 scale model for *Saxon Math 7/6,* Investigation 11.

Mary Burleson, who scheduled the project and coordinated efforts among the many participants.

Adriana Maxwell, who coordinated the production of the manuscripts.

Diane Blank, for her extensive and thorough analysis of state standards and numerous helpful suggestions for addressing those standards.

Nancy Crisler, for her contributions to the Teacher's Manuals.

Nancy Larson, for her generous help and insightful suggestions for the Teacher's Manuals.

Chris Braun, whose invaluable contributions as senior editor span the contents of the series.

John Saxon, whose unwavering focus on student success continues to inspire and guide.

Mary Hake, for her support, encouragement, and patience.

Stephen Hake
Temple City, California

LESSON

1 Sequences • Digits

WARM-UP

Facts Practice: 100 Addition Facts (Test A)†

Mental Math: Count by tens from 10 to 100. Count by hundreds from 100 to 1000.

a. 3 + 3 **b.** 30 + 30 **c.** 300 + 300

d. 40 + 50 **e.** 200 + 600 **f.** 50 + 50

g. 20 + 20 + 20 **h.** 500 + 500 + 500

Problem Solving:

The counting numbers are 1, 2, 3, 4, and so on. How many one-digit counting numbers are there?

NEW CONCEPTS

Sequences Counting is a math skill that we learn early in life. Counting by ones, we say the numbers

1, 2, 3, 4, 5, 6, ...

These numbers are called **counting numbers.** We can also count by a number other than one. Below we show the first five numbers for counting by twos and the first five numbers for counting by fives.

2, 4, 6, 8, 10, ...

5, 10, 15, 20, 25, ...

An ordered list of numbers forms a **sequence.** Each member of the sequence is a **term.** The three dots mean that the sequence continues even though the numbers are not written. We can study a sequence to discover its counting pattern, or rule. The rule can be used to find more terms in the sequence.

Example 1 What are the next three terms in this counting sequence?

3, 6, 9, 12, ____, ____, ____, ...

†For instructions on how to use the Warm-up activities, please consult the preface.

Solution The pattern is "Count up by threes." To find the next three terms, we may count up by threes, or we may count up by ones and emphasize every third term (one, two, *three,* four, five, *six,* ...). Either way, we find that the next three terms are **15, 18,** and **21.**

Example 2 Describe the rule for this counting sequence. What is the next term in the sequence?

56, 49, 42, _____, ...

Solution This sequence counts down. We find that the rule for this sequence is **"Count down by sevens."** Counting down by seven from 42 gives us **35.**

Digits There are ten **digits** in our number system. They are 0, 1, 2, 3, 4, 5, 6, 7, 8, and 9. The number 385 has three digits, and the last digit is 5. The number 148,567,896,094 has twelve digits, and the last digit is 4.

Example 3 The number 186,000 has how many digits?

Solution The number 186,000 has **six digits.**

Example 4 What is the last digit of 26,348?

Solution The number 26,348 has five digits. The last digit is **8.**

LESSON PRACTICE

Practice set Describe the rule for each counting sequence. Then write the next three terms in the sequence.

a. 6, 8, 10, _____, _____, _____, ...

b. 7, 14, 21, _____, _____, _____, ...

c. 4, 8, 12, _____, _____, _____, ...

d. 21, 18, 15, _____, _____, _____, ...

e. 45, 40, 35, _____, _____, _____, ...

f. 12, 18, 24, _____, _____, _____, ...

How many digits are in each of these numbers?

g. 36,756 **h.** 8002 **i.** 1,287,495

What is the last digit of each of these numbers?

j. 17 **k.** 3586 **l.** 654,321

MIXED PRACTICE

Problem set Write the next term in each counting sequence:

1. 10, 15, 20, _____, ...

2. 56, 49, 42, _____, ...

3. 8, 16, 24, _____, ...

4. 18, 27, 36, 45, _____, ...

5. 24, 21, 18, _____, ...

6. 32, 28, 24, 20, _____, ...

Write the missing term in each counting sequence:

7. 7, 14, _____, 28, 35, ...

8. 40, _____, 30, 25, 20, ...

9. 20, _____, 28, 32, 36, ...

10. 24, 32, _____, 48, ...

11. _____, 36, 30, 24, ...

12. 21, 28, _____, 42, ...

Describe the rule for each counting sequence, and write the next three terms:

13. 3, 6, 9, 12, _____, _____, _____, ...

14. 8, 16, 24, _____, _____, _____, ...

15. 6, 12, 18, _____, _____, _____, ...

16. 40, 35, 30, _____, _____, _____, ...

17. 18, 21, 24, _____, _____, _____, ...

18. 9, 18, 27, _____, _____, _____, ...

19. What word names an ordered list of numbers?

How many digits are in each number?

20. 186,000 **21.** 73,842 **22.** 30,004,091

What is the last digit of each number?

23. 26,348 **24.** 347 **25.** 9,675,420

LESSON

2 Even and Odd Numbers

WARM-UP

Facts Practice: 100 Addition Facts (Test A)

Mental Math: Count up and down by tens between 10 and 100. Count up and down by hundreds between 100 and 1000.

a. 6 + 6 **b.** 60 + 60 **c.** 600 + 600
d. 60 + 70 **e.** 70 + 80 **f.** 300 + 300 + 300
g. 90 + 90 **h.** 50 + 50 + 50 + 50

Problem Solving:

How many two-digit counting numbers are there?

NEW CONCEPT

Whole numbers are the counting numbers and the number 0.

0, 1, 2, 3, 4, 5, 6, ...

Counting by twos, we say the numbers

2, 4, 6, 8, 10, 12, 14, 16, 18, 20, ...

This is a special sequence. These numbers are **even numbers.** The number 0 is also an even number. The sequence of even numbers continues without end. The numbers 36 and 756 and 148,567,896,094 are all even. We can tell whether a whole number is even by looking at the last digit of the number. If the last digit is even, then the number is even. So even numbers end with 0, 2, 4, 6, or 8.

An even number of objects can be arranged in pairs. Twelve is an even number. Here we show 12 dots arranged in six pairs. Notice that every dot has a partner.

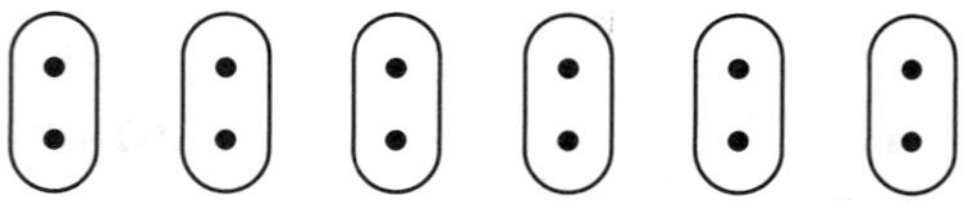

Next we show 13 dots arranged in pairs. We find that there is a dot that does not have a partner. So 13 is not even.

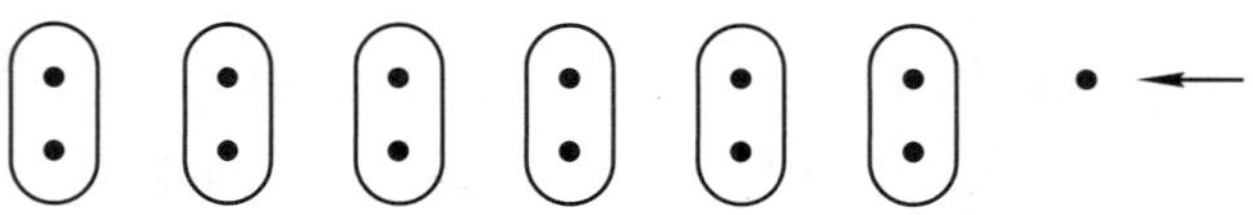

The whole numbers that are not even are **odd.** We can make a list of odd numbers by counting up by twos from the number 1. Odd numbers form this sequence:

1, 3, 5, 7, 9, 11, 13, 15, 17, ...

If the last digit of a number is 1, 3, 5, 7, or 9, then the number is odd. All whole numbers are either odd or even.

Example 1 Which of these numbers is even?

3586 2345 2223

Solution Even numbers are the numbers we say when counting by twos. We can see whether a number is odd or even by looking at the last digit of the number. If the last digit is even, then the number is even. The last digits of these three numbers are 6, 5, and 3, respectively. Since 6 is even and 5 and 3 are not, the only even number in the list is **3586.**

Example 2 Which of these numbers is not odd?

123,456 654,321 353,535

Solution All whole numbers are either odd or even. A number that is not odd is even. The last digits of these numbers are 6, 1, and 5, respectively. Since 6 is even (not odd), the number that is not odd is **123,456.**

Half of an even number is a whole number. We know this because an even number of objects can be separated into two equal groups. However, half of an odd number is not a whole number. If an odd number of objects is divided into two equal groups, then one of the objects will be broken in half. These two stories illustrate dividing an even number in half and dividing an odd number in half:

Sherry has 6 cookies to share with Leticia. If Sherry shares the cookies equally, each girl will have 3 cookies.

Herman has 5 cookies to share with Ivan. If Herman shares the cookies equally, each boy will have $2\frac{1}{2}$ cookies.

Activity: *Halves*

The table below lists the counting numbers 1 through 10. Below each counting number we have recorded half of the number. Continue the list of counting numbers and their halves for the numbers 11 through 20.

Counting number	1	2	3	4	5	6	7	8	9	10
Half of number	$\frac{1}{2}$	1	$1\frac{1}{2}$	2	$2\frac{1}{2}$	3	$3\frac{1}{2}$	4	$4\frac{1}{2}$	5

LESSON PRACTICE

Practice set Describe each number as odd or even:

a. 0 **b.** 1234 **c.** 20,001

d. 999 **e.** 3000 **f.** 391,048

g. All the students in the class separated into two groups. The same number of students were in each group. Was the number of students in the class an odd number or an even number?

h. Leticia has seven cookies to share with Willis. If Leticia shares the cookies equally, how many cookies will each person have?

MIXED PRACTICE

Problem set †**1.** If a whole number is not even, then what is it?
(2)

What is the last digit of each number?

2. 47,286,560 *(1)* **3.** 296,317 *(1)*

Describe each number as odd or even:

4. 15 *(2)* **5.** 196 *(2)* **6.** 3567 *(2)*

7. Which of these numbers is even?
(2)

3716 2345 2223

8. Which of these numbers is odd?
(2)

45,678 56,789 67,890

†The italicized numbers within parentheses underneath each problem number are called *lesson reference numbers.* These numbers refer to the lesson(s) in which the major concept of that particular problem is introduced. If additional assistance is needed, refer to the discussion, examples, or practice problems of that lesson.

9. Which of these numbers is not odd?
(2)

333,456 654,321 353,535

10. Which of these numbers is not even?
(2)

300 232 323

Write the next three terms in each counting sequence:

11. 9, 12, 15, _____, _____, _____, ...
(1)

12. 16, 24, 32, _____, _____, _____, ...
(1)

13. 120, 110, 100, _____, _____, _____, ...
(1)

14. 28, 24, 20, _____, _____, _____, ...
(1)

15. 55, 50, 45, _____, _____, _____, ...
(1)

16. 18, 27, 36, _____, _____, _____, ...
(1)

17. 36, 33, 30, _____, _____, _____, ...
(1)

18. 18, 24, 30, _____, _____, _____, ...
(1)

19. 14, 21, 28, _____, _____, _____, ...
(1)

20. 66, 60, 54, _____, _____, _____, ...
(1)

21. 48, 44, 40, _____, _____, _____, ...
(1)

22. 99, 90, 81, _____, _____, _____, ...
(1)

23. 88, 80, 72, _____, _____, _____, ...
(1)

24. 84, 77, 70, _____, _____, _____, ...
(1)

25. All the students in the class formed two lines. An equal number of students were in each line. Which of the following could not be the total number of students in the class?
(2)

A. 30 B. 31 C. 32

26. What number is half of 5?
(2)

27. Which of these numbers is a whole number?
(2)

A. half of 11 B. half of 12 C. half of 13

LESSON

3 Using Money to Illustrate Place Value

WARM-UP

Facts Practice: 100 Addition Facts (Test A)

Mental Math: Count up and down by tens between 10 and 200. Count up and down by hundreds between 100 and 2000.

a. 30 + 70	**b.** 20 + 300	**c.** 320 + 20
d. 340 + 200	**e.** 250 + 40	**f.** 250 + 400
g. 120 + 60	**h.** 600 + 120	

Problem Solving:

How many three-digit counting numbers are there?

NEW CONCEPT

Activity: *Place Value*

Materials needed:

- Activity Masters 1, 2, and 3 (two copies of each master per student; masters available in *Saxon Math 6/5 Assessments and Classroom Masters*)
- Activity Master 4 (one copy per student)
- locking plastic bags (one per student)
- paper clips (three per student)

Preparation:

Step 1: Cut out the $1, $10, and $100 money manipulatives on Activity Masters 1–3. A paper cutter, if available, will save time.

Step 2: Distribute the money manipulatives, copies of Activity Master 4 (Place-Value Template), bags, and paper clips to students.

Note: At the end of this lesson students may store their money manipulatives in the plastic bags, clipping together bills of the same denomination. This paper money will be useful in later lessons.

Place twelve $1 bills on the template in the ones position, as shown below.

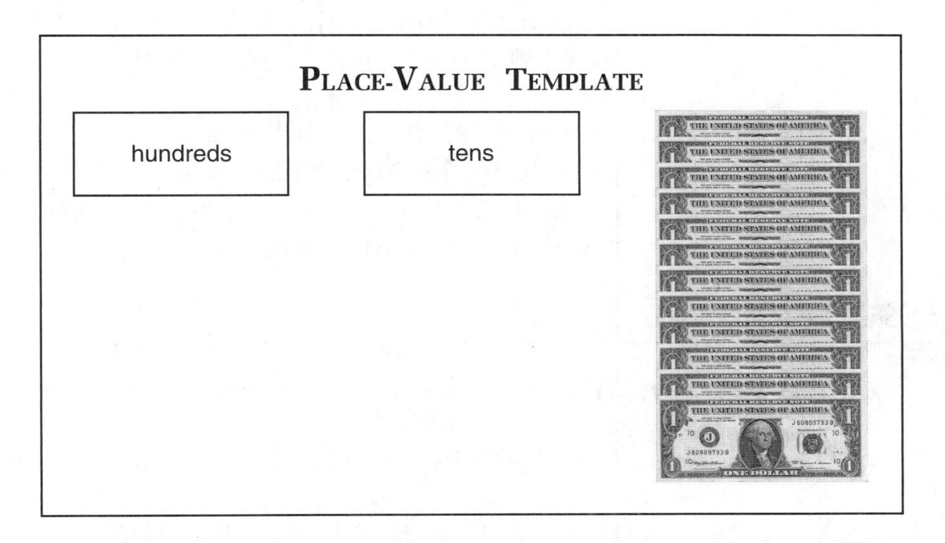

We can use fewer bills to represent $12 by exchanging ten $1 bills for one $10 bill. Remove ten $1 bills from the template, and replace them with one $10 bill in the tens position. You will get this arrangement of bills:

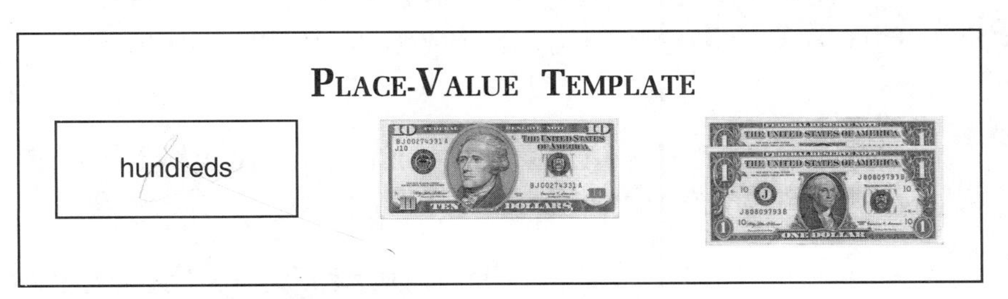

The bills on the template illustrate the **expanded form** of the number 12.

Expanded form: 1 ten + 2 ones

Now place $312 on the place-value template, using the fewest bills necessary. Use the bills to write 312 in expanded form.

From the template we see that the expanded form for 312 is

3 hundreds + 1 ten + 2 ones

Use the bills and place-value template to work these exercises:

1. Place twelve $10 bills on the place-value template. Then exchange ten of the bills for one $100 bill. Write the result in expanded form.

2. Place twelve $1 bills and twelve $10 bills on the template. Then exchange bills to show that amount of money using the least number of bills possible. Write the result in expanded form.

LESSON PRACTICE

Practice set

a. Which digit in 365 shows the number of tens?

b. Use digits to write the number for "3 hundreds plus 5 tens."

c. How much money is one $100 bill plus ten $10 bills plus fifteen $1 bills? You may use your bills to help find the answer.

MIXED PRACTICE

Problem set

1. (3) Use digits to write the number for "5 hundreds plus 7 tens plus 8 ones."

2. (3) Use digits to write the number for "2 hundreds plus 5 tens plus 0 ones."

3. (3) In 560, which digit shows the number of tens?

4. (3) In 365, which digit shows the number of ones?

5. (3) Ten $10 bills have the same value as one of what kind of bill?

6. (2) Which of these numbers is not odd?
   A. 365   B. 653   C. 536

7. (2) Which of these numbers is not even?
   A. 1234   B. 2345   C. 3456

8. (2) The greatest two-digit odd number is 99. What is the greatest two-digit even number?

9. (2) Two teams have an equal number of players. The total number of players on both teams could not be
   A. 22   B. 25   C. 50

10. We can count to 12 by 2's or by 3's. We do not count to 12 when counting by
(1)

A. 1's B. 4's C. 5's D. 6's

Write the next three terms in each counting sequence:

11. 9, 12, 15, _____, _____, _____, ...
(1)

12. 54, 48, 42, _____, _____, _____, ...
(1)

13. 8, 16, 24, _____, _____, _____, ...
(1)

14. 80, 72, 64, _____, _____, _____, ...
(1)

15. 16, 20, 24, _____, _____, _____, ...
(1)

16. 40, 36, 32, _____, _____, _____, ...
(1)

Describe the rule for each counting sequence, and find the next three terms:

17. 27, 36, 45, _____, _____, _____, ...
(1)

18. 81, 72, 63, _____, _____, _____, ...
(1)

19. 10, 20, 30, _____, _____, _____, ...
(1)

20. 33, 30, 27, _____, _____, _____, ...
(1)

21. What number equals four tens?
(3)

22. What number equals five hundreds?
(3)

23. How much money is two \$100 bills plus twelve \$10 bills plus fourteen \$1 bills? You may use your bills to help find the answer.
(3)

24. The number 80 means "eight tens." The number 800 means eight what?
(3)

25. The fifth term in the counting sequence below is 20. What is the ninth term in this sequence?
(1)

4, 8, 12, 16, ...

26. How much money is half of \$10?
(2)

27. How much money is half of \$5?
(2)

28. Is the greatest two-digit number an odd number or an even number?
(2)

LESSON

4 Comparing Whole Numbers

WARM-UP

Facts Practice: 100 Addition Facts (Test A)

Mental Math: Count up and down by tens between 0 and 200. Count up and down by hundreds between 0 and 2000.

a. 300 + 300 + 20 + 20 **b.** 250 + 50
c. 300 + 350 **d.** 320 + 320
e. 300 + 300 + 50 + 50 **f.** 250 + 60
g. 340 + 600 **h.** 240 + 320

Problem Solving:

The two-digit counting numbers that contain the digits 1 and 2 are 12 and 21. There are six three-digit counting numbers that contain the digits 1, 2, and 3. One of these numbers is 213. What are the other five numbers?

NEW CONCEPT

When we count from 1 to 10, we count in order of size from least to greatest.

1, 2, 3, 4, 5, 6, 7, 8, 9, 10

↑ least ↑ greatest

Of these numbers, the least is 1 and the greatest is 10. Since these numbers are arranged in order, we can easily see that 5 is greater than 4 and that 5 is less than 6.

We can use mathematical symbols to compare numbers. These symbols include the equal sign (=) and the "greater than/less than" symbol ($>$ or $<$).

$5 = 5$ is read "Five is **equal to** five."

$5 > 4$ is read "Five is **greater than** four."

$5 < 6$ is read "Five is **less than** six."

When using a "greater than/less than" symbol to compare two numbers, we place the symbol so that the smaller end points to the smaller number.

Example 1 Write the numbers 64, 46, and 54 in order from least to greatest.

Solution "From least to greatest" means "from smallest to largest." We write the numbers in this order:

46, 54, 64

Example 2 Complete each comparison by replacing the circle with the proper comparison symbol:

(a) $7 \bigcirc 7$ (b) $6 \bigcirc 4$ (c) $6 \bigcirc 8$

Solution When two numbers are equal, we show the comparison with an equal sign.

(a) $\mathbf{7 = 7}$

When two numbers are not equal, we place the greater than/less than symbol so that the smaller end points to the smaller number.

(b) $\mathbf{6 > 4}$ (c) $\mathbf{6 < 8}$

Example 3 Compare:

(a) $373 \bigcirc 47$ (b) $373 \bigcirc 382$

Solution (a) When comparing whole numbers, we know that numbers with more digits are greater than numbers with fewer digits.

$\mathbf{373 > 47}$

(b) When comparing whole numbers with the same number of digits, we consider the value place by place. The digits in the hundreds place are the same, but in the tens place 8 is greater than 7. So we have the following:

$\mathbf{373 < 382}$

Example 4 Use digits and a comparison symbol to write this comparison:

Six is less than ten.

Solution We translate the words into digits. The comparison symbol for "is less than" is "<".

$\mathbf{6 < 10}$

LESSON PRACTICE

Practice set

a. Write the numbers 324, 243, and 423 in order from least to greatest.

Complete each comparison by replacing the circle with the proper comparison symbol:

b. 36 ◯ 632 **c.** 110 ◯ 101

d. 90 ◯ 90 **e.** 112 ◯ 121

Write each comparison using mathematical symbols:

f. Twenty is less than thirty.

g. Twelve is greater than eight.

MIXED PRACTICE

Problem set Write each comparison using digits and a comparison symbol:

1. (4) Four is less than ten.

2. (4) Fifteen is greater than twelve.

Complete each comparison by replacing the circle with the proper comparison symbol:

3. (4) 97 ◯ 101 **4.** (4) 34 ◯ 43

5. (3) Use digits to write the number for "3 hundreds plus 6 tens plus 5 ones."

6. (3) Which digit in 675 shows the number of hundreds?

7. (3) Which digit in 983 shows the number of ones?

8. (3) One $100 bill equals ten of what kind of bill?

Describe each number as odd or even:

9. (2) 36,275 **10.** (2) 36,300 **11.** (2) 5,396,428

12. (2) The greatest two-digit odd number is 99. What is the greatest three-digit odd number?

13. We can count to 18 by 2's or by 3's. We do not count to 18
(1) when counting by

A. 1's B. 4's C. 6's D. 9's

14. Write the numbers 435, 354, and 543 in order from least
(4) to greatest.

15. The fourth term in the counting sequence below is 24.
(1) What is the ninth term in this sequence?

6, 12, 18, ...

16. What is the value of five $100 bills, thirteen $10 bills, and
(3) ten $1 bills? You may use your money manipulatives to help find the answer.

Write the next three terms in each counting sequence:

17. 20, 24, 28, ____, ____, ____, ...
(1)

18. 106, 104, 102, ____, ____, ____, ...
(1)

19. 0, 6, 12, ____, ____, ____, ...
(1)

20. 0, 7, 14, ____, ____, ____, ...
(1)

21. 40, 32, 24, ____, ____, ____, ...
(1)

22. 45, 36, 27, ____, ____, ____, ...
(1)

23. What number equals nine tens?
(3)

24. What number equals eleven tens?
(3)

25. What is the seventh term in this counting sequence?
(1)

8, 16, 24, ...

26. Is the eleventh term of this counting sequence odd or even?
(1, 2)

2, 4, 6, 8, ...

27. What number is half of 9?
(2)

28. In Room 12 there is one more boy than there are girls. Is
(2) the number of students in Room 12 odd or even?

LESSON

5 Naming Whole Numbers Through Hundreds • Dollars and Cents

WARM-UP

Facts Practice: 100 Addition Facts (Test A)

Mental Math: Count up and down by tens between 0 and 200. Count up and down by hundreds between 0 and 2000.

a. 200 + 60 + 300 | **b.** 20 + 600 + 30
c. 350 + 420 | **d.** 250 + 250
e. 400 + 320 + 40 | **f.** 30 + 330 + 100
g. 640 + 250 | **h.** 260 + 260

Problem Solving:

Write all the three-digit numbers that each have the digits 2, 3, and 4.

NEW CONCEPTS

Naming whole numbers through hundreds

If numbers are to be our "friends," we should learn their names. Naming numbers is not difficult if we pay attention to place values. In order to name larger numbers, we should first be able to name numbers that have three digits. Let's consider the number 365. Below we use expanded form to break the number into its parts. Then we show the name of each part.

three hundreds + six tens + five ones

"three hundred" "sixty" "five"

We will use words to name a number that we see and use digits to write a number that is named. Look at these examples:

18 eighteen

80 eighty

81 eighty-one

108 one hundred eight

821 eight hundred twenty-one

Notice that we do not use the word *and* when naming whole numbers. For example, we write the number 108 as "one hundred eight," **not** "one hundred *and* eight." Also notice that we use a hyphen when writing the numbers from 21 to 99 that do not end with zero. For example, we write 21 as "twenty-one," **not** "twenty one."

Dollars and cents Dollars and cents are written with a dollar sign and a decimal point. To name an amount of money, we first name the number of dollars, say "and," and then name the number of cents. The decimal point separates the number of dollars from the number of cents. For example, $324.56 is written as "three hundred twenty-four dollars and fifty-six cents."

LESSON PRACTICE

Practice set*†

a. Use words to name $563.45.

b. Use words to name 101.

c. Use words to name 111.

d. Use digits to write two hundred forty-five.

e. Use digits to write four hundred twenty.

f. Use digits to write five hundred three dollars and fifty cents.

MIXED PRACTICE

Problem set

1. (5) Use digits to write three hundred seventy-four dollars and twenty cents.

2. (5) Use words to name $623.15.

3. (5) Use digits to write two hundred five.

4. (5) Use words to name 109.

5. (4, 5) Write this comparison using digits and a comparison symbol:

One hundred fifty is greater than one hundred fifteen.

†The asterisk after "Practice set" indicates that additional practice problems intended for remediation are available in the appendix.

6. Compare: 346 ◯ 436
(4)

7. Use digits to write the number for "5 hundreds plus 7 tens plus 9 ones."
(3)

8. Arrange these four numbers in order from least to greatest:
(4)

462 624 246 426

9. Which digit in 567 shows the number of tens?
(3)

10. When counting up by tens, what number comes after 90?
(1)

Describe each number as odd or even:

11. 363,636 (2)
12. 36,363 (2)
13. 2000 (2)

14. The greatest three-digit odd number is 999. What is the greatest three-digit even number?
(2)

15. We can count to 20 by 2's or by 10's. We do not count to 20 when counting by
(1)

A. 1's B. 3's C. 4's D. 5's

16. There are equal numbers of boys and girls in the room. Which of the following could not be the number of students in the room?
(2)

A. 12 B. 29 C. 30 D. 44

17. What is the value of six $100 bills, nine $10 bills, and twelve $1 bills? You may use your money manipulatives to help find the answer.
(3)

Write the next four terms in each counting sequence:

18. 0, 9, 18, ____, ____, ____, ____, ...
(1)

19. 25, 30, 35, ____, ____, ____, ____, ...
(1)

20. 6, 12, 18, ____, ____, ____, ____, ...
(1)

State the rule for each counting sequence, and find the next four terms:

21. 100, 90, 80, _____, _____, _____, _____, ...
(1)

22. 90, 81, 72, _____, _____, _____, _____, ...
(1)

23. 88, 80, 72, _____, _____, _____, _____, ...
(1)

24. 7, 14, 21, _____, _____, _____, _____, ...
(1)

25. What is the ninth term in this counting sequence?
(1)

3, 6, 9, ...

26. Is the tenth term in this counting sequence odd or even?
(1, 2)

1, 3, 5, 7, 9, ...

27. Is the greatest three-digit whole number odd or even?
(2)

28. Tom and Jerry evenly shared the cost of a $7 pizza. How much did each person pay?
(2)

LESSON

6 Adding One-Digit Numbers • Using the Addition Algorithm

WARM-UP

Facts Practice: 100 Addition Facts (Test A)

Mental Math: Count up and down by 20's between 0 and 200. Count up and down by 200's between 0 and 2000.

a. 400 + 50 + 300 + 40 **b.** 320 + 300
c. 320 + 30 **d.** 320 + 330
e. 60 + 200 + 20 + 400 **f.** 400 + 540
g. 40 + 250 **h.** 450 + 450

Problem Solving:

Write the six three-digit numbers that have each of the digits 3, 4, and 5. Then arrange the six numbers in order from least to greatest.

NEW CONCEPTS

Adding one-digit numbers

Numbers that are added are called **addends.** The answer to an addition problem is the **sum.** We may add numbers in any order to find their sum. For example, 5 + 6 gives us the same sum as 6 + 5. This property of addition is called the **commutative property of addition.** When adding more than two numbers, this property allows us to add in any order we choose. Below we show three ways to add 6, 3, and 4. We point out the two numbers we added first.

$$\begin{array}{r} 6 \\ 3 \\ +\ 4 \\ \hline 13 \end{array} \quad (6+3=9) \qquad \begin{array}{r} 6 \\ 3 \\ +\ 4 \\ \hline 13 \end{array} \quad (3+4=7) \qquad \begin{array}{r} 6 \\ 3 \\ +\ 4 \\ \hline 13 \end{array} \quad (6+4=10)$$

As shown in the right-hand example above, we can sometimes find pairs of numbers that add up to 10. This makes the addition easier.

Example 1 What is the sum of 7, 4, 3, and 6?

Solution We add to find the sum. We may either add the numbers as they are written (horizontally) or align them in a column. Here we write the numbers in a column and add in an order that makes the work a little easier.

$$\begin{array}{r} 7 \\ 4 \\ 3 \\ +\ 6 \\ \hline \mathbf{20} \end{array} \quad (7+3=10,\ 4+6=10)$$

Example 2 Four one-digit whole numbers are added. Is the sum more than or less than 40? How do you know?

Solution We do not know the numbers, so we do not know the sum. However, we know that the sum is **less than 40,** because the greatest one-digit number is 9, and the sum of four 9's is only 36. The sum of the four whole numbers is actually any whole number less than 37—including zero (if the four numbers were all zero).

If zero is added to any number, the sum is identical to that number. Here are some examples:

$$2 + 0 = 2 \qquad 37 + 0 = 37 \qquad 999 + 0 = 999$$

This property of addition is called the **identity property of addition.**

Using the addition algorithm

In arithmetic we add, subtract, multiply, and divide numbers using *algorithms.* An **algorithm** is a procedure for getting an answer. Algorithms allow us to solve problems.

Adding money can help us understand the addition algorithm. Use your $100, $10, and $1 money manipulatives to act out the following example.

Example 3 Jamal had $462. Maria paid Jamal $58 rent when she landed on his property. Then how much money did Jamal have?

Solution First we will use bills to illustrate the story:

Jamal had $462.

4

6

2

Maria paid Jamal $58 rent.

5

8

When Jamal added Maria's rent money to the money he already had, he ended up with four $100 bills, eleven $10 bills, and ten $1 bills.

4

11

10

Jamal exchanged the ten $1 bills for one $10 bill. That gave Jamal twelve $10 bills. Then he exchanged ten $10 bills for one $100 bill. That gave Jamal five $100 bills and two $10 bills.

5

2

This exchange of bills shows that Jamal had **$520** after Maria paid him for rent.

Now we will show a pencil-and-paper solution that uses the addition algorithm. When using this addition algorithm, we are careful to line up digits that have the same place value.

Jamal had $462.	$462
Maria paid Jamal $58.	+ $ 58
Then Jamal had ...	

First we add the ones, then the tens, and then the hundreds.

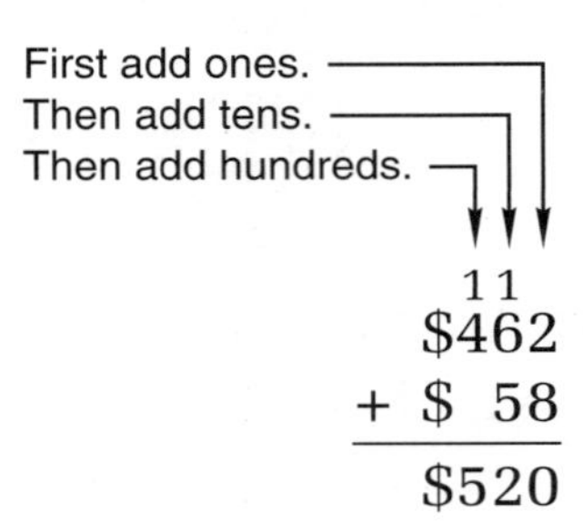

Notice we exchange 10 ones for 1 ten. Then we exchange 10 tens for 1 hundred.

LESSON PRACTICE

Practice set* Find each sum. When adding, look for combinations of numbers that add up to 10.

a. 8 + 6 + 2

b. 4 + 7 + 3 + 6

c. 9 + 6 + 4

d. 4 + 5 + 6 + 7

e. 7 + 3 + 4

f. 2 + 6 + 3 + 5

g. 6 + 7 + 5

h. 8 + 7 + 5 + 3

i. The sum of 5 one-digit whole numbers is certain to be

A. greater than 4   B. less than 50   C. an odd number

Use the addition algorithm to find each sum. When putting the numbers into columns, remember to line up the last digits.

j. \$463 + \$158   **k.** 674 + 555   **l.** \$323 + \$142 + \$365

m. 543 + 98   **n.** \$47 + \$485

MIXED PRACTICE

Problem set

1. (6) You may use money manipulatives to help answer the question in this story:

Jamal had \$520. After Hannah paid him \$86 rent, how much money did Jamal have?

2. (5) Use words to name \$212.50.

3. (3) In 274, which digit shows the number of hundreds?

Describe each number as odd or even:

4. (2) 1234   **5.** (2) 12,345   **6.** (2) 1,234,567

7. (5) Use digits to write five hundred eight dollars.

8. (5) Use words to name 580.

Find each sum. Look for combinations of 10.

9. (6) 1 + 6 + 9   **10.** (6) 7 + 6 + 4

11. (6) 8 + 3 + 1 + 7   **12.** (6) 4 + 5 + 6 + 7

13. (6) \$436 + \$527   **14.** (6) 592 + 408   **15.** (6) 963 + 79   **16.** (6) \$180 + \$747

17. (2) All the books were put into two piles. There was one more book in one pile than in the other pile. The total number of books in both piles could not be

A. 28   B. 29   C. 33   D. 55

Find the eighth term in each counting sequence:

18. (1) 10, 20, 30, ...   **19.** (1) 6, 12, 18, ...

20. (1) 7, 14, 21, ...   **21.** (1) 8, 16, 24, ...

22. Compare: nine hundred sixteen ◯ nine hundred sixty
(4)

23. Write this comparison using digits and a comparison symbol:
(4, 5)

Six hundred ninety is greater than six hundred nine.

24. Compare: $5 + 5 + 5$ ◯ $4 + 5 + 6$
(4, 6)

25. The smallest even two-digit whole number is 10. What is the smallest odd two-digit whole number?
(2)

26. Is the smallest three-digit number odd or even?
(2)

27. Is the 29th term in this counting sequence odd or even?
(1, 2)

2, 4, 6, 8, ...

28. Tabitha needs to read nine pages in her history book. If she wants to read half of those pages before dinner, how many pages does she need to read?
(2)

LESSON

7 Writing and Comparing Numbers Through Hundred Thousands • Ordinal Numbers

WARM-UP

Facts Practice: 100 Addition Facts (Test A)

Mental Math: Count up and down by 20's between 0 and 200. Count up and down by 200's between 0 and 2000.

a. \$25 + \$25	**b.** \$300 + \$450	**c.** \$250 + \$250
d. 30 + 450	**e.** \$75 + \$25	**f.** \$750 + \$250
g. \$50 + \$350	**h.** 360 + 360	

Problem Solving:

The sum of 12 and 21 is 33. What is the sum of the six three-digit numbers that each have the digits 1, 2, and 3? If the six numbers are arranged vertically, what is the sum of the digits in each column? Why is the sum of the digits in each column the same?

NEW CONCEPTS

Writing and comparing numbers through hundred thousands

We have practiced naming whole numbers with three or fewer digits. In this lesson we will begin naming whole numbers with four, five, and six digits.

The value of a digit depends upon its position in a number. The chart below lists the values of the first six whole-number places.

hundred thousands	ten thousands	thousands		hundreds	tens	ones
___	___	___	,	___	___	___

Commas are often used to write a whole number with many digits so that the number is easier to read. To place commas in a whole number, we count digits from the right-hand end of the number and insert a comma after every three digits.

54,321

The comma in this number marks the end of the thousands. To name this number, we read the number formed by the digits to the left of the comma and then say "thousand" at the comma. Finally, we read the number formed by the last three digits.

54,321

fifty-four thousand, three hundred twenty-one

Notice that we place a comma after the word *thousand* when we use words to name a number. Here we show some other examples:

$27,050 twenty-seven thousand, fifty dollars

125,000 one hundred twenty-five thousand

203,400 two hundred three thousand, four hundred

Whole numbers with four digits may be written with a comma, but in this book four-digit whole numbers will usually be written without a comma.

Example 1 Use words to name 52370.

Solution To help us read the number, we write it with a comma:

52,370

We name the number formed by the digits in front of the comma, write "thousand" and a comma, and then name the number formed by the digits after the comma. So 52,370 is **fifty-two thousand, three hundred seventy.**

Example 2 Use digits to write one hundred fifty thousand, two hundred thirty-four.

Solution We use digits to write "one hundred fifty" and write a comma for the word *thousand.* Then we use digits to write "two hundred thirty-four."

150,234

Example 3 Compare: 23,465 ◯ 23,654

Solution Since the digits in the ten-thousands place and thousands place match, we look to the hundreds place to make the comparison.

23,465 < 23,654

Example 4 Three of the longest underwater tunnels in North America are in New York City. The Brooklyn-Battery Tunnel is 9117 feet long, the Lincoln Tunnel is 8216 feet long, and the Holland Tunnel is 8558 feet long. Write the names and lengths of these tunnels in order from shortest to longest.

Solution Arranging the numbers in order from least to greatest arranges the tunnels in order from shortest to longest: **Lincoln Tunnel (8216 feet), Holland Tunnel (8558 feet), Brooklyn-Battery Tunnel (9117 feet).**

Ordinal numbers

Numbers used to name position or order are called **ordinal numbers.** This table shows two ways to write the first twelve ordinal numbers:

Ordinal Numbers for 1–12

1st	first
2nd	second
3rd	third
4th	fourth
5th	fifth
6th	sixth
7th	seventh
8th	eighth
9th	ninth
10th	tenth
11th	eleventh
12th	twelfth

Example 5 Tom was the fourth person in a line of ten people waiting for a movie. How many people were in front of Tom? How many people were behind Tom?

Solution We draw a picture to illustrate the problem.

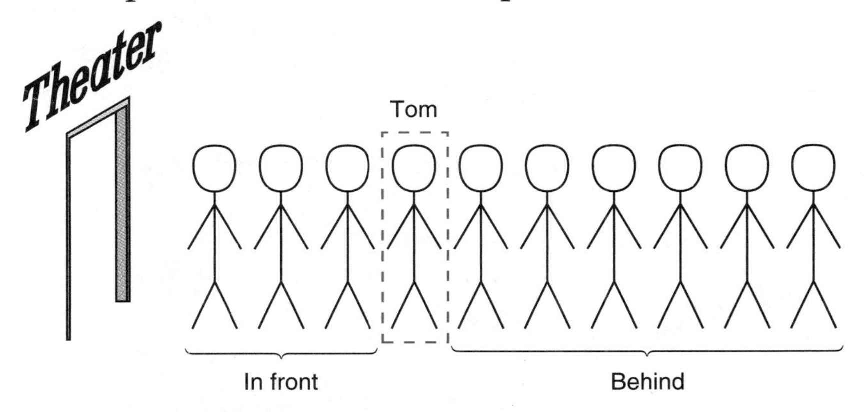

By counting people in our picture, we find that there were **three people in front** of Tom and **six people behind** him.

LESSON PRACTICE

Practice set* Use words to name each number. (*Hint:* Begin by writing the number with a comma.)

a. 36420

b. $12300

c. 4567

Use digits to write each number:

d. sixty-three thousand, one hundred seventeen

e. two hundred fifty-six thousand, seven hundred

f. fifty thousand, nine hundred twenty-four

g. seven hundred fifty thousand dollars

h. Christina was the sixth person in a line of ten people. How many people were in front of Christina, and how many people were behind her?

MIXED PRACTICE

Problem set

1. (6) You may use money manipulatives to help answer the question in this story:

Hannah had $462. After she was paid $88 rent, how much money did Hannah have?

2. (7) Which digit is in the tens place in 567?

3. (5) Use digits to write seven hundred seven.

4. (7) Mount Everest in Asia has the highest peak in the world. The peak is 29,035 feet above sea level. Use words to name this height.

5. (6) Find the sum of 54 and 246.

Find each sum:

6. (6) $463 + $364

7. (6) $286 + $414

8. (6) 709 + 314

Find the seventh term in each counting sequence:

9. (1) 10, 20, 30, ...

10. (1) 5, 10, 15, ...

11. 6, 12, 18, ...
(1)

12. 7, 14, 21, ...
(1)

13. 8, 16, 24, ...
(1)

14. 9, 18, 27, ...
(1)

15. Compare: two hundred fifty ◯ two hundred fifteen
(4)

16. Compare. (Try to answer the comparison without adding.)
(4, 6)

365 + 366 ◯ 365 + 365

Find each sum:

17. (6)
```
  $436
  $ 72
+ $ 54
```

18. (6)
```
  361
  493
+ 147
```

19. (6)
```
  506
   79
+ 434
```

20. Write this comparison using digits and a comparison symbol:
(4, 5)

Four hundred eight is less than four hundred eighty.

21. We can count to 24 by 2's or by 3's. We do not count to 24 when counting by
(1)

A. 4's B. 5's C. 6's D. 8's

Describe each number as odd or even:

22. 1969
(2)

23. 1492
(2)

24. 1776
(2)

25. The smallest even three-digit number is 100. What is the smallest odd three-digit number?
(2)

26. Of the twelve people in line, Rosario was fifth. How many people were in front of Rosario? How many were behind her?
(7)

27. Is the twentieth term in this counting sequence odd or even?
(2)

1, 3, 5, 7, ...

28. Five birds were perched on a branch. Could half of the birds fly away? Why or why not?
(2)

LESSON

8 Subtraction Facts • Fact Families

WARM-UP

Facts Practice: 100 Addition Facts (Test A)

Mental Math: Count up and down by 50's between 0 and 500. Count up and down by 500's between 0 and 5000.

a. 3000 + 3000 **b.** 5000 + 5000 **c.** 350 + 450
d. 370 + 580 **e.** \$275 + \$25 **f.** \$350 + \$500
g. 750 + 750 **h.** 250 + 750

Problem Solving:

The sum of the six numbers that have the digits 1, 2, and 3 is 1332. What is the sum of the six three-digit numbers that each have the digits 2, 4, and 6? What do you notice about the two sums?

NEW CONCEPTS

Subtraction facts Subtraction involves taking one number from another number. If five birds were perched on a branch and three flew away, then two birds would be left on the branch.

A number sentence for this story is

$$5 - 3 = 2$$

We read this number sentence, "Five minus three equals two." The dash (–) between the 5 and the 3 is called a **minus sign.** *The minus sign tells us to subtract the number to the right of the sign from the number to the left of the sign.* Order matters when we subtract. The answer to 5 – 3 does not equal the answer to 3 – 5. When we see 5 – 3, we must start with 5 and subtract 3.

When a subtraction problem is written in a column (with one number above the other) we start with the top number and subtract the bottom number. The two forms below mean the same thing. With both forms, we start with 5 and subtract 3.

$$5 - 3 = 2 \qquad \begin{array}{r} 5 \\ -\ 3 \\ \hline 2 \end{array}$$

The answer when we subtract is called a **difference.** We can say "the difference of 5 and 3 is 2."

Example 1 When 7 is subtracted from 12, what is the difference?

Solution We start with 12 and subtract 7. If we write the numbers horizontally, we write the 12 on the left. If we write the numbers in a column, we position the 12 on top and the 7 below the 2 in 12. This way, digits with the same place value are in the same column. We find that the difference of 12 and 7 is **5.**

$$12 - 7 = 5 \qquad \begin{array}{r} 12 \\ -\ 7 \\ \hline 5 \end{array}$$

Example 2 What is 8 minus 3?

Solution The word *minus* means "take away." For this problem, we take 3 away from 8. When we see the word *minus,* we may put a minus sign in its place. We find that 8 minus 3 equals **5.**

$$8 - 3 = 5 \qquad \begin{array}{r} 8 \\ -\ 3 \\ \hline 5 \end{array}$$

Fact families

Addition and subtraction are closely related. We say that addition and subtraction are **inverse operations** because one operation "undoes" the other. If we add 3 to 5, we get 8. If we then subtract 3 from 8, we again have 5. By subtracting 3, we undo the addition of 3.

For every addition fact we can form a subtraction fact. With the numbers 2, 3, and 5, for example, we can form two addition facts and two subtraction facts.

$$\begin{array}{r} 2 \\ +\ 3 \\ \hline 5 \end{array} \qquad \begin{array}{r} 5 \\ -\ 3 \\ \hline 2 \end{array} \qquad \begin{array}{r} 3 \\ +\ 2 \\ \hline 5 \end{array} \qquad \begin{array}{r} 5 \\ -\ 2 \\ \hline 3 \end{array}$$

We call the three numbers 2, 3, and 5 a **fact family.**

Example 3 Write two addition facts and two subtraction facts for the fact family 3, 4, and 7.

Solution **3 + 4 = 7   4 + 3 = 7   7 − 3 = 4   7 − 4 = 3**

LESSON PRACTICE

Practice set Subtract:

a. 17 − 9 **b.** 12 − 8 **c.** 15 − 9

d. 11 − 5 **e.** 17 − 8 **f.** 16 − 8

Write two addition facts and two subtraction facts for each fact family:

g. 7, 8, 15

h. 5, 7, 12

MIXED PRACTICE

Problem set

1. Which digit in 3654 is in the thousands place? *(7)*

2. List the five odd, one-digit numbers. *(2)*

3. When seven is subtracted from 15, what is the difference? *(8)*

4. When 56 is added to 560, what is the sum? *(6)*

5. What is seven minus four? *(8)*

6. What is sixty-four plus two hundred six? *(6)*

7. Use words to name $812,000. *(7)*

8. Use digits to write eight hundred two. *(5)*

9. Write a two-digit odd number using 5 and 6. *(2)*

10. Use words to name the number for "4 hundreds plus 4 tens plus 4 ones." *(3)*

Describe the rule for each counting sequence, and write the ninth term.

11. (1) 6, 12, 18, ...

12. (1) 3, 6, 9, ...

13. (8) Write two addition facts and two subtraction facts for the fact family 4, 8, and 12.

14. (2) Think of two odd numbers and add them. Is the sum odd or even?

Subtract to find each difference:

15. (8) 18 − 9

16. (8) 15 − 7

17. (8) 12 − 5

18. (8) 11 − 8

19. (8) 14 − 6

20. (8) 13 − 9

Add to find each sum:

21. (6) \$36 + \$403 + \$97

22. (6) 572 + 386 + 38

23. (6) 47 + 135 + 70

24. (6) \$590 + \$306 + \$75

25. (2, 4, 6) If the greatest odd number in the list below is added to the smallest even number in the list, then what is the sum?

364 287 428 273

26. (2) Write the smallest four-digit whole number. Is the number odd or even?

27. (2) Half of the 18 students were girls. How many girls were there?

28. (2) From Tom's house to school and back is five miles. How far is it from Tom's house to school?

LESSON

9 Practicing the Subtraction Algorithm

WARM-UP

Facts Practice: 100 Subtraction Facts (Test B)

Mental Math: Count up and down by 50's between 0 and 500. Count up and down by 500's between 0 and 5000.

a. $250 + $250 **b.** 6000 + 6000 **c.** $75 + $125
d. 750 + 750 **e.** 60 − 20 **f.** 600 − 200
g. 6000 − 2000 **h.** 860 + 70

Problem Solving:

The letters P, T, and A can be arranged in six different orders. Write the six possible orders, and circle the ones that spell words.

NEW CONCEPT

We may find a subtraction answer by counting, by using objects, or by remembering combinations. When subtracting larger numbers, it is helpful to have a method. Recall from Lesson 6 that a method for solving a problem is an *algorithm.* In this lesson we will practice an algorithm for subtraction. We will use a money example to help us understand the algorithm. It may be helpful to act out the story in this money example using your $100, $10, and $1 bills.

Maria has $524. She must pay Jamal $58 for rent. After she pays Jamal, how much money will she have?

We will use five $100 bills, two $10 bills, and four $1 bills to show how much money Maria has.

5

2

4

From $524, Maria must pay Jamal $58, which is five $10 bills and eight $1 bills. Maria has enough money to pay Jamal, but she doesn't have enough $10 bills and $1 bills to pay him the exact amount. Before Maria pays Jamal, she must exchange one \$10 bill for ten \$1 bills. Then she will have enough ones.

Maria still does not have enough tens, so she must also exchange one \$100 bill for ten \$10 bills.

Now Maria can pay Jamal with five \$10 bills and eight \$1 bills. Taking away 5 tens and 8 ones leaves this much:

After she pays Jamal, Maria will have \$466.

We exchanged bills to show the subtraction. We also use exchange with the pencil-and-paper algorithm. We write the subtraction problem and begin by subtracting the ones.

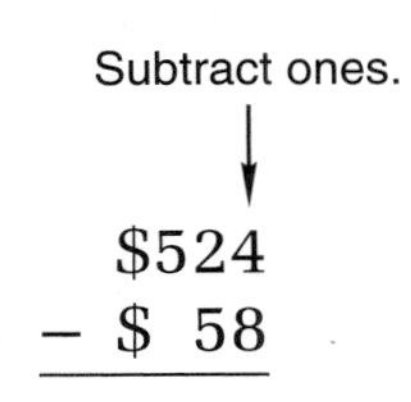

We cannot subtract \$8 from \$4. We need more ones. We look at the tens column and see 2 tens. We exchange 1 ten for 10 ones, which gives us 1 ten and 14 ones. Now we can subtract the ones.

$$\begin{array}{r} \$5\,\overset{1}{\cancel{2}}\,{}^{1}4 \\ -\ \$\ \ 5\,8 \\ \hline 6 \end{array}$$

Next we subtract the tens. We cannot subtract 5 tens from 1 ten, so again we will exchange. This time we exchange 1 hundred for 10 tens, which gives us 4 hundreds and 11 tens. Now we finish subtracting.

$$\begin{array}{r} \$\overset{4}{\cancel{5}}\,\overset{{}^{1}1}{\cancel{2}}\,{}^{1}4 \\ -\ \$\ \ 5\ 8 \\ \hline \$4\ 6\ 6 \end{array}$$

Since the value of every column is 10 times the value of the column to its right, we can use this method any time we come to a column in which we cannot subtract.

Example Use the subtraction algorithm to find each difference:

(a) $\begin{array}{r} \$346 \\ -\ \$264 \\ \hline \end{array}$ (b) 219 − 73 (c) $\begin{array}{r} 600 \\ -\ 123 \\ \hline \end{array}$

Solution (a) $\begin{array}{r} \$\overset{2}{\cancel{3}}\,{}^{1}4\ 6 \\ -\ \$2\ 6\ 4 \\ \hline \mathbf{\$\ \ 8\ 2} \end{array}$ (b) $\begin{array}{r} \overset{1}{\cancel{2}}\,{}^{1}1\ 9 \\ -\ \ \ 7\ 3 \\ \hline \mathbf{1\ 4\ 6} \end{array}$

(c) $\begin{array}{r} \overset{5}{\cancel{6}}\,\overset{{}^{1}9}{\cancel{0}}\,{}^{1}0 \\ -\ 1\ 2\ 3 \\ \hline \mathbf{4\ 7\ 7} \end{array}$ or $\begin{array}{r} \overset{5\ 9}{\cancel{6\ 0}}\,{}^{1}0 \\ -\ 1\ 2\ 3 \\ \hline \mathbf{4\ 7\ 7} \end{array}$

Notice part (c). When we try to exchange 1 ten for 10 ones, we find that there are zero tens in the tens column. So we must go to the hundreds column to create some tens. We show two ways to do this. In the first method we exchange 1 hundred for 10 tens, and then we exchange 1 of those tens for 10 ones. In the second method we think of 600 as 60 tens. Taking 1 of the tens leaves 59 tens. Some people think this method of subtracting across zeros is easier and neater than the first.

LESSON PRACTICE

Practice set* Subtract:

a. $\begin{array}{r} \$496 \\ -\ \$157 \\ \hline \end{array}$ **b.** $\begin{array}{r} 400 \\ -\ 136 \\ \hline \end{array}$ **c.** $\begin{array}{r} \$315 \\ -\ \$264 \\ \hline \end{array}$

d. $\begin{array}{r} \$500 \\ -\ \$\ 63 \\ \hline \end{array}$

e. $\begin{array}{r} 435 \\ -\ 76 \\ \hline \end{array}$

f. $\begin{array}{r} 800 \\ -\ 406 \\ \hline \end{array}$

g. 86 − 48

h. $132 − $40

i. 203 − 47

MIXED PRACTICE

Problem set

1. (9) You may use money manipulatives to help answer the question in this story:

Hannah had $550. After she paid a tax of $75, how much money did Hannah have?

2. (2) List the five even, one-digit numbers.

3. (3) Which digit in 596 shows the number of tens?

4. (3) One hundred is equal to how many tens?

5. (8) When seven is subtracted from 15, what is the difference?

6. (8) Write two addition facts and two subtraction facts for the fact family 7, 8, and 15.

7. (5, 6) What is the sum of one hundred ninety and one hundred nineteen?

8. (4, 5) Write this comparison using digits and a comparison symbol:

Five hundred forty is greater than five hundred fourteen.

9. (7) Yosemite National Park in California is one of the oldest national parks in the United States. Yosemite covers 761,266 acres and became a national park in the year 1890. Use words to name the number of acres in Yosemite National Park.

10. (2, 4) Write a three-digit even number less than 200 using the digits 1, 2, and 3.

11. (9)
$$\begin{array}{r} \$346 \\ -\ \$178 \\ \hline \end{array}$$

12. (9)
$$\begin{array}{r} 56 \\ -\ 38 \\ \hline \end{array}$$

13. (9)
$$\begin{array}{r} \$219 \\ -\ \$\ \ 73 \\ \hline \end{array}$$

14. (9)
$$\begin{array}{r} 600 \\ -\ 321 \\ \hline \end{array}$$

15. (9)
$$\begin{array}{r} 300 \\ -\ 124 \\ \hline \end{array}$$

16. (9)
$$\begin{array}{r} \$500 \\ -\ \$246 \\ \hline \end{array}$$

17. (9)
$$\begin{array}{r} 608 \\ -\ 314 \\ \hline \end{array}$$

18. (9)
$$\begin{array}{r} 415 \\ -\ 378 \\ \hline \end{array}$$

19. (6)
$$\begin{array}{r} \$787 \\ \$156 \\ +\ \$324 \\ \hline \end{array}$$

20. (6)
$$\begin{array}{r} 573 \\ 90 \\ +\ 438 \\ \hline \end{array}$$

21. (6)
$$\begin{array}{r} \$645 \\ \$489 \\ +\ \$\ \ 65 \\ \hline \end{array}$$

22. (6)
$$\begin{array}{r} 429 \\ 85 \\ +\ 671 \\ \hline \end{array}$$

Write the ninth term in each counting sequence:

23. (1) 7, 14, 21, ...

24. (1) 9, 18, 27, ...

25. (1) 8, 16, 24, ...

26. (2, 5) Is three hundred seventy an odd number or an even number?

27. (4, 9) Compare. (Try to answer the comparison before subtracting. Then subtract and compare.)

$$31 - 12 \bigcirc 31 - 15$$

28. (2) Half of 20 is 10. What number is half of 21?

LESSON

10 Missing Addends

WARM-UP

Facts Practice: 100 Subtraction Facts (Test B)

Mental Math: Count up and down by 25's between 0 and 200. (Think quarters.) Count up and down by 20's between 0 and 200.

a. 5000 + 4500 **b.** 6000 − 4000 **c.** \$750 + \$250

d. 380 + 90 **e.** 500 − 400 **f.** 125 + 125

g. 640 + 260 **h.** 6 + 6 − 2 + 5

Problem Solving:

Arrange the letters R, T, and A in six different orders. Circle the arrangements that spell words.

NEW CONCEPT

In the number sentence below, there is a missing addend. A letter is used to stand for the missing addend. The letter W is used, but any uppercase or lowercase letter may be used.

$$8 + W = 15$$

A number sentence with an equal sign is often called an **equation.** Since eight plus seven equals 15, we know that the missing addend in this equation is seven. Notice that we can find a missing addend by subtracting. For the number sentence $8 + W = 15$, we subtract eight from 15 to find the missing number:

$$15 - 8 = 7$$

Example 1 Find the missing addend:

$$\begin{array}{r} 24 \\ + \ M \\ \hline 37 \end{array}$$

Solution There are two addends and the sum.

$$\begin{array}{rl} 24 & \text{addend} \\ +\ M & \text{addend} \\ \hline 37 & \text{sum} \end{array}$$

One of the addends is 24. The sum is 37. We subtract 24 from 37 and find that the missing addend is **13.** Then we substitute 13 into the original problem to be sure the answer is correct.

$$\begin{array}{r} 37 \\ -\ 24 \\ \hline 13 \end{array} \nearrow \begin{array}{r} 24 \\ +\ 13 \\ \hline 37 \end{array}$$

Example 2 Find the missing addend: $15 + 20 + 6 + W = 55$

Solution In this equation there are four addends and the sum. The known addends are 15, 20, and 6. Their total is 41.

$$\begin{array}{r} \left.\begin{array}{r} 15 \\ 20 \\ 6 \end{array}\right\} 41 \\ +\ W \quad \\ \hline 55 \quad \end{array}$$

So 41 plus W equals 55. We can find the missing addend by subtracting 41 from 55, which gives us **14.** Then we check the answer.

$$\begin{array}{r} 55 \\ -\ 41 \\ \hline 14 \end{array} \nearrow \begin{array}{r} 15 \\ 20 \\ 6 \\ +\ 14 \\ \hline 55 \end{array}$$

We see that the answer is correct.

LESSON PRACTICE

Practice set Find each missing addend:

a. $35 + m = 67$

b. $n + 27 = 40$

c. $5 + 7 + 9 + f = 30$

d. $15 + k + 10 + 25 = 70$

MIXED PRACTICE

Problem set

1. You may use money manipulatives to help answer the question in this story: (6)

   Maria won $200 in an essay contest. If she had $467 before she won the contest, how much money did she have after she won the contest?

2. For the fact family 4, 5, and 9, write two addition facts and two subtraction facts. (8)

3. Use the digits 4, 5, and 6 to write a three-digit odd number that is greater than 500. (2, 4)

4. Write this comparison using digits and a comparison symbol: (4, 5)

   Six hundred thirteen is less than six hundred thirty.

5. $34 + m = 61$ (10)

6. What is five hundred ten minus fifty-one? (5, 9)

7. Which digit in 325,985 shows the number of hundreds? (3)

8. We can count to 30 by 3's or by 10's. We do not count to 30 when counting by (1)

   A. 2's B. 4's C. 5's D. 6's

9. Think of one odd number and one even number and add them. Is the sum odd or even? (2)

10. Compare. (Try to answer the comparison before subtracting.) (4, 9)

$$100 - 10 \bigcirc 100 - 20$$

11. (9) $\begin{array}{r} \$363 \\ -\ \$179 \\ \hline \end{array}$ **12.** (9) $\begin{array}{r} 400 \\ -\ 176 \\ \hline \end{array}$ **13.** (9) $\begin{array}{r} \$570 \\ -\ \$\ \ 91 \\ \hline \end{array}$ **14.** (9) $\begin{array}{r} 504 \\ -\ 175 \\ \hline \end{array}$

15. (6) $\begin{array}{r} \$367 \\ \$\ \ 48 \\ +\ \$135 \\ \hline \end{array}$ **16.** (6) $\begin{array}{r} 179 \\ 484 \\ +\ 201 \\ \hline \end{array}$ **17.** (6) $\begin{array}{r} \$305 \\ \$897 \\ +\ \$725 \\ \hline \end{array}$ **18.** (6) $\begin{array}{r} 32 \\ 248 \\ +\ 165 \\ \hline \end{array}$

19. (9) $463 − $85

20. (6) 432 + 84 + 578

21. (10) $18 + w = 42$

22. (10) $12 + r = 80$

Write the next four terms in each counting sequence:

23. (1) 3, 6, 9, 12, ...

24. (1) 4, 8, 12, 16, ...

25. (1) 6, 12, 18, 24, ...

26. (3, 7) How many $100 bills are needed to total $1000?

27. (2) Sabrina folded an $8\frac{1}{2}$-by-11-inch piece of paper in half as shown. The folded paper made a rectangle that was $8\frac{1}{2}$ inches by how many inches?

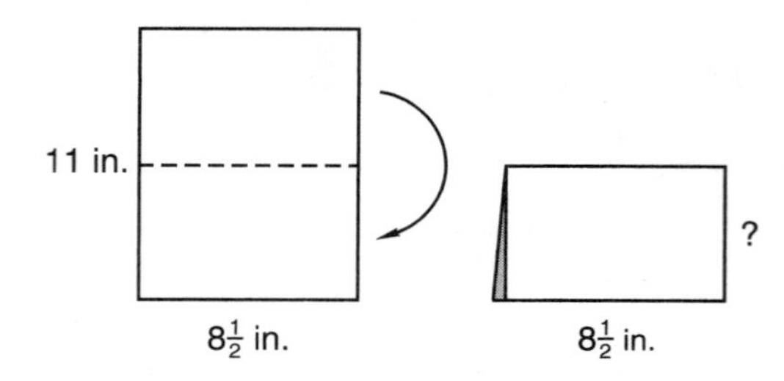

28. (2) Is half of 37,295 a whole number? Why or why not?

INVESTIGATION 1

Focus on

Story Problems

Materials needed for each student:

- 1 copy each of Activity Masters 5, 6, 7, and 8 (available in *Saxon Math 6/5 Assessments and Classroom Masters*)

In this investigation we will study four kinds of math stories: stories about **combining,** stories about **separating,** stories about **equal groups,** and stories about **comparing.** We will see one example of each type of story. All the stories contain three numbers. A story becomes a story problem when one of its numbers is replaced with a question. We will make three different story problems for each story in this investigation by replacing the numbers with questions. In later lessons we will practice solving story problems.

Stories about combining

We combine two (or more) quantities by adding them together. We start with some and add some more. Here is a story about combining:

(a) The troop hiked 8 miles in the morning.

(b) The troop hiked 7 miles in the afternoon.

(c) Altogether, the troop hiked 15 miles.

Notice that there are three numbers. The numbers in (a) and (b) add up to the number in (c). If we know any two of the numbers, we can figure out the third number. The story is written in three sentences. Suppose sentence (a) were missing. Read sentences (b) and (c); then write a question that asks for the number in sentence (a). Start the question with the words, "How many miles ..."

Now suppose sentence (b) were missing from the story. Read sentences (a) and (c); then write a question that asks for the number in sentence (b). Start with the words, "How many miles ..."

Finally, suppose sentence (c) were missing. Read sentences (a) and (b); then write a question that asks for the number in (c). This time start the question, "Altogether, how many miles ..."

Stories about separating

We separate one quantity from a larger quantity by taking some away (by subtracting). Here is a story about separating:

(d) Jack went to the store with $28.

(e) Jack spent $12 at the store.

(f) Jack left the store with $16.

This is a story about Jack's money. Jack had some money; then some money "went away" at the store. There are three numbers in the story. If one of the numbers were missing, we could figure out the missing number. Suppose sentence (d) were missing. Read sentences (e) and (f); then write a question that asks for the number in sentence (d). Start with the words, "How much money ..."

Now suppose sentence (e) were missing. Read sentences (d) and (f); then write a question that asks for the number in sentence (e). Start with the words, "How much money ..."

Finally, suppose sentence (f) were missing. Read (d) and (e); then write a question that asks for the number in sentence (f).

Stories about equal groups

Some stories are about items that are clustered in groups of equal size. These stories might describe the number of groups, the number in each group, and/or the total number in all groups. By multiplying the number in each group by the number of groups, we can find the total in all groups. Here is an example of an "equal groups" story:

At Lincoln School there are the same number of students in each fifth-grade class.

(g) At Lincoln School there are 4 classes of fifth-grade students.

(h) There are 30 students in each fifth-grade class.

(i) Altogether, there are 120 fifth-grade students at Lincoln School.

Again we see three numbers in the story. If we know two of the numbers, we can figure out the third number. Suppose sentence (g) were missing. Read sentences (h) and (i); then write a question that asks for the number in sentence (g). Start with the words, "How many classes ..."

Now suppose sentence (h) were missing. Read sentences (g) and (i); then write a question that asks for the number in sentence (h). Start with the words, "How many students ..."

Finally, suppose sentence (i) were missing. Read sentences (g) and (h); then write a question that asks for the number in sentence (i). Start with the words, "Altogether, how many ..."

Stories about comparing

One way to compare two numbers is to find how much larger or how much smaller one number is than the other. By subtracting the smaller number from the larger number, we find the difference of the numbers. Consider this story about comparing:

(j) Abe is 5 years old.

(k) Gabe is 11 years old.

(l) Gabe is 6 years older than Abe.

A comparison may be stated two ways. For example, sentence (l) could have been written, "Abe is 6 years younger than Gabe."

Once again, our story has three numbers. If we know two of the numbers, we can figure out the third number. Suppose sentence (j) were missing. Read sentences (k) and (l); then write a question that asks for the number in sentence (j).

Now suppose sentence (k) were missing. Read sentences (j) and (l); then write a question that asks for the number in sentence (k).

Finally, suppose sentence (l) were missing. Read sentences (j) and (k); then write a question that asks for the number in sentence (l). You should be able to phrase the question two different ways.

LESSON

11 Story Problems About Combining

WARM-UP

Facts Practice: 100 Subtraction Facts (Test B)

Mental Math: Count up and down by 25's between 0 and 200.
Count up and down by 250's between 0 and 2000.

a. 6000 + 3200 **b.** 5000 – 3000 **c.** 375 + 125
d. 570 + 250 **e.** 350 – 300 **f.** 540 – 140
g. 7 + 6 – 3 + 4 **h.** 10 – 3 + 7 + 10

Problem Solving:

Find three different letters that can be arranged to spell at least two different words. Can you think of three different letters that can be arranged to spell three different words?

NEW CONCEPT

In action stories a heroic character often risks danger to rescue someone who is in trouble. The details and characters of the stories might differ, but the underlying idea, the plot, is the same.

Although we consider many stories in mathematics, there are only a few *kinds* of stories. One kind of math story has a "some plus some more" plot. Here are three "some plus some more" stories:

The troop hiked 8 miles in the morning and 7 miles in the afternoon. Altogether, how many miles did the troop hike?

After Jamal paid Hannah $120 rent, Hannah had $645. How much money did Hannah have before Jamal paid Hannah for rent?

Cheryl counted 18 children on the playground before the bell rang. After the bell rang, more children ran onto the playground. Then Cheryl counted 98 children. How many children ran onto the playground after the bell rang?

In each of these stories there was "some." Then "some more" was added to make a total. The total results from *combining* the two quantities. In mathematics the process of combining is called *addition*. Here are two ways to write the addition pattern:

$$\begin{array}{r} \text{Some} \\ +\ \text{Some more} \\ \hline \text{Total} \end{array} \qquad \text{Some} + \text{some more} = \text{total}$$

In each story we are given two numbers and are asked to find a third number. We can write an addition equation to find the third number.

To answer the question in a "some plus some more" story, we will:

1. Write the equation.
2. Find the missing number.
3. Check whether our answer is reasonable and our arithmetic is correct.

Example 1 The troop hiked 8 miles in the morning and 7 miles in the afternoon. Altogether, how many miles did the troop hike?

Solution This is a story about combining some miles hiked in the morning with some more miles hiked in the afternoon. We follow three steps.

Step 1: We write an equation for the story that follows the addition pattern.

$$\text{Some} + \text{some more} = \text{total}$$

$$8 \text{ mi} + 7 \text{ mi} = t$$

Step 2: We find the missing number. In this story the missing number is the total, so we add 8 miles and 7 miles.

$$8 \text{ mi} + 7 \text{ mi} = 15 \text{ mi}$$

Step 3: We check to be sure the answer is reasonable and the arithmetic is correct. It is reasonable that the total distance hiked is more than the distance hiked during either the morning or the afternoon. Also, the sum of 8 and 7 is correct. So the troop hiked a total of **15 miles.**

Example 2 After Jamal paid Hannah $120 rent, Hannah had $645. How much money did Hannah have before Jamal paid Hannah for rent?

Solution Hannah had some money. Jamal paid her some more money. We write an equation. This time we will write the equation vertically.

Hannah had some money.	S
Jamal paid Hannah $120.	+ $120
Then Hannah had $645.	$645

The missing number represents how much money Hannah had before Jamal paid her. We can find a missing addend by subtracting.

$$\begin{array}{r} \$645 \\ -\ \$120 \\ \hline \$525 \end{array}$$

Now we check whether the answer is reasonable and correct. Is it reasonable that Hannah had less money before Jamal paid her? To check the answer, we return to the original problem and add.

Hannah had some money.	$525
Jamal paid Hannah $120.	+ $120
Then Hannah had $645.	$645

Before Jamal paid her, Hannah had **$525.**

Example 3 Cheryl counted 18 children on the playground before the bell rang. After the bell rang, more children ran onto the playground. Then Cheryl counted 98 children. How many children ran onto the playground after the bell rang?

Solution We write an equation for the story. There were 18 children at first. Then some more children arrived, making the total 98 children.

$$\text{Some} + \text{some more} = \text{total}$$
$$18 + m = 98$$

The missing number is one of the addends. We can find a missing addend by subtracting.

$$\begin{array}{r} 98\text{ children} \\ -\ 18\text{ children} \\ \hline 80\text{ children} \end{array}$$

We check whether the answer is reasonable and correct. Is it reasonable that 80 children ran onto the playground after the bell rang? Is the arithmetic correct? We add to check the answer:

$$\begin{array}{r} 18 \text{ children} \\ +\ 80 \text{ children} \\ \hline 98 \text{ children} \end{array}$$

After the bell rang, **80 children** ran onto the playground.

LESSON PRACTICE

Practice set In problems **a** and **b**, write an equation for the story. Then find the missing number and check the answer.

a. Tammy wants to buy a camera. She has \$24. The camera costs \$41. How much more money does Tammy need?

b. Ling was swimming laps when her mother came to watch. Her mother watched Ling swim her final 16 laps. If Ling swam 30 laps in all, how many laps did Ling swim before her mother arrived?

c. Write a story about combining for this equation. Then answer the question in your story problem.

$$\$12 + \$24 = T$$

MIXED PRACTICE

Problem set For problems 1–3, write an equation and find the answer.

1. (11) Nia scored 21 points in the game. If she scored 13 points in the first half of the game, how many points did she score in the second half?

2. (11) Nia's team scored 62 points to win the game. If the team scored 29 points in the second half, how many points did the team score in the first half?

3. (11) The Lees traveled 397 miles one day and 406 miles the next day. Altogether, how many miles did the Lees travel in two days?

4. (8) For the fact family 8, 9, and 17, write two addition facts and two subtraction facts.

5. (2) What is the greatest three-digit even number that can be written using the digits 1, 2, and 3?

6. (4, 6) Compare. (Try to answer the comparison before adding.)

$$8 + 7 + 6 \bigcirc 6 + 7 + 8$$

7. (7) Write this comparison using digits and a comparison symbol:

Eighty thousand is greater than eighteen thousand.

8. (5, 8) Write the following sentence using digits and symbols:

Forty minus fourteen equals twenty-six.

9. (2, 6) Think of two odd numbers and one even number. Add them all together. Is the sum odd or even?

10. (5) Use digits to write four hundred eight dollars and seventy cents.

11. (9) $\begin{array}{r} \$872 \\ -\ \$\ 56 \\ \hline \end{array}$

12. (9) $\begin{array}{r} 706 \\ -\ 134 \\ \hline \end{array}$

13. (9) $\begin{array}{r} \$800 \\ -\ \$139 \\ \hline \end{array}$

14. (9) $\begin{array}{r} 365 \\ -\ 285 \\ \hline \end{array}$

15. (10) $\begin{array}{r} 578 \\ +\quad A \\ \hline 600 \end{array}$

16. (6) $\begin{array}{r} \$640 \\ \$152 \\ +\ \$749 \\ \hline \end{array}$

17. (6) $\begin{array}{r} 365 \\ 294 \\ +\ 716 \\ \hline \end{array}$

18. (6) $\begin{array}{r} \$475 \\ \$233 \\ +\ \$\ 76 \\ \hline \end{array}$

19. (9) $317 − $58

20. (10) $433 + 56 + Q = 497$

21. (10) $7 + w = 15$

22. (10) $15 + y = 70$

Write the next four terms in each counting sequence:

23. (1) 9, 18, 27, 36, ...

24. (1) 8, 16, 24, 32, ...

25. (1) Find the tenth term of this counting sequence:

7, 14, 21, ...

26. (1) Below is a different kind of sequence. Notice how the figure turns from one term of the sequence to the next.

◺, ◸, ◹, ...

What is the next term in the sequence?

A. ◿ B.  ⌂ C.  ⊕

27. (2) Every morning Kayla runs to the tracks and back. If the round trip is 3 miles, how far is it to the tracks?

28. (2) Find half of 10. Then find half of half of 10. Write both answers.

29. (11) Michael has $18. His brother has $15. Use this information to write a story problem about combining. Then answer the question in your story problem.

LESSON

12 Lines • Number Lines • Tally Marks

WARM-UP

Facts Practice: 100 Subtraction Facts (Test B)

Mental Math: Count up and down by 25's between 0 and 300.
Count up and down by 50's between 0 and 500.

a. 6500 + 500 **b.** 1000 − 500 **c.** 75 + 75
d. 750 + 750 **e.** 460 − 400 **f.** 380 − 180
g. 20 + 30 − 5 **h.** 16 − 8 + 4 − 2 + 1

Problem Solving:

Lance, Molly, and José lined up side by side for a picture. Then they changed their order. Then they changed their order again. List all the possible side-by-side arrangements Lance, Molly, and José could make.

NEW CONCEPTS

Lines

In mathematics we study numbers. We also study shapes such as circles, squares, and triangles. The study of shapes is called **geometry.** The simplest figures in geometry are the **point** and the **line.** A line does not end. Part of a line is called a **line segment** or just a *segment.* A line segment has two endpoints. A **ray** (sometimes called a *half line*) begins at a point and continues without end. Here we illustrate a point, a line, a segment, and a ray. The arrowheads on the line and the ray show the directions in which those figures continue without end.

Point Line Line segment Ray

Lines, rays, and segments can be **horizontal, vertical,** or **oblique.** The term *horizontal* comes from the word *horizon.* When we look into the distance, the horizon is the line where the earth and sky seem to meet. A horizontal line is level with the horizon, extending left and right. A vertical line extends up and down.

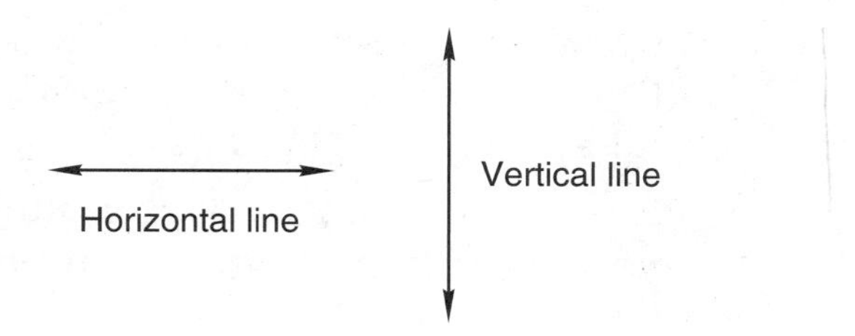

A line or segment that is neither horizontal nor vertical is oblique. An oblique line appears to be slanted.

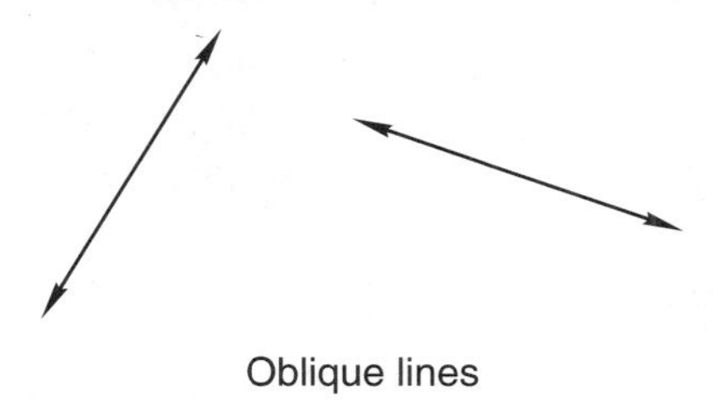

Oblique lines

Number lines By carefully marking and numbering a line, we can make a **number line. A number line shows numbers at a certain distance from zero.** On the following number line, the distance from 0 to 1 is a segment of a certain length, which we call a *unit segment.* The distance from 0 to 5 is five unit segments. The arrowheads show that the number line continues in both directions. Numbers to the left of zero are called **negative numbers.** We read the minus sign by saying "negative," so we read –3 as "negative three." The small marks above each number are **tick marks.**

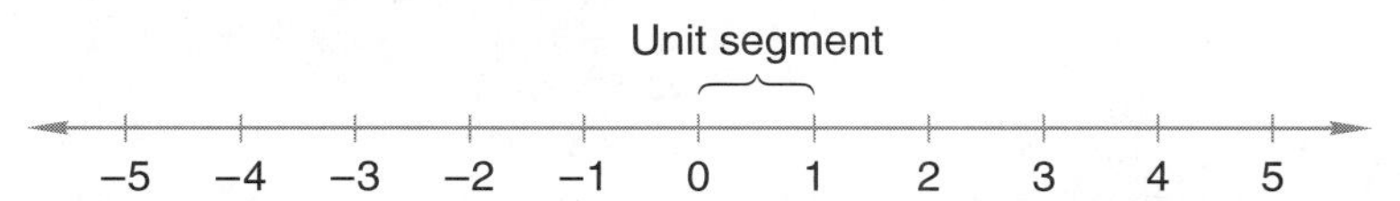

The numbers shown on the number line above are called **integers.** Integers include all the counting numbers, the negatives of all the counting numbers, and the number zero.

Example 1 This sequence counts down by ones. Write the next six numbers in the sequence, and say the numbers aloud as a class.

5, 4, 3, ...

Solution The next six numbers in the sequence are

2, 1, 0, –1, –2, –3

We read these numbers as "two, one, zero, negative one, negative two, negative three."

Example 2 Draw a number line marked with whole numbers from 0 to 5.

Solution Begin by drawing a line segment. An arrowhead should be drawn on each end of the segment to show that the number line continues without end. Make a tick mark for zero and label it "0." Make *equally spaced* tick marks to the right of zero for the numbers 1, 2, 3, 4, and 5. Label those tick marks. When you are finished, your number line should look like this:

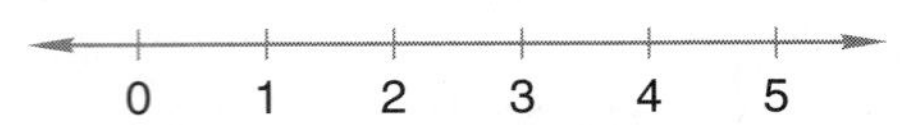

To count on a number line, it is important to focus our attention more on the **segments** than on the tick marks. To help us concentrate on the segments, we will answer questions such as the following:

Example 3 How many unit segments are there from 2 to 5 on the number line in example 2?

Solution On the number line above, the distance from 0 to 1 is one unit segment. We see one unit segment from 2 to 3, another from 3 to 4, and a third from 4 to 5. Thus, the number of unit segments from 2 to 5 is **three.**

Example 4 On the number line below, arrows (a) and (b) indicate integers. Write the two integers, using a comparison symbol to show which integer is greater.

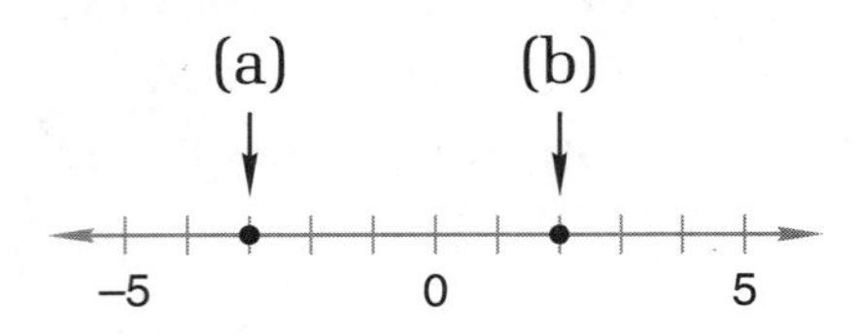

Solution Arrow (a) indicates –3 and arrow (b) indicates 2. Numbers to the right on the number line are greater than numbers to the left. We may write the comparison two ways.

$$\mathbf{-3 < 2} \quad \text{or} \quad \mathbf{2 > -3}$$

Tally marks

Tally marks are used to keep track of a count. Each tally mark counts as one. Here we show the tallies for the numbers one through six.

Notice that the tally mark for five is a diagonal mark crossing four vertical marks.

Example 5 What number is represented by this tally? 𝍸 𝍸 𝍸 ||

Solution We see three groups of five, which is 15, and we see two more tally marks, which makes **17.**

LESSON PRACTICE

Practice set

a. Which of these represents a line segment?

A. ———— B. ←————→ C. •————→

b. Draw a vertical line.

c. Draw a horizontal segment.

d. Draw an oblique ray.

e. Draw a number line marked with integers from –3 to 3.

f. How many unit segments are there from –2 to 3 on the number line you drew in problem **e?**

g. What are the next five numbers in this counting sequence?

10, 8, 6, ...

h. Write the two integers indicated on this number line, using a comparison symbol between the integers to show which is greater.

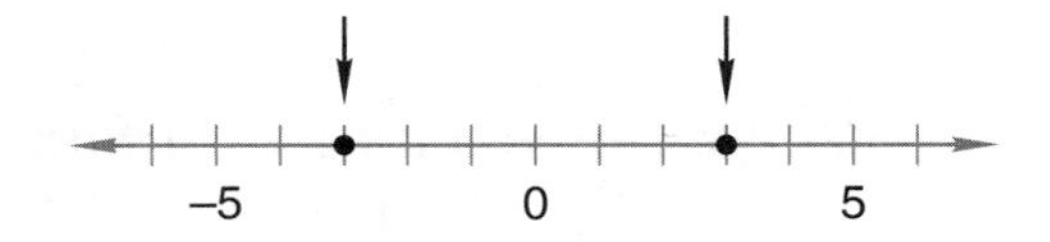

i. What whole number is six unit segments to the right of –4 on the number line above?

j. What number is represented by this tally? 𝍸 𝍸 ||||

MIXED PRACTICE

Problem set

1. (12) How many unit segments are there from 2 to 7 on the number line?

2. (12) Use tally marks to show the number 7.

For problems 3 and 4, write an equation and find the answer.

3. (11) Gilbert weighs 94 pounds. Andy weighs 86 pounds. If they both stand on one scale, what will they weigh together?

4. (11) Andy weighs 86 pounds. With his book bag on, Andy weighs 110 pounds. How much does Andy's book bag weigh?

5. (9) $\begin{array}{r} 862 \\ -\ 79 \\ \hline \end{array}$

6. (9) $\begin{array}{r} \$420 \\ -\ \$137 \\ \hline \end{array}$

7. (9) $\begin{array}{r} 508 \\ -\ 96 \\ \hline \end{array}$

8. (9) $\begin{array}{r} \$500 \\ -\ \$136 \\ \hline \end{array}$

9. (6) $\begin{array}{r} \$248 \\ \$514 \\ +\ \$\ 18 \\ \hline \end{array}$

10. (6) $\begin{array}{r} 907 \\ 45 \\ +\ 653 \\ \hline \end{array}$

11. (6) $\begin{array}{r} \$367 \\ \$425 \\ +\ \$740 \\ \hline \end{array}$

12. (10) $\begin{array}{r} W \\ +\ 427 \\ \hline 568 \end{array}$

13. (10) $38 + 427 + P = 475$

14. (9) $580 − $94

15. (4) The number 57 is between which pair of numbers?

A. 40 and 50 B. 50 and 60 C. 60 and 70

16. (7) Write this comparison using digits and a comparison symbol:

Eighteen thousand is less than eighty thousand.

17. (8) Write two addition facts and two subtraction facts for the fact family 4, 6, and 10.

18. (2, 8) Think of an odd number and an even number. Subtract the smaller number from the larger number. Is the answer odd or even?

In problems 19 and 20, find the missing number that makes the equation true.

19. (10) $18 + m = 150$

20. (10) $12 + y = 51$

21. (4, 6) In this problem the letters x and y are each one-digit numbers. Compare: $x + y \bigcirc 19$

Write the next six terms in each counting sequence:

22. 2, 4, 6, ...
(1)

23. 3, 6, 9, ...
(1)

24. 4, 8, 12, ...
(1)

25. 30, 25, 20, ...
(1, 12)

26. Use words to write 5280.
(7)

27. Is the 99th term of this counting sequence odd or even?
(1, 2)

2, 4, 6, 8, ...

28. Grace has $7.00 in her wallet and $4.37 in a coin jar. Use this information to write a story problem about combining, and answer the question in your story problem.
(11)

29. Mark bought a box of pencils, two pens, and a notebook. What was the total cost of the items?
(6, 11)

Item	Price
Notebook	$1.79
Box of pencils	$1.50
Pen	$0.65

LESSON

13 Multiplication as Repeated Addition • Adding and Subtracting Dollars and Cents

WARM-UP

Facts Practice: 100 Subtraction Facts (Test B)

Mental Math: Count by 25¢ from 25¢ to $3.00 and from $3.00 to 25¢.

a. 6500 − 500 **b.** 2000 − 100 **c.** 225 + 225
d. 750 + 750 **e.** 360 − 200 **f.** 425 − 125
g. 50 + 50 − 25 **h.** 8 + 8 − 1 + 5 − 2

Problem Solving:

Copy this addition problem and fill in the missing digits:

```
  3_4
+ 23_
 ----
  _03
```

NEW CONCEPTS

Multiplication as repeated addition

Consider this word problem:

There are 5 rows of desks with 6 desks in each row. How many desks are there in all?

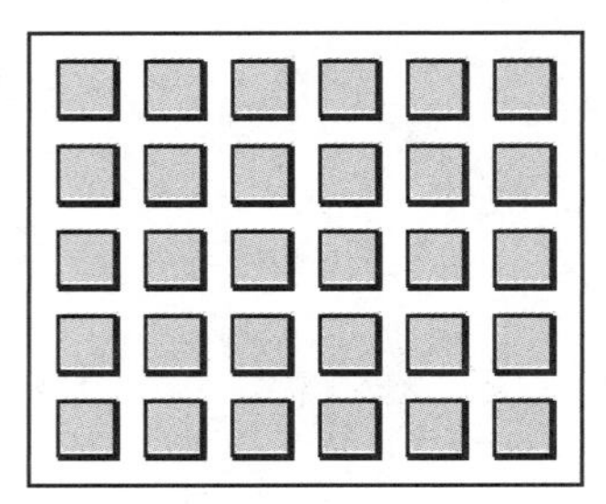

There are many ways to find an answer to this problem. One way is to count the desks one at a time. Another way is to count the desks in one row and then count the number of rows. Since there are 5 rows of 6 desks, we can add five 6's together, as we show here:

$$6 + 6 + 6 + 6 + 6$$

We can also **multiply** to find the number of desks. Whenever we need to add the same number over and over, we may multiply. To find the sum of five 6's, we multiply 5 by 6. We show two ways to write this.

$$5 \times 6 \qquad \begin{array}{r} 6 \\ \times\ 5 \\ \hline \end{array}$$

The × is called a **times sign.** We read 5 × 6 by saying "five times six." "Five times six" means the sum of five 6's. Multiplication is a way to repeatedly add a number. In the example above, the second number (6) is added the number of times shown by the first number (5).

In the picture on the previous page, we see 5 rows of desks with 6 desks in each row. However, if we turn the book sideways, we see 6 rows of desks with 5 desks in each row. So we see that six 5's is the same as five 6's.

> Five 6's (5 × 6) means 6 + 6 + 6 + 6 + 6, which equals 30.
>
> Six 5's (6 × 5) means 5 + 5 + 5 + 5 + 5 + 5, which also equals 30.

We see that the answer to 6 × 5 is the same as the answer to 5 × 6. This shows us that we may multiply in any order.

Example 1 Change this addition problem to a multiplication problem:

$$7 + 7 + 7 + 7$$

Solution We are asked to change the addition problem into a multiplication problem. We are not asked to find the sum. We see four 7's added together. We can write four 7's as the multiplication **4 × 7.** We note that four 7's is also equal to 7 × 4.

Adding and subtracting dollars and cents

To add or subtract dollars and cents, we align the decimal points so that we add or subtract digits with the same place value.

Example 2 (a) \$3.45 + \$6.23 + \$0.50 (b) \$4.50 − \$3.80

Solution (a)

$$\begin{array}{r} \overset{1}{\$3}.45 \\ \$6.23 \\ +\ \$0.50 \\ \hline \mathbf{\$10.18} \end{array}$$

(b)

$$\begin{array}{r} \$\overset{3}{\cancel{4}}.\overset{1}{5}0 \\ -\ \$3.80 \\ \hline \mathbf{\$0.70} \end{array}$$

The answer to (b) is 70 cents. The zero to the left of the decimal point shows that there are no dollars.

A number of dollars may be written either with or without a decimal point. Both of these mean "five dollars":

$5 $5.00

When adding or subtracting dollars and cents, it is a good idea to write all whole-dollar amounts with a decimal point and two zeros, as we show below.

Example 3 Add: $5 + $8.75 + $10 + $0.35

Solution We rewrite the problem so that the whole-dollar amounts contain a decimal point and two zeros.

$5.00 + $8.75 + $10.00 + $0.35

Next, we set up the problem so that the decimal points line up vertically. Then we add. We place the decimal point in the answer in line with the decimal points in the problem.

```
  $ 5.00
  $ 8.75
  $10.00
+ $ 0.35
--------
  $24.10
```

LESSON PRACTICE

Practice set* Write a multiplication problem for each of these addition problems:

a. 8 + 8 + 8 + 8 **b.** 25 + 25 + 25

c. Write a multiplication problem that shows how to find the number of X's.

X X X X X X
X X X X X X
X X X X X X

Find each sum or difference:

d.
```
  $5.26
+ $8.92
-------
```

e.
```
  $3.27
- $2.65
-------
```

f. $10 + $3.75 + $2 **g.** $5 − $1.87

MIXED PRACTICE

Practice set

1. (12) Draw a number line marked with whole numbers from 0 to 8. How many unit segments are there from 3 to 7?

2. (12) Use tally marks to show the number 9.

For problems 3 and 4, write an equation and find the answer.

3. (11) Corina hiked 33 miles in one day. If she hiked 14 miles after noon, how many miles did she hike before noon?

4. (11) Of the 23 students in the classroom, 11 were boys. How many girls were in the classroom?

5. (8) Write two addition facts and two subtraction facts for the fact family 3, 7, and 10.

6. (11, 13) Jim paddled the canoe down the river 25 miles each day for 5 days. How far did he travel in 5 days?

7. (9) $\begin{array}{r} 300 \\ -\ 114 \\ \hline \end{array}$

8. (13) $\begin{array}{r} \$5.60 \\ -\ \$2.84 \\ \hline \end{array}$

9. (9) $\begin{array}{r} 203 \\ -\ \ 87 \\ \hline \end{array}$

10. (9) $\begin{array}{r} \$512 \\ -\ \$123 \\ \hline \end{array}$

11. (10) $\begin{array}{r} 432 \\ +\ \ B \\ \hline 683 \end{array}$

12. (13) $\begin{array}{r} \$2.54 \\ \$5.36 \\ +\ \$0.75 \\ \hline \end{array}$

13. (6) $\begin{array}{r} 387 \\ 496 \\ +\ 874 \\ \hline \end{array}$

14. (6) $\begin{array}{r} \$\ 97 \\ \$436 \\ +\ \$468 \\ \hline \end{array}$

15. (4, 8) Compare: fifteen minus five ◯ fifteen minus six

16. (2, 6) Think of two even numbers and one odd number. Add them together. Is the sum odd or even?

17. (13) \$4.56 + \$13.76

18. (9) \$5127 − \$49

19. (10) $N + 27 + 123 = 153$

20. (9) 2510 − 432

21. (13) \$5 − \$3.36

22. (13) \$5 + \$3.36 + \$0.54

Write the next six terms in each counting sequence:

23. (1) 6, 12, 18, ...

24. (1) 7, 14, 21, ...

25. (1) 8, 16, 24, ...

26. (1) 9, 18, 27, ...

27. (4, 6) Compare: $3 + 3 + 3 + 3$ ◯ $4 + 4 + 4$

28. (13) Change this addition problem to a multiplication problem:

$$7 + 7 + 7 + 7 + 7 + 7$$

29. (13) Write a multiplication problem that shows how to find the number of X's.

X X X X X
X X X X X
X X X X X
X X X X X

LESSON

14 Missing Numbers in Subtraction

WARM-UP

Facts Practice: 100 Subtraction Facts (Test B)

Mental Math: Count by 25¢ from 25¢ to $3.00 and from $3.00 to 25¢. Count by 50¢ from 50¢ to $5.00 and from $5.00 to 50¢.

a. 2500 + 500 **b.** 2500 − 500 **c.** 390 + 450

d. $7.50 + $2.50 **e.** 10 + 10 − 5 + 10 − 5

f. How much money is 2 quarters? ... 3 quarters? ... 4 quarters?

Problem Solving:

Copy this addition problem and fill in the missing digits:

```
  52_
+ _94
 _0_2
```

NEW CONCEPT

The numbers in an addition fact can be read in reverse direction to form a subtraction fact.

READING DOWN		READING UP
Three plus four is seven. ↓	$\begin{array}{r} 3 \\ +\ 4 \\ \hline 7 \end{array}$	↑ Seven minus four is three.

Likewise, the numbers in a subtraction fact can be read in reverse direction to form an addition fact.

READING DOWN		READING UP
Nine minus five is four. ↓	$\begin{array}{r} 9 \\ -\ 5 \\ \hline 4 \end{array}$	↑ Four plus five is nine.

Example 1 Reverse the order of these numbers to make an addition equation:

$$68 - 45 = 23$$

Solution We write the numbers in reverse order.

23 45 68

Then we insert a plus sign and an equal sign to make an equation.

$$\mathbf{23 + 45 = 68}$$

Example 2 Reverse the order of these numbers to make an addition equation:

$$\begin{array}{r} 77 \\ -\ 23 \\ \hline 54 \end{array}$$

Solution We write the numbers in reverse order, using a plus sign instead of a minus sign.

$$\begin{array}{r} \mathbf{54} \\ +\ \mathbf{23} \\ \hline \mathbf{77} \end{array}$$

In this lesson we will practice finding missing numbers in subtraction problems. There are three numbers in a subtraction problem. Any one of the three numbers may be missing. Sometimes, changing a subtraction problem to an addition problem can help us find the missing number.

Example 3 Find the missing number:

$$\begin{array}{r} F \\ -\ 15 \\ \hline 24 \end{array}$$

Solution We need to find the first number in this subtraction problem. When 15 is subtracted from F, the difference is 24. So F must be more than 24. We will read this subtraction problem in reverse direction to form an addition problem.

READING DOWN ↓		READING UP ↑
F minus fifteen is twenty-four.	$\begin{array}{r} F \\ -\ 15 \\ \hline 24 \end{array}$	Twenty-four plus fifteen is F.

Reading up, we see that 24 plus 15 is F. This means we can find F by adding 24 and 15.

$$\begin{array}{r} 24 \\ +\ 15 \\ \hline 39 \end{array}$$

We find that F is **39.** To check our work, we replace F with 39 in the original problem.

REPLACE F WITH 39.

$$\begin{array}{r} F \\ -\ 15 \\ \hline 24 \end{array} \longrightarrow \begin{array}{r} 39 \\ -\ 15 \\ \hline 24 \end{array} \quad \text{This is correct.}$$

From the previous example, we know the following:

$\begin{array}{r} F \\ -\ S \\ \hline T \end{array}$ ←	The first number of a subtraction problem can be found by adding the other two numbers.

Example 4 Find the missing number: $\begin{array}{r} 45 \\ -\ S \\ \hline 21 \end{array}$

Solution We need to find the second number in this subtraction problem. When S is subtracted from 45, the difference is 21. So S must be less than 45. We will read this problem in both directions.

Reading Down		Reading Up
Forty-five minus S is twenty-one. ↓	$\begin{array}{r} 45 \\ -\ S \\ \hline 21 \end{array}$	↑ Twenty-one plus S is forty-five.

Reading up, we see that 21 plus S is 45. When the problem is read this way, S is a missing addend. We find a missing addend by subtracting.

$$\begin{array}{r} 45 \\ -\ 21 \\ \hline 24 \end{array}$$

We find that S is **24.** Now we replace S with 24 in the original problem to check the answer.

Replace S with 24.

$$\begin{array}{r} 45 \\ -\ S \\ \hline 21 \end{array} \longrightarrow \begin{array}{r} 45 \\ -\ 24 \\ \hline 21 \end{array}$$ This is correct.

Here is another fact to remember for finding missing numbers in subtraction problems:

$\begin{array}{r} F \\ -\ S \\ \hline T \end{array}$ ←	The second or third number of a subtraction problem can be found by subtracting.

LESSON PRACTICE

Practice set Reverse the order of the numbers to change each subtraction equation to an addition equation:

a. $34 - 12 = 22$

b. $\begin{array}{r} 56 \\ -\ 29 \\ \hline 27 \end{array}$

Find the missing number in each subtraction problem:

c. $w - 8 = 6$

d. $23 - y = 17$

e. $\begin{array}{r} N \\ -\ 24 \\ \hline 48 \end{array}$

f. $\begin{array}{r} 63 \\ -\ P \\ \hline 20 \end{array}$

g. $\begin{array}{r} Q \\ -\ 36 \\ \hline 14 \end{array}$

h. $\begin{array}{r} 42 \\ -\ R \\ \hline 24 \end{array}$

MIXED PRACTICE

Problem set

1. (12) Draw a number line marked with integers from –5 to 5. How many unit segments are there from 1 to 5?

2. (5) Use words to name $4.48.

3. (7) Use digits to write eight hundred eighteen thousand, eighty.

4. (12) John used tally marks to keep track of the number of votes each candidate received. The winner received 11 votes. Use tally marks to show the number 11.

For problems 5 and 6, write an equation and find the answer.

5. (11) Janet is reading a 260-page book. She has read 85 pages. How many more pages does she have left to read?

6. (11) Esmerelda mixed 32 ounces of soda with 24 ounces of juice to make the punch. How many ounces of punch did she make?

7. (4, 5) Write this comparison using digits and a comparison symbol:

Fifty-six is less than sixty-five.

8. (2) Write the greatest three-digit even number that contains the digits 1, 2, and 3.

9. (13) $\begin{array}{r} \$43.10 \\ - \$\ 1.54 \\ \hline \end{array}$

10. (13) $\begin{array}{r} \$3.01 \\ - \$1.03 \\ \hline \end{array}$

11. (14) $\begin{array}{r} 600 \\ - \quad M \\ \hline 364 \end{array}$

12. (9) $\begin{array}{r} 4625 \\ - 1387 \\ \hline \end{array}$

13. (13) $\begin{array}{r} \$3.67 \\ \$4.12 \\ + \$5.01 \\ \hline \end{array}$

14. (6) $\begin{array}{r} \$573 \\ \$\ 96 \\ + \$427 \\ \hline \end{array}$

15. (6) $\begin{array}{r} 68 \\ 532 \\ + 176 \\ \hline \end{array}$

16. (10) $\begin{array}{r} 436 \\ + \quad Y \\ \hline 634 \end{array}$

17. (14) $100 - N = 48$

18. (13) $31.40 − $13.40

19. (10) $6 + 48 + 9 + W = 100$

20. (6) $3714 + 56 + 459$

21. (14) Reverse the order of the numbers to change this subtraction equation to an addition equation:

$$50 - 18 = 32$$

22. (1, 12) This sequence counts down by threes. What are the next six terms in the sequence?

12, 9, 6, ...

23. (8) Write two addition facts and two subtraction facts for the fact family 2, 8, and 10.

24. (14) $\begin{array}{r} N \\ - 17 \\ \hline 12 \end{array}$

25. (14) $\begin{array}{r} P \\ - 175 \\ \hline 125 \end{array}$

26. (13) Change this addition problem to a multiplication problem:

$$10 + 10 + 10 + 10$$

27. (2) In a class of 23 students, there are 12 girls. Do girls make up more than or less than half the class?

28. (12) Draw a horizontal segment and a vertical ray.

29. (11) Some stories about combining have more than two addends. The story below has four addends. Answer the question in this story problem.

The football team scored 6 points in the first quarter, 13 points in the second quarter, 7 points in the third quarter, and 6 points in the fourth quarter. How many points did the team score in all four quarters?

LESSON

15 Making a Multiplication Table

WARM-UP

Facts Practice: 100 Subtraction Facts (Test B)

Mental Math: Count up and down by 50's between 0 and 500.

a. 50 + 50 + 50
b. 500 + 500 + 500
c. 24 + 26
d. 240 + 260
e. 480 − 200
f. 30 + 15
g. 270 + 280
h. $4.50 − $1.25
i. How much money is 3 quarters? ... 5 quarters? ... 6 quarters?
j. 10 + 6 − 1 + 5 + 10

Problem Solving:

Billy, Ricardo, and Sherry finished first, second, and third in the race, though not necessarily in that order. List the different orders in which they could have finished.

NEW CONCEPT

Below we list several sequences of numbers. Together, these sequences form an important pattern.

Zeros	0	0	0	0	0	0
Ones	1	2	3	4	5	6
Twos	2	4	6	8	10	12
Threes	3	6	9	12	15	18
Fours	4	8	12	16	20	24
Fives	5	10	15	20	25	30
Sixes	6	12	18	24	30	36

This pattern is sometimes called a **multiplication table.** A multiplication table usually lists the first ten or more **multiples** of the first ten or more whole numbers. On a multiplication table, we can find the answer to questions such as, "How much is three 4's?" We do this by using the rows and columns on the table. (Rows run left to right, and columns run top to bottom.)

To find how much three 4's equals, we locate the row that begins with 3 and the column that begins with 4. Then we look across the row and down the column for the number where the row and column meet.

Column ↓

	0	1	2	3	4	5	6
0	0	0	0	0	0	0	0
1	0	1	2	3	4	5	6
2	0	2	4	6	8	10	12
Row → 3	0	3	6	9	(12)	15	18
4	0	4	8	12	16	20	24
5	0	5	10	15	20	25	30
6	0	6	12	18	24	30	36

We find that three 4's equals 12.

Numbers that are multiplied together are called **factors.** The factors in the problem above are 3 and 4. The answer to a multiplication problem is called a **product.** From the table we see that the product of 3 and 4 is 12. The table shows us that 12 is also the product of 4 and 3. (Find the number where row 4 and column 3 meet.) Thus, the **commutative property** applies to multiplication as well as to addition. We may choose the order of factors when we multiply.

Notice the row and column of zeros. When we multiply by zero, the product is zero. This fact is called the **property of zero for multiplication.** We may think of 2×0 or 10×0 or 100×0 as two 0's or ten 0's or one hundred 0's added together. In each case the sum is zero.

Notice also that 1 times any number is that number. For example, one 5 is 5 and five 1's is 5. The fact that one times a number equals the number might seem obvious, but it is also very powerful. This fact is called the **identity property of multiplication.** It is used in every math course you will take from now until the end of high school.

Activity: *Multiplication Table*

Materials needed by each student:

- 1 copy of Activity Master 9 (masters available in *Saxon Math 6/5 Assessments and Classroom Masters*)

The multiplication table above has 7 columns and 7 rows. Using Activity Master 9, make a multiplication table with 11 columns and 11 rows. Be sure to line up the numbers carefully. Use your multiplication table to answer the Lesson Practice problems below.

LESSON PRACTICE

Practice set In your multiplication table, find where the indicated row and column meet. Write that number as your answer.

a. column 4, row 5 → ?

b. column 2, row 6 → ?

c. column 6, row 3 → ?

d. column 10, row 8 → ?

Find each product:

e. 6×7

f. 8×9

g. 8×4

h. 3×10

i. 50×0

j. 25×1

k. The answer to a multiplication problem is called the *product.* What do we call the numbers that are multiplied together?

MIXED PRACTICE

Problem set

1. (12) Draw a number line marked with integers from –3 to 10. How many unit segments are there from 3 to 8?

2. (7) Kobe was the ninth person in line. How many people were in front of him?

3. (12) Mary used tally marks to count the number of trucks, cars, and motorcycles that drove by her house. Thirteen cars drove by her house. Use tally marks to show the number 13.

4. Write two addition facts and two subtraction facts for the fact family 1, 9, and 10.
(8)

For problems 5 and 6, write an equation and find the answer. (*Hint:* Problem 6 has three addends.)

5. During the first four weeks of school, Gretta read two books totaling 429 pages. If the first book was 194 pages long, how long was the second book?
(11)

6. Chang read three books. The first was 172 pages, the second 168 pages, and the third 189 pages. What was the total number of pages in the three books?
(11)

7. 3×6
(15)

8. 4×8
(15)

9. 7×9
(15)

10. 9×10
(15)

11. (14) $\begin{array}{r} A \\ -\ 819 \\ \hline 100 \end{array}$

12. (13) $\begin{array}{r} \$6.00 \\ -\ \$5.43 \\ \hline \end{array}$

13. (9) $\begin{array}{r} \$501 \\ -\ \$256 \\ \hline \end{array}$

14. (14) $\begin{array}{r} 510 \\ -\ Q \\ \hline 256 \end{array}$

15. (6) $\begin{array}{r} \$564 \\ \$796 \\ +\ \$287 \\ \hline \end{array}$

16. (10) $\begin{array}{r} N \\ +\ 96 \\ \hline 432 \end{array}$

17. (6) $\begin{array}{r} 608 \\ 930 \\ +\ 762 \\ \hline \end{array}$

18. (13) $\begin{array}{r} \$4.36 \\ \$2.18 \\ +\ \$3.94 \\ \hline \end{array}$

19. $360 + 47 + B = 518$
(10)

20. $\$10 - \$9.18$
(13)

21. Write the smallest three-digit even number that has the digits 1, 2, and 3.
(2)

22. Compare. (Try to answer the comparison without adding or multiplying.)
(4, 15)

$$5 + 5 + 5 \bigcirc 3 \times 5$$

23. Use digits and symbols to write "Twelve equals ten plus two."
(4, 6)

24. What term is missing in this counting sequence?
(1)

..., 32, 40, 48, _____, 64, ...

25. *(5)* Use digits to write eight hundred eighty dollars and eight cents.

26. *(7)* Compare: 346,129 ◯ 346,132

27. *(2)* A dozen is 12. How many is half of half a dozen?

28. *(13)* Write a multiplication problem that shows how to find the number of circles in this pattern.

29. *(4, 12)* Two integers are indicated by arrows on this number line. Write the two integers, using a comparison symbol to show which number is greater and which is less.

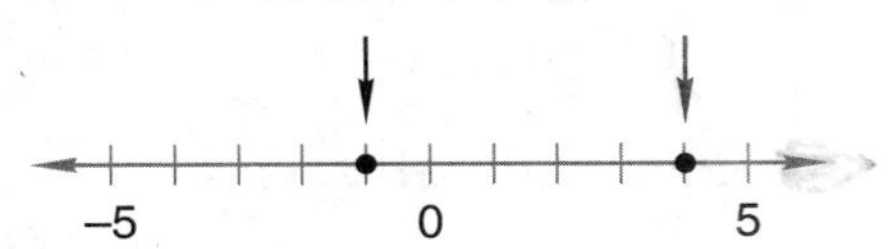

LESSON

16 Story Problems About Separating

WARM-UP

Facts Practice: 100 Multiplication Facts (Test C)

Mental Math:

Follow the pattern shown below to answer problems **a–f.**

$$3 \times 4 = 12$$
$$3 \times 40 = 120$$
$$3 \times 400 = 1200$$

a. 3 × 5	**b.** 3 × 50	**c.** 3 × 500
d. 5 × 6	**e.** 5 × 60	**f.** 5 × 600
g. \$3.75 − \$2.50	**h.** 140 + 16	**i.** 30 + 12
j. 20 + 15 − 5 + 10 + 4		

Problem Solving:

Copy this addition problem and fill in the missing digits:

```
   _6_
+  37_
------
  _248
```

NEW CONCEPT

Stories about combining have an addition pattern. In this lesson we will consider stories about separating. Stories about separating have a subtraction pattern. Below are examples of stories about separating. The plot of these stories is "some went away."

Jack had \$28. After he spent \$12, how much money did Jack have?

After losing 234 pounds, Jumbo still weighed 4368 pounds. How much did Jumbo weigh before he lost the weight?

Four hundred runners started the race, but many dropped out along the way. If 287 runners finished the race, then how many dropped out of the race?

Here are two ways to write the pattern for stories about separating:

```
  Some
− Some went away
----------------
  What is left
```

Some − some went away = what is left

Example 1 Jack had \$28. After he spent \$12, how much money did Jack have?

Solution Jack had some money, \$28. Then he spent some of his money, so some went away. He still had some money left. We write the equation using a subtraction pattern.

$$S - A = L$$

$$\$28 - \$12 = L$$

To find what is left, we subtract.

$$\begin{array}{r} \$28 \\ -\ \$12 \\ \hline \$16 \end{array}$$

Then we check whether our answer is sensible and our arithmetic is correct. The answer is sensible because Jack has less money than he started with. We can check the arithmetic by "adding up."

$$\begin{array}{r} \$28 \\ -\ \$12 \\ \hline \$16 \end{array}$$

Add Up

\$16 plus \$12 is \$28.
The answer is correct.

After spending \$12, Jack had **\$16** left.

Example 2 After losing 234 pounds, Jumbo still weighed 4368 pounds. How much did Jumbo weigh before he lost the weight?

Solution We write a "some went away" pattern.

Before, Jumbo weighed ...	W pounds
Then Jumbo lost ...	− 234 pounds
Jumbo still weighed ...	4368 pounds

To find the first number of a subtraction problem, we add.

$$\begin{array}{r} \scriptstyle 1\,1 \\ 4368 \text{ pounds} \\ +\ \ 234 \text{ pounds} \\ \hline 4602 \text{ pounds} \end{array}$$

Now we check the answer. Is it sensible? Yes. (We expect Jumbo's weight before to be greater than his weight after.) Is the arithmetic correct? We can check the arithmetic by using the answer in the original equation.

$$\begin{array}{r} W \\ -\ 234 \\ \hline 4368 \end{array} \longrightarrow \begin{array}{r} \scriptstyle 5\ 9\ 1 \\ 4\,\not{6}\,\not{0}\,2 \\ -\ \ 2\,3\,4 \\ \hline 4\,3\,6\,8 \end{array} \quad \text{This is correct.}$$

Before losing weight, Jumbo weighed **4602 pounds.**

Example 3 Four hundred runners started the race, but many dropped out along the way. If 287 runners finished the race, then how many dropped out of the race?

Solution We write a "some went away" pattern.

400 runners started ...	400 runners
Some dropped out ...	$-$ D runners
287 runners finished ...	287 runners

We find the missing number by subtracting.

$$\begin{array}{r} \overset{3}{\cancel{4}}\,\overset{9}{\cancel{0}}\,{}^{1}0 \text{ runners} \\ -\ 2\ 8\ 7 \text{ runners} \\ \hline 1\ 1\ 3 \text{ runners} \end{array}$$

Now we check whether the answer is reasonable and the arithmetic is correct. Since we expect the number of runners who dropped out to be less than the number who started, our answer is reasonable. We check the arithmetic as follows:

$$\begin{array}{r} 400 \\ -\ \ D \\ \hline 287 \end{array} \longrightarrow \begin{array}{r} \overset{3}{\cancel{4}}\,\overset{9}{\cancel{0}}\,{}^{1}0 \\ -\ 1\ 1\ 3 \\ \hline 2\ 8\ 7 \end{array} \quad \text{This is correct.}$$

There were **113 runners** who dropped out of the race.

LESSON PRACTICE

Practice set In problems **a–c,** write an equation for the story problem. Then answer the question.

a. Five hundred runners started the race. Only 293 finished the race. How many runners dropped out of the race?

b. After paying $85 rent, Jamal still had $326. How much money did Jamal have before he paid the rent?

c. The 26 members of the posse split into two groups. Fourteen members rode into the mountains, while the rest rode down to the river. How many members rode down to the river?

d. For the following equation, write a story problem about separating. Then answer the question in your story problem.

$$\$20 - \$12 = L$$

MIXED PRACTICE

Problem set

1. (11) The price went up from \$26 to \$32. By how many dollars did the price increase? (Use an addition pattern to solve the problem.)

2. (12) Use tally marks to show the number 15.

3. (5) Use words to name \$205.50.

4. (8) For the fact family 6, 8, and 14, write two addition facts and two subtraction facts.

For problems 5–7, write an equation and find the answer.

5. (11) Pink ink spots were here, and pink ink spots were there. Four hundred spots here, six hundred spots there, how many ink spots were everywhere?

6. (16) The custodian put away 24 chairs, leaving 52 chairs in the room. How many chairs were in the room before the custodian put some away? Use a subtraction pattern to solve the problem.

7. (16) Jill had \$24. She spent \$8. Then how much money did Jill have left? Use a subtraction pattern to solve the problem.

8. (15) 3×7 **9.** (15) 6×7 **10.** (15) 3×8 **11.** (15) 7×10

12. (14) $\begin{array}{r} B \\ -\ 256 \\ \hline 56 \end{array}$ **13.** (14) $\begin{array}{r} 900 \\ -\ \ \ C \\ \hline 90 \end{array}$ **14.** (13) $\begin{array}{r} \$4.18 \\ -\ \$2.88 \\ \hline \end{array}$ **15.** (9) $\begin{array}{r} \$406 \\ -\ \$278 \\ \hline \end{array}$

16. (6) $\begin{array}{r} \$357 \\ \$946 \\ +\ \$130 \\ \hline \end{array}$ **17.** (10) $\begin{array}{r} G \\ +\ 843 \\ \hline 1000 \end{array}$ **18.** (6) $\begin{array}{r} 365 \\ 52 \\ +\ 548 \\ \hline \end{array}$ **19.** (13) $\begin{array}{r} \$3.15 \\ \$2.87 \\ +\ \$1.98 \\ \hline \end{array}$

20. (2, 15) Think of two one-digit odd numbers. Multiply them. Is the product odd or even?

21. Which of these is a horizontal line?
(12)

A. B. C.

22. Use digits and a comparison symbol to write this comparison:
(4)

Eight hundred forty is greater than eight hundred fourteen.

23. What number is missing in this counting sequence?
(1)

..., 24, 30, 36, _____, 48, 54, ...

24. Compare: $4 \times 3 \bigcirc 2 \times 6$
(4, 15)

25. The letter y stands for what number in this equation?
(10)

$$36 + y = 63$$

26. What word is used to describe a line that goes straight up and down?
(12)

27. How many cents is half a dollar? How many cents is half of half a dollar?
(2)

28. Write a multiplication problem that shows how to find the number of small squares in this rectangle.
(13)

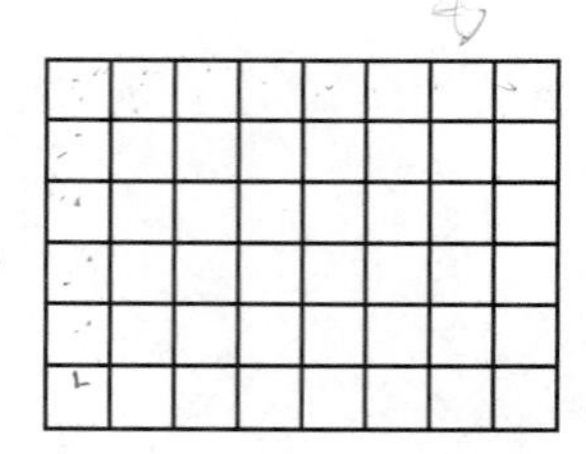

29. Josefina had \$32. She spent \$15. Use this information to write a story problem about separating. Then answer the question in your story problem.
(16)

LESSON

17 Multiplying by One-Digit Numbers

WARM-UP

Facts Practice: 100 Multiplication Facts (Test C)

Mental Math: Count up and down by 5's between 1 and 51 (1, 6, 11, 16, ...). Count by 50¢ to \$5.00 and from \$5.00 to 50¢.

a. 4×6	**b.** 4×60	**c.** 4×600
d. 5×8	**e.** 5×80	**f.** 5×800
g. $80 + 12$	**h.** $160 + 24$	**i.** $580 - 60$

j. $5 \times 6 + 12 - 2 + 10 - 1$

Problem Solving:

Quarters, dimes, nickels, and pennies are often put into paper or plastic rolls to make their value easier to identify. Quarters are put into rolls of 40 quarters. Dimes are put into rolls of 50 dimes. One roll of quarters has the same value as how many rolls of dimes?

NEW CONCEPT

We may solve the following problem either by adding or by multiplying:

A ticket to the basketball game costs \$24. How much would three tickets cost?

To find the answer by adding, we add the price of three tickets.

$$\begin{array}{r} {}^{1} \\ \$24 \\ \$24 \\ +\ \$24 \\ \hline \$72 \end{array}$$

To find the answer by multiplying, we multiply \$24 by 3. First we multiply the 4 ones by 3. This makes 12 ones, which is the same as 1 ten and 2 ones. We write the 2 ones below the line and the 1 ten above the tens column.

$$\begin{array}{r} {}^{1} \\ \$24 \\ \times\ \ \ 3 \\ \hline 2 \end{array}$$

Next we multiply the 2 tens by 3, making 6 tens. Then we add the 1 ten to make 7 tens.

$$\begin{array}{r} \overset{1}{\$24} \\ \times \quad 3 \\ \hline \$72 \end{array}$$

Example 1 Use dimes and pennies to illustrate $3 \times \$0.14$.

Solution One dime and four pennies is $0.14. We lay out three sets of these coins to show $3 \times \$0.14$.

We see that the total is 3 dimes and 12 pennies. Since 12 pennies is more than a dime, we trade ten pennies for a dime. The result is 4 dimes and 2 pennies. So $3 \times \$0.14$ is $0.42.

Example 2 Find the product: $\$0.25 \times 6$

Solution Mentally, we can find that 6 quarters is $1.50. However, we will use this problem to practice the pencil-and-paper algorithm for multiplication. Think of $0.25 as 2 dimes and 5 pennies. First we multiply 5 pennies by 6, which makes 30 pennies. Since 30 pennies equals 3 dimes and 0 pennies, we write "0" below the line and "3" above the dimes column.

$$\begin{array}{r} \$0.\overset{3}{2}5 \\ \times \quad 6 \\ \hline 0 \end{array}$$

Next we multiply 2 dimes by 6, making 12 dimes. Then we add the 3 dimes to make 15 dimes. Fifteen dimes equals 1 dollar and 5 dimes, so we write "5" below the line and "1" above the dollars column.

$$\begin{array}{r} {}^{1\ 3} \\ \$0.25 \\ \times \quad 6 \\ \hline 50 \end{array}$$

There are no dollars to multiply, so we write the 1 in the dollars place below the line. Finally, we insert the decimal point two places from the right-hand end and write the dollar sign.

$$\begin{array}{r} {}^{1\ 3} \\ \$0.25 \\ \times \quad 6 \\ \hline \mathbf{\$1.50} \end{array}$$

Example 3 Find the product: 6×325

Solution We follow the same method as in example 2. We multiply 5 ones by 6 and get 30. We write "0" below the line and "3" above the tens column. Next we multiply 2 tens by 6, making 12 tens, and add the 3 tens to get 15 tens. We write "5" below the line and "1" above the next digit. Then we multiply 3 hundreds by 6 and add the 1 hundred. The product of 6×325 is **1950.**

$$\begin{array}{r} {}^{1\,3} \\ 325 \\ \times \quad 6 \\ \hline 1950 \end{array}$$

Example 4 A ticket to the show costs \$7. How much would four tickets cost?

Solution We may find the answer by adding or by multiplying.

$$\$7 + \$7 + \$7 + \$7 = \$28 \qquad 4 \times \$7 = \$28$$

Four tickets would cost **\$28.**

LESSON PRACTICE

Practice set* Find each product:

a. $\$36 \times 5$ **b.** 50×8 **c.** $7 \times \$0.43$

d. $\begin{array}{r} 340 \\ \times \quad 8 \\ \hline \end{array}$ **e.** $\begin{array}{r} \$7.68 \\ \times \quad 4 \\ \hline \end{array}$ **f.** $\begin{array}{r} 506 \\ \times \quad 6 \\ \hline \end{array}$

g. $\begin{array}{r} \$394 \\ \times \quad 7 \\ \hline \end{array}$ **h.** $\begin{array}{r} 607 \\ \times \quad 9 \\ \hline \end{array}$ **i.** $\begin{array}{r} \$9.68 \\ \times \quad 3 \\ \hline \end{array}$

j. Each tire costs $42. At that price, how much would four tires cost? Show how to find the total cost by adding and by multiplying.

k. Nathan had five quarters in his pocket. Write and solve a multiplication problem that shows the value of the quarters in Nathan's pocket.

l. Devon bought three bottles of milk for $2.14 each. Altogether, how much did the milk cost? Find the answer by multiplying.

MIXED PRACTICE

Problem set

1. (12) Draw a vertical line segment.

2. (17) Cedric read 3 books. Each book had 120 pages. How many pages did Cedric read? Find the answer once by adding and again by multiplying.

For problems 3 and 4, write an equation and find the answer.

3. (11) The spider spun its web for 6 hours the first night and for some more hours the second night. If the spider spent a total of 14 hours spinning its web those two nights, how many hours did the spider spend the second night?

4. (16) After buying a notebook for $1.45, Carmela had $2.65. How much money did Carmela have before she bought the notebook?

5. (17) $\begin{array}{r} 24 \\ \times \ 3 \\ \hline \end{array}$ **6.** (17) $\begin{array}{r} \$36 \\ \times \quad 4 \\ \hline \end{array}$ **7.** (17) $\begin{array}{r} 45 \\ \times \ 5 \\ \hline \end{array}$ **8.** (17) $\begin{array}{r} \$56 \\ \times \quad 6 \\ \hline \end{array}$

9. (17) $\begin{array}{r} \$3.25 \\ \times \quad 6 \\ \hline \end{array}$ **10.** (17) $\begin{array}{r} 432 \\ \times \quad 9 \\ \hline \end{array}$ **11.** (17) $\begin{array}{r} \$2.46 \\ \times \quad 7 \\ \hline \end{array}$ **12.** (17) $\begin{array}{r} 364 \\ \times \quad 8 \\ \hline \end{array}$

13. (10) $\begin{array}{r} C \\ +\ 147 \\ \hline 316 \end{array}$ **14.** (13) $\begin{array}{r} \$4.20 \\ -\ \$3.75 \\ \hline \end{array}$ **15.** (14) $\begin{array}{r} 604 \\ -\quad W \\ \hline 406 \end{array}$ **16.** (14) $\begin{array}{r} M \\ -\ 73 \\ \hline 800 \end{array}$

17. (10) $3 + N + 15 + 9 = 60$

18. $\$90 + \$6.75 + \$7.98 + \$0.02$
(13)

19. Doreen bought five pens for $0.24 each. Altogether, how much did the pens cost? Find the answer to the problem by changing this addition problem into a multiplication problem:
(13, 17)

$$\$0.24 + \$0.24 + \$0.24 + \$0.24 + \$0.24$$

20. Find the product: 26×7
(17)

21. Think of two one-digit even numbers. Multiply them. Is the product odd or even?
(2, 15)

22. Compare: $12 \times 1 \bigcirc 24 \times 0$
(4, 15)

23. Use digits and a comparison symbol to write this comparison:
(7)

Five hundred four thousand is less than five hundred fourteen thousand.

24. What number is missing in this counting sequence?
(1)

..., 21, 28, 35, ______, 49, 56, ...

25. Which digit in 375 shows the number of hundreds?
(3)

26. These tally marks represent what number? 卌 卌 |||
(12)

27. Is the 100th term of this counting sequence odd or even?
(1, 2)

1, 3, 5, 7, ...

28. Write a multiplication problem that shows how to find the number of small squares in this rectangle.
(13)

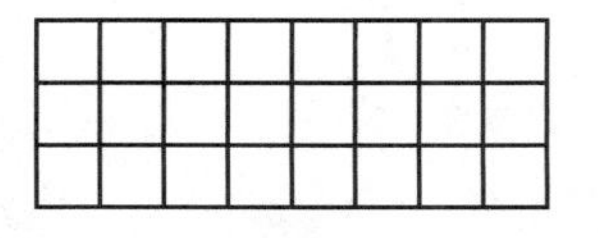

29. Use the commutative property to change the order of factors. Then multiply. Show your work.
(15)

$$\begin{array}{r} 5 \\ \times\ 24 \\ \hline \end{array}$$

LESSON 18 Multiplying Three Factors • Missing Numbers in Multiplication

WARM-UP

Facts Practice: 100 Multiplication Facts (Test C)

Mental Math: Count up and down by 5's between 1 and 51. Count up and down by 200's between 0 and 2000.

a. 3×30 plus 3×2 **b.** 4×20 plus 4×3
c. 5×30 plus 5×4 **d.** 6×700
e. $1000 - 100$ **f.** $320 + 32$
g. $3.75 − $1.25 **h.** $6 \times 4 + 1 + 10 - 5 + 3$

Problem Solving:

All of the digits 1 through 9 are used in this addition problem. Copy the problem and fill in the missing digits.

```
   3__
 + 452
 -----
   ___
```

NEW CONCEPTS

Multiplying three factors

In this lesson we will learn how to multiply three numbers together. Remember that numbers multiplied together are called *factors.* In the problem below we see three factors.

$$9 \times 8 \times 7$$

To multiply three factors, we first multiply two of the factors together. Then we multiply the product we get by the third factor.

First we multiply 9 by 8 to get 72. $9 \times 8 \times 7 =$

Then we multiply 72 by 7 to get 504. $72 \times 7 = 504$

Since multiplication is commutative, we may multiply numbers in any order. Sometimes, changing the order of the factors can make a problem easier.

Example 1 Find the product: $6 \times 3 \times 5$

Solution To find the product of three factors, we first multiply two of the factors. Then we multiply the product we get by the third factor. We may choose an order of factors that makes the problem easier. In this problem we choose to multiply 6 and 5 first; then we multiply the resulting product by 3.

$$6 \times 3 \times 5$$

$$30 \times 3 = \mathbf{90}$$

Example 2 Find the product: $5 \times 7 \times 12$

Solution The order in which we choose to multiply can affect the difficulty of the problem. If we multiply 5 by 7 first, then we must multiply 35 by 12. But if we multiply 5 by 12 first, then we would multiply 7 by 60 next. The second way is easier and can be done mentally. We rearrange the factors to show the order in which we choose to multiply.

$$5 \times 12 \times 7$$

$$60 \times 7 = \mathbf{420}$$

Example 3 How many blocks were used to build this shape?

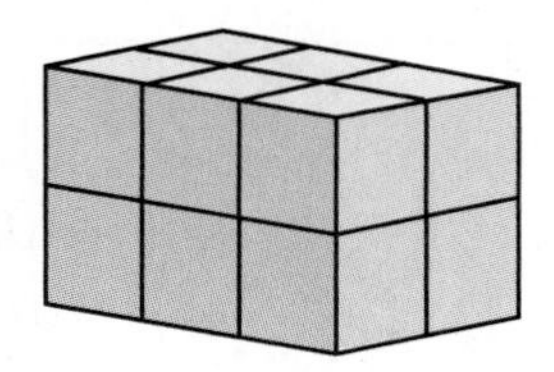

Solution We may count all the blocks, or we may multiply three numbers. We can see that the top layer has 2 rows of 3 blocks. So we know there are 2×3 blocks in each layer. Since there are two layers, we multiply the number in each layer by 2.

$$2 \times 3 \times 2 = 12$$

The shape was built with **12 blocks.**

Missing numbers in multiplication Now we will practice finding missing factors in multiplication problems. In this type of problem we are given one factor and a product.

Example 4 Find each missing factor:

(a)
$$\begin{array}{r} W \\ \times\ 3 \\ \hline 18 \end{array}$$

(b) $3N = 24$

(c) $6 \times 5 = 3 \times A$

Solution Before we start, we must understand what each equation means. In (b) the expression $3N$ means "3 times N." In (c), if we multiply 6 and 5, we see that the equation means $30 = 3 \times A$. Now we are ready to find the missing factors. There are many ways to do this. In these examples we could count how many 3's add up to 18, to 24, and to 30. We could also use a multiplication table. In the table below, look across the 3's row to 18, 24, and 30, and then look to the top of each column for the missing factor. We see that the missing factors are 6, 8, and 10.

Columns ↓ (6, 8, 10)

	0	1	2	3	4	5	6	7	8	9	10
0	0	0	0	0	0	0	0	0	0	0	0
1	0	1	2	3	4	5	6	7	8	9	10
2	0	2	4	6	8	10	12	14	16	18	20
Row → 3	0	3	6	9	12	15	(18)	21	(24)	27	(30)

The fastest way to find missing factors is to recall the multiplication facts. Since $3 \times 6 = 18$ and $3 \times 8 = 24$ and $3 \times 10 = 30$, we know the missing factors are (a) **6**, (b) **8**, and (c) **10**.

LESSON PRACTICE

Practice set* For problems **a–d**, copy each problem and then multiply. Show which numbers you chose to multiply first.

a. $5 \times 7 \times 6$

b. $10 \times 9 \times 8$

c. $3 \times 4 \times 25$

d. $4 \times 3 \times 2 \times 1 \times 0$

e. How many blocks were used to build this figure? Write a multiplication problem that provides the answer.

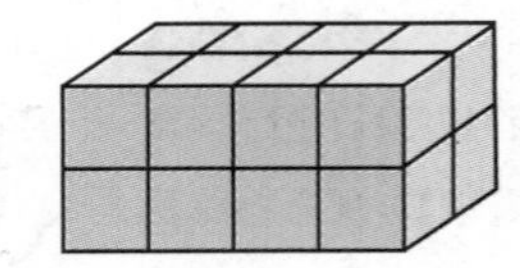

Find each missing factor:

f. $5M = 30$

g. $3B = 21$

h. $3 \times 4 = N \times 2$

i. $\begin{array}{r} P \\ \times\ 4 \\ \hline 24 \end{array}$

j. $\begin{array}{r} 9 \\ \times\ Q \\ \hline 81 \end{array}$

k. $\begin{array}{r} W \\ \times\ 9 \\ \hline 0 \end{array}$

MIXED PRACTICE

Problem set

1. (12) Draw a horizontal line and a vertical line. Then write the words *horizontal* and *vertical* to label each line.

For problems 2–4, write an equation and find the answer.

2. (16) Once he started exercising regularly, Reggie's weight dropped from 223 pounds to 195 pounds. How many pounds did Reggie lose?

3. (11) In one class there are 33 students. Fourteen of the students are boys. How many girls are in the class?

4. (11) In another class there are 17 boys and 14 girls. How many students are in the class?

For problems 5–8, try to find each product mentally before using pencil and paper.

5. (18) $6 \times 4 \times 5$

6. (18) $5 \times 6 \times 12$

7. (18) $5 \times 10 \times 6$

8. (18) $9 \times 7 \times 10$

9. (17) $\begin{array}{r} \$407 \\ \times\ \quad 8 \\ \hline \end{array}$

10. (17) $\begin{array}{r} 375 \\ \times\ \quad 6 \\ \hline \end{array}$

11. (17) $\begin{array}{r} \$4.86 \\ \times\ \quad 9 \\ \hline \end{array}$

12. (17) $\begin{array}{r} 308 \\ \times\ \quad 7 \\ \hline \end{array}$

13. (18) $9G = 36$

14. (17) $\begin{array}{r} \$573 \\ \times\ \quad 9 \\ \hline \end{array}$

15. (18) $8H = 48$

16. (17) $\begin{array}{r} \$7.68 \\ \times\ \quad 4 \\ \hline \end{array}$

17. (10) $456 + 78 + F = 904$

18. (6) $34 + 75 + 123 + 9$

19. (13) $\$36.70 - \$7.93$

20. (14) $H - 354 = 46$

21. (13, 17) Luis bought four folders for $0.37 each. Altogether, how much money did the folders cost?

22. (2, 15) Think of a one-digit odd number and a one-digit even number. Multiply them. Is the product odd or even?

23. (18) Find the missing factor: $6 \times 4 = 8 \times N$

24. (4, 15) Use digits and symbols to write this comparison:

Eight times eight is greater than nine times seven.

25. (8) For the fact family 7, 8, and 15, write two addition facts and two subtraction facts.

26. (13) Write a multiplication fact that shows the number of squares in this rectangle.

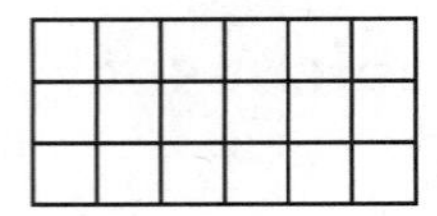

27. (18) Write a three-factor multiplication fact that shows the number of blocks in this figure.

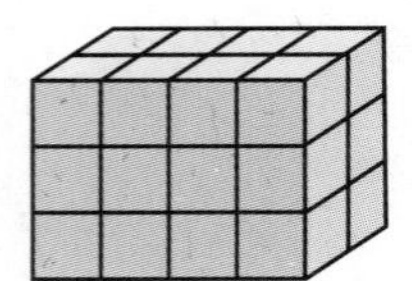

28. (1, 12) What are the next three integers in this counting sequence?

8, 6, 4, 2, ...

29. (2) As a "rule of thumb," the height of a child on his or her second birthday is about half the height the child will be as an adult. When Kurt turned two years old, he was 3 feet tall. If the rule works for Kurt, about how tall will he be as an adult?

LESSON

19 Division Facts

WARM-UP

Facts Practice: 100 Multiplication Facts (Test C)

Mental Math: Count up and down by 5's between 2 and 52 (2, 7, 12, 17, ...).

a. $2 \times 5 \times 6$ **b.** $6 \times 5 \times 3$

c. 6×30 plus 6×2 **d.** 4×60 plus 4×5

e. $1000 - 800$ **f.** $640 + 24$

g. $\$5.00 - \$0.50$ **h.** $9 \times 9 - 1 + 10 + 10$

Problem Solving:

For breakfast, lunch, and dinner, Bethany ate soup and eggs and ham, one for each meal, but not necessarily in that order. (a) List all the possible arrangements of meals Bethany could have eaten. (b) If Bethany never eats eggs for lunch, how many arrangements of meals are possible?

NEW CONCEPT

Searching for a missing factor is called **division.** A division problem is like a miniature multiplication table. The product is shown inside a symbol called a *division box* ($\overline{)\quad}$). The two factors are outside the box. One factor is in front, and the other is on top. In the problem below, the factor on top is missing.

$$3\overset{?}{\overline{)12}}$$

To solve this problem, we need to know what number times 3 equals 12. Since $3 \times 4 = 12$, we know that the missing factor is 4. We write our answer this way:

$$3\overset{4}{\overline{)12}}$$

Example 1 What is the missing number in this problem? $4\overset{?}{\overline{)20}}$

Solution To find the missing number, we think, "Four times what number equals 20?" We find that the missing number is 5, which we write above the division box:

$$4\overset{\mathbf{5}}{\overline{)20}}$$

Example 2 Divide: $3\overline{)18}$

Solution This is the way division problems are often written. We need to find the number that goes above the box. We think, "Three times what number equals 18?" We remember that $3 \times 6 = 18$, so the answer to the division problem is 6.

$$\begin{array}{r} \mathbf{6} \\ 3\overline{)18} \end{array}$$

Example 3 Divide: $10\overline{)80}$

Solution We think, "How many 10's make 80?" Since $8 \times 10 = 80$, the answer is **8.**

Multiplication and division are inverse operations. One operation undoes the other. If we start with 5 and multiply by 6, we get a product of 30. If we then divide 30 by 6, the result is 5, the number we started with. The division by 6 undid the multiplication by 6. Using the commutative property and inverse operations, we find that the three numbers that form a multiplication fact also form division facts.

Example 4 Write two multiplication facts and two division facts for the fact family 5, 6, and 30.

Solution

$$\begin{array}{r} \mathbf{6} \\ \times\ \mathbf{5} \\ \hline \mathbf{30} \end{array} \qquad \begin{array}{r} \mathbf{5} \\ \times\ \mathbf{6} \\ \hline \mathbf{30} \end{array} \qquad \begin{array}{r} \mathbf{6} \\ \mathbf{5\overline{)30}} \end{array} \qquad \begin{array}{r} \mathbf{5} \\ \mathbf{6\overline{)30}} \end{array}$$

LESSON PRACTICE

Practice set Find the missing number in each division fact:

a. $2\overline{)16}$ **b.** $4\overline{)24}$ **c.** $6\overline{)30}$ **d.** $8\overline{)56}$

e. $3\overline{)21}$ **f.** $10\overline{)30}$ **g.** $7\overline{)28}$ **h.** $9\overline{)36}$

i. Write two multiplication facts and two division facts for the fact family 3, 8, and 24.

MIXED PRACTICE

Problem set For problems 1 and 2, write an equation and find the answer.

1. (16) The \$45 dress was marked down to \$29. By how many dollars had the dress been marked down?

2. (11) Room 15 collected 243 aluminum cans. Room 16 collected 487 cans. Room 17 collected 608 cans. How many cans did the three rooms collect in all?

3. (11, 13) There are 5 rows of desks with 6 desks in each row. How many desks are there in all? Find the answer once by adding and again by multiplying.

4. (5, 7) Use words to name \$4,587.20.

5. (19) For the fact family 7, 8, and 56, write two multiplication facts and two division facts.

6. (19) $3\overline{)24}$ **7.** (19) $6\overline{)18}$ **8.** (19) $4\overline{)32}$ **9.** (19) $10\overline{)40}$

10. (17) $\begin{array}{r} \$4.83 \\ \times \quad 7 \\ \hline \end{array}$ **11.** (17) $\begin{array}{r} 659 \\ \times \quad 8 \\ \hline \end{array}$ **12.** (17) $\begin{array}{r} \$706 \\ \times \quad 4 \\ \hline \end{array}$ **13.** (18) $9M = 54$

14. (18) $8 \times 10 \times 7$ **15.** (18) $9 \times 8 \times 5$

16. (13) $\begin{array}{r} \$65.40 \\ - \$19.18 \\ \hline \end{array}$ **17.** (14) $\begin{array}{r} 4000 \\ - \quad T \\ \hline 1357 \end{array}$ **18.** (14) $\begin{array}{r} R \\ - 1915 \\ \hline 269 \end{array}$

19. (6) $\begin{array}{r} 907 \\ 415 \\ + 653 \\ \hline \end{array}$ **20.** (13) $\begin{array}{r} \$3.67 \\ \$4.25 \\ + \$7.40 \\ \hline \end{array}$ **21.** (10) $\begin{array}{r} 427 \\ + \quad K \\ \hline 813 \end{array}$

22. (10) $356 + L + 67 = 500$

23. (10) $86 + w = 250$

24. (18) Find the missing factor: $6 \times 6 = 4N$

25. (4, 15) Use digits and symbols to write this comparison:

Eight times six is less than seven times seven.

26. (2) Ed cut a 15-inch long piece of licorice in half. How long was each half?

27. (13) Write a multiplication fact that shows how many squares cover this rectangle.

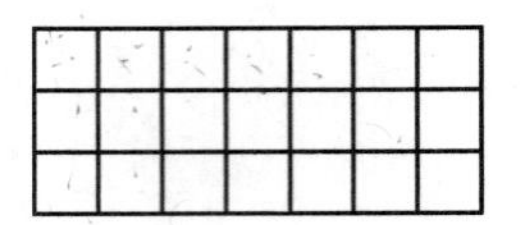

28. (18) Write a three-factor multiplication fact that shows how many blocks form this figure.

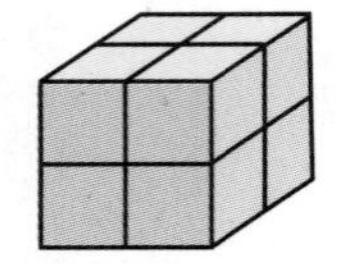

29. (7) The Mississippi River begins in Minnesota. From there it flows 2340 miles to the Gulf of Mexico. The Missouri River is 2315 miles long and begins in Montana. The Colorado River is the longest river in the U.S. west of the Rocky Mountains. It starts in the Rocky Mountains and flows 1450 miles to the Gulf of California. Write the names of the three rivers in order from shortest to longest.

LESSON

20 Three Ways to Show Division

WARM-UP

Facts Practice: 90 Division Facts (Test D)

Mental Math: Count up and down by 5's between 2 and 52.

a. 4×32 equals 4×30 plus 4×2. Find 4×32.

b. $5 \times 8 \times 3$

c. $4 \times 5 \times 6$

d. 4×23

e. 4×54

f. $1000 - 990$

g. $7 \times 7 + 1 + 25 + 25$

Problem Solving:

Copy this addition problem and fill in the missing digits:

```
   _3_
+   _1
------
  __3_
```

NEW CONCEPT

We use different ways to show division. Here are three ways to show "twelve divided by four":

$$4\overline{)12} \qquad 12 \div 4 \qquad \frac{12}{4}$$

In the first form we use a *division box.* In the second we use a *division sign.* In the third we use a *division bar.* To solve longer division problems, we usually use the first form. In later math courses we will use the third form more often. We should be able to read and solve division problems in each form and to change from one form to another.

Three numbers are involved in every division problem:

1. The number being divided: $\mathbf{15} \div 3 = 5$
2. The number by which it is divided: $15 \div \mathbf{3} = 5$
3. The answer to the division: $15 \div 3 = \mathbf{5}$

These numbers are called the **dividend, divisor,** and **quotient.** In the example above, the dividend is 15, the divisor is 3, and the quotient is 5. The location of these numbers in each form is shown below.

Positions of Divisor, Dividend, and Quotient

Division box	$\text{divisor}\overline{)\text{dividend}}$ with quotient above
Division sign	dividend ÷ divisor = quotient
Division bar	$\frac{\text{dividend}}{\text{divisor}}$ = quotient

Example 1 Use words to show how each problem is read:

(a) $12 \div 6$ (b) $\frac{12}{6}$ (c) $6\overline{)12}$

Solution For all three division symbols, we say "divided by." The division in (a) is read from left to right: "twelve divided by six."

The division in (b) is read from top to bottom: "twelve divided by six." (This may also be read as "twelve over six.")

The division in (c) is written with a division box. We read the number inside the box first: "twelve divided by six."

We see that all three problems are read the same. Each problem shows the same division, **"twelve divided by six."**

Example 2 Write this division problem in two other forms. $15 \div 3$

Solution We read this problem, "fifteen divided by three." Fifteen is the dividend.

To show division with a division bar, we write the dividend on top. $\mathbf{\frac{15}{3}}$

To show division with a division box, we write the dividend inside the box. $\mathbf{3\overline{)15}}$

Example 3 Divide: $\frac{15}{5}$

Solution The bar is a way to show division. We think, "Five times what number is 15?" The answer is **3.**

Example 4 In the following equation, which number is the divisor, which number is the dividend, and which number is the quotient?

$$\frac{56}{7} = 8$$

Solution The **dividend, 56,** is divided by the **divisor, 7.** The answer is the **quotient, 8.**

LESSON PRACTICE

Practice set

a. Show "10 divided by 2" in three different forms.

b. Use three different division forms to show "24 divided by 6."

Use words to show how each division problem is read:

c. $3\overline{)21}$ **d.** $12 \div 6$ **e.** $\frac{30}{5}$

Rewrite each division problem with a division box:

f. $63 \div 7$ **g.** $\frac{42}{6}$ **h.** 30 divided by 6

i. Identify the quotient, dividend, and divisor in this equation:

$$63 \div 9 = 7$$

Find the answer (quotient) to each division problem:

j. $\frac{60}{10}$ **k.** $\frac{42}{7}$ **l.** $28 \div 4$ **m.** $36 \div 6$

n. Compare: $24 \div 4 \bigcirc 24 \div 6$

MIXED PRACTICE

Problem set

1. (12) Draw a horizontal number line marked with even integers from –6 to 6.

2. (19) Write two multiplication facts and two division facts for the fact family 4, 9, and 36.

3. (12) Use tally marks to show the number 16.

4. (11, 17) Jim reads 40 pages per day. How many pages does Jim read in 4 days? Find the answer once by adding and again by multiplying.

5. (11) There are 806 students at Gidley School. If there are 397 girls, how many boys are there? Write an equation and find the answer.

6. (5, 6) What is the sum of five hundred twenty-six and six hundred eighty-four?

Use words to show how each problem is read:

7. (20) $6\overline{)24}$

8. (20) $15 \div 3$

9. (4, 20) Compare: $\frac{15}{3} \bigcirc \frac{15}{5}$

10. (18) $8M = 24$

11. (20) $10\overline{)90}$

12. (20) $\frac{27}{3}$

13. (17) $\begin{array}{r} \$23.18 \\ \times \quad 6 \\ \hline \end{array}$

14. (17) $\begin{array}{r} 4726 \\ \times \quad 8 \\ \hline \end{array}$

15. (17) $\begin{array}{r} \$34.09 \\ \times \quad 7 \\ \hline \end{array}$

16. (4, 18) Compare. (Try to answer the comparison before multiplying.)

$$5 \times 6 \times 7 \bigcirc 7 \times 6 \times 5$$

17. (13, 17) Snider bought five notebooks for \$3.52 each. What was the total cost of the five notebooks? Change this addition problem to a multiplication problem and find the total:

$$\$3.52 + \$3.52 + \$3.52 + \$3.52 + \$3.52$$

18. (13) $\begin{array}{r} \$40.00 \\ - \$24.68 \\ \hline \end{array}$

19. (14) $\begin{array}{r} 1207 \\ - \quad R \\ \hline 943 \end{array}$

20. (14) $\begin{array}{r} Z \\ - 1358 \\ \hline 4444 \end{array}$

21. (6) $\begin{array}{r} 3426 \\ 1547 \\ + 2684 \\ \hline \end{array}$

22. (10) $\begin{array}{r} 4318 \\ + \quad M \\ \hline 4343 \end{array}$

23. (13) $\begin{array}{r} \$13.06 \\ \$ \ 4.90 \\ + \$60.75 \\ \hline \end{array}$

24. (4) Use digits and symbols to write this comparison:

Ten times two is greater than ten plus two.

25. What are the next three terms in this counting sequence?
(1, 12)

24, 18, 12, 6, ...

26. In this equation, which number is the divisor?
(20)

$$27 \div 3 = 9$$

27. Write a multiplication equation that shows the number of squares in this rectangle.
(13)

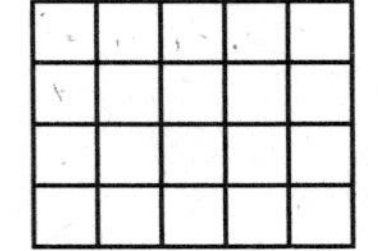

28. Sarah went to the store with $35 and came home with $9. Use this information to write a story problem about separating. Then answer the question in your story problem.
(16)

29. Arrange these years in order from earliest to latest:
(4)

1620, 1789, 1492, 1776

INVESTIGATION 2

Focus on

Fractions: Halves, Fourths, and Tenths

A **fraction** describes part of a whole. The "whole" may be a single thing, such as a whole pie or a whole inch, or the "whole" may be a group, such as a whole class of students or a whole bag of marbles.

We use two numbers to write a fraction. The bottom number, the **denominator**, shows the number of *equal parts* in the whole. The top number, the **numerator**, shows how many of the equal parts are counted.

$$\frac{1}{2}$$

1 ← Numerator
2 ← Denominator

We read fractions from top to bottom, as shown below.

$\frac{1}{2}$	"one half" (sometimes we just say "half")
$\frac{1}{4}$	"one fourth" or "one quarter"
$\frac{3}{4}$	"three fourths" or "three quarters"
$\frac{1}{10}$	"one tenth"
$\frac{9}{10}$	"nine tenths"

Many fraction problems are "equal groups" problems. The denominator of the fraction shows the number of equal groups. We divide the total by the denominator to find the number in each group.

Example 1 Half of the 18 students in the class are girls. How many girls are in the class?

Solution The word *half* means "one of two equal groups." In this story, one group is girls and the other is boys. We find the number in each group by dividing by 2.

$$2\overline{)18}\quad 9$$

In each group there are 9 students. This means there are **9 girls** in the class.

Example 2 (a) How many cents is one fourth of a dollar?

(b) How many cents is three fourths of a dollar?

Solution The word *fourth* means that the whole dollar is divided into four equal parts. Since four quarters equals a dollar, one fourth of a dollar equals a quarter, which is twenty-five cents. Three fourths of a dollar (three quarters) equals seventy-five cents.

(a) One fourth of a dollar is **twenty-five cents.**

(b) Three fourths of a dollar is **seventy-five cents.**

Example 3 One tenth of the 30 students earned an A on the test. How many students earned an A?

Solution One tenth means "one of ten equal parts." We can find one tenth of 30 by dividing 30 by 10.

$$10\overline{)30}\quad 3$$

One tenth of 30 is 3. So **3 students** earned an A.

Use this information to answer problems 1–4:

There were 20 pumpkins in the garden. One fourth of the pumpkins were too small, one tenth were too large, and one half were just the right size. The rest of the pumpkins were not yet ripe.

1. How many pumpkins were too small?

2. How many pumpkins were too large?

3. How many pumpkins were just the right size?

4. How many pumpkins were not yet ripe?

Use this information to answer problems 5 and 6:

> *During the hike Sam found that he could carry a pack $\frac{1}{2}$ his weight for half a mile without resting. He could carry a pack $\frac{1}{4}$ his weight for two miles without resting. Sam weighs 80 pounds.*

5. How heavy a pack could Sam carry for 2 miles without resting?

6. How heavy a pack could Sam carry for only half a mile without resting?

Activity: *Using Fraction Manipulatives*

Materials needed:

- photocopies of Activity Masters 10, 11, and 12 (1 copy of each master per student; masters available in *Saxon Math 6/5 Assessments and Classroom Masters*)
- scissors
- envelopes or locking plastic bags

Note: Color-coding the fraction manipulatives makes sorting easier. If you wish to color-code the manipulatives, photocopy each master on a different color of construction paper, or, before cutting, have students color both sides of the fraction circles using different colors for different masters. Following the activity, each student may store the fraction manipulatives in an envelope or plastic bag for use in later lessons.

Preparation:

Distribute materials. Have students separate the fraction manipulatives by cutting out the circles and cutting apart the fraction slices along the lines.

Use your fraction manipulatives to help with each exercise below.

7. Show that two quarters equals one half.

8. Two quarters of a circle is what percent of a whole circle?

9. How many tenths equal one half?

10. Is one quarter plus two tenths more or less than one half?

11. One fourth of a circle plus two tenths of a circle is what percent of a whole circle?

12. One half of a circle plus four tenths of a circle is what percent of a whole circle?

Two half circles can be put together to form a whole circle. This equation states that two halves equal a whole:

$$\frac{1}{2} + \frac{1}{2} = 1$$

Another way to write this is

$$\frac{2}{2} = 1$$

One half circle and two quarter circles can also be put together to form a whole circle.

$$\frac{1}{2} + \frac{1}{4} + \frac{1}{4} = 1$$

Another way to write this is

$$\frac{1}{2} + \frac{2}{4} = 1$$

13. Use your fraction manipulatives to find other ways to form a whole circle. Write an equation for each way you find.

14. One fourth of a circle plus one tenth of a circle is what percent of a whole circle?

15. Two fourths of a circle plus two tenths of a circle is what percent of a whole circle?

16. What percent of a circle is one half plus one fourth plus one tenth?

17. What fraction piece covers one half of a half circle?

Compare:

18. $\frac{3}{4} \bigcirc 75\%$

19. $\frac{3}{10} \bigcirc 25\%$

20. $50\% \bigcirc \frac{1}{4} + \frac{1}{4}$

21. $30\% \bigcirc \frac{1}{4} + \frac{1}{10}$

22. $\frac{1}{2} + \frac{1}{4} + \frac{1}{10} \bigcirc 90\%$

23. $\frac{1}{2} + \frac{1}{4} + \frac{1}{4} \bigcirc 100\%$

24. $\frac{1}{4} + \frac{1}{10} + \frac{1}{10} \bigcirc \frac{1}{2}$

25. $\frac{1}{10} + \frac{1}{10} + \frac{1}{10} \bigcirc \frac{1}{4}$

LESSON

21 Problems About Equal Groups

WARM-UP

Facts Practice: 90 Division Facts (Test D or E)

Mental Math: Count up and down by 25's between 0 and 200. Count up and down by 250's between 0 and 2000.

a. 3×40 plus 3×5

b. 4×50 plus 4×4

c. 4×45

d. 4×54

e. $120 + 70$

f. $560 - 200$

g. $210 + 35$

h. Start with 5, $\times$ 6, + 2, $\div$ 4, + 1, $\div$ 3[†]

Problem Solving:

The uppercase letter A encloses one area, B encloses two areas, and C does not enclose an area. List all the uppercase and lowercase letters that enclose at least one area.

NEW CONCEPT

Stories about combining have an addition pattern. Stories about separating have a subtraction pattern. Stories about **equal groups** have a multiplication pattern. Here are three "equal groups" stories:

At Lincoln School there are 4 classes of fifth graders with 30 students in each class. Altogether, how many students are in the four classes?

The coach separated the 48 players into 6 teams with the same number of players on each team. How many players were on each team?

Monifa raked up 28 bags of leaves. On each trip she could carry away 4 bags. How many trips did it take Monifa to carry away all the bags?

[†]As a shorthand, we will use commas to separate operations to be performed sequentially from left to right. In this case, $5 \times 6 = 30$, then $30 + 2 = 32$, then $32 \div 4 = 8$, then $8 + 1 = 9$, then $9 \div 3 = 3$. The answer is 3.

There are three numbers in a completed "equal groups" story—the number of groups, the number in each group, and the total number in all the groups. These numbers are related by multiplication. Here we show the multiplication pattern written two ways:

$$\text{Number of groups} \times \text{number in each group} = \text{total}$$

$$\begin{array}{r} \text{Number in each group} \\ \times\ \text{Number of groups} \\ \hline \text{Total} \end{array}$$

The number of groups is one factor, and the "in each" number is the other factor. The total number in all groups is the product.

In an "equal groups" story problem, one of the numbers is missing. If the total is missing, we multiply to find the missing number. If the "in each" number or the number of groups is missing, we divide.

Example 1 At Lincoln School there are 4 classes of fifth graders with 30 students in each class. Altogether, how many students are in the 4 classes?

Solution This story is about equal groups. We are given the number of groups (4 classes) and the number in each group (30 students). We write an equation.

$$\text{Number of groups} \times \text{number in each group} = \text{total}$$

$$4 \times 30 = T$$

We multiply to find the missing number.

$$4 \times 30 = 120$$

We check whether the answer is reasonable. There are many more students in four classes than in one class, so 120 is reasonable. There are **120 students** in all 4 classes.

Example 2 The coach separated 48 players into 6 teams with the same number of players on each team. How many players were on each team?

Solution This is an "equal groups" story. The groups are teams. We are given the number of groups (6 teams) and the total number of players (48 players). We are asked to find the number of players on each team. We write an equation.

$$\begin{array}{rl} & N \text{ players on each team} \\ \times & 6 \text{ teams} \\ \hline & 48 \text{ players on all 6 teams} \end{array}$$

We find the missing number, a factor, by dividing.

$$6\overline{)48}\quad = 8$$

There were **8 players** on each team. The answer is reasonable because 6 teams of 8 players is 48 players in all.

Example 3 Monifa raked up 28 bags of leaves. On each trip she could carry away 4 bags. How many trips did it take Monifa to carry away all the bags?

Solution The objects are bags, and the groups are trips. The missing number is the number of trips. We show two ways to write the equation.

$$\begin{array}{rl} & 4 \text{ bags in each trip} \\ \times & N \text{ trips} \\ \hline & 28 \text{ bags in all the trips} \end{array}$$

$$4N = 28$$

The missing number is a factor, which we find by dividing.

$$28 \div 4 = 7$$

Monifa took **7 trips** to carry away all 28 bags.

LESSON PRACTICE

Practice set For problems **a–d,** write an equation and find the answer.

a. On the shelf were 4 cartons of eggs. There were 12 eggs in each carton. How many eggs were in all four cartons?

b. Thirty desks are arranged in 6 equal rows. How many desks are in each row?

c. Twenty-one books are stacked in piles with 7 books in each pile. How many piles are there?

d. If 56 zebus were separated into 7 equal herds, then how many zebus would be in each herd?

e. Write an "equal groups" story problem for this equation. Then answer the question in your story problem.

$$6 \times \$0.75 = T$$

MIXED PRACTICE

Problem set For problems 1–3, write an equation and find the answer.

1. (21) The coach separated the PE class into 8 teams with the same number of players on each team. If there are 56 students in the class, how many are on each team? Use a multiplication pattern.

2. (16) Tony opened a bottle containing 32 ounces of milk and poured 8 ounces of milk into a bowl of cereal. How many ounces of milk remained in the bottle?

3. (11) The set of drums costs eight hundred dollars. The band has earned four hundred eighty-seven dollars. How much more must the band earn in order to buy the drums?

4. (12) Draw an oblique line.

5. (19) Write two multiplication facts and two division facts for the fact family 6, 7, and 42.

6. (19) $8\overline{)72}$

7. (18) $6N = 42$

8. (19) $9\overline{)36}$

9. (18) $6N = 48$

10. (20) $56 \div 7$

11. (20) $\frac{70}{10}$

12. (4, 20) Compare: $24 \div 4 \bigcirc 30 \div 6$

13. (17) $\begin{array}{r} 367 \\ \times \quad 8 \\ \hline \end{array}$

14. (17) $\begin{array}{r} \$5.04 \\ \times \quad 7 \\ \hline \end{array}$

15. (17) $\begin{array}{r} 837 \\ \times \quad 9 \\ \hline \end{array}$

16. (18) $6 \times 8 \times 10$

17. (18) $7 \times 20 \times 4$

18. $40 − $29.34
(13)

19. $R - 4568 = 6318$
(14)

20. $5003 - W = 876$
(14)

21.
(10)
$$\begin{array}{r} 268 \\ + \quad M \\ \hline 687 \end{array}$$

22.
(13)
$$\begin{array}{r} \$9.65 \\ \$2.43 \\ + \ \$1.45 \\ \hline \end{array}$$

23.
(6)
$$\begin{array}{r} 382 \\ 96 \\ + \ 182 \\ \hline \end{array}$$

24. If a dozen items are divided into two equal groups, how many will be in each group?
(21)

25. What are the next three terms in this counting sequence?
(1)

..., 50, 60, 70, 80, 90, _____, _____, _____, ...

26. Use words to show how this problem is read: $\frac{10}{2}$
(20)

27. What number is the dividend in this equation?
(20)

$$60 \div 10 = 6$$

28. Below is a story problem about equal groups. After you find the answer to the question, use it to rewrite the last sentence as a statement instead of a question.
(21)

The books arrived in 5 boxes. There were 12 books in each box. How many books were in all 5 boxes?

29. The fraction $\frac{1}{2}$ is equivalent to what percent?
(Inv. 2)

LESSON

22 One-Digit Division with a Remainder • Divisibility by 2, 5, and 10

WARM-UP

Facts Practice: 64 Multiplication Facts (Test F)

Mental Math: Count up and down by 50's between 0 and 500. Count up and down by 500's between 0 and 5000.

a. 10×5 **b.** 10×25

c. 5×50 plus 7×5 **d.** 4×56

e. 3×56 **f.** $150 + 25$

g. $180 + 30$ **h.** $850 - 150$

i. Start with 6, $\times$ 6, $-$ 1, $\div$ 5, $+$ 1, $\div$ 2

Problem Solving:

Copy this subtraction problem and fill in the missing digits:

$$\begin{array}{r} 4\_6 \\ -\ \_1\_ \\ \hline 237 \end{array}$$

NEW CONCEPTS

One-digit division with a remainder

Division and multiplication are inverse operations. We can use division to find a missing factor. Then we can use multiplication to check our division. We show this below:

$$5\overline{)35}^{\,7} \qquad \begin{array}{r} 7 \\ \times\ 5 \\ \hline 35 \end{array}\ \text{check}$$

Instead of writing a separate multiplication problem, we can show the multiplication as part of the division problem. After dividing to get 7, we multiply 7 by 5 and write the product under the 35. This shows that there are exactly 7 fives in 35.

$$\begin{array}{r} 7 \\ 5\overline{)35} \\ 35 \end{array}$$

Not all division problems have a whole-number quotient. Consider this question:

If 16 pennies are divided among 5 children, how many pennies will each child receive?

If we try to divide 16 into 5 equal groups, we find that there is no whole number that is an exact answer.

$$5\overline{)16}\text{ ?}$$

To answer the question, we think, "What number of fives is close to but not more than 16?" We answer that question with the number 3. We write "3" above the box and multiply to show that 3 fives is 15. Each child will get 3 pennies.

$$\begin{array}{r} 3 \\ 5\overline{)16} \\ 15 \end{array}$$

Now we subtract 15 from 16 to show how many pennies are left over. The amount left over is called the **remainder.** Here the remainder is 1, which means that one penny will be left over.

$$\begin{array}{r} 3 \\ 5\overline{)16} \\ -\ 15 \\ \hline 1 \end{array}$$

How we deal with remainders depends upon the question we are asked. For now, when we answer problems written with digits and division symbols, we will write the remainder at the end of our answer, with the letter "R" in front, as we show here.

$$\begin{array}{r} 3\text{ R }1 \\ 5\overline{)16} \\ -\ 15 \\ \hline 1 \end{array}$$

Example 1 Divide: $50 \div 8$

Solution We rewrite the problem with a division box. We think, "What number of eights is close to but not more than 50?" We answer "6" and then multiply 6 by 8 to get 48. We subtract to find the amount left over and write this remainder at the end of the answer.

$$\begin{array}{r} \mathbf{6\text{ R }2} \\ 8\overline{)50} \\ -\ 48 \\ \hline 2 \end{array}$$

Divisibility by 2, 5, and 10 For some division problems, we can decide whether there will be a remainder before we begin dividing. Here we show three rows from a multiplication table. We show the rows for twos, fives, and tens. In each row all the numbers can be divided by the first number of the row without leaving a remainder.

	1	2	3	4	5	6	7	8	9	10
twos	2	4	6	8	10	12	14	16	18	20
fives	5	10	15	20	25	30	35	40	45	50
tens	10	20	30	40	50	60	70	80	90	100

Notice that all the numbers in the twos row are even. If an even number is divided by 2, there will be no remainder. If an odd number is divided by 2, a remainder of 1 will result.

All the numbers in the fives row end in 5 or 0. If a whole number ending in 5 or 0 is divided by 5, there will be no remainder. If a whole number divided by 5 does not end in 5 or 0, there will be a remainder.

All numbers in the tens row end in zero. If a whole number ending in zero is divided by 10, there will be no remainder. If a whole number divided by 10 does not end in zero, there will be a remainder.

Example 2 Without dividing, decide which of these division problems will have a remainder. (There might be more than one.)

A. $2\overline{)16}$ B. $5\overline{)40}$ C. $10\overline{)45}$ D. $2\overline{)15}$

Solution **Problem C.** will have a remainder because 45 does not end in zero. Only numbers ending in zero can be divided by 10 without a remainder.

Problem D. will have a remainder because 15 is not even. Only even numbers can be divided by 2 without a remainder.

LESSON PRACTICE

Practice set* Divide. Write each answer with a remainder.

a. $5\overline{)23}$ **b.** $6\overline{)50}$ **c.** $37 \div 8$

d. $4\overline{)23}$ **e.** $7\overline{)50}$ **f.** $40 \div 6$

g. $10\overline{)42}$ **h.** $9\overline{)50}$ **i.** $34 \div 9$

j. Without dividing, decide which of these division problems will have a remainder.

$10\overline{)60}$ $5\overline{)44}$ $2\overline{)18}$

k. Which of these numbers can be divided by 2 without a remainder?

25 30 35

MIXED PRACTICE

Problem set

1. (12) Draw two horizontal lines, one above the other.

For problems 2–4, write an equation and find the answer.

2. (21) Huck collected 32 night crawlers for fishing. If he put an equal number in each of his 4 pockets, how many night crawlers did he put in each pocket?

3. (16) Julissa started a marathon, a race of approximately 26 miles. After running 9 miles, about how far did Julissa still have to run to finish the race?

4. (11) Eight hundred forty mice came in the front door. Four hundred eighteen mice came in the back door. Altogether, how many came in through the front and back doors? Use an addition pattern.

5. (22) $56 \div 10$

6. (22) $20 \div 3$

7. (22) $7\overline{)30}$

8. (18) $3 \times 7 \times 10$

9. (18) $2 \times 3 \times 4 \times 5$

10. (17) $\begin{array}{r} \$394 \\ \times \quad 8 \\ \hline \end{array}$

11. (17) $\begin{array}{r} 678 \\ \times \quad 4 \\ \hline \end{array}$

12. (17) $\begin{array}{r} \$6.49 \\ \times \quad 9 \\ \hline \end{array}$

13. (20) $\frac{63}{7}$

14. (20) $\frac{56}{8}$

15. (20) $\frac{42}{6}$

16. (17) $\begin{array}{r} \$4.08 \\ \times \quad 7 \\ \hline \end{array}$

17. (17) $\begin{array}{r} 3645 \\ \times \quad 6 \\ \hline \end{array}$

18. (17) $\begin{array}{r} 3904 \\ \times \quad 4 \\ \hline \end{array}$

19. (15, 18) $8 \times 0 = 4N$

20. (14) $C - 462 = 548$

21. (13) $\$36.15 - \$29.81$

22. (10) $963 + a = 6000$

23. (20) Use words to show how this problem is read: $4\overline{)12}$

24. (2, 22) Think of an odd number. Multiply it by 2. If the product is divided by 2, will there be a remainder?

25. What are the next three terms in this counting sequence?
(1, 12)

50, 40, 30, 20, 10, ...

26. Grandpa has 10 quarters. If he gives each of his 3 grandchildren 3 quarters, how many quarters will he have left?
(22)

27. Compare: 46,208 ◯ 46,028
(7)

28. How many $\frac{1}{4}$ circles equal a half circle?
(Inv. 2)

29. The fraction $\frac{1}{4}$ is equivalent to what percent?
(Inv. 2)

LESSON

23 Recognizing Halves

WARM-UP

Facts Practice: 90 Division Facts (Test D or E)

Mental Math: Count up by 5's from 1 to 51 (1, 6, 11, 16, ...). Count up and down by 3's between 0 and 36.

a. 10×7
b. 10×75
c. 7×30 plus 7×5
d. 5×35
e. 6×35
f. $280 + 14$
g. $240 + 12$
h. $960 - 140$
i. $6 \times 4, + 1, \div 5, + 1, \div 2$

Problem Solving:

Behind curtains A, B, and C were three prizes: a car, a boat, and a pogo stick. One prize was behind each curtain. List all the possible arrangements of prizes behind the curtains.

NEW CONCEPT

Many fractions equal one half. Here we show five fractions equal to one half:

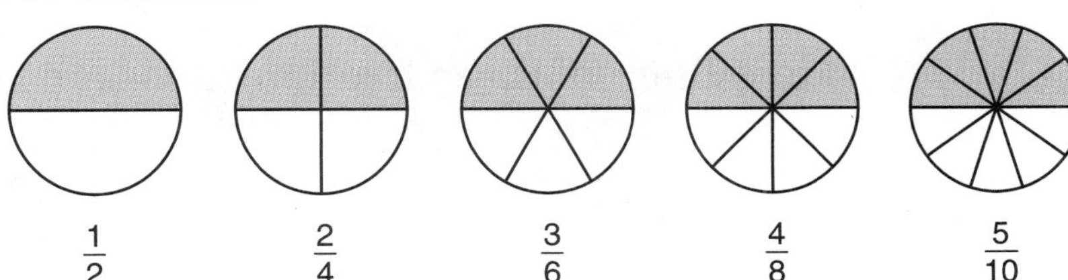

$\frac{1}{2}$ $\frac{2}{4}$ $\frac{3}{6}$ $\frac{4}{8}$ $\frac{5}{10}$

Notice that the numerator of each fraction is half the denominator.

$\frac{2}{4}$ Two is half of four.

$\frac{3}{6}$ Three is half of six.

$\frac{4}{8}$ Four is half of eight.

$\frac{5}{10}$ Five is half of ten.

A fraction is equal to $\frac{1}{2}$ if the numerator is half the denominator. A fraction is less than $\frac{1}{2}$ if the numerator is less than half the denominator. A fraction is greater than $\frac{1}{2}$ if the numerator is more than half the denominator.

Example 1 Which fraction is not equal to $\frac{1}{2}$?

A. $\frac{9}{18}$ B. $\frac{10}{25}$ C. $\frac{25}{50}$ D. $\frac{50}{100}$

Solution In each choice, the numerator is half the denominator, except for **B.** $\mathbf{\frac{10}{25}}$.

Example 2 Compare: $\frac{5}{12} \bigcirc \frac{4}{8}$

Solution The denominator of $\frac{5}{12}$ is 12, and half of 12 is 6. Since 5 is less than half of 12, $\frac{5}{12}$ is less than $\frac{1}{2}$. The other fraction, $\frac{4}{8}$, equals $\frac{1}{2}$. So $\frac{5}{12}$ is less than $\frac{4}{8}$.

$$\mathbf{\frac{5}{12} < \frac{4}{8}}$$

LESSON PRACTICE

Practice set

a. Think of a counting number. Double it. Then write a fraction equal to $\frac{1}{2}$ using your number and its double.

b. Which of these fractions does not equal $\frac{1}{2}$?

A. $\frac{7}{14}$ B. $\frac{8}{15}$ C. $\frac{9}{18}$ D. $\frac{21}{42}$

c. Compare: $\frac{5}{8} \bigcirc \frac{5}{12}$

d. Compare: $\frac{12}{24} \bigcirc \frac{6}{12}$

MIXED PRACTICE

Problem set For problems 1–4, write an equation and find the answer.

1. (16) It cost $3.48 to rent the movie. Leo gave the clerk $5.00. How much money should Leo get back? Use a subtraction pattern.

2. (11) The burger cost $1.45 and the fries cost $0.95. What was the cost of the burger and fries together? Use an addition pattern.

3. A week is 7 days. How many days is 52 weeks? Use a multiplication pattern.
(21)

4. Sumiko, Hector, and Julie divided the money equally. If there was $24 to start with, how much money did each receive? Use a multiplication pattern.
(21)

5. One half of the 20 students had finished the book. One fourth had not yet started the book.
(Inv. 2)

(a) How many students had finished the book?

(b) How many students had not started the book?

6. Compare: $\frac{3}{10} \bigcirc \frac{3}{6}$
(23)

7. $40 \div 6$
(22)

8. $3\overline{)20}$
(22)

9. $60 = N \times 10$
(18)

10. $\begin{array}{r} \$3.08 \\ \times \quad 7 \\ \hline \end{array}$
(17)

11. $\begin{array}{r} 2514 \\ \times \quad 3 \\ \hline \end{array}$
(17)

12. $\begin{array}{r} 697 \\ \times \quad 8 \\ \hline \end{array}$
(17)

13. Use words to show how this problem is read: $7\overline{)35}$
(20)

14. $4 \times 3 \times 10$
(18)

15. $12 \times 2 \times 10$
(18)

16. $\begin{array}{r} 4035 \\ - \quad S \\ \hline 3587 \end{array}$
(14)

17. $\begin{array}{r} M \\ - \ 1056 \\ \hline 5694 \end{array}$
(14)

18. $\begin{array}{r} \$70.00 \\ - \ \$ \ 7.53 \\ \hline \end{array}$
(13)

19. $5.00 + $8.75 + $10.00 + $0.35
(13)

20. $6.25 + $0.85 + $4.00 + D = $20.00
(10, 13)

21. Write two multiplication facts and two division facts for the fact family 7, 9, and 63.
(19)

22. Write the numbers 48, 16, and 52 in order from greatest to least.
(4)

23. Draw two vertical lines side by side.
(12)

24. Use words to name the number 212,500.
(7)

25. Write two addition facts and two subtraction facts for the fact family 7, 9, and 16.
(8)

26. Which fraction below does not equal $\frac{1}{2}$?
(23)

A. $\frac{10}{20}$ B. $\frac{20}{40}$ C. $\frac{40}{80}$ D. $\frac{80}{40}$

27. The fraction $\frac{3}{4}$ is equivalent to what percent?
(Inv. 2)

28. Mary has nine quarters in her coin purse. Write and answer a multiplication problem that shows the value of the nine quarters.
(17)

29. Write an "equal groups" story problem for this equation. Then answer the question in your story problem.
(21)

$$3 \times 12 = P$$

LESSON

24 Parentheses • Associative Property

WARM-UP

Facts Practice: 64 Multiplication Facts (Test F)

Mental Math: Count up by 5's from 2 to 52. Count up and down by 3's between 0 and 36.

a. 10×12 **b.** 10×20
c. 8×40 plus 8×2 **d.** 7×42
e. 6×42 **f.** $\frac{1}{2}$ of 12
g. $\frac{1}{4}$ of 12 **h.** $\frac{1}{10}$ of 30
i. $1000 - 100$
j. 6×3, $+ 2$, $\div 2$, $- 2$, $\div 2$

Problem Solving:

Copy this subtraction problem and fill in the missing digits:

$$\begin{array}{r} \_4\_ \\ -\ 3\_2 \\ \hline 58 \end{array}$$

NEW CONCEPTS

Parentheses

The **operations of arithmetic** are addition, subtraction, multiplication, and division. When there is more than one operation in a problem, **parentheses** can show us the order for doing the operations. Parentheses separate a problem into parts. We do the part inside the parentheses first. In the problem below, the parentheses tell us to add 5 and 4 before we multiply by 6.

$$6 \times \underbrace{(5 + 4)} =$$
$$6 \times \quad 9 \quad = 54$$

Example 1 Simplify: $8 - (4 + 2)$

Solution It takes two steps to find the answer to this problem. The parentheses show us which step to take first. We add 4 and 2 to get 6. Then we subtract 6 from 8 and get **2.**

$$8 - \underbrace{(4 + 2)} =$$
$$8 - \quad 6 \quad = 2$$

Example 2 Compare: $2 \times (3 + 4) \bigcirc (2 \times 3) + 4$

Solution The numbers and operations on both sides are the same, but the order for doing the operations is different. We follow the proper order on both sides and find that the left-hand side is greater than the right-hand side.

$$2 \times \underbrace{(3 + 4)} \bigcirc \underbrace{(2 \times 3)} + 4$$
$$2 \times 7 \bigcirc 6 + 4$$
$$\mathbf{14} > \mathbf{10}$$

Associative property When performing the operations of arithmetic, we perform one operation at a time. So if we have three numbers to add, we decide which two numbers to add first. Suppose we wish to find $4 + 5 + 6$. We may find $4 + 5$ first and then add 6, or we may find $5 + 6$ first and then add 4. Either way, the sum is 15.

$$(4 + 5) + 6 = 4 + (5 + 6)$$

Whichever way we group the addends, the result is the same. This property is called the **associative property of addition.**

The associative property also applies to multiplication, but not to subtraction or division. Below we illustrate the **associative property of multiplication.** Whichever way we group the factors, the product is the same.

$$(2 \times 3) \times 4 \bigcirc 2 \times (3 \times 4)$$
$$6 \times 4 \bigcirc 2 \times 12$$
$$24 = 24$$

LESSON PRACTICE

Practice set* Solve each problem by following the proper order of operations:

a. $6 - (4 - 2)$ **b.** $(6 - 4) - 2$

c. $(8 \div 4) \div 2$ **d.** $8 \div (4 \div 2)$

e. $12 \div (4 - 1)$ **f.** $(12 \div 4) - 1$

g. Name the four operations of arithmetic.

For each problem, write the proper comparison symbol, and state whether the associative property applies.

h. $(8 \div 4) \div 2 \bigcirc 8 \div (4 \div 2)$

i. $(8 - 4) - 2 \bigcirc 8 - (4 - 2)$

j. $(8 \times 4) \times 2 \bigcirc 8 \times (4 \times 2)$

MIXED PRACTICE

Problem set

1. (Inv. 2) How much money is one half of a dollar plus one fourth of a dollar?

For problems 2–4, write an equation and find the answer.

2. (21) How many horseshoes are needed to shoe 25 horses? Use a multiplication pattern.

3. (16) Inez removed some eggs from a carton of one dozen eggs. If nine eggs remained in the carton, how many eggs did Inez remove?

4. (11) The auditorium had nine hundred fifty-six seats. Only ninety-eight seats were occupied. How many seats were not occupied? Which pattern did you use?

5. (19) Write two multiplication facts and two division facts for the fact family 5, 10, and 50.

6. (24) Compare: $3 \times (4 + 5) \bigcirc (3 \times 4) + 5$

7. (24) $30 - (20 + 10)$

8. (24) $(30 - 20) + 10$

9. (24) Compare: $4 \times (6 \times 5) \bigcirc (4 \times 6) \times 5$

10. (22) $60 \div 7$

11. (22) $50 \div 6$

12. (22) $10\overline{)44}$

13. (17) $\begin{array}{r} \$50.36 \\ \times \quad 4 \\ \hline \end{array}$

14. (17) $\begin{array}{r} 7408 \\ \times \quad 6 \\ \hline \end{array}$

15. (17) $\begin{array}{r} 4637 \\ \times \quad 9 \\ \hline \end{array}$

16. (13, 14) $\begin{array}{r} W \\ - \$9.62 \\ \hline \$14.08 \end{array}$

17. (14) $\begin{array}{r} 4730 \\ - \quad J \\ \hline 2712 \end{array}$

18. (13) $\begin{array}{r} \$30.00 \\ - \$\ 0.56 \\ \hline \end{array}$

19. (13) $\$3.54 + \$12 + \$1.66$

20. (13) $\$20 - \$16.45$

21. Write two addition facts and two subtraction facts for the fact family 9, 5, and 14.
(8)

22. Which digit in 256 shows the number of hundreds?
(3)

23. Freda bought four postcards for \$0.35 each. Altogether, how much did the postcards cost? Change this addition problem to a multiplication problem and find the total cost.
(13, 17)

\$0.35 + \$0.35 + \$0.35 + \$0.35

24. What is the tenth term of this counting sequence?
(1)

3, 6, 9, 12, 15, ...

25. When any of these odd numbers is divided by 2, there is a remainder of 1. Which of these odd numbers can be divided by 5 without a remainder?
(22)

A. 23 B. 25 C. 27 D. 29

26. Draw two vertical lines.
(12)

27. Write two multiplication facts and two division facts for the fact family 7, 8, and 56.
(19)

28. Compare: (8 + 4) + 2 ◯ 8 + (4 + 2)
(24)

Based on your answer, does the associative property apply to addition?

29. (a) What number is half of 14?
(2, 23)

(b) Write a fraction equal to $\frac{1}{2}$ using 14 and its half.

LESSON

25 Listing the Factors of Whole Numbers

WARM-UP

Facts Practice: 90 Division Facts (Test D or E)

Mental Math: Count up by 5's from 3 to 53 (3, 8, 13, 18, ...). Count by 7's from 0 to 77. (A calendar can help you start.)

a. 10×10 **b.** 10×100 **c.** 6×24
d. 6×42 **e.** 7×42 **f.** $\frac{1}{2}$ of 40
g. $\frac{1}{4}$ of 40 **h.** $\frac{1}{10}$ of 40 **i.** $365 - 100$
j. 6×2, $- 2$, $\times 2$, $+ 1$, $\div 3$

Problem Solving:

Tom was thinking of a two-digit even number. Tom hinted that you say the number when counting by 3's and when counting by 7's, but not when counting by 4's. What number was Tom thinking of?

NEW CONCEPT

The **factors** of a number are all the whole numbers that can divide it without leaving a remainder. For example, the factors of 6 are 1, 2, 3, and 6 because each of these numbers divides 6 without leaving a remainder.

Example 1 List the factors of 20.

Solution We look for all the whole numbers that divide 20 without leaving a remainder. Which numbers can be put into this box to give us an answer without a remainder?

$\boxed{?}\overline{)20}$

One way to find out is to start with 1 and to try each whole number up to 20. If we do this, we find that the numbers that divide 20 evenly are **1, 2, 4, 5, 10,** and **20.** These are the factors of 20. All other whole numbers leave a remainder.

We can cut our search for factors in half if we record the quotient when we find a factor.

$$1\overline{)20}\ \text{quotient } 20$$ Both **1** and **20** are factors.

$$2\overline{)20}\ \text{quotient } 10$$ Both **2** and **10** are factors.

$$4\overline{)20}\ \text{quotient } 5$$ Both **4** and **5** are factors.

Example 2 List the factors of 23.

Solution The only factors of 23 are **1** and **23.** Every number greater than 1 has at least two factors: the number 1 and itself.

Sometimes we can discover some factors of a number just by looking at one or two of its digits. For example, a factor of every even number is 2, and any whole number ending in 0 or 5 has 5 as a factor. Since 20 is even and ends with zero, we know that both 2 and 5 are factors of 20.

Example 3 Which of these numbers is not a factor of 30?

A. 2 B. 3 C. 4 D. 5

Solution We see that 30 is an even number ending in zero. So 2 and 5 are factors. And we quickly see that 30 can be divided by 3 without a remainder. The only choice that is not a factor is **C. 4.**

LESSON PRACTICE

Practice set List the factors of each of these numbers:

a. 4 **b.** 3 **c.** 6 **d.** 5

e. 8 **f.** 11 **g.** 9 **h.** 12

i. 1 **j.** 14 **k.** 2 **l.** 15

m. Two is not a factor of which of these numbers?

A. 236 B. 632 C. 362 D. 263

n. Five is not a factor of which of these numbers?

A. 105 B. 150 C. 510 D. 501

o. Which of these numbers is not a factor of 40?

A. 2 B. 5 C. 6 D. 10

MIXED PRACTICE

Problem set For problems 1–3, write an equation and find the answer.

1. (21) The Christmas-tree farm planted 9 rows of trees, with 24 trees in each row. How many trees were planted?

2. (16) The haircut cost \$6.75. Mila paid for it with a \$10 bill. How much money should she get back?

3. (21) Donna bought four cartons of milk for \$1.12 each. Altogether, how much did Donna spend?

4. (25) List the factors of 30.

5. (25) List the factors of 13.

6. (24) Compare: $4 \times (6 \times 10) \bigcirc (4 \times 6) \times 10$

7. (24) Which property of multiplication is illustrated in problem 6?

8. (24) $6 \times (7 + 8)$

9. (24) $(6 \times 7) + 8$

10. (19) Write two multiplication facts and two division facts for the fact family 10, 12, and 120.

11. (18) $9n = 54$

12. (22) $55 \div 8$

13. (17) $\begin{array}{r} 1234 \\ \times \quad 5 \\ \hline \end{array}$

14. (17) $\begin{array}{r} \$5.67 \\ \times \quad 8 \\ \hline \end{array}$

15. (17) $\begin{array}{r} 987 \\ \times \quad 6 \\ \hline \end{array}$

16. (13, 14) $W - \$13.55 = \$5$

17. (14) $2001 - R = 1002$

18. (6) $4387 + 124 + 96$

19. (6) $3715 + 987 + 850$

20. (10, 13) $\$6.75 + \$8 + \$1.36 + P = \$20$

21. (Inv. 2) How much money is $\frac{1}{2}$ of a dollar plus $\frac{1}{4}$ of a dollar plus $\frac{1}{10}$ of a dollar?

22. (7) Use words to name the number 894,201.

23. (20) Which number is the divisor in this equation? $6\overline{)42}$ with quotient 7

24. What is the tenth term in this counting sequence?
(1)

5, 10, 15, 20, ...

25. Think of a whole number. Multiply it by 2. Is the answer odd or even?
(2, 15)

26. Two is not a factor of which of these numbers?
(25)

A. 456 B. 465 C. 654 D. 564

27. Which property of addition is illustrated by this equation?
(24)

$$(6 + 7) + 8 = 6 + (7 + 8)$$

28. Write a multiplication equation that shows the number of blocks used to build this figure.
(18)

29. The fraction $\frac{1}{10}$ is equivalent to what percent?
(Inv. 2)

LESSON

26 Division Algorithm

WARM-UP

Facts Practice: 64 Multiplication Facts (Test F)

Mental Math: Count up by 5's from 4 to 54. Count by 7's from 0 to 77.

a. How many cents is 1 quarter? ... 2 quarters? ... 3 quarters?

b. 10×34 **c.** 5×34 **d.** $\frac{1}{2}$ of 8

e. $\frac{1}{4}$ of 8 **f.** $\frac{3}{4}$ of 8 **g.** $640 + 32$

h. 5×8, $+ 2$, $\div 6$, $\times 3$, $- 1$, $\div 2$

Problem Solving:

Use each of the digits 5, 6, 7, 8, and 9 to complete this addition problem:

$$\begin{array}{r} \_\,\_ \\ +\ \ \_ \\ \hline \_\,\_ \end{array}$$

NEW CONCEPT

A *division algorithm* is a method for solving division problems whose answers have not been memorized. A division algorithm breaks large division problems into a series of smaller division problems that are easier to do. In each of the smaller problems we follow four steps: **divide, multiply, subtract,** and **bring down.** As we do each step, we write a number. Drawing this division chart a few times will help us remember the steps:

Division Chart

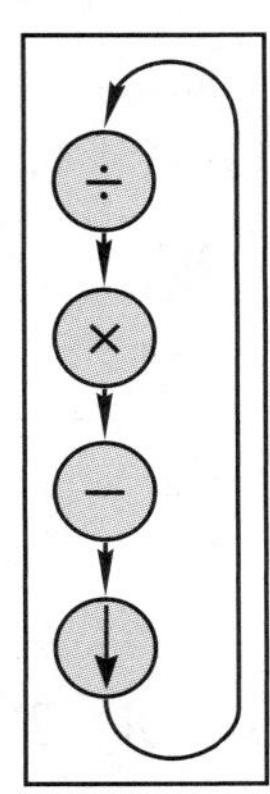

Step 1: Divide and write a number.

Step 2: Multiply and write a number.

Step 3: Subtract and write a number.

Step 4: Bring down the next digit.

Every time we bring down a digit, we divide again, even if the answer when we divide is zero. We continue to divide, multiply, subtract, and bring down until there are no digits left to bring down.

Example 1 Divide: $3\overline{)\$8.52}$

Solution We begin by breaking the division problem into a smaller problem. Our first division problem in this example is $3\overline{)\$8}$.

$$3\overline{)\$8.52}$$

We divide and write "\$2" above the \$8. Then we multiply \$2 by 3, which is \$6. We write "6" below the \$8. We subtract and get \$2. Then we bring down the next digit, which is 5.

$$\begin{array}{r} \$2 \\ 3\overline{)\$8.52} \\ -\ 6\downarrow \\ \hline 2\ 5 \end{array}$$

Now we begin a new division problem, $3\overline{)25}$. The 2 of 25 is in the dollars place, and the 5 is in the dimes place. So this division is like dividing 25 dimes by 3. The answer is 8 dimes, which we write above the 5. We multiply 8 dimes by 3, which is 24 dimes. We write "24" below the 25. Then we subtract, get 1 dime, and bring down the 2 cents.

$$\begin{array}{r} \$2\ 8 \\ 3\overline{)\$8.52} \\ -\ 6 \\ \hline 2\ 5 \\ -\ 2\ 4\downarrow \\ \hline 12 \end{array}$$

We are ready to begin the last division, $3\overline{)12}$. The "12" is 12¢. We divide and write "4" for 4¢ above the 2. Then we multiply and subtract. There are no digits to bring down. There is no remainder. We write the decimal point in the answer above the decimal point in the division box to get **\$2.84.**

$$\begin{array}{r} \$2.84 \\ 3\overline{)\$8.52} \\ -\ 6 \\ \hline 2\ 5 \\ -\ 2\ 4 \\ \hline 12 \\ -\ 12 \\ \hline 0 \end{array}$$

We can check a division answer by multiplying. We multiply \$2.84 by 3 and get \$8.52. The three numbers of the multiplication answer should match the three numbers in the division box.

$$\begin{array}{r} {}^{2\ 1} \\ \$2.84 \\ \times\qquad 3 \\ \hline \$8.52 \end{array} \quad \text{check}$$

Example 2 Divide: $5\overline{)234}$

Solution Since we cannot divide 2 by 5, we begin with the division $5\overline{)23}$. We divide and write "4" above the 3 of 23. Then we multiply, subtract, and bring down.

$$\begin{array}{r} 4 \\ 5\overline{)234} \\ -\ 20\downarrow \\ \hline 34 \end{array}$$

Now we begin the new division, $5\overline{)34}$. We divide and write "6" above the 4. Then we multiply and subtract. Since there is no other number to bring down, we are finished. The remainder is 4. Thus, the answer is **46 R 4.** This means that 234 equals 46 fives plus 4.

$$\begin{array}{r} 46 \text{ R } 4 \\ 5\overline{)234} \\ -\ 20 \\ \hline 34 \\ -\ 30 \\ \hline 4 \end{array}$$

Checking a division answer with a remainder takes two steps. First we multiply. Then we add the remainder to the product we get. To check our answer to this division, we multiply 46 by 5 and then add 4.

$$\begin{array}{rl} 46 & \\ \times\ \ 5 & \\ \hline 230 & \\ +\ \ \ 4 & \text{remainder} \\ \hline 234 & \text{check} \end{array}$$

Example 3 Solve: $5N = 365$

Solution Two numbers are multiplied, 5 and N. The product is 365. We can find an unknown factor by dividing the product by the known factor. We divide 365 by 5 and find that N is **73.**

$$\begin{array}{r} 73 \\ 5\overline{)365} \\ -\ 35 \\ \hline 15 \\ -\ 15 \\ \hline 0 \end{array}$$

LESSON PRACTICE

Practice set* Divide:

a. $4\overline{)\$5.56}$ **b.** $9\overline{)375}$ **c.** $3\overline{)\$4.65}$ **d.** $5\overline{)645}$

e. $7\overline{)\$3.64}$ **f.** $7\overline{)365}$ **g.** $10\overline{)546}$ **h.** $4\overline{)\$4.56}$

i. Show how to check this division answer. $\begin{array}{r} 12 \text{ R } 3 \\ 6\overline{)75} \end{array}$

Find each missing factor:

j. $3x = 51$ **k.** $4y = 92$ **l.** $6z = 252$

MIXED PRACTICE

Problem set For problems 1–3, write an equation and find the answer.

1. (16) The bicycle tire cost $2.98. Jen paid for the tire with a $5 bill. How much should she get back in change?

2. (21) Sarita sent 3 dozen cupcakes to school for a party. How many cupcakes did she send? Use a multiplication pattern.

3. (11) When three new students joined the class, the number of students increased to 28. How many students were in the class before the new students arrived?

4. (2, 23) (a) What is the smallest two-digit even number?

(b) What is half of the number in part (a)?

(c) Use the answers to parts (a) and (b) to write a fraction equal to $\frac{1}{2}$.

5. (25) List the factors of 16.

6. (26) $5\overline{)\$3.75}$

7. (26) $4\overline{)365}$

8. (26) $6m = 234$

9. (26) $\$4.32 \div 6$

10. (26) $\frac{123}{3}$

11. (26) $\frac{576}{6}$

12. (17) $\$7.48 \times 4$

13. (17) 609×8

14. (18) $7 \times 8 \times 10$

15. (15, 18) $7 \times 8 \times 0$

16. (14) $9374 - M = 4938$

17. (13) $\$10 - \$6.24$

18. (10) $L + 427 + 85 = 2010$

19. (13) $\$12.43 + \$0.68 + \$10$

20. (4, 18) Compare. (Try to answer the comparison without multiplying.)

$$3 \times 40 \bigcirc 3 \times 4 \times 10$$

21. (18) $8 \times 90 = 8 \times 9 \times N$

22. Write two multiplication facts and two division facts for
(19) the fact family 8, 9, and 72.

23. A checkerboard has 64 squares. The squares are in 8
(21) equal rows. How many squares are in each row? Use a multiplication pattern.

24. How much money is $\frac{3}{4}$ of a dollar plus $\frac{3}{10}$ of a dollar?
(Inv. 2)

25. What number is halfway between 400 and 600?
(12)

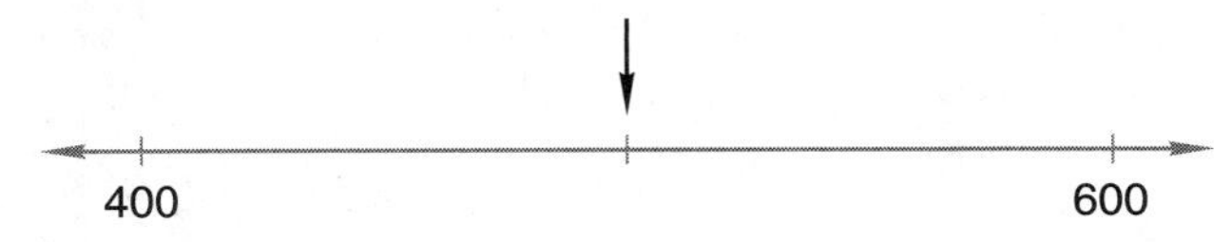

26. This equation shows that 7 is a factor of 91. Which other factor of 91 is shown by this equation?
(25)

$$7\overline{)91}\text{ with quotient }13$$

27. What is the sum of three hundred forty-seven and eight
(5, 6) hundred nine?

28. Here is Todd's answer to a division problem. Show how to check the answer. Is Todd's answer correct? Why or why not?
(26)

$$4\overline{)75}\text{ with quotient }16\text{ R }3$$

29. Which of these numbers is not a factor of 15?
(25)

A. 1 B. 2 C. 3 D. 5

LESSON

27 Reading Scales

WARM-UP

Facts Practice: 64 Multiplication Facts (Test F)

Mental Math: How many days are in a common year? ... a leap year? Count by 12's from 12 to 60.

a. How many days are in 2 weeks? ... 3 weeks? ... 4 weeks?

b. 10×24 **c.** 6×24 **d.** $\frac{1}{2}$ of 100

e. $\frac{1}{4}$ of 100 **f.** $\frac{3}{4}$ of 100 **g.** $\frac{1}{10}$ of 100

h. $\frac{7}{10}$ of 100 **i.** $6 \times 6,\ -\ 1,\ \div\ 5$

Problem Solving:

Use each of the digits 5, 6, 7, 8, and 9 to complete this subtraction problem:

```
  __
-  _
----
  __
```

NEW CONCEPT

Number lines can be horizontal, vertical, or even curved. It is not necessary to show every whole number on a number line. Some number lines show only even numbers or numbers we say when counting by 5's. The locations of unlabeled numbers must be figured out.

One use of a number line is as a **scale** for measuring temperature. Two commonly used temperature scales are the **Fahrenheit** (F) scale and the **Celsius** (C) scale. On the Fahrenheit scale, water freezes at 32°F and boils at 212°F. The Celsius scale is a **centigrade** scale, meaning there are one hundred gradations, or degrees, between the freezing and boiling points of water. On the Celsius scale, water freezes at 0°C and boils at 100°C.

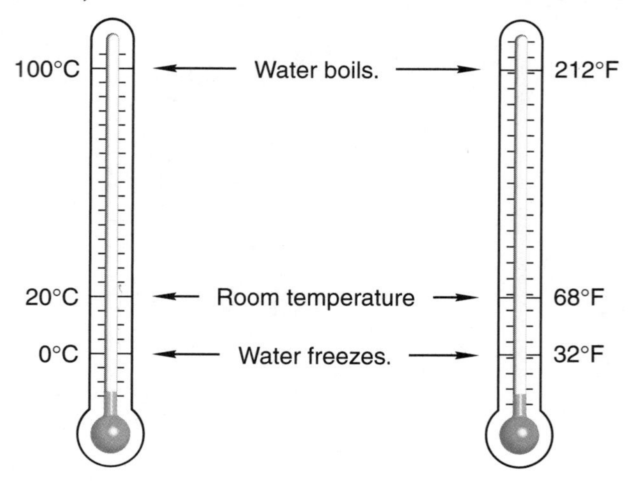

Example 1 What temperature is shown by this thermometer?

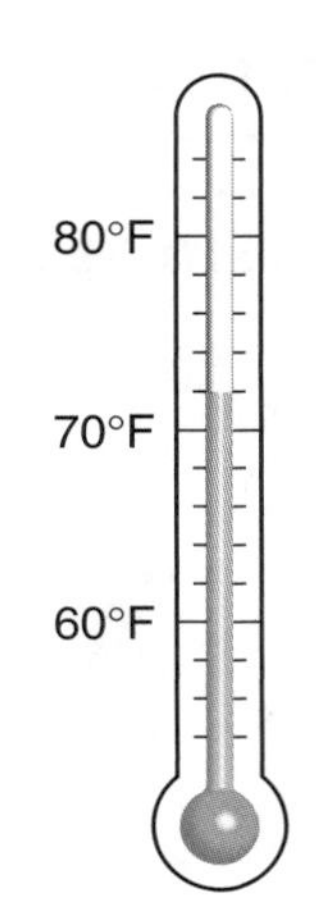

Solution This thermometer indicates the temperature in degrees Fahrenheit, which is abbreviated "°F." On the scale, only every 10° is labeled. There are five spaces between every 10°. That means every space equals 2°. One space up from 70° is 72°. The thermometer shows a temperature of **72°F.**

Example 2 To what number on this scale is the arrow pointing?

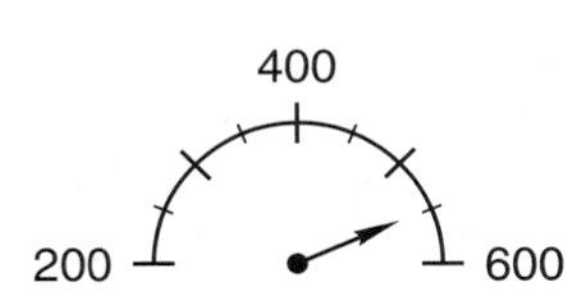

Solution As we move along the curve toward the right, we see that the numbers grow larger. The arrow points to a location past the 400 mark and near the 600 mark. Halfway between the 400 and 600 marks is a long mark that stands for 500. The arrow points halfway between the 500 and 600 marks, so it points to **550.**

Example 3 Draw a horizontal number line from 0 to 500 with only zero and hundreds marked and labeled.

Solution We draw a horizontal number line and make marks for 0, 100, 200, 300, 400, and 500. These marks should be evenly spaced. We then label the marks. Our number line should look like this:

LESSON PRACTICE

Practice set

a. Draw a number line from 0 to 100 with only zero and tens marked and labeled.

b. On the Celsius scale, what temperature is five degrees less than the freezing point of water?

c. Points A and B on this number line indicate two numbers. Write the two numbers, using a comparison symbol to show which is greater and which is less.

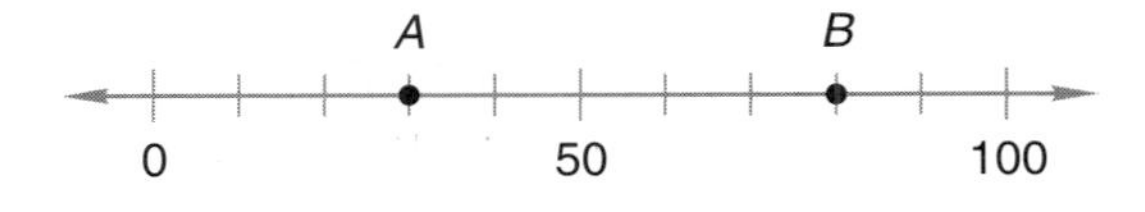

MIXED PRACTICE

Problem set For problems 1–4, write an equation and find the answer.

1. (11) On the first 3 days of their trip, the Smiths drove 408 miles, 347 miles, and 419 miles. Altogether, how far did they drive in 3 days?

2. (21) Tom is 5 feet tall. There are 12 inches in 1 foot. How many inches tall is Tom?

3. (16) Fifteen minutes after the store opened, only seven autographed footballs remained in the store. If customers had purchased 27 autographed footballs during the first 15 minutes, how many autographed footballs were in the store when it opened?

4. (17, 21) Lucy sold 9 cups of lemonade for $0.15 each. How much money did Lucy collect by selling lemonade?

5. (2) Andrew's age is half of David's age. If David is 12 years old, then how old is Andrew?

6. (26) $864 \div 5$

7. (26) $\$2.72 \div 4$

8. (26) $608 \div 9$

9. (24, 26) $378 \div (18 \div 3)$

10. (27) What temperature is shown by this thermometer?

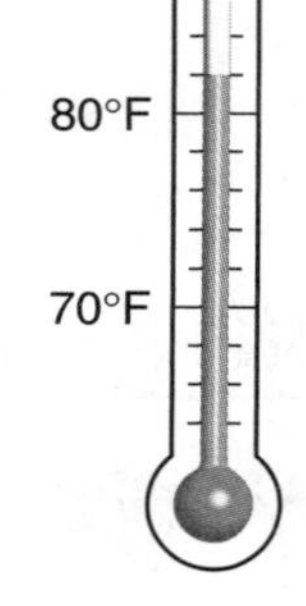

11. (17) $\begin{array}{r} \$52.60 \\ \times \quad 7 \\ \hline \end{array}$

12. (17) $\begin{array}{r} 3874 \\ \times \quad 6 \\ \hline \end{array}$

13. (17) $\begin{array}{r} 9063 \\ \times \quad 8 \\ \hline \end{array}$

14. (27) To what number on this scale is the arrow pointing?

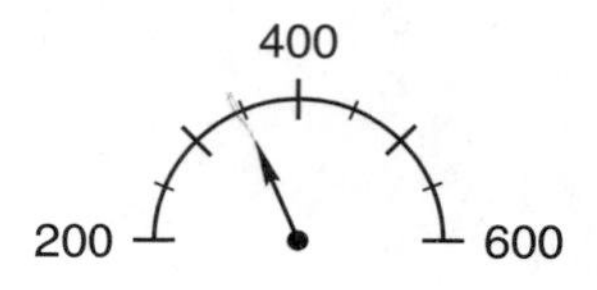

15. (10) $386 + 4287 + 672 + M = 5350$

16. (27) Draw a horizontal number line from 0 to 50 with only zero and tens marked and labeled.

17. The number 78 is between which of these pairs of numbers?
(4)

A. 60 and 70 B. 70 and 80 C. 80 and 90

18. List the factors of 30.
(25)

19. When three hundred ninety-seven is subtracted from four hundred five, what is the difference?
(5, 9)

20. In Khadija's class there is one more boy than there are girls. Which could not be the number of students in Khadija's class?
(2)

A. 25 B. 27 C. 28 D. 29

21. On the Celsius scale, what temperature is ten degrees below the freezing point of water?
(27)

22. What are the next three terms in this counting sequence?
(1)

..., 160, 170, 180, _____, _____, _____, ...

23. Which digit in 537 shows the number of hundreds?
(3)

24. Use words to name 327,040.
(7)

25. To what number is the arrow pointing?
(27)

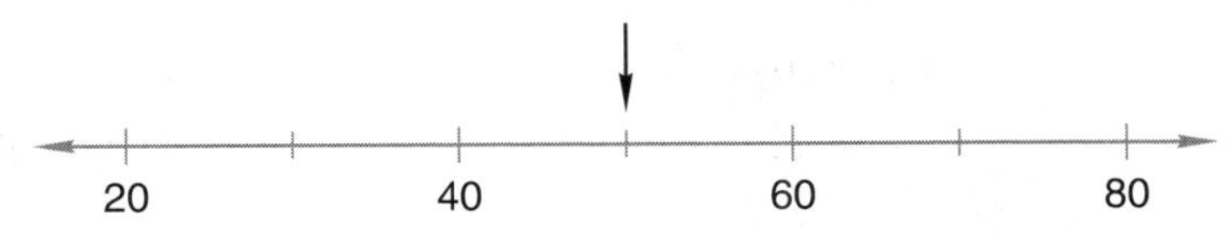

26. Show three ways to write "24 divided by 3" with digits and division symbols.
(20)

27. Here is Madeline's answer to a division problem. Show how to check the division. Is Madeline's answer correct? Why or why not?
(26)

$$7\overline{)100}\quad 14\text{ R }2$$

28. Compare: $12 \div (6 \div 2) \bigcirc (12 \div 6) \div 2$
(20, 24)

Does the associative property apply to division?

29. The fraction $\frac{3}{10}$ is equivalent to what percent?
(Inv. 2)

LESSON

28 Measuring Time

WARM-UP

Facts Practice: 90 Division Facts (Test D or E)

Mental Math: Count by 12's from 12 to 72. Count by 5's from 2 to 52.

a. How many days is 2 weeks? ... 3 weeks? ... 4 weeks?

b. 100×25 **c.** 7×25 **d.** $\frac{1}{2}$ of 40

e. $\frac{1}{4}$ of 40 **f.** $\frac{3}{4}$ of 40 **g.** $\frac{1}{10}$ of 40

h. $\frac{9}{10}$ of 40 **i.** $7 \times 7, + 1, \div 5, \div 5$

Problem Solving:

Half of the students in the room were girls. Half of the girls had brown hair. Half the brown-haired girls wore ponytails. If 4 brown-haired girls were wearing ponytails, how many students were in the room?

NEW CONCEPT

We measure the passage of time by the movement of Earth. A **day** is the length of time it takes Earth to spin around on its axis once. We divide a day into 24 equal parts called **hours.** Each hour is divided into 60 equal lengths of time called **minutes,** and each minute is divided into 60 **seconds.**

Besides spinning on its axis, Earth also moves on a long journey around the Sun. The time it takes to travel around the Sun is a **year.** It takes Earth about $365\frac{1}{4}$ days to travel once around the Sun. To make the number of days in every year a whole number, we have three years in a row that have 365 days each. These years are called **common years.** Then we have one year that has 366 days. A year with 366 days is called a **leap year.**†

A year is divided into 12 **months.** The month February has 28 days in common years and 29 days in leap years. Four months have 30 days each. All the rest have 31 days each. Seven days in a row is called a **week.** We may refer to a calendar to see which day of the week a particular day of the month falls on.

†Sometimes there are seven years in a row without a leap year. This happens around "century years" that cannot be divided evenly by 400. For example, the seven-year span 1897–1903 contains no leap years, because 1900 cannot be divided evenly by 400.

To identify longer spans of time, we may use the terms **decade, century,** and **millennium.** A decade is a period of ten years, and a century is a period of 100 years. A millennium is a period of 1000 years.

Example 1 A century is how many decades?

Solution A century is 100 years. A decade is 10 years. Since 10 tens equals 100, a century is **10 decades.**

Example 2 According to this calendar, June 8, 2014, is what day of the week?

JUNE 2014						
S	M	T	W	T	F	S
1	2	3	4	5	6	7
8	9	10	11	12	13	14
15	16	17	18	19	20	21
22	23	24	25	26	27	28
29	30					

Solution Many calendars are designed so that the first day of the week is Sunday. On this calendar the letters at the top of each column stand for "Sunday," "Monday," "Tuesday," "Wednesday," "Thursday," "Friday," and "Saturday." Thus, June 8, 2014, is a **Sunday,** the second Sunday of the month.

The time of day can be shown by a clock. A clock can be either *digital* or *analog.* Analog clocks show time with hands that point to places on a circular number line. An analog clock actually contains two number lines in one. One number line is the hour scale. It has 12 marks, usually numbered, that show the hours of the day. The other number line is the minute scale. It has 60 smaller marks, usually unnumbered, that show the minutes of the hour. The two scales are wrapped into a circle so that the ends are joined. A full day is 24 hours long, but most clocks show only 12 hours.

We "tell time" by reading the locations on the number line to which the hands are pointing. With the short hand we read from the hour scale, and with the long hand we read from the minute scale.

When writing the time of day, we write the hour, a colon, and two digits to show the number of minutes after the hour. This format is also used by many digital clocks to show the time of day. The time shown by the clock above is 1:45.

The 24 hours of a day are divided into **a.m.** hours and **p.m.** hours. The time 12:00 a.m. is called *midnight* and is the beginning of each day. The time 12:00 p.m. is called *noon* and is the midpoint of each day. The 12 hours before noon are the "a.m." hours. The 12 hours after noon are the "p.m." hours. When stating the time of day, we will use the labels "a.m." and "p.m." to prevent confusion.

When naming the time of day, we sometimes refer to fractions of an hour, such as "a quarter after two" or "a quarter to four" or "half past seven." A quarter of an hour is 15 minutes, so a quarter after two is 2:15, and a quarter to four is 3:45. Half past seven is 7:30.

Example 3 If it is morning, what time is shown by this clock? What will the time be in three hours?

Solution The clock shows 5 minutes after the ninth hour. The proper form is hour, colon, two digits for the minutes, and then "a.m." or "p.m." The time indicated is **9:05 a.m.** In three hours the time will be after noon, so the "a.m." will switch to "p.m." The time will be **12:05 p.m.**

LESSON PRACTICE

Practice set*

a. Four centuries is how many years?

b. According to the calendar in example 2, what is the date of the third Thursday in June 2014?

c. A leap year has how many days?

d. What is the name for $\frac{1}{10}$ of a century?

e. Write the time that is 2 minutes after eight in the evening.

f. Write the time that is a quarter to nine in the morning.

g. Write the time that is 20 minutes after noon.

h. Write the time that is 30 minutes after midnight.

i. Write the time that is a quarter after nine in the morning.

j. If it is morning, what time is shown by the clock?

k. What time would be shown by the clock 2 hours later?

MIXED PRACTICE

Problem set For problems 1–3, write an equation and find the answer.

1. (16) After Luis paid Jenny $600 for rent, he had $1267 remaining. How much money did Luis have before paying rent?

2. (11) Mae-Ying had $1873. She earned $200 more for baby-sitting. How much money did she then have? Use an addition pattern.

3. (21) Dan separated 52 cards into 4 equal piles. How many cards were in each pile? Write a multiplication pattern.

4. (28) One half of a decade is how many years?

5. (25) Which factors of 18 are also factors of 24?

6. (26) $\frac{\$5.43}{3}$

7. (26) $\frac{\$6.00}{8}$

8. (24, 26) $528 \div (28 \div 7)$

9. (26) $6w = 696$

10. (28) It is evening. What time is shown by this clock? What will be the time in three hours?

11. (28) Write the time that is half past noon.

12. (Inv. 2) How much money is $\frac{1}{2}$ of a dollar plus $\frac{5}{10}$ of a dollar?

13. (28) According to this calendar, May 10, 2042, is what day of the week?

MAY 2042

S	M	T	W	T	F	S
				1	2	3
4	5	6	7	8	9	10
11	12	13	14	15	16	17
18	19	20	21	22	23	24
25	26	27	28	29	30	31

14. (2) What is the largest three-digit even number that has the digits 5, 6, and 7?

15. (6)
$$\begin{array}{r} 4387 \\ 2965 \\ +\ 4943 \\ \hline \end{array}$$

16. (13)
$$\begin{array}{r} \$63.75 \\ -\ \$46.88 \\ \hline \end{array}$$

17. (14)
$$\begin{array}{r} 4010 \\ -\ \quad F \\ \hline 563 \end{array}$$

18. (17)
$$\begin{array}{r} 3408 \\ \times\ \quad 7 \\ \hline \end{array}$$

19. (17)
$$\begin{array}{r} \$3.56 \\ \times\ \quad 8 \\ \hline \end{array}$$

20. (17)
$$\begin{array}{r} 487 \\ \times\ \ 9 \\ \hline \end{array}$$

21. (28) What time is 5 minutes before nine in the morning?

22. (19) Write two multiplication facts and two division facts for the fact family 10, 2, and 20.

23. (26) Show how to check this division answer. Is the answer correct?

$$\begin{array}{r} 22 \text{ R } 2 \\ 9\overline{)200} \end{array}$$

24. (1) What are the next three terms in this counting sequence?

..., 400, 500, 600, 700, _____, _____, _____, ...

25. (27) To what number is the arrow pointing?

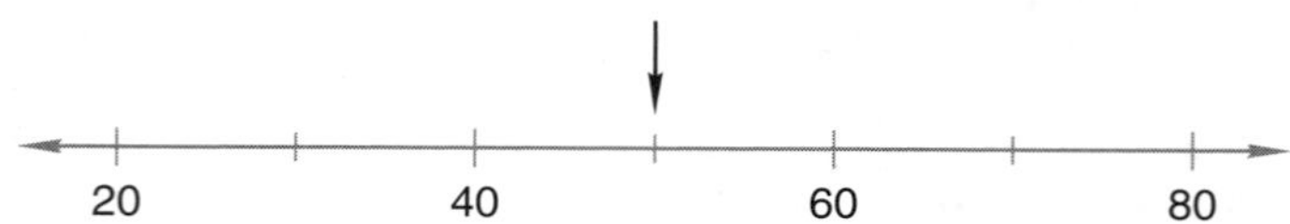

26. (13) Which multiplication fact shows the number of small squares in this rectangle?

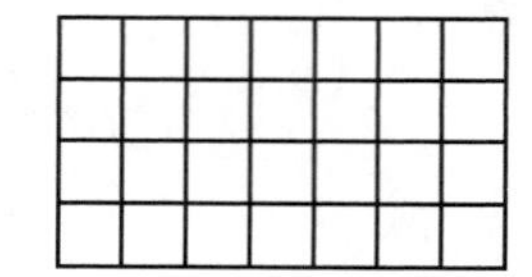

27. (28) How many centuries equal a millennium?

28. (Inv. 2) How many quarter circles equal a whole circle?

29. (23, 28)
(a) How many minutes are in an hour?

(b) How many minutes are in half an hour?

(c) Use the numbers in the answers to parts (a) and (b) to write a fraction equal to one half.

LESSON

29 Multiplying by Multiples of 10 and 100

WARM-UP

Facts Practice: 64 Multiplication Facts (Test F)

Mental Math: How many months are in a year? How many weeks are in a year? Name the months of the year in order.

a. How many months are in 2 years? ... 3 years? ... 4 years?

b. 10 × 50 **c.** 6 × 43 **d.** $\frac{1}{2}$ of 50

e. $\frac{1}{10}$ of 50 **f.** $\frac{5}{10}$ of 50 **g.** 750 − 250

h. 9 × 9, − 1, ÷ 2, + 2, ÷ 6

Problem Solving:

Copy this multiplication problem and fill in the missing digits:

$$\begin{array}{r} 36 \\ \times \ \_ \\ \hline \_\_2 \end{array}$$

NEW CONCEPT

The multiples of a number are the answers we get when we multiply the number by 1, 2, 3, 4, and so on. **Multiples of 10** all end in zero.

10, 20, 30, 40, 50, 60, ...

Any multiple of 10 can be written as a number times 10. For example:

20 = 2 × 10

30 = 3 × 10

40 = 4 × 10

Multiples of 100 all end with at least two zeros.

100, 200, 300, 400, 500, 600, ...

Any multiple of 100 can be written as a number times 100. For example:

200 = 2 × 100

300 = 3 × 100

400 = 4 × 100

When we multiply by a multiple of 10, we may multiply by the digit(s) in front of the zero and then multiply by 10. We will show this by multiplying 25 by 30.

The problem:	$25 \times 30 =$
We think:	$25 \times 3 \times 10 =$
We multiply 25 by 3:	$75 \times 10 =$
Then we multiply 75 by 10:	$75 \times 10 = 750$

The last step placed a zero after the 75. So when we multiply by a multiple of 10, we may multiply by the digit(s) in front of the zero and then place a zero on the end of that answer.

This can be shown when we write a problem vertically. We may write the numbers so that the multiple of 10 is on the bottom and the zero "hangs out" to the right. Here we write 25 times 30 vertically. We multiply 25 by 3. Then we bring down the zero (multiply by 10) and find that 25×30 is 750.

$$\begin{array}{r} \overset{1}{2}5 \\ \times \;\; 3\,0 \\ \hline 750 \end{array}$$

We may use a similar method to multiply by multiples of 100. When we multiply by a multiple of 100, we can write the problem so that *two* zeros "hang out" to the right. We show this by multiplying 25 by 300.

We write the problem with 300 on the bottom and its zeros out to the right. We multiply 25 by 3 hundreds and get 75 hundreds. We write 7500.

$$\begin{array}{r} \overset{1}{2}5\;\; \\ \times \;\; 300 \\ \hline 7500 \end{array}$$

Example 1 Multiply: 37×40

Solution We write the problem so that the multiple of 10 is on the bottom. We let the zero "hang out" to the right. Then we multiply.

$$\begin{array}{r} \overset{2}{3}7 \\ \times \;\; 4\,0 \\ \hline \mathbf{1480} \end{array}$$

Example 2 Multiply: $3.75 × 10

Solution When multiplying whole numbers by 10, we may simply attach a zero. The zero shifts all other digits one place to the left. However, when multiplying dollars and cents by 10, attaching a zero does not shift the other digits from their places:

$3.750 is the same as $3.75

This is because the decimal point sets the place values, and attaching a zero does not change the position of the decimal point. When multiplying dollars and cents by whole numbers, we position the decimal point in the answer so that there are two digits to the right of the decimal point.

$$\begin{array}{r} \$3.75 \\ \times \quad 10 \\ \hline \mathbf{\$37.50} \end{array}$$

LESSON PRACTICE

Practice set* Multiply:

a. 34 × 20

b. 50 × 48

c. 34 × 200

d. 500 × 36

e. 55 × 30

f. $1.25 × 30

g. 55 × 300

h. $1.25 × 300

i. 60 × 45

j. $2.35 × 40

k. 400 × 37

l. $1.43 × 200

MIXED PRACTICE

Problem set For problems 1–3, write an equation and find the answer.

1. (21) Laura, Lesley, and Tracey equally shared 1 dozen cookies. Each of the girls had how many cookies?

2. (16) Michael had $841 before he had to pay a $75 luxury tax. After paying the tax, how much money did he have?

3. (21) The sheet of stamps had 10 rows of stamps with 10 stamps in each row. How many stamps were on the sheet?

4. (28) What year came one century after 1776?

5. (25) List the factors of 10.

6. (29) 37×60

7. (18, 29) $37 \times 6 \times 10$

8. (29) 50×46

9. (29) $60 \times \$0.73$

10. (24, 29) $50 \times (1000 - 200)$

11. (3) What is the place value of the 5 in 356?

12. (28) Write the time that is 30 minutes before noon.

13. (Inv. 2) How much money is $\frac{1}{2}$ of a dollar plus $\frac{3}{4}$ of a dollar plus $\frac{3}{10}$ of a dollar?

14. (5, 29) What is the product of thirty-eight and forty?

15. (7) Use words to name the number 944,000.

16. (6)
$$\begin{array}{r} 4637 \\ 2843 \\ +\ 6464 \\ \hline \end{array}$$

17. (9)
$$\begin{array}{r} 4618 \\ -\ 2728 \\ \hline \end{array}$$

18. (13)
$$\begin{array}{r} \$60.00 \\ -\ \$\ 7.63 \\ \hline \end{array}$$

19. (26) $364 \div 10$

20. (26) $7w = 364$

21. (26) $\dfrac{364}{7}$

22. (2, 15) Think of a whole number. Multiply it by 2. Now add 1. Is the final answer odd or even?

23. (28) According to this calendar, what is the date of the third Sunday in May 1957?

MAY 1957

S	M	T	W	T	F	S
			1	2	3	4
5	6	7	8	9	10	11
12	13	14	15	16	17	18
19	20	21	22	23	24	25
26	27	28	29	30	31	

24. The number 356 is between which pair of numbers?
(4)

A. 340 and 350 B. 350 and 360 C. 360 and 370

25. What are the next three terms in this counting sequence?
(1)

..., 600, 700, 800, ______, ______, ______, ...

26. Which of these numbers has both 2 and 5 as factors?
(25)

A. 205 B. 502 C. 250

27. Show how to check this division answer. Is the answer correct?
(26)

$$7\overline{)300}\quad 43 \text{ R } 1$$

28. Compare: 12 − (6 − 2) ◯ (12 − 6) − 2
(24)

Does the associative property apply to subtraction?

29. Five tenths of a circle equals what percent of a circle?
(Inv. 2)

LESSON

30 Interpreting Pictures of Fractions and Percents

WARM-UP

Facts Practice: 90 Division Facts (Test D or E)

Mental Math: How many years are in a century? ... in a decade? Count by 12's from 12 to 60.

a. How many months are in 1 year? ... 2 years? ... 3 years?

b. $35 + 47$ **c.** $370 + 50$ **d.** 100×40

e. 4×36 **f.** $\frac{1}{2}$ of 12 **g.** $\frac{1}{4}$ of 12

h. $4 \times 7, - 1, \div 3, + 1, \times 10$

Problem Solving:

Bob flipped a coin three times. It landed heads up twice and tails up once, but not necessarily in that order. List the possible orders of the three coin flips.

NEW CONCEPT

A picture can help us understand the meaning of a fraction. This circle is divided into six equal parts. One of the parts is shaded. So $\frac{1}{6}$ of the circle is shaded.

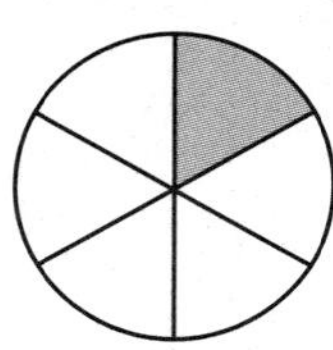

Five of the six parts are not shaded. So $\frac{5}{6}$ of the circle is not shaded.

Example 1 What fraction of this group of circles is shaded?

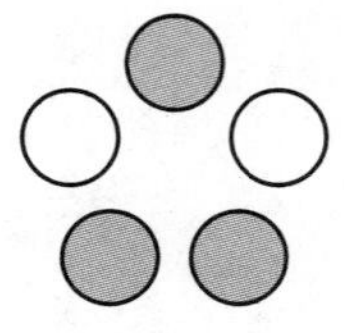

Solution We see a group of five circles. Three of the five circles are shaded. So $\frac{3}{5}$ of the group is shaded.

Example 2 What fraction of this circle is not shaded?

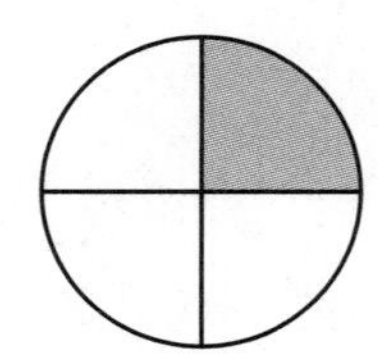

Solution The circle is divided into four equal parts. One part is shaded and three parts are not shaded. The fraction that is not shaded is $\frac{3}{4}$.

Fractions and **percents** are two ways to describe parts of a whole. A whole is 100 percent, which we abbreviate as 100%. So half of a whole is half of 100%, which is 50%.

All of this rectangle is shaded.

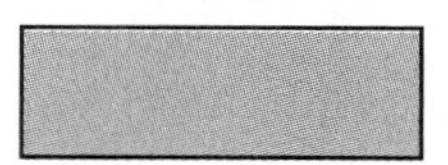

100% of this rectangle is shaded.

Half of this rectangle is shaded.

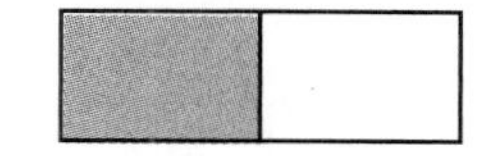

50% of this rectangle is shaded.

Thinking about cents as part of a dollar can help us understand percents. Just as one cent is one hundredth of a whole dollar, one percent is one hundredth of a whole.

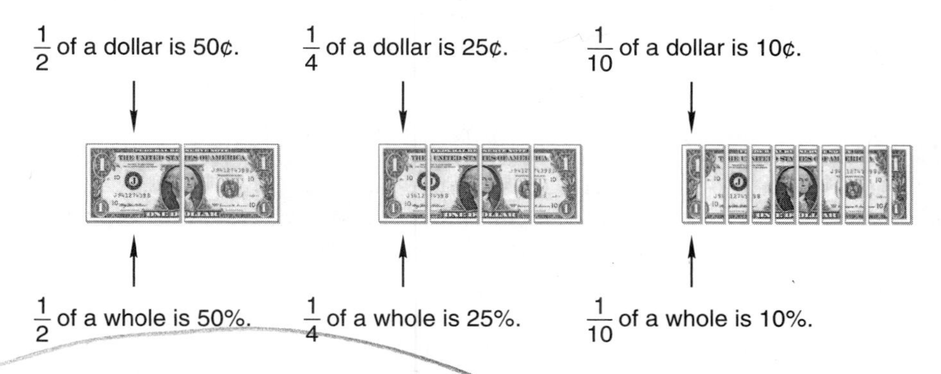

Example 3 What percent of this square is shaded?

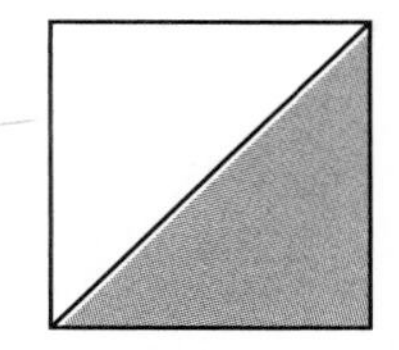

Solution One half of the square is shaded. The whole square is 100%, so one half of the square is **50%.**

Example 4 Three quarters plus a dime is what percent of a dollar?

Solution Three quarters plus a dime is 85¢, which is 85 hundredths of a dollar. This amount is **85%** of one dollar.

LESSON PRACTICE

Practice set Refer to the triangle to answer problems **a–d.**

a. What fraction of the triangle is shaded?

b. What percent of the triangle is shaded?

c. What fraction of the triangle is not shaded?

d. What percent of the triangle is not shaded?

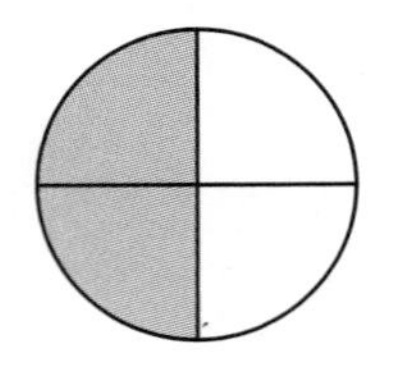

e. What are two fractions that name the shaded part of this circle?

f. What percent of the circle is shaded?

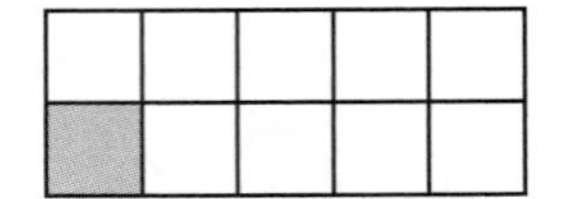

g. What fraction of this rectangle is shaded?

h. What percent of the rectangle is shaded?

i. A quarter plus a nickel is what percent of a dollar?

In the tables below, name the percent of a dollar represented by the number of coins stated.

Number of quarters	Percent of a dollar
4 quarters	**j.**
3 quarters	**k.**
2 quarters	**l.**
1 quarter	**m.**

Number of dimes	Percent of a dollar
10 dimes	**n.**
9 dimes	**o.**
8 dimes	**p.**
7 dimes	**q.**
6 dimes	**r.**
5 dimes	**s.**
4 dimes	**t.**
3 dimes	**u.**
2 dimes	**v.**
1 dime	**w.**

MIXED PRACTICE

Problem set For problems 1–4, write an equation and find the answer.

1. *(16)* There were 100 stamps on the sheet. Thai has used 36 of them. How many stamps are left?

2. *(16, 28)* The first month of the year is January, which has 31 days. After January, how many days are left in a common year?

3. *(21)* Each quart of juice could fill 4 cups. How many quarts of juice were needed to fill 28 cups?

4. Lorna used five $0.45 stamps to mail the heavy envelope. What was the total value of the stamps on the envelope?
(21)

5. Draw two vertical lines that stay the same distance apart.
(12)

6. List the factors of 25.
(25)

7. (30)

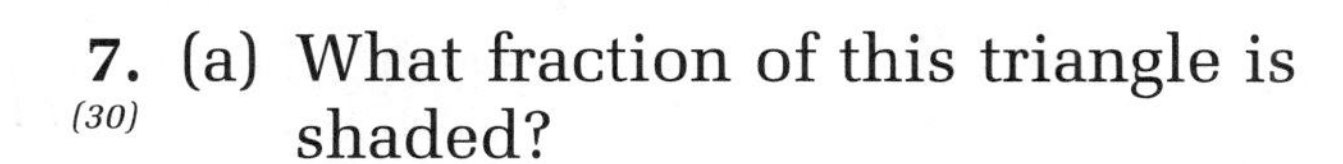

(a) What fraction of this triangle is shaded?

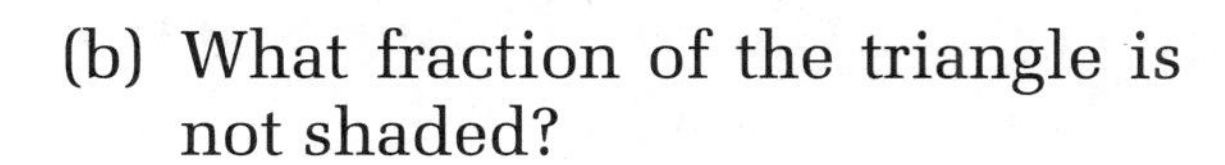

(b) What fraction of the triangle is not shaded?

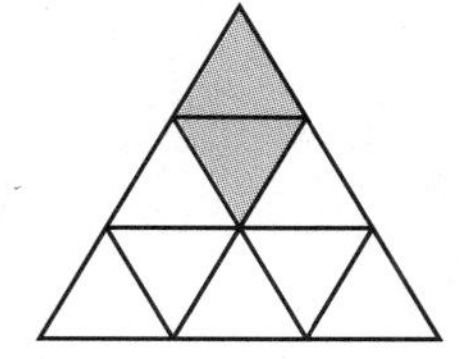

8. What number is the denominator in the fraction $\frac{2}{3}$?
(Inv. 2)

9. Write the time that is a quarter to eight in the morning.
(28)

10. (13, 14)
$$\begin{array}{r} W \\ -\ \$19.46 \\ \hline \$28.93 \end{array}$$

11. (9)
$$\begin{array}{r} 3010 \\ -\ 1342 \\ \hline \end{array}$$

12. (6)
$$\begin{array}{r} 28 \\ 54 \\ 75 \\ 91 \\ +\ 26 \\ \hline \end{array}$$

13. (29)
$$\begin{array}{r} 764 \\ \times\ \ 30 \\ \hline \end{array}$$

14. (29)
$$\begin{array}{r} \$9.08 \\ \times\ \ \ 60 \\ \hline \end{array}$$

15. $6\overline{)\$7.44}$
(26)

16. 362 ÷ 10
(26)

17. $4\overline{)898}$
(26)

18. $42.37 + $7.58 + $0.68 + $15
(13)

19. (48 × 6) − 9
(24)

20. 6 × 30 × 12
(18)

21. From February 1 to September 1 is how many months?
(28)

22. What is the sum of six hundred five and five hundred ninety-seven?
(5, 6)

23. Which of these numbers is between 360 and 370?
(4)

A. 356 B. 367 C. 373

24. What are the next three terms in this counting sequence?
(1)

..., 250, 260, 270, 280, ______, ______, ______, ...

25. It was the hottest day of the year. What temperature is shown on this Celsius thermometer?
(27)

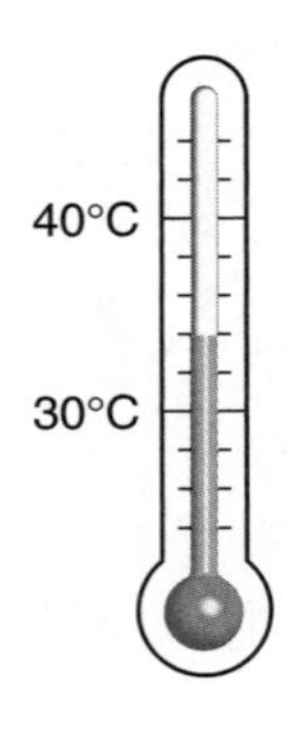

26. What year came one decade after 1802?
(28)

27. A quarter plus a dime is what percent of a dollar?
(30)

28. Show how to check this division answer. Is the answer correct?
(26)

$$100 \div 7 = 14 \text{ R } 2$$

29. Compare. (Try to answer the comparison before dividing.)
(4, 26)

$$100 \div 4 \bigcirc 100 \div 5$$

INVESTIGATION 3

Focus on

Fractions: Thirds, Fifths, and Eighths

Recall from Investigation 2 that we can use fractions to describe part of a group.

Example One third of the 24 students earned an A on the test. How many students earned an A on the test?

Solution We see the word "third," so we divide the group of students into three equal parts. The number in one part is the number of students who earned an A, since one third earned an A.

$$3\overline{)24}\quad 8$$

We find that **8 students** earned an A on the test.

Use this information to answer problems 1–8:

> *Students were given two hours to finish a 120-question final exam. One third of the questions on the test were true-false. One fifth of the questions were fill-in-the-blank. One eighth of the questions were short answer. The rest of the questions were multiple choice. Stephanie answered half of the questions in the first hour.*

1. How many questions did Stephanie answer in the first hour?

2. How many questions were true-false?

3. How many questions were fill-in-the-blank?

4. How many questions were short answer?

5. How many questions were multiple choice?

6. Did the multiple-choice questions make up more than or less than $\frac{1}{3}$ of the questions on the test?

7. Together, did the true-false and fill-in-the-blank questions make up more than or less than half of the test?

8. Together, did the true-false and short-answer questions make up more than or less than half of the test?

Activity: *Using Fraction Manipulatives*

In this activity we will make and use fraction manipulatives for thirds, fifths, and eighths.

Materials needed:

- photocopies of Activity Masters 13, 14, and 15 (1 copy of each master per student; masters available in *Saxon Math 6/5 Assessments and Classroom Masters*)
- fraction manipulatives from Investigation 2
- scissors

Note: Color-coding the fraction manipulatives makes sorting easier. If you wish to color-code the manipulatives, photocopy each master on a different color of construction paper, or, before cutting, have students color both sides of the fraction circles using different colors for different masters. Following the activity, each student may store the fraction manipulatives for use in later lessons.

Preparation:

Distribute materials. Have students separate the fraction manipulatives by cutting out the circles and cutting apart the fraction slices along the lines.

Use all your fraction manipulatives (halves, thirds, fourths, fifths, eighths, and tenths) to work the exercises in this investigation.

9. Show that four eighths equals one half.

10. Show that a fifth equals two tenths.

11. How many eighths equal a fourth?

12. Is two fifths more or less than one half?

13. One third of a circle is what percent of a circle?

14. Three fifths of a circle is what percent of a circle?

15. Four eighths of a circle is what percent of a circle?

16. Sarah has a half circle, a quarter circle, and an eighth of a circle. How much more does she need to have a whole circle?

17. Can you make half of a circle using only thirds?

18. Can you make half of a circle using only fifths?

19. If you had fraction pieces for sevenths, do you think you could make half a circle using only sevenths? Why or why not?

20. What fraction is $\frac{1}{2}$ of $\frac{1}{2}$?

21. What fraction is $\frac{1}{2}$ of $\frac{1}{4}$?

22. What fraction is $\frac{1}{2}$ of $\frac{1}{5}$?

23. What fraction do you suppose is $\frac{1}{2}$ of $\frac{1}{3}$?

24. What single fraction piece equals $\frac{2}{8}$?

25. Two eighths of a circle equals what percent of a circle?

Use your fraction manipulatives to illustrate these additions and subtractions. Write a complete number sentence for each.

26. $\frac{1}{5} + \frac{2}{5}$

27. $\frac{3}{8} + \frac{5}{8}$

28. $\frac{2}{3} - \frac{1}{3}$

29. $\frac{5}{8} - \frac{2}{8}$

Compare. Use your fraction manipulatives to assist you.

30. $\frac{1}{3} + \frac{1}{5} \bigcirc \frac{1}{2}$

31. $\frac{1}{3} + \frac{1}{8} \bigcirc 50\%$

32. $\frac{1}{3} + \frac{1}{3} \bigcirc 67\%$

33. $\frac{1}{3} + \frac{1}{5} + \frac{1}{8} \bigcirc 1$

34. Arrange these fractions in order from least to greatest:

$$\frac{1}{2}, \frac{1}{8}, \frac{1}{5}, \frac{1}{3}, \frac{1}{10}, \frac{1}{4}$$

35. Arrange these percents in order from least to greatest:

$$33\frac{1}{3}\%, 12\frac{1}{2}\%, 10\%, 50\%, 25\%, 20\%$$

LESSON

31 Pairs of Lines

WARM-UP

Facts Practice: 64 Multiplication Facts (Test F)

Mental Math: How many days are in a leap year? ... in a common year? Count by 12's from 12 to 84. Count by 5's from 3 to 53.

a. How many is 2 dozen? ... 3 dozen? ... 4 dozen?

b. 48 + 25 **c.** 1200 + 340 **d.** 50% of 20

e. 25% of 20 **f.** 10% of 20 **g.** 7 × 32

h. 4 × 9, − 1, ÷ 5, + 1, × 4

Problem Solving:

Copy this multiplication problem and fill in the missing digits:

$$\begin{array}{r} 45 \\ \times\ \_ \\ \hline \_0 \end{array}$$

NEW CONCEPT

When lines cross we say that they **intersect.** If we draw two straight lines on the same flat surface, then those lines either intersect at some point or they do not intersect at all. Lines that go in the same directions and do not intersect are called **parallel lines.** Parallel lines always stay the same distance apart. Thinking of train tracks can give us the idea of parallel lines. Here are pairs of parallel lines and parallel line segments:

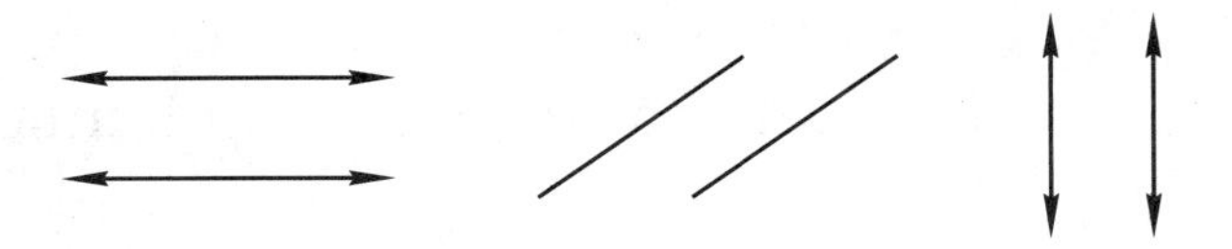

Lines on the same surface that are not parallel are called **intersecting lines.** Here are pairs of intersecting lines and intersecting line segments:

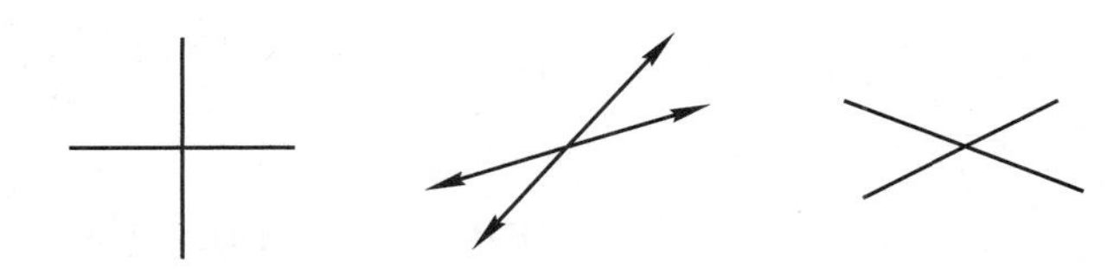

The pair of segments on the left are **perpendicular.** Perpendicular lines and segments intersect to form "square corners." The other two pairs of lines and segments are **oblique.** Oblique lines and segments are neither parallel nor perpendicular.

Activity: *Parallel and Perpendicular Segments*

For this activity, work with a partner. Draw a line segment. Then have your partner draw two line segments, one parallel to your segment and the other perpendicular to it. Repeat the activity, switching roles with your partner.

Example 1 Draw a pair of oblique lines.

Solution We draw two lines that intersect but that do not form square corners. Many arrangements are possible.

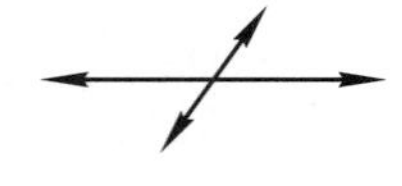

Example 2 Which of the following figures does *not* appear to contain perpendicular segments?

A. B. C. D.

Solution Perpendicular segments intersect to form square corners. The segments in A appear to be perpendicular. (You may need to turn your book slightly to help you see this.) The segments in B and D also appear to be perpendicular. The segments that do not appear to be perpendicular are those in choice **C.**

LESSON PRACTICE

Practice set

a. Draw two parallel segments.

b. Draw two perpendicular lines.

c. Draw two oblique segments.

MIXED PRACTICE

Problem set

1. Draw a pair of intersecting lines that are perpendicular.
(31)

2. Lani bought a kaleidoscope for \$4.19. If she paid for it with a \$10 bill, how much money should she get back? Write a subtraction pattern and solve the problem.
(16)

3. How many hours are there in 7 days?
(21, 28)

4. From 6:00 a.m. to 4:00 p.m. the temperature rose 23° to 71°F. What was the temperature at 6:00 a.m.? Write an addition pattern and solve the problem.
(11)

5. What fraction of this group is shaded?
(30)

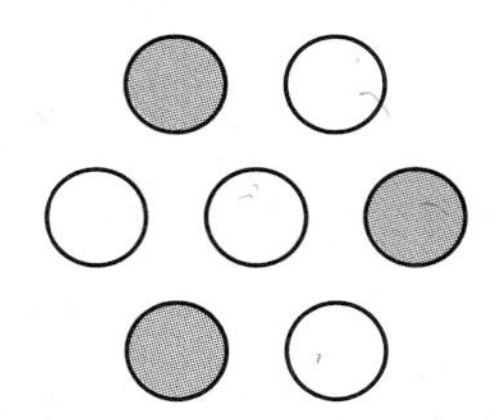

6. List the factors of 19.
(25)

7. *(13)* $\begin{array}{r} \$16.38 \\ -\ \$\ 9.47 \\ \hline \end{array}$

8. *(14)* $\begin{array}{r} 1000 \\ -\quad Q \\ \hline 576 \end{array}$

9. *(26)* $5n = 280$

10. *(29)* $\begin{array}{r} 476 \\ \times\ \ 80 \\ \hline \end{array}$

11. *(29)* $\begin{array}{r} \$9.68 \\ \times\ \ 60 \\ \hline \end{array}$

12. *(26)* $\dfrac{\$19.44}{8}$

13. Write the time that is thirty minutes before midnight.
(28)

14. Compare: $\frac{1}{10}$ of 100 ◯ $\frac{1}{2}$ of 20
(Inv. 2)

15. Jackson bought five boxes of his favorite cereal for \$2.87 each. Altogether, how much did the five boxes of cereal cost? Change this addition problem to a multiplication problem and find the total:
(13, 17)

$$\$2.87 + \$2.87 + \$2.87 + \$2.87 + \$2.87$$

16. $\$96 + \$128.13 + \$27.49 + W = \$300$
(10, 13)

17. $328 \div (32 \div 8)$
(24, 26)

18. $648 - (600 + 48)$
(24)

19. Think of an odd number. Multiply it by 2. Now add 1. Is the final answer odd or even?
(2)

20. Which of these numbers has neither 2 nor 5 as a factor?
(25)

A. 125 B. 251 C. 512 D. 215

21. It is afternoon. What time is shown by this clock?
(28)

22. What number is the numerator of the fraction $\frac{2}{3}$?
(Inv. 2)

23. Use words to name the number 123,400.
(7)

24. What percent of a circle is $\frac{2}{3}$ of a circle?
(Inv. 3)

25. Copy this number line, and draw an arrow that points to the location of the number 75.
(27)

26. Show how to check this division answer. Is the answer correct?
(26)

$$8\overline{)300}\quad 37\text{ R }6$$

27. (a) How many years is a century?
(23, 28)

(b) How many years is half a century?

(c) Use the numbers in the answers to parts (a) and (b) to write a fraction equal to $\frac{1}{2}$.

28. One fourth of an hour is how many minutes?
(Inv. 2, 28)

29. Write the next four terms in this counting sequence:
(1, 12)

27, 18, 9, ...

LESSON

32 Angles • Polygons

WARM-UP

Facts Practice: 90 Division Facts (Test D or E)

Mental Math: How many years are in a decade? ... in a century? Count by 12's from 12 to 84. Count by 5's from 4 to 54.

a. How many hours are in 1 day? ... 2 days?
b. 39 + 45 **c.** 680 − 400 **d.** 10 × 60
e. 50% of 30 **f.** 10% of 30 **g.** 8 × 45
h. 6 × 7, − 2, ÷ 5, × 2, − 1, ÷ 5

Problem Solving:

Jan flipped a coin three times. It landed heads up once and tails up twice. List the possible orders of the three coin flips.

NEW CONCEPTS

Angles

When lines or segments intersect, angles are formed. Here we show four angles:

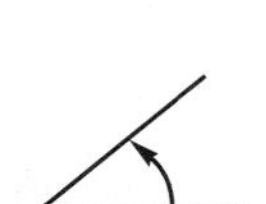
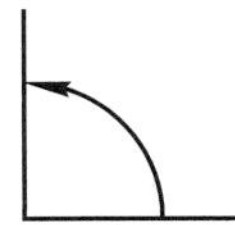
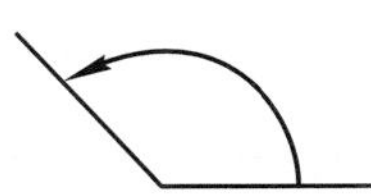
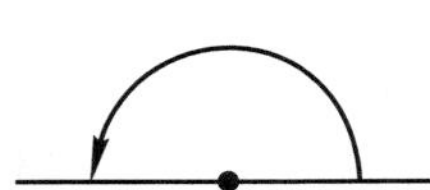

An **angle** is an "opening" between intersecting lines or segments. We see in the figures above that the amount of opening can vary; angles may be more open or less open than other angles. We have different names for angles depending upon how open they are.

An angle that is like the corner of a square is called a **right angle.** *Right angle* does not mean the angle opens to the right. A right angle may open in any direction. *Right angle* simply means "square corner." Sometimes we draw a small square in the angle to indicate that it is a right angle.

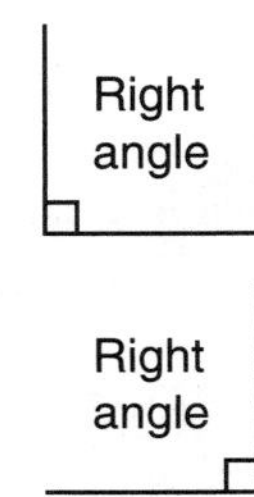

An angle whose opening is less than a right angle is an **acute angle.** Some remember this as "a cute" little angle.

Acute angle

An angle whose opening is more than a right angle is an **obtuse angle.**

Obtuse angle

An angle open to form a straight line is a **straight angle.** The angle formed by the hands of a clock at 6 o'clock is an example of a straight angle.

Straight angle

Example 1 Which of these angles appears to be a right angle?

A. B. C.

Solution A right angle looks like the corner of a square. Angle A is open too wide, and angle C is not open wide enough. The only angle that appears to be a right angle is **angle B.**

Polygons

A **plane** is a flat surface that extends without end. The classroom floor is part of a plane that extends beyond the walls. The wall surfaces in the room are parts of other planes. Planes can contain flat shapes such as triangles, squares, and circles. Some of these flat shapes are **polygons.**

A polygon is a flat shape formed by line segments that close in an area. Each of these shapes is a polygon:

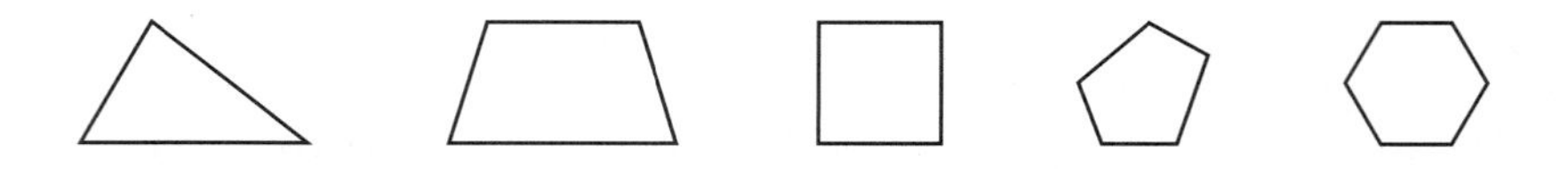

The line segments that form a polygon are called **sides.** Two sides meet at a **vertex** (plural: *vertices*) to form an angle. A polygon may have three or more sides and it has as many vertices and angles as it has sides.

Polygons do not have any curved sides. These figures are not polygons:

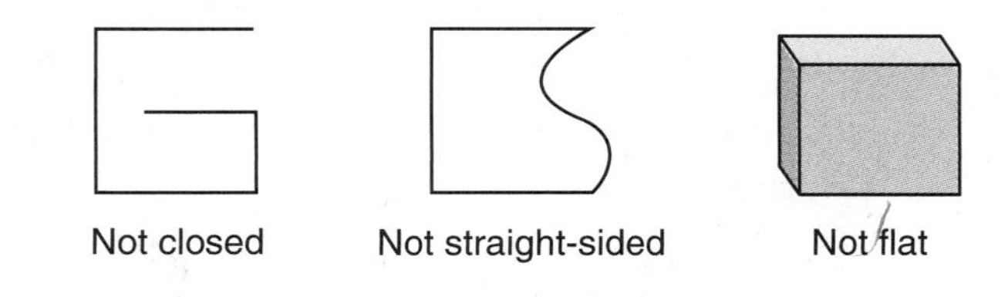

Polygons are named by the number of sides they have. The table below names some common polygons.

Polygons

Polygon	Number of Sides	Example
triangle	3	
quadrilateral	4	
pentagon	5	
hexagon	6	
heptagon	7	
octagon	8	
decagon	10	
dodecagon	12	
n-gon	*n*	

Notice that a four-sided polygon is a **quadrilateral.** There are different kinds of quadrilaterals, such as squares, rectangles, parallelograms, and trapezoids. We will study these classifications in more detail later.

Example 2 Which of these shapes is not a quadrilateral?

A. B. C. D.

Solution A quadrilateral is a polygon with four sides. The shape that does not have four sides is choice **C.**

Sometimes we enclose an area by using smooth curves. A circle is one example of an area that is enclosed by a smooth curve. Since a circle does not enclose an area with line segments, a circle is not a polygon.

Example 3 Which of these shapes is not a polygon?

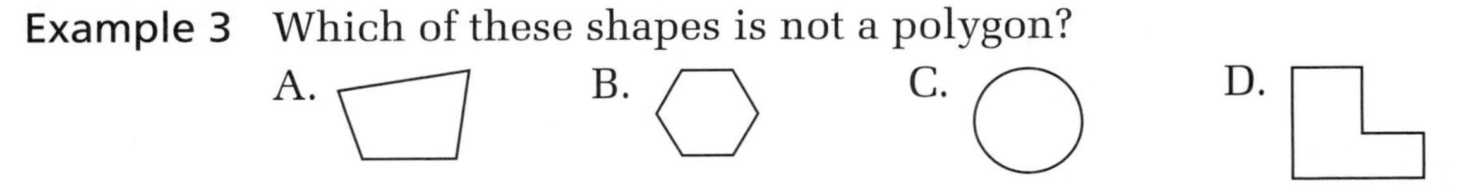

Solution A polygon is formed by line segments. A circle is a smooth curve. So the shape that is not a polygon is choice **C.**

Example 4 Name each of these polygons.

(a)  (b)  (c)

Solution (a) The polygon has six sides. It is a **hexagon.**

(b) This 12-sided polygon is a **dodecagon.**

(c) The block-letter T has 8 sides and is an **octagon.**

Figures that have the same size and shape are **congruent.** The three triangles below are congruent even though they have been flipped and turned to different positions.

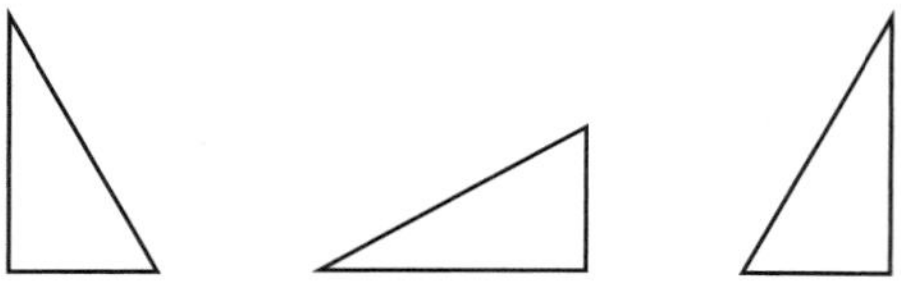

Example 5 Which two rectangles below are congruent?

A. 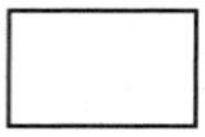 B. 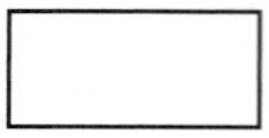 C. 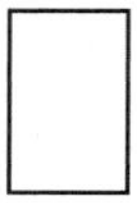 D.

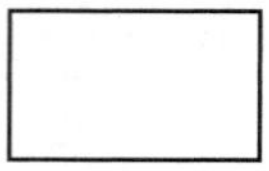

Solution Rectangles **A.** and **C.** have the same size and shape, so they are congruent figures.

Congruent figures are also **similar.** Similar figures have the same shape. They may or may not be the same size. When looking at two similar figures that are not the same size, the larger figure will look like a magnified version of the smaller figure. These two triangles are similar but not congruent.

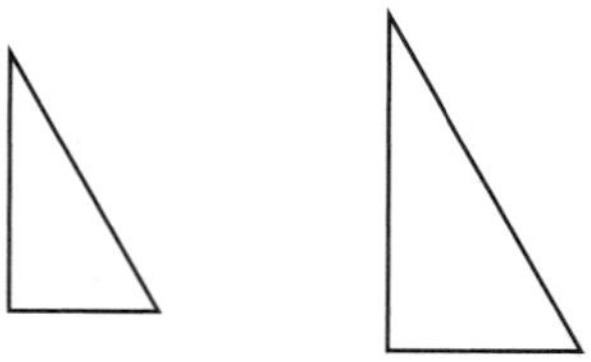

Example 6 Which two triangles below are similar?

A. 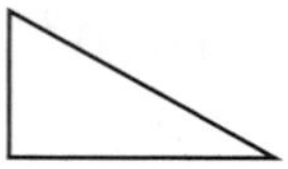 B. 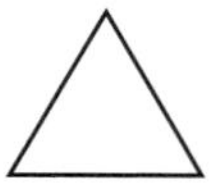 C. 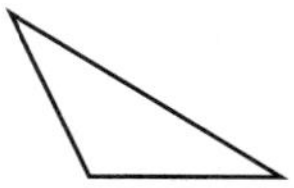 D.

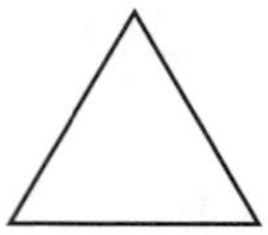

Solution Triangles **B.** and **D.** have the same shape, so they are similar figures.

LESSON PRACTICE

Practice set Draw an example of each angle:

a. acute angle **b.** obtuse angle **c.** right angle

Describe each angle shown as acute, obtuse, right, or straight:

d. 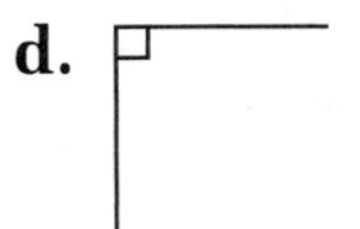 **e.** **f.** **g.**

h. Draw a triangle with two perpendicular sides.

i. A quadrilateral is a polygon with how many sides?

j. Draw a quadrilateral that has one pair of parallel sides.

k. Draw a quadrilateral with two pairs of parallel sides.

l. Draw a quadrilateral that has no parallel sides. (Begin by drawing two nonparallel segments; then connect those with two other nonparallel segments.)

Name each shape:

m. 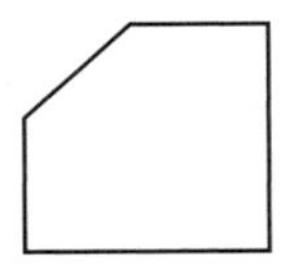 **n.** 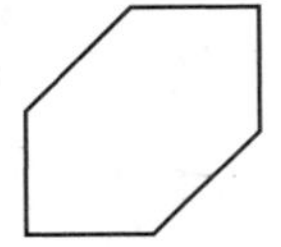 **o.** 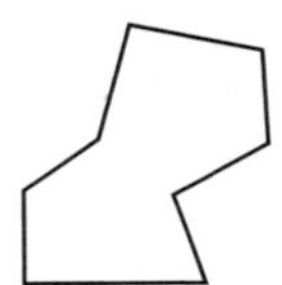

p. Draw a polygon shaped like the block letter F. What type of polygon did you draw?

q. Draw two triangles that are congruent.

MIXED PRACTICE

Problem set

1. *(Inv. 2)* Suki took \$20 to the carnival. She spent $\frac{1}{2}$ of her money on rides, $\frac{1}{4}$ of her money on food, and $\frac{1}{10}$ of her money on parking. How much did Suki spend on rides? ... on food? ... on parking?

For problems 2–4, write an equation and find the answer.

2. *(21)* Hank says that the horse trough holds 18 buckets of water. If a bucket holds 3 gallons, how many gallons does the trough hold?

3. (21) Terrell chopped a tree that was 52 feet tall into four logs of equal length. How many feet long was each log?

4. (16) After 20 minutes Shelly had answered 17 of the 45 questions on the test. How many questions remained for Shelly to answer?

5. (28, 29) How many seconds are in 1 hour?

6. (13)
$$\begin{array}{r} \$56.37 \\ \$34.28 \\ +\ \$\ 9.75 \\ \hline \end{array}$$

7. (14)
$$\begin{array}{r} 5286 \\ -\quad K \\ \hline 4319 \end{array}$$

8. (13)
$$\begin{array}{r} \$40.00 \\ -\ \$39.56 \\ \hline \end{array}$$

9. (10)
$$\begin{array}{r} 67 \\ 72 \\ 43 \\ 91 \\ 48 \\ 19 \\ 648 \\ +\quad M \\ \hline 996 \end{array}$$

10. (24, 26) $936 \div (36 \div 9)$

11. (29)
$$\begin{array}{r} 596 \\ \times\quad 600 \\ \hline \end{array}$$

12. (26) $\dfrac{\$46.56}{8}$

13. (29) $\$4.07 \times 80$

14. (15, 18) $9 \times 12 \times 0$

15. (26) $936 \div 7$

16. (Inv. 3) Compare: $\frac{1}{3}$ of 60 ◯ $\frac{1}{5}$ of 100

17. (32) Which of these angles does not appear to be a right angle?

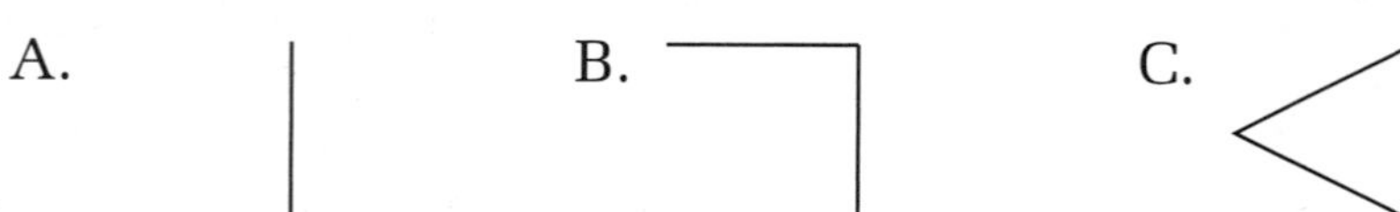

18. (25) List the factors of 18.

19. (30) What fraction of the rectangle is shaded? What percent of the rectangle is shaded?

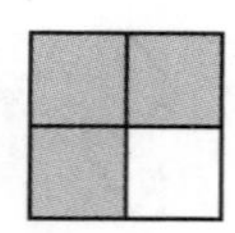

20. (31) Draw a horizontal line segment and a vertical line segment that intersect.

21. (28) According to this calendar, July 17, 2025, is what day of the week?

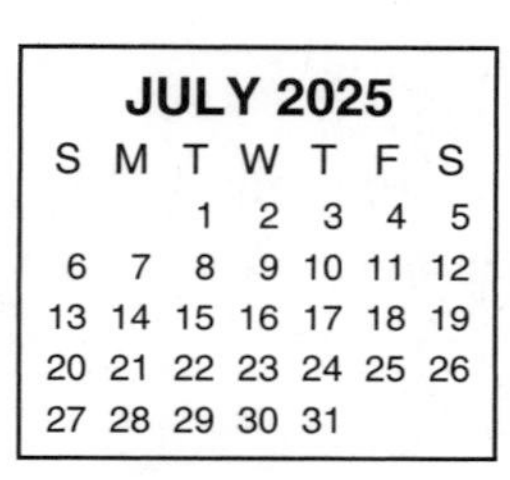

JULY 2025

S	M	T	W	T	F	S
		1	2	3	4	5
6	7	8	9	10	11	12
13	14	15	16	17	18	19
20	21	22	23	24	25	26
27	28	29	30	31		

22. What is the name for the bottom number of a fraction?
(Inv. 2)

23. Write two multiplication facts and two division facts for the fact family 9, 10, and 90.
(19)

24. What are the next three terms in this counting sequence?
(1)

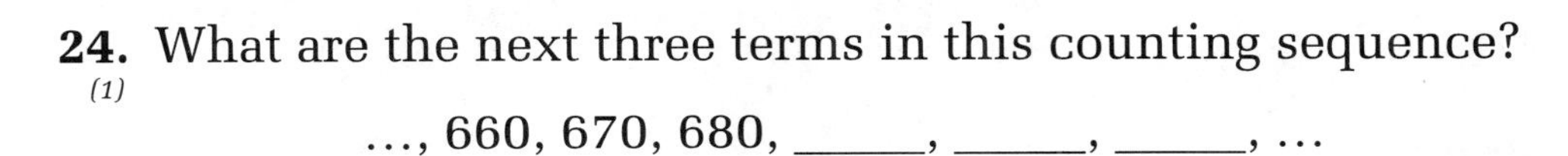

..., 660, 670, 680, ______, ______, ______, ...

25. To what number on this scale is the arrow pointing?
(27)

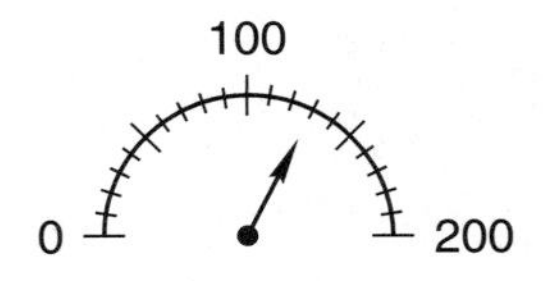

26. Draw a polygon shaped like a block letter H. What type of polygon did you draw?
(32)

27. Show how to check this division answer. Is the answer correct?
(26)

$$7\overline{)400}\quad 57\text{ R }1$$

28. Use the digits 0, 2, and 5 to make a three-digit number that has both 2 and 5 as factors.
(25)

29. (a) How many is a dozen?
(2, 23)

(b) How many is half a dozen?

(c) Use the answers to parts (a) and (b) to write a fraction equal to $\frac{1}{2}$.

LESSON

33 Rounding Numbers Using a Number Line

WARM-UP

Facts Practice: 64 Multiplication Facts (Test F)

Mental Math: How many months are in half a year? ... a year and a half? Count by 12's from 12 to 96. Count by 6's from 6 to 96.

a. How much money is 4 quarters? ... 5 quarters? ... 6 quarters?

b. 9 × 42 **c.** 25 × 10 **d.** 50% of 40

e. 25% of 40 **f.** 10% of 40 **g.** 840 − 140

h. 8 × 8, − 1, ÷ 9, × 3, − 1, ÷ 4

Problem Solving:

Copy this multiplication problem and fill in the missing digits:

$$\begin{array}{r} \_\_ \\ \times\ 8 \\ \hline \_6 \end{array}$$

NEW CONCEPT

Two of the following statements use **exact numbers,** while the other two statements use **round numbers.** Can you tell which statements use round numbers?

About 600 people attended the homecoming game.

The attendance at the game was 614.

The price of the shoes was $48.97.

The shoes cost about $50.

The first and last statements use round numbers. The term *round numbers* usually refers to numbers ending with one or more zeros. Round numbers are often used in place of exact numbers because they are easy to understand and to work with.

When we **round** a number, we find another number to which the number is near. One way we can do this is with a number line. To round 67 to the nearest ten, for example, we find the multiple of ten that is nearest to 67. On the number line below we see that 67 falls between the multiples 60 and 70.

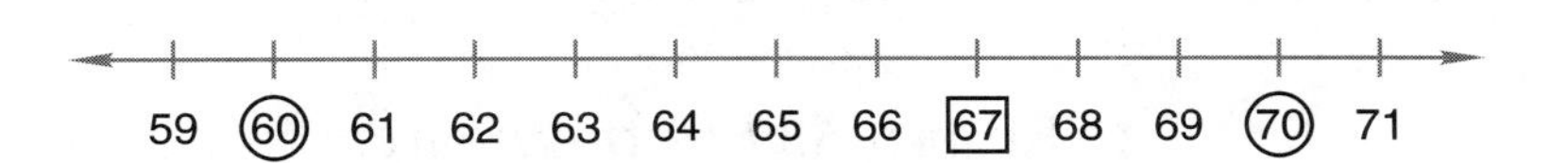

Since 67 is nearer to 70 than to 60, we *round up* to 70.

When the number we are rounding is halfway between two round numbers, we usually round up to the larger round number. Sixty-five is halfway between 60 and 70. We would round 65 to 70. Likewise, 450 is halfway between 400 and 500, so we would round 450 to 500.

Example Round 523 to the nearest hundred.

Solution When we round a number to the nearest hundred, we find the multiple of 100 to which it is nearest. Recall that the multiples of 100 are the numbers we say when we count by hundreds: 100, 200, 300, 400, and so on. We use a number line marked and labeled with hundreds to picture this problem.

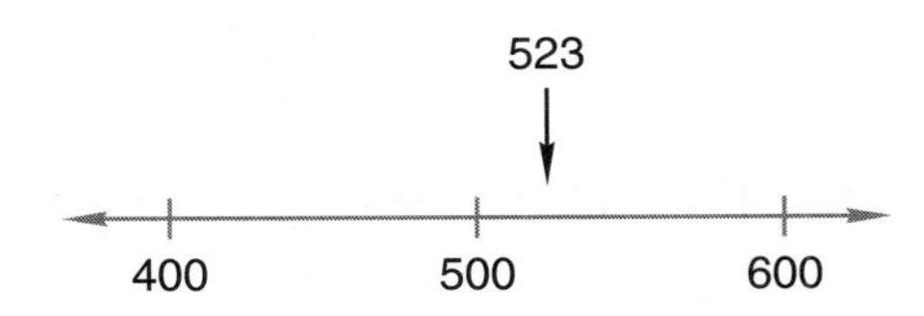

Placing 523 on the number line, we see that it falls between the multiples 500 and 600. Since 523 is nearer to 500 than to 600, we *round down* to **500**.

LESSON PRACTICE

Practice set* Round to the nearest ten. For each problem, you may draw a number line to help you round.

a. 72 **b.** 87 **c.** 49 **d.** 95

Round to the nearest hundred. For each problem, you may draw a number line to help you round.

e. 685 **f.** 420 **g.** 776 **h.** 350

MIXED PRACTICE

Problem set

1. Draw a pair of vertical parallel lines. (31)

2. Round 537 to the nearest hundred. (33)

3. Round 78 to the nearest ten. (33)

4. Forty animals were brought to the pet show. One half were mammals, one fourth were fish, one tenth were reptiles, and the rest were birds. How many mammals were brought to the pet show? How many fish? How many reptiles? How many birds? (Inv. 2)

5. Keisha was standing in a line that had 10 people in it, including herself. If 5 people were in front of her, how many people were behind her? (7)

For problems 6–8, write an equation and find the answer.

6. Seven hours is how many minutes? (21, 28)

7. After paying $7.50 for a movie ticket, Salvador still had $3.75. How much money did Salvador have before paying for the ticket? (16)

8. The Smiths set the trip odometer at zero when they started their trip. By the end of the first day, the Smiths had traveled 427 miles. By the end of the second day, they had traveled a total of 902 miles. How far did the Smiths travel the second day? (11)

9. (13)
$$\begin{array}{r} \$34.28 \\ \$\ 9.76 \\ +\ \$20.84 \\ \hline \end{array}$$

10. (14)
$$\begin{array}{r} 3526 \\ -\quad V \\ \hline 1617 \end{array}$$

11. (13)
$$\begin{array}{r} \$10.00 \\ -\ \$\ 0.86 \\ \hline \end{array}$$

12. (6)
$$\begin{array}{r} 499 \\ 25 \\ 43 \\ 756 \\ 67 \\ 94 \\ +\ \ 32 \\ \hline \end{array}$$

13. (29)
$$\begin{array}{r} 563 \\ \times\quad 90 \\ \hline \end{array}$$

14. (29)
$$\begin{array}{r} \$2.86 \\ \times\quad 70 \\ \hline \end{array}$$

15. (29)
$$\begin{array}{r} 479 \\ \times\ \ 800 \\ \hline \end{array}$$

16. $3\overline{)1122}$ (26)

17. $6m = \$5.76$ (26)

18. $10\overline{)2735}$ (26)

19. $64.23 + $5.96 + $17 + ($1 − $0.16) (13, 24)

20. From March 1 to December 1 is how many months?
(28)

21. What fraction of the circle is shaded? What percent of the circle is shaded? Is more than or less than 50% of the circle shaded?
(30, Inv. 3)

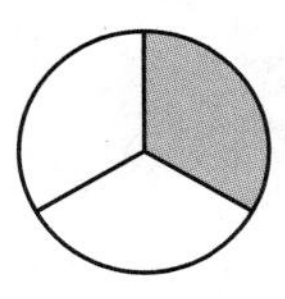

22. Which word means "parallel to the horizon"?
(31)

A. vertical B. oblique C. horizontal

23. Write the time that is a quarter after one in the afternoon.
(28)

24. Draw a horizontal number line from 0 to 50 with only zero and tens marked and labeled.
(27)

25. What is the tenth term of this counting sequence?
(1)

7, 14, 21, ...

26. Draw an acute angle.
(32)

27. List the factors of 7.
(25)

28. At which of these times are the hands of a clock perpendicular?
(28, 31)

A. 6:00 B. 12:30 C. 9:00

29. Main Street and Allen Street intersect at a traffic light. The two streets form square corners where they meet.
(31)

(a) Draw segments to show how Main Street and Allen Street meet.

(b) Which of these words best describes the segments in your drawing?

A. parallel

B. perpendicular

C. oblique

LESSON

34 Division with Zeros in the Quotient

WARM-UP

Facts Practice: 90 Division Facts (Test D or E)

Mental Math: How many years is half a century? ... half a decade?

a. Round 48 to the nearest ten.

b. 50 + 80 **c.** 50 × 8 **d.** 4 × 27

e. 50% of 50 **f.** 10% of 50 **g.** 1420 + 300

h. 3 × 8, + 1, × 2, − 1, ÷ 7, ÷ 7

Problem Solving:

If a coin is flipped twice, it may land heads then heads, or heads then tails, or tails then heads, or tails then tails. If a coin is flipped three times, what are the possible orders in which it could land?

NEW CONCEPT

Recall that the answer to a division problem is called a **quotient.** Sometimes when we divide, one or more of the digits in the quotient is a zero. When this happens, we continue to follow the four steps in the division algorithm: divide, multiply, subtract, and bring down.

Example 1 Divide: $6\overline{)365}$

Solution We begin by breaking the division problem into a smaller problem: $6\overline{)36}$.

$$6\overline{)365}$$

Then we divide, multiply, subtract, and bring down. When we subtract, we get zero, which we may or may not write, and we bring down the 5. Since there is a number to bring down, we divide again. The new division is $6\overline{)5}$.

$$\begin{array}{r} 6 \\ 6\overline{)365} \\ -\,36\downarrow \\ \hline 05 \end{array}$$

Since we cannot divide 5 by 6 even once, we write a zero in the quotient, multiply, and subtract. Since there is no other number to bring down, the division is finished and the remainder is 5. Our answer is **60 R 5.**

$$\begin{array}{r} 60 \text{ R } 5 \\ 6\overline{)365} \\ -\,36 \\ \hline 5 \\ -\,0 \\ \hline 5 \end{array}$$

Example 2 Divide: $6\overline{)635}$

Solution We break the division problem into smaller problems. We can find $6\overline{)6}$, so we divide, multiply, subtract, and bring down. The next division is $6\overline{)3}$.

$$\begin{array}{r} 1 \\ 6\overline{)635} \\ -6\downarrow \\ \hline 03 \end{array}$$

Since the dividend (number we are dividing) is less than the divisor (number we are dividing by), we write a zero in the quotient. Then we multiply, subtract, and bring down. The next division is $6\overline{)35}$.

$$\begin{array}{r} 10 \\ 6\overline{)635} \\ -6 \\ \hline 3 \\ -0 \\ \hline 35 \end{array}$$

We divide 35 by 6, multiply, and subtract. Since there is no other number to bring down, the division is finished and the remainder is 5. When we divide 635 into 6 equal parts, there are 105 in each part with 5 "left over." Our answer is **105 R 5.**

$$\begin{array}{r} 105 \text{ R } 5 \\ 6\overline{)635} \\ -6 \\ \hline 3 \\ -0 \\ \hline 35 \\ -30 \\ \hline 5 \end{array}$$

Again, we check a division answer by multiplying the quotient by the divisor and then adding the remainder to this product.

$$\begin{array}{r} 105 \\ \times \quad 6 \\ \hline 630 \\ + \quad 5 \\ \hline 635 \end{array}$$

Since the result, 635, equals the dividend, we can be confident that our answer is correct.

LESSON PRACTICE

Practice set* Divide:

a. $3\overline{)61}$ **b.** $6\overline{)242}$ **c.** $3\overline{)121}$ **d.** $4\overline{)1628}$

e. $4\overline{)122}$ **f.** $5\overline{)\$5.25}$ **g.** $2\overline{)\$6.18}$ **h.** $6\overline{)4981}$

i. $10\overline{)301}$ **j.** $4\overline{)\$8.24}$ **k.** $7\overline{)\$5.60}$ **l.** $8\overline{)4818}$

m. Show how to check this division answer. Is the answer correct?

$$\begin{array}{r} 108 \text{ R } 2 \\ 6\overline{)650} \end{array}$$

MIXED PRACTICE

Problem set

1. (31) Draw a horizontal line. Draw another line that is perpendicular to the horizontal line.

2. (Inv. 2) One hundred students named their favorite vegetable. One half named "beans," one fourth named "broccoli," one tenth named "peas," and the rest named "spinach." How many students named beans? ... broccoli? ... peas? ... spinach?

3. (28) What year was one century after 1849?

For problems 4–6, write an equation and find the answer.

4. (21, 28, 29) How many minutes are in one day?

5. (21) In one year Henrietta laid 10 dozen eggs. How many eggs is that? Use a multiplication pattern.

6. (16) When Morgan finished reading page 127 of a 300-page book, he still had how many pages left to read?

7. (34) $6\overline{)365}$

8. (34) $6\overline{)\$6.36}$

9. (34) $5\overline{)536}$

10. (26) $10\overline{)653}$

11. (34) $4\overline{)\$4.36}$

12. (29) 95×500

13. (33) Round 83 to the nearest ten.

14. (10) $345 + 57 + 760 + 398 + 762 + 584 + W = 3000$

15. (24) $3004 - (3000 - 4)$

16. (29) $\$5.93 \times 40$

17. (Inv. 3) Compare: $\frac{1}{3}$ of 12 ◯ $\frac{1}{8}$ of 24

18. (13) $\$12 + \$8.75 + \$0.96$

19. (13) $\$20 - \$12.46$

20. (18, 29) $8 \times 30 \times 15$

21. (18) $6 \times 7 \times 8 \times 9$

22. (1) What are the next three terms in this counting sequence?

..., 460, 470, 480, _____, _____, _____, ...

23. (30) What fraction of this square is shaded? What percent of the square is shaded?

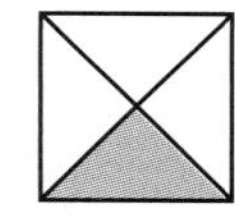

24. *(31, 32)* If two segments that intersect are perpendicular, then what kind of angle do they form?

A. acute B. right C. obtuse

25. *(28)* It is morning. What time is shown by this clock?

26. *(32)* Which two triangles appear to be congruent?

A. 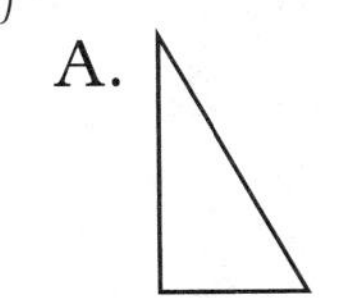 B. C. 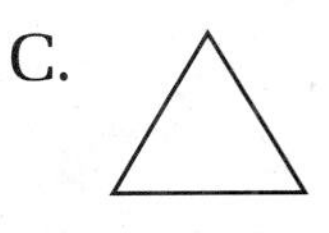 D. 

27. *(26)* Show how to check this division answer. Is the answer correct?

$$\begin{array}{r} 84 \text{ R } 8 \\ 9\overline{)764} \end{array}$$

28. *(24)* Abigail calculated the number of blocks in this figure by finding (2 × 3) × 4. Moe found the number of blocks by finding 2 × (3 × 4). Who was correct and why?

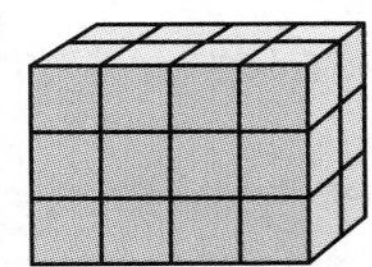

29. *(22)* Without dividing, decide which of these division problems will not have a remainder.

A. $\frac{49}{2}$ B. $\frac{52}{5}$ C. $\frac{600}{10}$

LESSON

35 Problems About Comparing • Problems About Elapsed Time

WARM-UP

Facts Practice: 64 Multiplication Facts (Test F)

Mental Math: Name the months of the year. How many months are in 3 years? ... 4 years? ... 5 years?

a. Round 285 to the nearest hundred.

b. 300 + 800 **c.** 300 × 8 **d.** 42 × 5

e. $\frac{1}{2}$ of 42 **f.** 50% of $8.00 **g.** 25% of $8.00

h. 3 × 9, + 1, ÷ 7, + 1, × 5, − 1, ÷ 4

Problem Solving:

List the possible arrangements of the letters A, E, and R. What percent of the possible arrangements spell words?

NEW CONCEPTS

Problems about comparing

Numbers are used to describe the quantity of objects.

There were 11 football players on the team.

Numbers are also used to describe the size of objects.

The biggest player weighed 245 pounds.

Some stories compare numbers of objects or sizes of objects.

The biggest player weighed 245 pounds. The smallest player weighed 160 pounds. The biggest player weighed how much more than the smallest player?

In comparison stories one number is larger and another number is smaller. Drawing a sketch can help us understand a comparison story. We will draw two rectangles, one taller than the other. Then we will draw an arrow from the top of the shorter rectangle to extend as high as the taller rectangle. The length of the arrow shows the difference in height between the two rectangles. The two rectangles and the arrow each have a circle for a number. For this story, the rectangles stand for the weights of the two players.

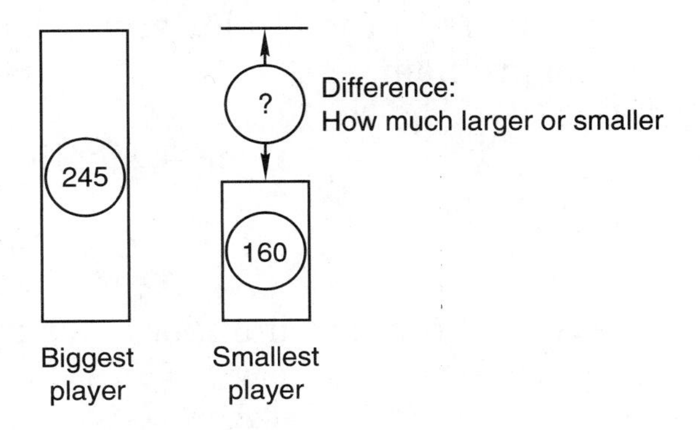

A comparison story may be solved by using a subtraction pattern. If we subtract the smaller number from the larger number, we find the difference between the two numbers. Here we show two ways to write a comparison equation:

$$\begin{array}{r} \text{Larger} \\ -\ \text{Smaller} \\ \hline \text{Difference} \end{array} \qquad \text{Larger} - \text{smaller} = \text{difference}$$

In this story the number missing is the difference, which we find by subtracting.

$$\begin{array}{r} \overset{1}{\not{2}}\,{}^{1}4\,5 \text{ pounds} \\ -\ 1\,6\,0 \text{ pounds} \\ \hline 8\,5 \text{ pounds} \end{array}$$

We find that the biggest player weighs 85 pounds more than the smallest player.

Example 1 Abe is 6 years younger than his brother Gabe. Abe is 11 years old. How old is Gabe?

Solution We will draw two rectangles to illustrate the story. The rectangles stand for the boys' ages. Since Abe is younger, his rectangle is shorter.

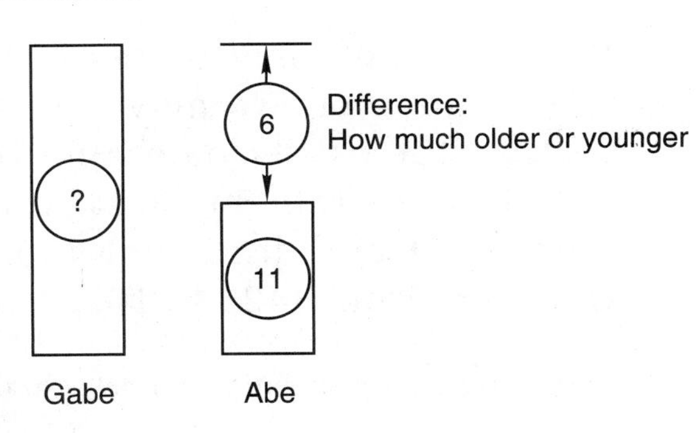

From the story we know that Abe is 11 and that the difference between his age and his brother's age is 6 years. We write the numbers in the circles and use a subtraction pattern to solve the problem.

$$\text{Larger} - \text{smaller} = \text{difference}$$

$$G - 11 = 6$$

We find the first number of a subtraction problem by adding.

$$\begin{array}{r} G \\ -\ 11 \\ \hline 6 \end{array} \quad \text{so} \quad \begin{array}{r} 6 \\ +\ 11 \\ \hline G \end{array}$$

Adding 6 and 11, we find that Gabe is **17 years old.**

Problems about elapsed time Elapsed-time problems are like comparison problems. **Elapsed time** is the amount of time between two points in time.

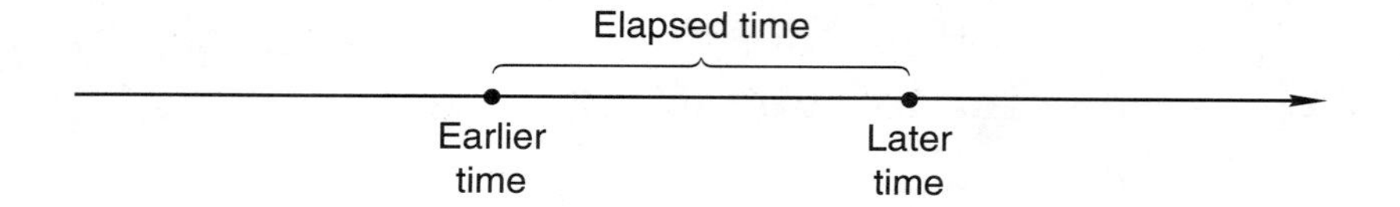

Your age is an example of elapsed time. Your age is the difference between the present time and the time of your birth. To calculate elapsed time, we subtract the earlier time from the later time. Below are two forms of the equation. We use the word *difference* for elapsed time.

$$\begin{array}{l} \quad\text{Later} \\ -\ \text{Earlier} \\ \hline \quad\text{Difference} \end{array} \qquad \text{Later} - \text{earlier} = \text{difference}$$

Example 2 How many years were there from 1492 to 1620?[†]

Solution To find the number of years from one date to another, we may subtract. We subtract the earlier date from the later date. In this problem we subtract 1492 from 1620 and find that there were **128 years** from 1492 to 1620.

$$\begin{array}{r} 1620 \\ -\ 1492 \\ \hline 128 \end{array}$$

[†]Unless otherwise specified, all dates in this book are A.D.

LESSON PRACTICE

Practice set For problems **a–e,** write an equation and solve the problem.

a. There were 4 more boys than girls in the class. If there were 17 boys in the class, how many girls were there?

b. The Mackinac Bridge spans 3800 feet, which is 400 feet less than the span of the Golden Gate Bridge. What is the span of the Golden Gate Bridge?

c. From Rome to Paris is 1120 kilometers. From Rome to London is 1448 kilometers. The distance from Rome to London is how much greater than the distance from Rome to Paris?

d. How many years were there from 1066 to 1776?

e. How many years were there from 1776 to 1787?

MIXED PRACTICE

Problem set

1. Draw a pair of intersecting oblique lines.
(12, 31)

For problems 2–5, write an equation and find the answer.

2. In three games Sherry's bowling scores were 109, 98, and 135. What was her total score for all three games?
(11)

3. Santiago is 8 inches taller than Samantha. If Santiago is 63 inches tall, how tall is Samantha?
(35)

4. How many years were there from 1886 to 1986?
(35)

5. The toll for one car to cross the bridge was $1.50. In ten minutes, 40 cars crossed the bridge. What was the total toll for the 40 cars?
(21, 29)

6. What is the product of nine hundred nineteen and ninety?
(29)

7. Which two quadrilaterals appear to be similar?
(32)

A. B. C. D.

8. List the factors of 28.
(25)

9. $4m = 432$
(26, 34)

10. $423 \div 6$
(34)

11. $243 \div 8$
(34)

12. $2001 \div 4$
(34)

13. $1020 \div 10$
(34)

14. $420 \div (42 \div 6)$
(24, 34)

15. Round 468 to the nearest hundred.
(33)

16. (6)
$$\begin{array}{r} 4657 \\ 285 \\ +\ 1223 \\ \hline \end{array}$$

17. (9)
$$\begin{array}{r} 3165 \\ -\ 1635 \\ \hline \end{array}$$

18. (13)
$$\begin{array}{r} \$10.00 \\ -\ \$\ 8.93 \\ \hline \end{array}$$

19. (29)
$$\begin{array}{r} 436 \\ \times\ \ 70 \\ \hline \end{array}$$

20. (17)
$$\begin{array}{r} \$8.57 \\ \times\ \ \ 7 \\ \hline \end{array}$$

21. (29)
$$\begin{array}{r} 600 \\ \times\ 900 \\ \hline \end{array}$$

22. What fraction of this rectangle is shaded? What percent of the rectangle is shaded? Is more than or less than 50% of the rectangle shaded?
(30, Inv. 3)

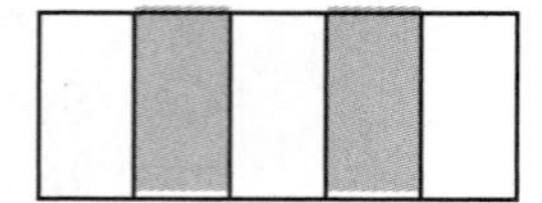

23. What time is a quarter to three in the afternoon?
(28)

24. From November 1 of one year to March 1 of the next year is how many months?
(28)

25. What are the next three terms in this counting sequence?
(1)

..., 1900, 2000, 2100, _____, _____, _____, ...

26. Show how to check this division answer. Is the answer correct?
(26)

$$6\overline{)432}\ \ 72$$

27. In John's class there are half as many girls as boys. There are 14 boys. How many girls are there?
(2)

28. Use words to name the number 68,200.
(7)

29. (a) Draw a right angle.
(32)

(b) Draw an acute angle whose opening is half the size of a right angle.

LESSON

36 Classifying Triangles

WARM-UP

Facts Practice: 90 Division Facts (Test D or E)

Mental Math: How many is half a dozen? ... one and a half dozen? ... two and a half dozen?

a. Round 73 to the nearest ten.
b. 70 + 80
c. 70 × 8
d. 8 × 73
e. $\frac{1}{2}$ of 24
f. 50% of $12
g. 25% of $12
h. 360 + 200
i. 9 × 6, + 2, ÷ 7, + 1, × 4, ÷ 6

Problem Solving:

There are three crosswalk signals between Julie's home and school. When Julie comes to a signal, she either walks through the crosswalk or waits for the signal to turn. List the eight possible patterns of signals for Julie's walk to school. Use the words "walk" and "wait."

NEW CONCEPT

In this lesson we will learn the names of different kinds of triangles. All triangles have three angles and three sides, but we can **classify,** or sort, triangles by the size of their angles and by the relative lengths of their sides.

Recall from Lesson 32 that three types of angles are acute angles, right angles, and obtuse angles.

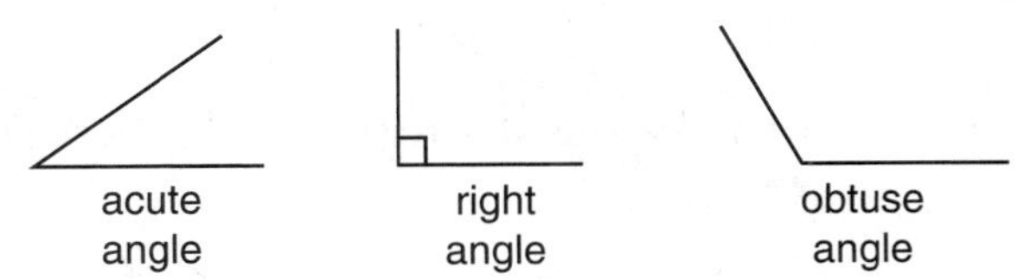

Triangles that contain these angles can be classified as acute, right, or obtuse.

Triangles Classified by Angles

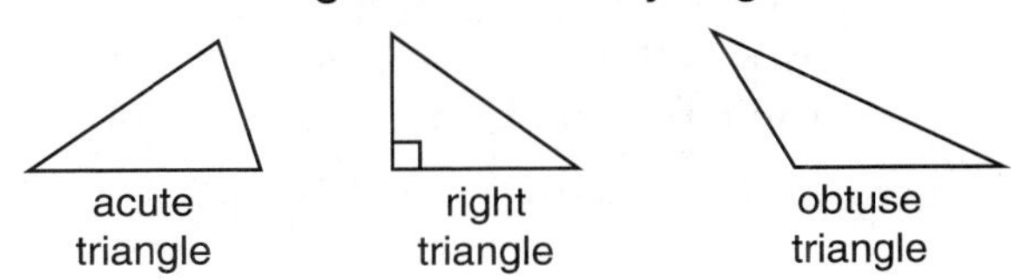

Every triangle has at least two acute angles. If all three angles are acute, the triangle is an **acute triangle.** If one of the angles is a right angle, the triangle is a **right triangle.** If one of the angles is obtuse, the triangle is an **obtuse triangle.**

We can also classify triangles by the comparative lengths of their sides.

Triangles Classified by Sides

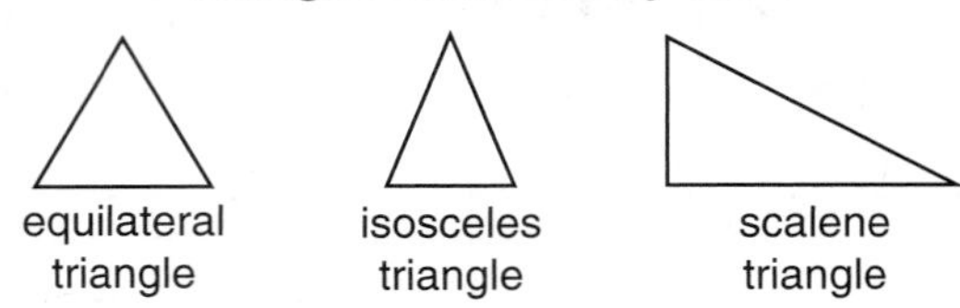

The three sides of an **equilateral triangle** have equal lengths. At least two sides of an **isosceles triangle** have equal lengths. All three sides of a **scalene triangle** have different lengths.

Every triangle can be classified both by angles **and** by sides. Notice that the scalene triangle illustrated above also appears to be a right triangle, while the isosceles and equilateral triangles are also acute triangles.

Example 1 All three sides of this triangle have the same length. Which of the following terms does not describe the triangle?

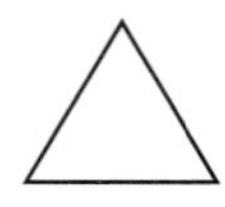

A. equilateral B. acute C. isosceles D. right

Solution The triangle is an equilateral triangle and an acute triangle. It is also an isosceles triangle, because *at least* two of the sides have equal lengths. Because none of the angles of the triangle is a right angle, the correct answer is **D. right.**

Example 2 Draw three angles with sides that are segments of equal length. Make the first angle acute, the second right, and the third obtuse. Then, for each angle, draw a segment between the endpoints to form a triangle. Classify each triangle by sides and by angles.

Solution We draw each angle with two segments that have a common endpoint.

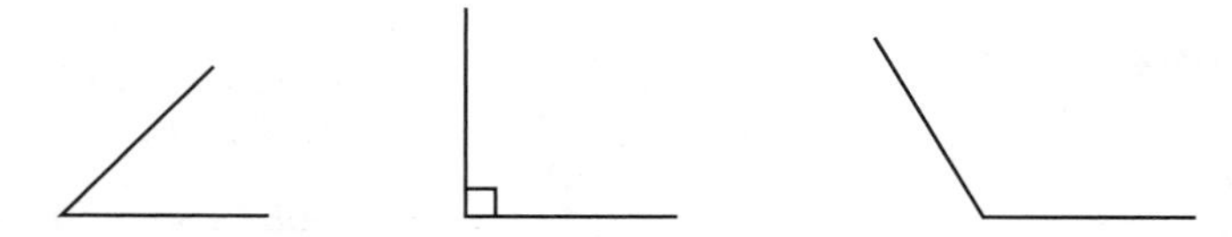

Then we draw segments to form three triangles. Since two sides of each triangle have equal length, all three triangles are isosceles. Here are the classifications for each triangle:

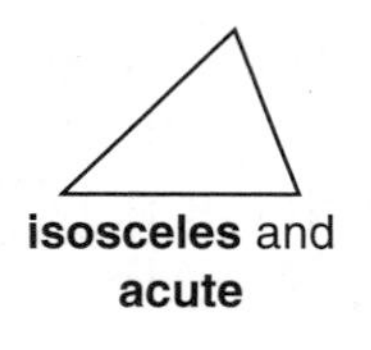

isosceles and **acute**

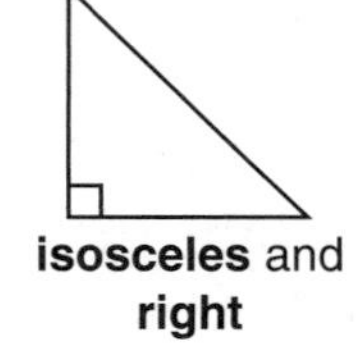

isosceles and **right**

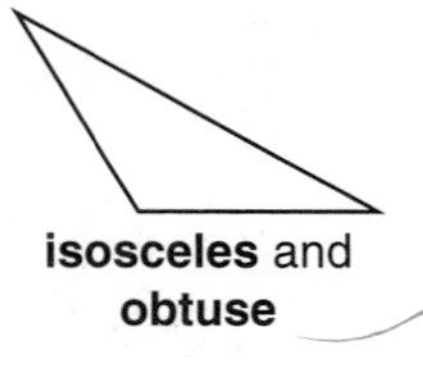

isosceles and **obtuse**

LESSON PRACTICE

Practice set Classify each triangle by angles:

a. **b.** **c.**

Classify each triangle by sides:

d. **e.** 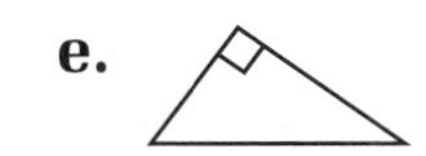 **f.** 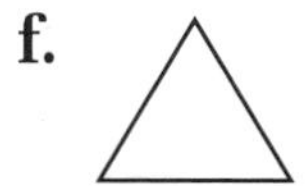

g. Draw a right triangle.

h. Draw an equilateral triangle.

i. Draw a right angle with sides that are segments of equal length. Then draw a segment between the endpoints to form a triangle. What type of right triangle did you draw?

j. If you were to draw a diagonal segment on your paper between opposite corners (vertices), you would divide your rectangular paper into two congruent triangles. Classify by sides and by angles the triangles that would be formed.

MIXED PRACTICE

Problem set

1. *(31)* Draw a pair of horizontal parallel line segments. Make both segments the same length.

For problems 2–4, write an equation and find the answer.

2. *(35)* Jason is reading a book that has 336 pages. Willis is reading a book that has 402 pages. Willis's book has how many more pages than Jason's book?

3. *(21)* Jason has one week to read a 336-page book. How many pages should he read each day to finish the book on time? Use a multiplication pattern.

4. *(21)* A fortnight is 2 weeks. How many days is a fortnight? Use a multiplication pattern.

5. Round 780 to the nearest hundred.
(33)

6. Which triangle has one obtuse angle?
(36)

A. B.  C.

7. How many years were there from 1776 to 1976?
(35)

8. When the students voted for president, Jeremy received 119 votes and Juanita received 142 votes. Juanita won by how many votes?
(35)

9. What is the name for the top number of a fraction?
(Inv. 2)

10. Which of these shapes is not a polygon? Why?
(32)

A. B. C. D.

11. Cindy has two fourths of a circle and three tenths of a circle. What does she need to make a whole circle?
(Inv. 2)

12. *(29)* $\begin{array}{r} 763 \\ \times \quad 800 \\ \hline \end{array}$

13. *(17)* $\begin{array}{r} \$24.08 \\ \times \quad 6 \\ \hline \end{array}$

14. *(29)* $\begin{array}{r} 976 \\ \times \quad 40 \\ \hline \end{array}$

15. *(29)* $\begin{array}{r} 400 \\ \times \quad 50 \\ \hline \end{array}$

16. *(14)* $\begin{array}{r} 5818 \\ - \quad M \\ \hline 4747 \end{array}$

17. *(13)* $\begin{array}{r} \$98.98 \\ \$36.25 \\ \$\ 4.97 \\ + \ \$87.64 \\ \hline \end{array}$

18. *(9)* $\begin{array}{r} 1010 \\ - \quad 918 \\ \hline \end{array}$

19. $7w = \$7.63$
(18, 34)

20. $368 \div 9$
(34)

21. $6\overline{)4248}$
(34)

22. $8\overline{)\$10.00}$
(26)

23. What are the next three terms in this counting sequence?
(1)

..., 2700, 2800, 2900, ______, ______, ______, ...

24. What fraction of this hexagon is shaded? Is more than or less than 25% of the hexagon shaded? Is more than or less than 10% of the hexagon shaded?
(30)

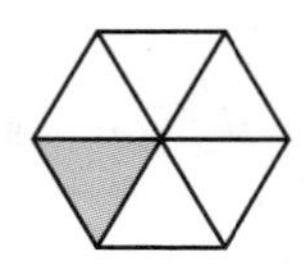

25. To what number is the arrow pointing?
(27)

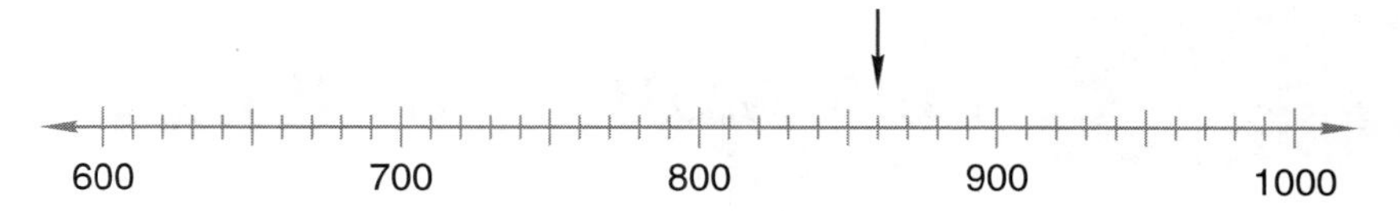

26. Show how to check this division answer. Is the answer correct?
(26)

$$784 \div 6 = 13 \text{ R } 4$$

27. Write a multiplication fact that shows how to find the number of small squares in this rectangle.
(13)

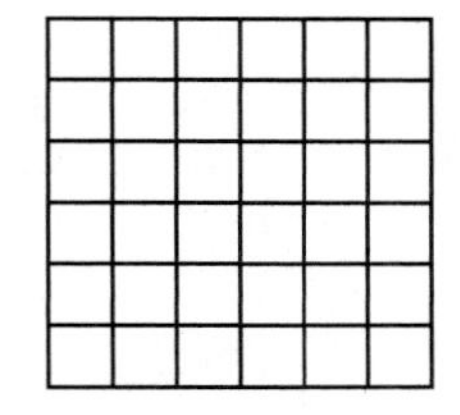

28. Seven tenths of a circle is what percent of a circle?
(Inv. 2)

29. Draw an isosceles right triangle.
(36)

LESSON

37 Drawing Pictures of Fractions

WARM-UP

Facts Practice: 64 Multiplication Facts (Test F)

Mental Math: Count by 12's from 12 to 96. How many months are in 5 years? ... 6 years?

a. Round 890 to the nearest hundred.

b. 900 + 900

c. 900 × 4

d. 4 × 89

e. 4 × 90 minus 4 × 1

f. 9 × 9, − 9, ÷ 9

g. 50% of 60¢

h. 25% of 60¢

i. 10% of 60¢

Problem Solving:

The license plates of a certain state have three letters followed by three digits. One license plate reads CAR 123. How many different license plates from the state could begin "CAR" and end with any arrangement of all the digits 1, 2, and 3? List the possible license plates.

NEW CONCEPT

A picture can help us understand the meaning of a fraction.

Example 1 Draw three squares and shade $\frac{1}{2}$ of each square a different way.

Solution The denominator of $\frac{1}{2}$ tells us to cut each square into two equal parts. The numerator of the fraction tells us to shade one of the parts. There are many ways to do this. Here we show three different ways:

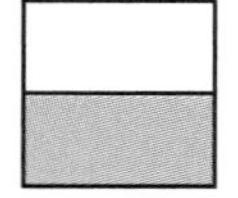 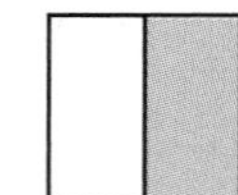 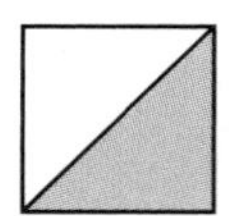

When drawing pictures of fractions, we must always be careful to divide the pictures into **equal** parts. The square below has been cut into two parts, but the parts are not equal. Therefore, the square has not been divided into halves.

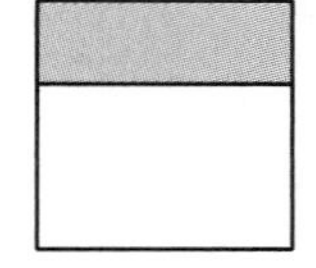

This does not represent $\frac{1}{2}$.

Example 2 Draw a rectangle and shade $\frac{1}{3}$ of it.

Solution After we draw the rectangle, we must divide it into three equal parts. If we begin by dividing it in half, we will not be able to divide it into three equal parts.

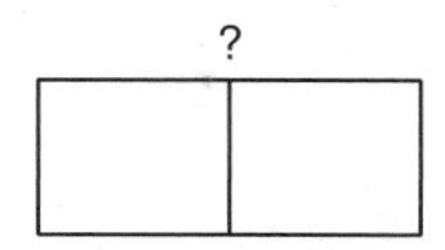

We must plan ahead. To form 3 parts, we draw 2 equally spaced segments. Here we show two different ways to shade $\frac{1}{3}$ of a rectangle:

LESSON PRACTICE

Practice set* **a.** Draw a circle and shade $\frac{1}{4}$ of it.

b. Draw a rectangle and shade $\frac{2}{3}$ of it.

c. The shaded portion of this square represents the fraction $\frac{3}{4}$. Show another way to shade $\frac{3}{4}$ of a square.

d. The shaded portion of this circle represents the fraction $\frac{1}{3}$. Draw a circle and shade $\frac{2}{3}$ of it.

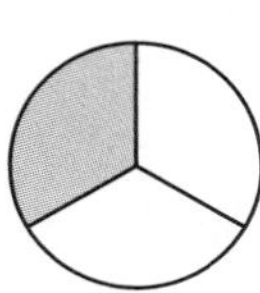

MIXED PRACTICE

Problem set **1.** (12, 31) Draw a pair of horizontal parallel line segments. Make the lower segment longer than the upper segment.

2. (30, 37) Draw three rectangles and shade $\frac{1}{2}$ of each rectangle a different way. What percent of each rectangle is shaded?

For problems 3–5, write an equation and find the answer.

3. (11) When Bill cleaned his room, he found 39 marbles, 20 baseball cards, a toothbrush, 4 pencils, and a peanut butter sandwich. How many items did he find?

4. (21) There are 12 inches in a foot. How many inches are in 3 feet?

5. (35) How many years were there from 1517 to 1620?

6. (25) List the factors of 40.

7. (30, Inv. 3) What fraction of this octagon is not shaded? Is more than or less than 50% of the octagon not shaded? What percent of the octagon is shaded?

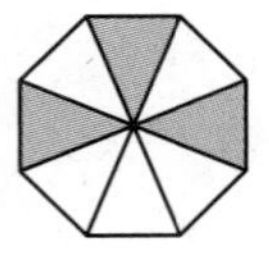

8. (28) From May 1 of one year to February 1 of the next year is how many months?

9. (33) Round 46 to the nearest ten.

10. (36) Draw a right triangle.

11. (13) $$\begin{array}{r} \$36.51 \\ \$74.15 \\ +\ \$25.94 \\ \hline \end{array}$$

12. (14) $$\begin{array}{r} 3040 \\ -\ \ W \\ \hline 2950 \end{array}$$

13. (13) $$\begin{array}{r} \$90.00 \\ -\ \$20.30 \\ \hline \end{array}$$

14. (29) $$\begin{array}{r} 592 \\ \times\ \ 90 \\ \hline \end{array}$$

15. (29) $$\begin{array}{r} \$4.75 \\ \times\ \ \ 80 \\ \hline \end{array}$$

16. (10) $$\begin{array}{r} 43 \\ C \\ 29 \\ 467 \\ +\ 94 \\ \hline 700 \end{array}$$

17. (4, 20, 34) Compare: $\frac{840}{8} \bigcirc \frac{460}{4}$

18. (29) 720×400

19. (26, 34) $6w = \$12.24$

20. (24, 34) $1000 \div (100 \div 10)$

21. (24, 26, 29) $60 \times (235 \div 5)$

22. (18, 29) $42 \times 30 \times 7$

23. (13, 24) $\$20 - (\$3.48 + \$12 + \$4.39)$

24. (Inv. 2) Duncan fit one half of a circle, one fourth of a circle, and one tenth of a circle together. What percent of the circle was missing?

25. (32) Which of these shapes is not a polygon?

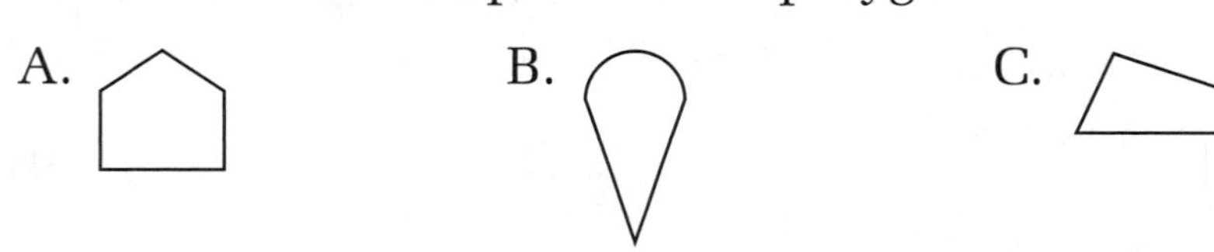

26. Draw a rectangle and shade $\frac{1}{3}$ of it.
(37)

27. What year was one decade before 1932?
(28)

28. (a) An octagon has how many angles?
(23, 32)

(b) A quadrilateral has how many angles?

(c) Use the answers to parts (a) and (b) to write a fraction equal to $\frac{1}{2}$.

29. Which of these is not shaped like a quadrilateral?
(32)

A.

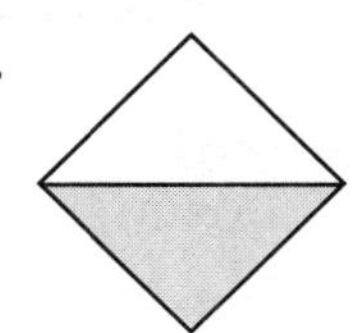

B. ELM AVE.

C.

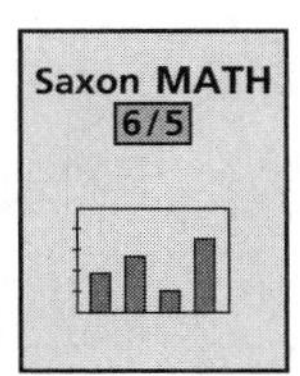

D.

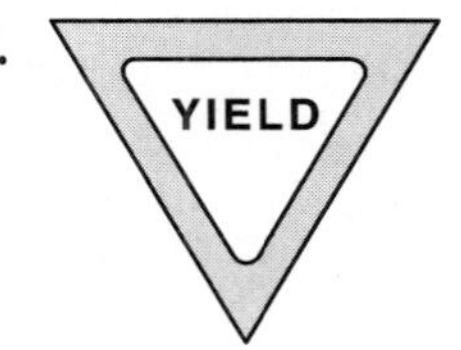

LESSON

38 Fractions and Mixed Numbers on a Number Line

WARM-UP

Facts Practice: 90 Division Facts (Test D or E)

Mental Math: Name the relationship: Diana's mother's father is Diana's ______.
Count by 12's from 12 to 96. How many is 5 dozen? ... 6 dozen? ... 7 dozen?

a. Round 615 to the nearest hundred. **b.** 700 + 800
c. 10 × 70 **d.** 5 × 24 **e.** $\frac{1}{2}$ of 44
f. 50% of 80¢ **g.** 25% of 80¢ **h.** 10% of 80¢
i. 6 × 6, − 6, ÷ 6, + 1, − 6

Problem Solving:

Copy this addition problem and fill in the missing digits:

```
  __
+  1
 ___
```

NEW CONCEPT

A number line is made up of a series of points. The points on the line represent numbers. On the number line below whole numbers are labeled. However, there are many numbers on the line that are not labeled. We mark some unlabeled numbers with arrows:

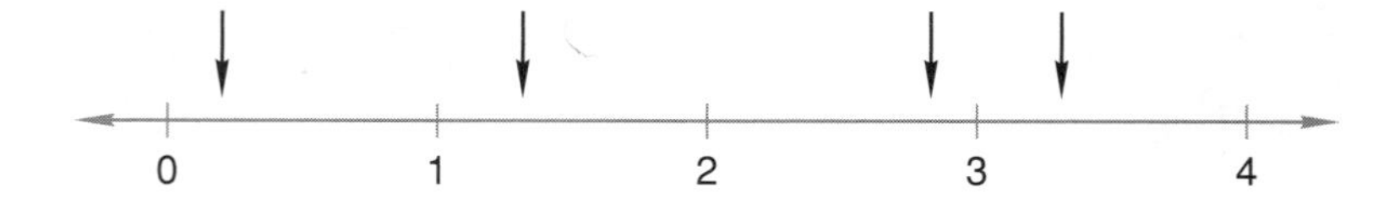

Many of the unlabeled points on a number line can be named with fractions and **mixed numbers.** Mixed numbers are numbers like $1\frac{1}{2}$ (one and one half) that are a whole number and a fraction together.

To identify a fraction or mixed number on a number line, we need to count the divisions between the whole numbers. On the number line below, the distance between every two whole numbers has been divided into three sections (into thirds). Thus, each small section is one third $\left(\frac{1}{3}\right)$. (Be careful to count the *sections* of the number line and not the marks that separate the sections.)

A point on a number line is named by its distance from zero. The location of the point marked by arrow A is given by the whole number 1 plus the length of one section. So the number for that point is $1\frac{1}{3}$. The point marked with arrow B is the whole number 3 plus the length of two sections. The number for point B is $3\frac{2}{3}$.

When reading from number lines with sections smaller than 1, follow these steps:

1. Find the whole-number distance from zero up to (but not past) the point to be named. This is the whole-number part of the answer.
2. Next, count the number of sections between whole numbers. This number is the denominator of the fraction.
3. Then count the number of sections past the whole number to the point being named. This is the numerator of the fraction.

Example 1 Name the fraction or mixed number marked by each arrow on these number lines:

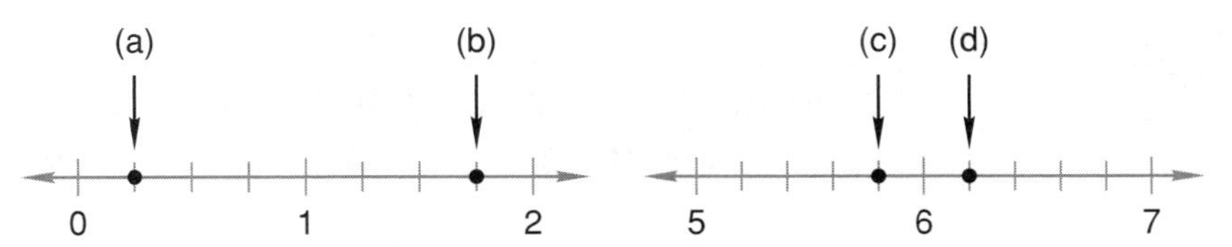

Solution Point (a) is between 0 and 1, so it is named by a fraction and not by a mixed number. The distance between whole numbers on this number line is divided into fourths. Point (a) is one section from zero, which is $\mathbf{\frac{1}{4}}$.

The distance from zero to point (b) is 1 plus the length of three sections, or $\mathbf{1\frac{3}{4}}$.

The distance from zero to point (c) is 5 plus a fraction. The distance between whole numbers on this number line is divided into fifths. Point (c) is four sections from 5, which is $\mathbf{5\frac{4}{5}}$.

The distance from zero to point (d) is 6 plus the length of one section, or $\mathbf{6\frac{1}{5}}$.

Example 2 Here we show two number lines. On one number line the fraction $\frac{2}{3}$ is graphed. On the other number line $\frac{3}{4}$ is graphed.

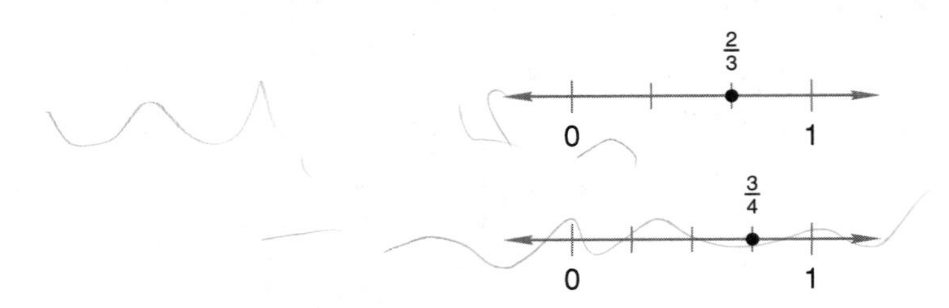

Refer to these number lines to compare the fractions $\frac{2}{3}$ and $\frac{3}{4}$.

$$\frac{2}{3} \bigcirc \frac{3}{4}$$

Solution Both $\frac{2}{3}$ and $\frac{3}{4}$ are greater than 0 but less than 1. Numbers to the right on the number line are greater than numbers to the left. So $\frac{3}{4}$ is greater than $\frac{2}{3}$.

$$\mathbf{\frac{2}{3} < \frac{3}{4}}$$

LESSON PRACTICE

Practice set* Name the fraction or mixed number marked by each arrow on these number lines:

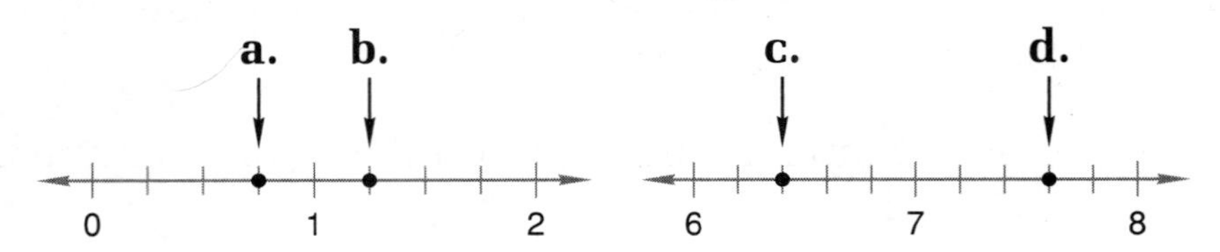

Three fractions are graphed on the number line below. Refer to the number line to compare the fractions in problems **e, f,** and **g.**

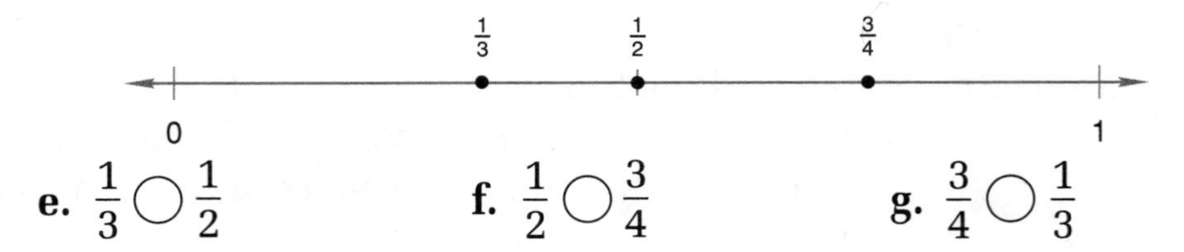

e. $\frac{1}{3} \bigcirc \frac{1}{2}$ **f.** $\frac{1}{2} \bigcirc \frac{3}{4}$ **g.** $\frac{3}{4} \bigcirc \frac{1}{3}$

MIXED PRACTICE

Problem set

1. Draw a pair of horizontal parallel line segments. Make the upper segment longer than the lower segment.
(12, 31)

2. Sam scored $\frac{1}{4}$ of the team's 28 points. How many points did Sam score?
(Inv. 2)

For problems 3–6, write an equation and find the answer.

3. Tickets to the matinee were $4.75 each. Mr. Jones bought four tickets. What was the total cost of the tickets?
(21)

4. The used-car dealer bought a car for $725 and sold it for $1020. How much profit did the dealer make on the car? Use a subtraction pattern.
(35)

5. In 2 hours the 3 boys picked a total of 1347 cherries. If they share the cherries evenly, then each boy will get how many cherries?
(21)

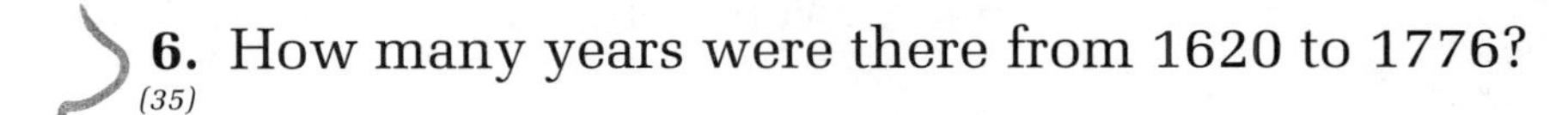

6. How many years were there from 1620 to 1776?
(35)

7. Which triangle has three acute angles?
(36)

A. 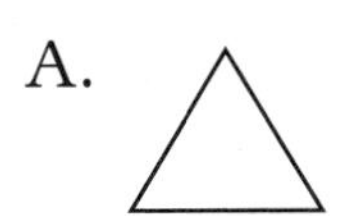   B. 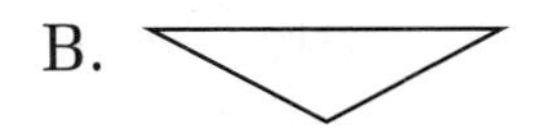   C.

8. Draw a circle and shade $\frac{3}{4}$ of it. What percent of the circle is shaded?
(30, 37)

9. How many days are in a leap year?
(28)

10. A stop sign has the shape of an octagon. An octagon has how many sides?
(32)

11. $3647 + 92 + 429$
(6)

12. $3518 - 1853$
(9)

13. $4 \times 6 \times 8 \times 0$
(15, 18)

14. $3518 \div 10$
(26)

15. $\$4.76 + \$12 + \$0.97 + W = \$20$
(10, 13)

16. $\$100 - \$87.23$
(13)

17. 786×900
(29)

18. $\$63.18 \div 9$
(34)

19. $375 \times (640 \div 8)$
(24, 29, 34)

20. Compare: $(3 \times 5) \times 7 \bigcirc 3 \times (5 \times 7)$
(24)

21. Every four-sided polygon is which of the following?
(32)

A. square   B. rectangle   C. quadrilateral

22. What are the next three terms in this counting sequence?
(1)

..., 1800, 1900, 2000, 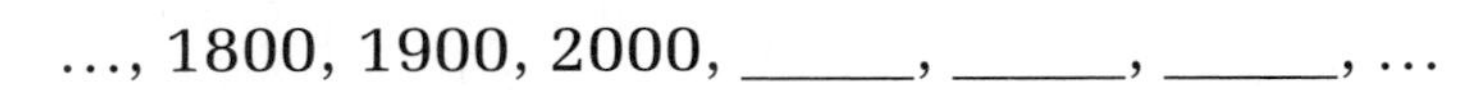
______, ______, ______, ...

23. To what mixed number is the arrow pointing?
(38)

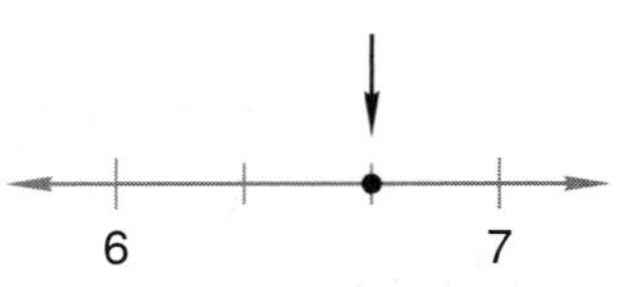

24. To what fraction is the arrow pointing?
(38)

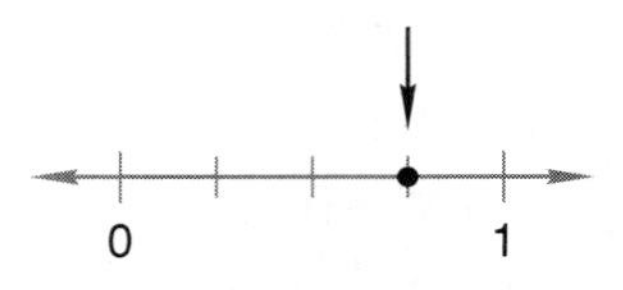

25. It is 9:45 a.m. What time will it be in 4 hours?
(28)

26. Round 649 to the nearest hundred.
(33)

27. If the divisor is 6 and the quotient is 3, then what is the dividend?
(20)

28. (a) Are all squares congruent?
(32)

(b) Are all squares similar?

29. Refer to the number line below to complete the comparison.
(38)

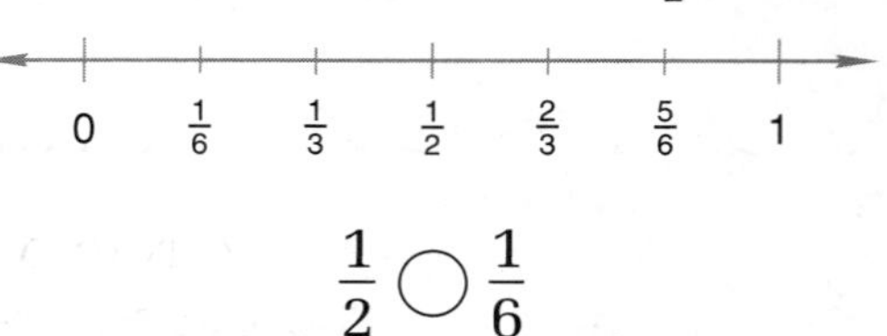

$$\frac{1}{2} \bigcirc \frac{1}{6}$$

LESSON

39 Comparing Fractions by Drawing Pictures

WARM-UP

Facts Practice: 64 Multiplication Facts (Test F)

Mental Math: Name the relationship: Gilbert's father's sister is Gilbert's ______.
Count by 7's from 7 to 84. How many days is 3 weeks? ... 4 weeks? ... 6 weeks?

a. Round 78 to the nearest ten. **b.** 830 − 200
c. 600 × 4 **d.** 6 × 24 **e.** 10 × 100
f. $\frac{1}{2}$ of 21 **g.** 50% of \$2.00 **h.** 25% of \$2.00
i. 5 × 5, + 5, ÷ 5, − 5

Problem Solving:

Which number do you say when you count by 6's from 6 to 36 **and** when you count by 8's from 8 to 48?

NEW CONCEPT

One fourth of the circle on the left is shaded. One half of the circle on the right is shaded.

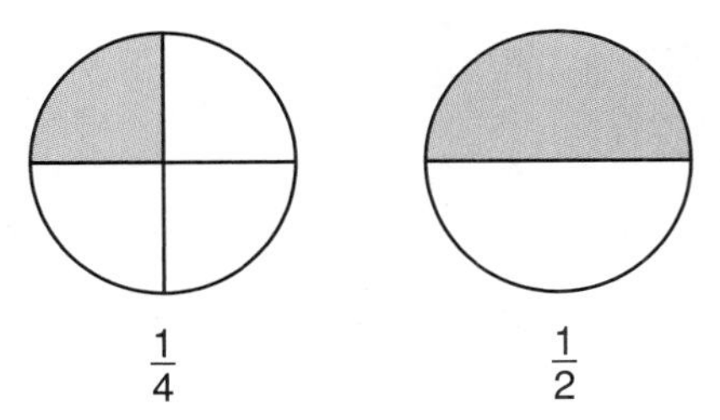

$\frac{1}{4}$ $\frac{1}{2}$

We see that less of the circle on the left is shaded. This is because $\frac{1}{4}$ is a smaller fraction than $\frac{1}{2}$. We can write this comparison using a comparison symbol as

$$\frac{1}{4} < \frac{1}{2}$$

In this lesson we will begin comparing fractions by drawing pictures of the fractions and comparing the pictures.

Example Draw pictures to compare these fractions: $\frac{1}{2} \bigcirc \frac{1}{3}$

Solution We might think that $\frac{1}{3}$ is greater than $\frac{1}{2}$ because 3 is greater than 2. However, by drawing pictures, we will see that $\frac{1}{3}$ is actually less than $\frac{1}{2}$. If an object is divided into 3 parts, each part will be smaller than if the object were divided into 2 parts.

To begin, we draw two *congruent* shapes. We choose to draw two equal-sized rectangles, and we label the rectangles $\frac{1}{2}$ and $\frac{1}{3}$. Next, we divide the rectangles into the number of parts shown by the denominator, and we shade the number of parts shown by the numerator.

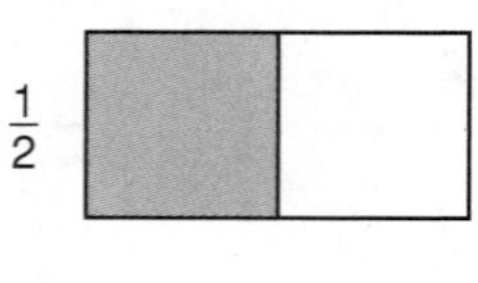

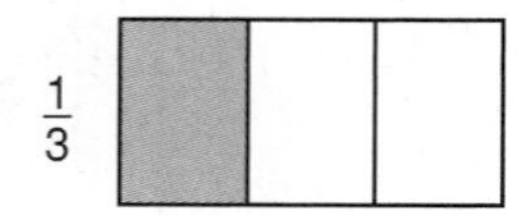

Then we compare the shaded areas. We see that more of the rectangle is shaded when $\frac{1}{2}$ is shaded than when $\frac{1}{3}$ is shaded. So our answer is

$$\mathbf{\frac{1}{2} > \frac{1}{3}}$$

LESSON PRACTICE

Practice set Draw pictures to compare each pair of fractions. When drawing pictures of any two fractions, be sure to draw the shapes the same size.

a. $\frac{1}{2} \bigcirc \frac{2}{3}$

b. $\frac{1}{2} \bigcirc \frac{2}{4}$

c. $\frac{1}{3} \bigcirc \frac{1}{4}$

d. $\frac{2}{3} \bigcirc \frac{3}{4}$

MIXED PRACTICE

Problem set

1. *(31, 32)* Draw a pair of horizontal parallel line segments of the same length. Form a quadrilateral by connecting the ends of the segments.

For problems 2–4, write an equation and find the answer.

2. *(21, 28)* How many years is five centuries?

3. *(35)* Paloma is 6 years older than her sister. If Paloma is 13 years old, then how old is her sister? Use a subtraction pattern.

4. *(21)* Joe walked 488 feet going to the end of the pier and back. How long is the pier?

5. (39) Draw pictures to compare these fractions: $\frac{1}{4} \bigcirc \frac{1}{3}$

6. (2) What number is half of 23?

7. (Inv. 2) Emily's cat ate $\frac{1}{4}$ of a dozen fish fillets. How many fish fillets did Emily's cat eat?

8. (33) Round 84 to the nearest ten.

9. (25) List the factors of 35.

10. (13)
$$\begin{array}{r} \$93.18 \\ \$42.87 \\ + \$67.95 \\ \hline \end{array}$$

11. (13)
$$\begin{array}{r} \$30.00 \\ - \$\ 8.75 \\ \hline \end{array}$$

12. (6)
$$\begin{array}{r} 46 \\ 23 \\ 97 \\ 15 \\ 24 \\ 55 \\ + 55 \\ \hline \end{array}$$

13. (14)
$$\begin{array}{r} 4304 \\ - \quad B \\ \hline 3452 \end{array}$$

14. (29)
$$\begin{array}{r} \$6.38 \\ \times \quad 60 \\ \hline \end{array}$$

15. (29)
$$\begin{array}{r} 640 \\ \times \quad 700 \\ \hline \end{array}$$

16. (34) $\frac{640}{8}$

17. (26) $\frac{720}{10}$

18. (34) $\frac{\$6.24}{6}$

19. (34) $\frac{1236}{4}$

20. (34) $563 \div 7$

21. (26) $4718 \div 9$

22. (26) $8m = 3000$

23. (28) What time is 20 minutes before midnight?

24. (Inv. 2, Inv. 3) A quarter of a circle plus an eighth of a circle is what percent of a circle?

25. (28) According to this calendar, what is the date of the third Saturday in April 1901?

APRIL 1901

S	M	T	W	T	F	S
	1	2	3	4	5	6
7	8	9	10	11	12	13
14	15	16	17	18	19	20
21	22	23	24	25	26	27
28	29	30				

26. Refer to the number line below to answer parts (a)–(c).
(38)

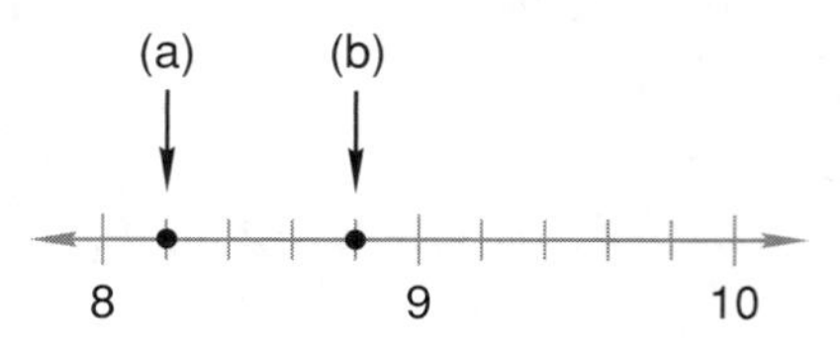

(a) To what mixed number is arrow (a) pointing?

(b) To what mixed number is arrow (b) pointing?

(c) Write your answers to (a) and (b), using a comparison symbol to show which mixed number is greater and which is less.

27. What is the product of four hundred sixteen and sixty?
(5, 29)

28. (a) How many hours are in a day?
(23, 28)

(b) How many hours are in half a day?

(c) Use the numbers in the answers to parts (a) and (b) to write a fraction equal to $\frac{1}{2}$.

29. A full turn is 360 degrees. How many degrees is $\frac{1}{2}$ of a turn?
(Inv. 2, 23, 34)

LESSON

40 Pictures of Mixed Numbers • Writing Quotients as Mixed Numbers, Part 1

WARM-UP

Facts Practice: 90 Division Facts (Test D or E)

Problem Solving:

Use the information in this paragraph to complete the statements that follow. Drawing a diagram may help you with the problem.

Alice and Bob are the mother and father of Carol and George. Carol and her husband Donald have a son, Edward. George and his wife Fiona have a daughter, Heather.

a. Alice is Edward's ______.

b. Heather is Bob's ______.

c. George is Edward's ______.

d. Heather is Carol's ______.

e. Donald is Bob's ______.

f. Alice is Fiona's ______.

g. Edward is Heather's ______.

NEW CONCEPTS

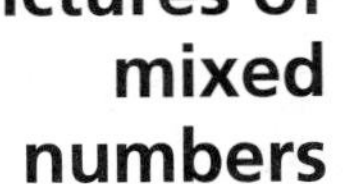

Pictures of mixed numbers

The picture below shows some pies on a shelf.

We see two whole pies and one half of another pie. There are two and one half pies on the shelf. Using digits, we write "two and one half" this way:

$$2\frac{1}{2}$$

Example 1 Use a mixed number to name the number of shaded circles shown here.

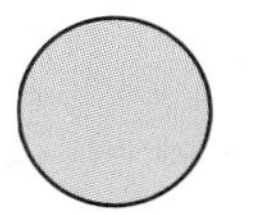

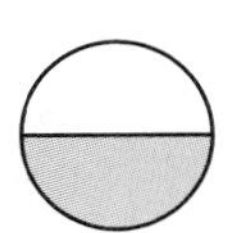

Solution We see two circles. The completely shaded circle represents the whole number 1. Half of the second circle is shaded. It represents the fraction $\frac{1}{2}$. Together, the number of shaded circles is one and one half.

$$\mathbf{1\frac{1}{2}}$$

Writing quotients as mixed numbers, part 1

Some problems have answers that are mixed numbers. For example, what is the width of the rectangle formed by folding a sheet of notebook paper in half as shown?

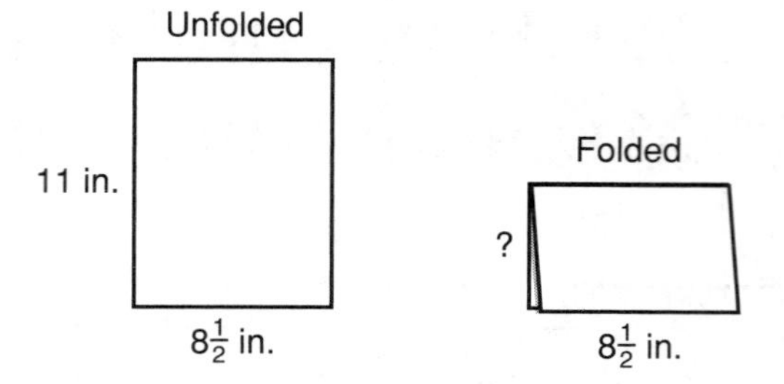

The width of the rectangle is half of 11 inches, which is $5\frac{1}{2}$ inches.

Example 2 Peter, Edmund, and Lucy will equally share seven chicken potpies. How many pies are there for each person?

Solution First we will use a diagram to explain the solution. We need to divide the pies into three equal groups. We can arrange six of the pies into three groups of two pies:

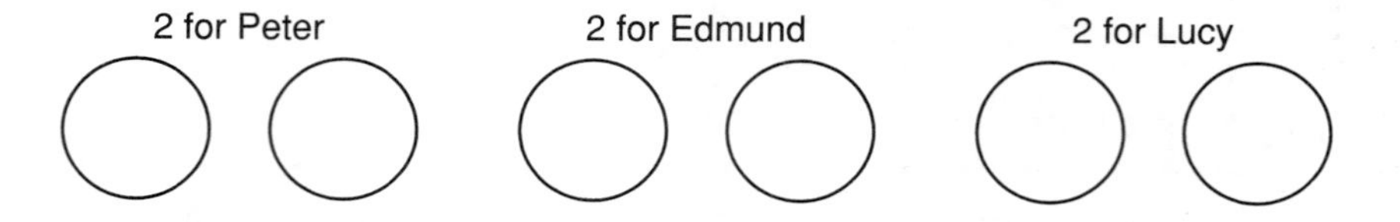

However, there are seven pies. So there is still one pie to be divided. We divide the remaining pie into thirds:

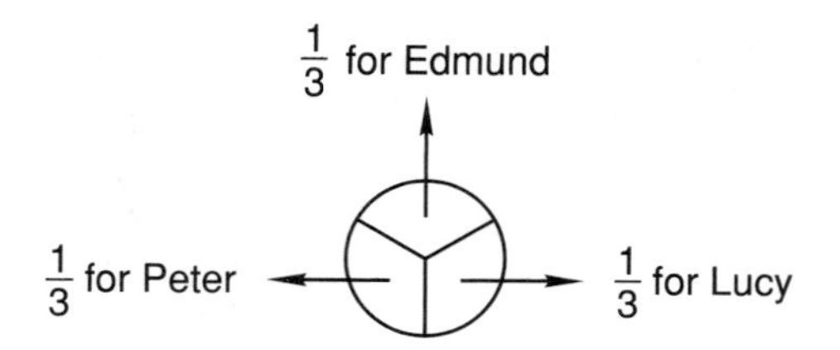

We find that there are $2\frac{1}{3}$ **pies** for each person.

Now we will show how to find the answer using a pencil-and-paper algorithm. To divide seven pies into three equal groups, we divide 7 by 3.

$$\begin{array}{r} 2 \\ 3\overline{)7} \\ -\ 6 \\ \hline 1 \end{array}$$

The quotient is 2, which means "2 whole pies." The remainder is 1, which means one pie has not been divided. Now we divide the remaining pie by three.

One divided by three is the fraction one third. We write "$\frac{1}{3}$" after the whole number above the division box.

$$\begin{array}{r} 2\frac{1}{3} \\ 3\overline{)7\phantom{\frac{1}{3}}} \\ -\,6\phantom{\frac{1}{3}} \\ \hline 1\phantom{\frac{1}{3}} \end{array}$$

This answer means that each person will get **$2\frac{1}{3}$ pies.**

LESSON PRACTICE

Practice set Write a mixed number to name the number of shaded circles in each diagram:

a. 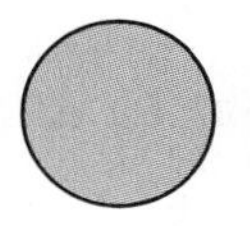 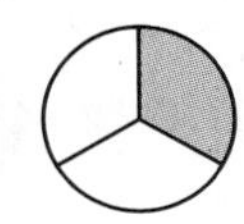

b. 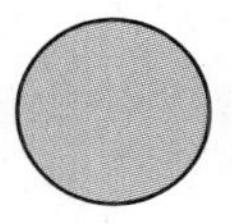  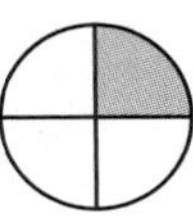

Draw and shade circles to represent each mixed number:

c. three and one half

d. one and three fourths

e. Use a diagram to explain the answer to this story problem. Then show how to find the answer using pencil and paper. Follow the example in the lesson.

Taro, Susan, Edmund, and Lucy will equally share nine chicken potpies. How many pies are there for each person?

MIXED PRACTICE

Problem set

1. (31, 32) Draw a pair of horizontal parallel line segments. Make the lower segment longer than the upper segment. Connect the endpoints of the segments to form a quadrilateral.

2. (37) If 1 pie is shared equally by 6 people, then each person will get what fraction of the pie?

3. (40) Dan, Juan, and Thuy are sharing 4 oranges equally. How many oranges does each person have?

For problems 4–6, write an equation and find the answer.

4. (21) One hundred forty students were divided equally into 5 classes. How many students were in each class?

5. (35) Khanh weighs 105 pounds. Sammy weighs 87 pounds. Sammy weighs how many pounds less than Khanh?

6. (35) The first flag of the United States had 13 stars. How many more stars does the current flag have?

7. (2, Inv. 3) What percent is half of 25%?

8. (32) A hexagon has how many more sides than a pentagon?

9. (Inv. 2) One half of a circle plus one fourth of a circle is what percent of a whole circle?

10. (38) Refer to the number line below to answer parts (a)–(c).

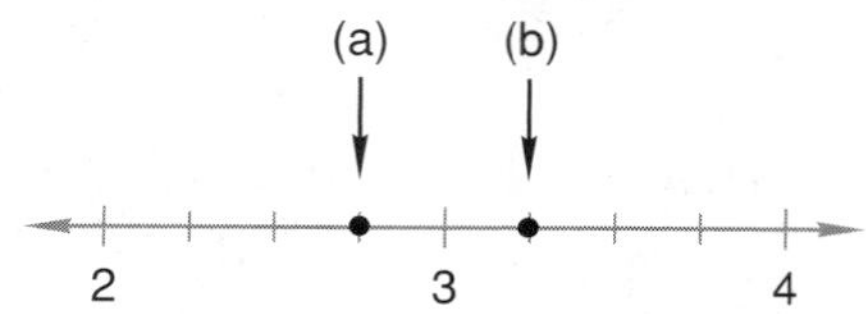

(a) To what mixed number is arrow (a) pointing?

(b) To what mixed number is arrow (b) pointing?

(c) Write your answers to (a) and (b), using a comparison symbol to show which number is greater and which is less.

11. (14) $M - 345 = 534$

12. (6)
$$\begin{array}{r} 785 \\ 964 \\ 287 \\ +\ 846 \\ \hline \end{array}$$

13. (9)
$$\begin{array}{r} 7106 \\ -\ 3754 \\ \hline \end{array}$$

14. (29)
$$\begin{array}{r} \$3.84 \\ \times\quad 60 \\ \hline \end{array}$$

15. (29) 769×800

16. (34) $\dfrac{\$24.48}{8}$

17. (34) $\dfrac{4320}{9}$

18. (13, 24) $\$20 - (\$1.45 + \$6.23 + \$8)$

19. (13, 17) $3742 + 3742 + 3742 + 3742 + 3742$

20. How many circles are shaded?
(40)

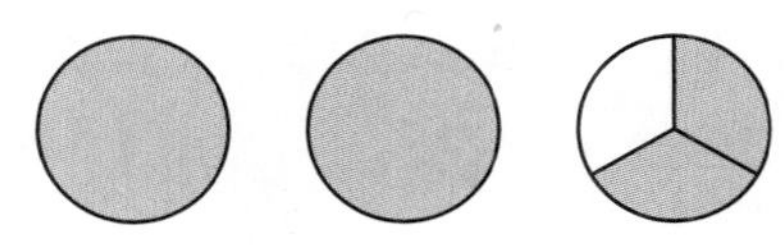

21. Round 650 to the nearest hundred.
(33)

22. A year is what fraction of a decade? A year is what percent of a decade?
(28, 30)

23. Which of these angles appears to be an obtuse angle?
(32)

A. B. C.

24. What are the next three terms in this counting sequence?
(1)

..., 60, 70, 80, ______, ______, ______, ...

25. To what number on this scale is the arrow pointing?
(27)

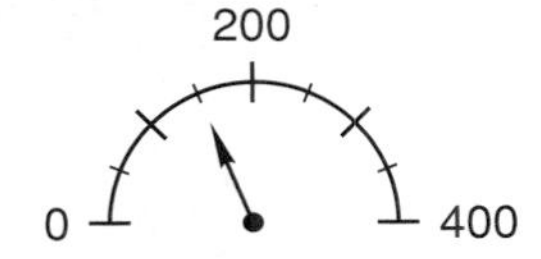

26. Draw two circles of the same size. Shade $\frac{1}{4}$ of one circle and $\frac{1}{3}$ of the other circle. Then compare these fractions:
(39)

$$\frac{1}{4} \bigcirc \frac{1}{3}$$

27. Which of these fractions is greater than one half?
(23)

A. $\frac{5}{12}$ B. $\frac{3}{5}$ C. $\frac{7}{14}$ D. $\frac{10}{21}$

28. A full turn is 360 degrees. How many degrees is $\frac{1}{4}$ of a turn?
(Inv. 2, 34)

29. How many years were there from 1776 to 1789?
(35)

INVESTIGATION 4

Focus on

Measuring Angles

One way to measure an angle is with **degrees.** Here we show four angles and their measures in degrees. (Read 30° as "thirty degrees.")

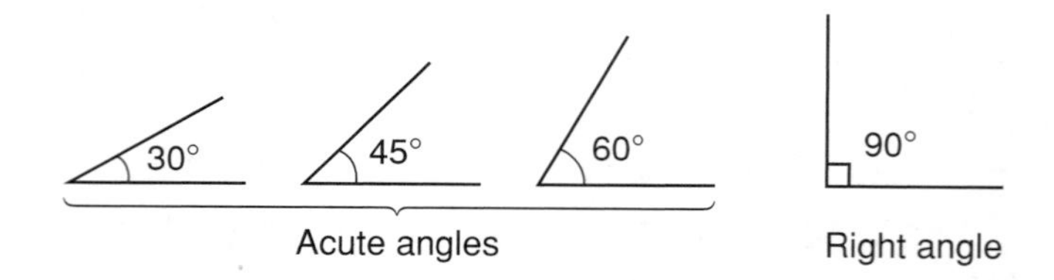

Note that a right angle measures 90°, and acute angles measure less than 90°. Obtuse angles measure more than 90° and less than a straight angle, which measures 180°.

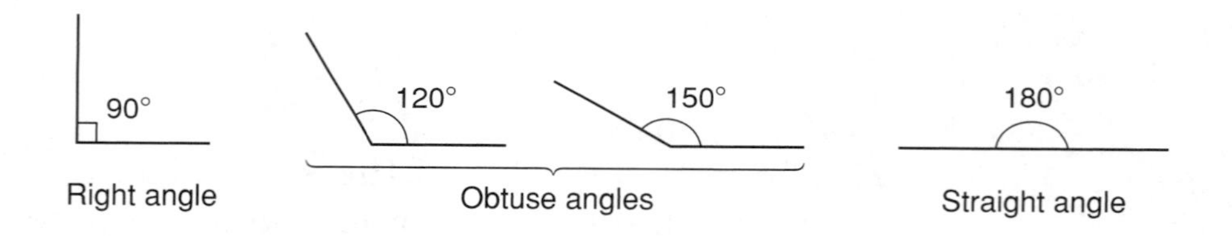

A full circle contains 360°, as demonstrated in the following activity.

Activity: *Angle Exercise*

1. Beginning with your arms extended forward at 0°, raise one arm to form a 90° angle.

2. Beginning with your arms extended forward at 0°, raise one arm up, around, and halfway down to form a 180° angle.

3. Beginning with your arms extended forward at 0°, move one arm up through 90°, down through 180°, and continue around to 360°.

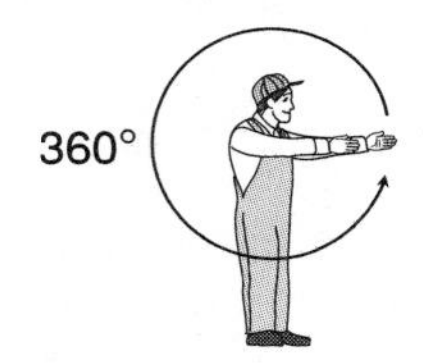

A **protractor** is a tool for measuring angles. Here is a picture of a typical protractor:

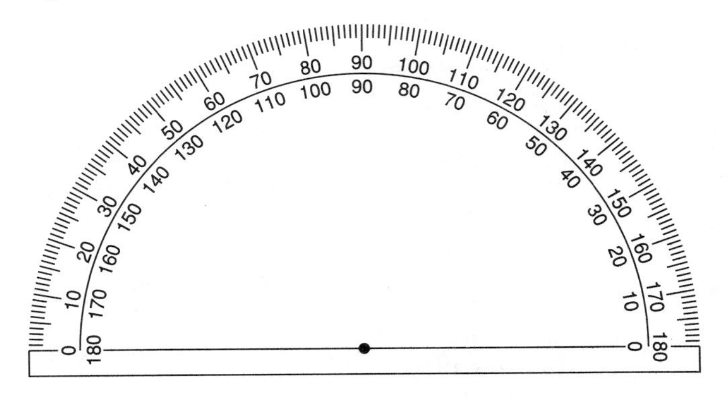

A protractor usually has two scales, one ranging from 0° to 180° from left to right, the other ranging from 0° to 180° from right to left. By paying attention to whether you are measuring an acute angle or an obtuse angle, you will know which scale to read. The angle being measured below is an acute angle, so we know that its measure is 60° and not 120°.

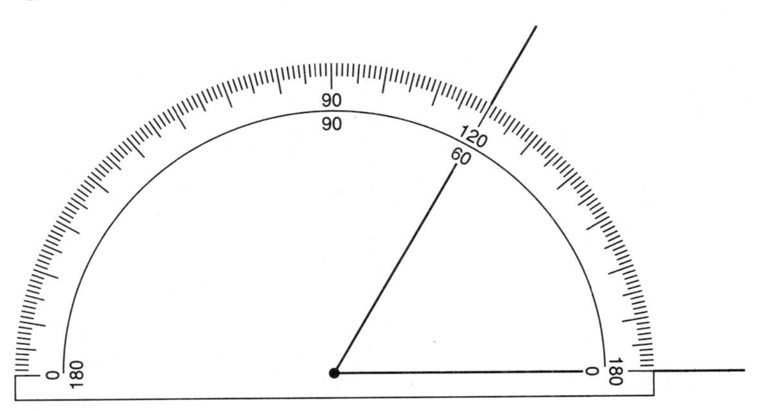

To measure an angle, we follow these three steps. (Refer to the illustration below.)

Step 1: Position the center of the protractor curve on the vertex of the angle.

Step 2: Also position one of the 0° marks on one side of the angle.

Step 3: Check that both steps 1 and 2 have been done correctly; then read the scale where the other side of the angle passes through the scale.

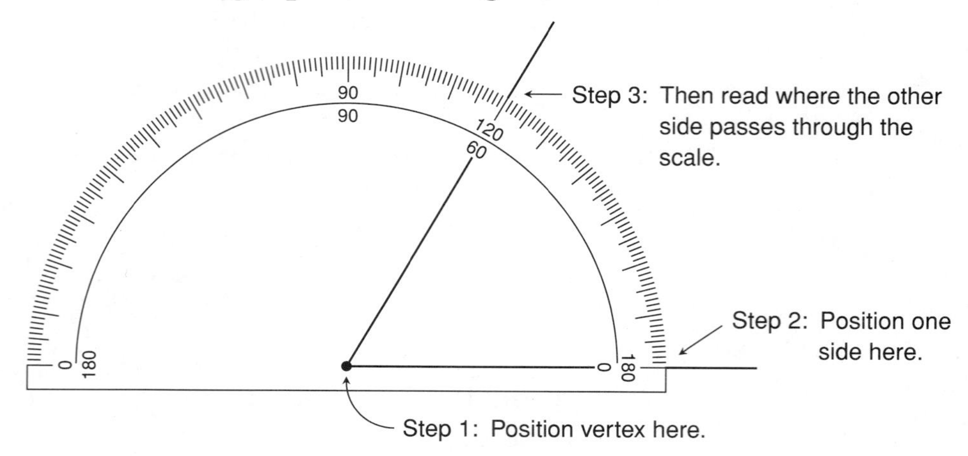

Activity: *Measuring Angles*

Materials needed for each student:

- protractor
- copy of Activity Master 16 (masters available in *Saxon Math 6/5 Assessments and Classroom Masters*)

4. With your protractor, measure each angle on Activity Master 16.

Activity: *Drawing Angles*

Materials needed for each student:

- protractor
- plain paper
- pencil

We can use a protractor to help us draw an angle of a specific size. Follow these steps:

Step 1: Using the straight edge of your protractor, draw a segment long enough to extend from the center of the protractor through the scale.

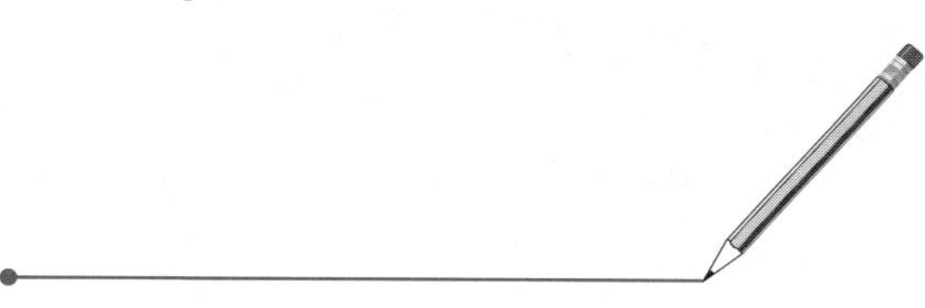

Step 2: Position the protractor so that the center is over one endpoint of the segment (the intended vertex) and the segment passes through a 0° mark.

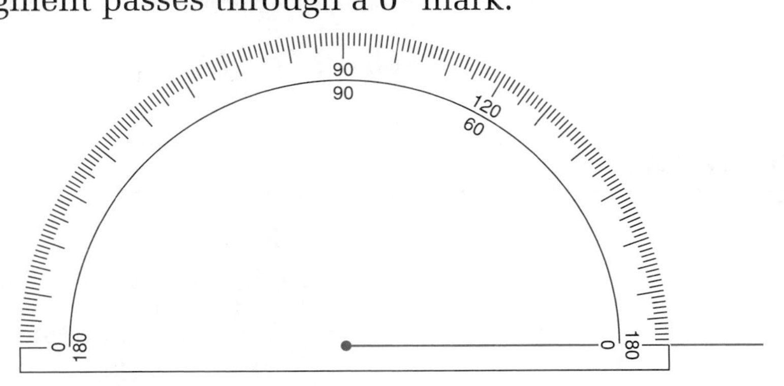

Step 3: Find the number on the protractor that matches the size of the angle you wish to draw. (Be sure you are reading from the correct scale.) Make a dot on your paper even with the scale mark on the protractor. (We have shown a mark for a 60° angle.)

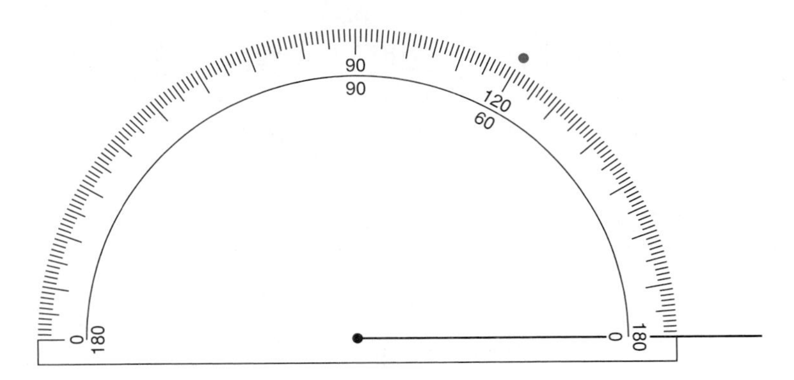

Step 4: Remove the protractor and draw the remaining side of the angle from the vertex through the dot.

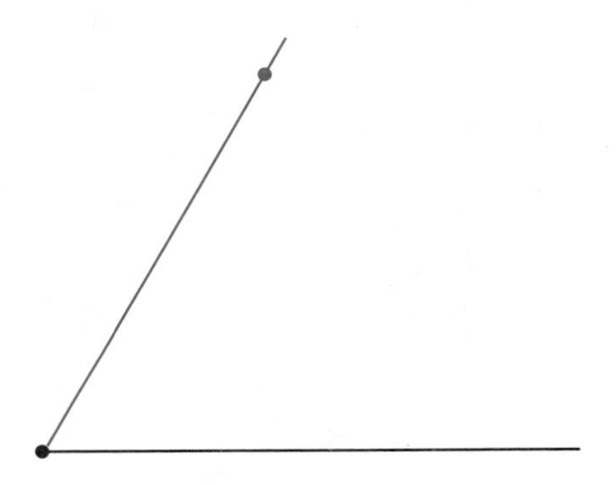

Follow steps 1–4 above to draw angles with these measures:

5. 30° **6.** 90° **7.** 110° **8.** 70°

LESSON

41 Adding and Subtracting Fractions with Common Denominators

WARM-UP

Facts Practice: 64 Multiplication Facts (Test F)

Mental Math: Name the relationship: Javier's mother's sister's son is Javier's ______.
How many is $\frac{1}{2}$ dozen? ... 2 dozen? ... $2\frac{1}{2}$ dozen?

a. Round 380 to the nearest hundred. **b.** 860 − 240
c. 24 × 7 **d.** 8 × 800 **e.** 10 × 25
f. $\frac{1}{2}$ of 15 **g.** 50% of 10¢ **h.** 10% of 10¢
i. 6 × 7, − 2, ÷ 5, + 1, ÷ 3, − 3

Problem Solving:

Copy this division problem and fill in the missing digits: $4\overline{)\_\_}$ with quotient 24

NEW CONCEPT

We may use fraction manipulatives to help us add and subtract fractions.

Example 1 Use your fraction manipulatives to illustrate this addition. Then write a number sentence for the addition.

$$\frac{2}{4} + \frac{1}{4}$$

Solution Using the manipulatives, we form the fractions $\frac{2}{4}$ and $\frac{1}{4}$.

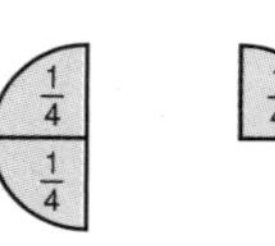

$\frac{2}{4}$ $\frac{1}{4}$

To add the two fractions, we combine them. We see that $\frac{2}{4}$ plus $\frac{1}{4}$ makes $\frac{3}{4}$.

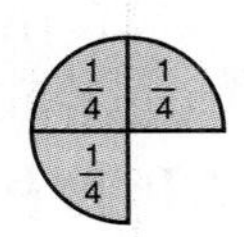

$$\mathbf{\frac{2}{4} + \frac{1}{4} = \frac{3}{4}}$$

Notice that the denominators of the fractions we added, $\frac{2}{4}$ and $\frac{1}{4}$, are the same. Fractions with the same denominators are said to have **common denominators.** When fractions have common denominators, we can add or subtract the fractions by simply adding or subtracting the numerators. We do not add or subtract the denominators.

$$\frac{2}{4} + \frac{1}{4} = \frac{3}{4}$$

Add the numerators.

Leave the denominators unchanged.

Example 2 Use your fraction manipulatives to illustrate this subtraction. Then write a number sentence for the subtraction.

$$\frac{7}{10} - \frac{4}{10}$$

Solution We form the fraction $\frac{7}{10}$.

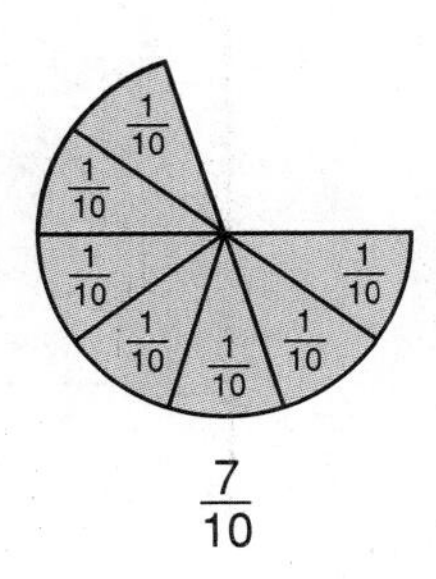

$\frac{7}{10}$

Then we remove $\frac{4}{10}$. We see that $\frac{3}{10}$ remains.

Start with $\frac{7}{10}$. − Remove $\frac{4}{10}$. = $\frac{3}{10}$ remains.

$$\mathbf{\frac{7}{10} - \frac{4}{10} = \frac{3}{10}}$$

Example 3 Add: $1\frac{1}{4} + 1\frac{2}{4}$

Solution To add mixed numbers, we add whole numbers to whole numbers and fractions to fractions. The whole numbers in this addition are 1 and 1. We add them and get 2. The fractions are $\frac{1}{4}$ and $\frac{2}{4}$. We add them and get $\frac{3}{4}$.

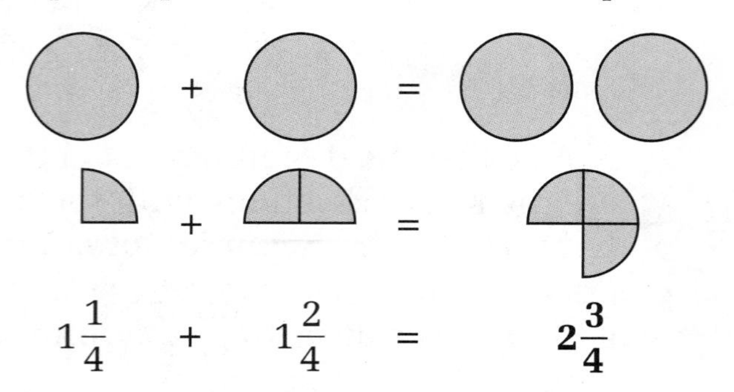

Example 4 Subtract: $2\frac{1}{2} - 1\frac{1}{2}$

Solution We start with $2\frac{1}{2}$.

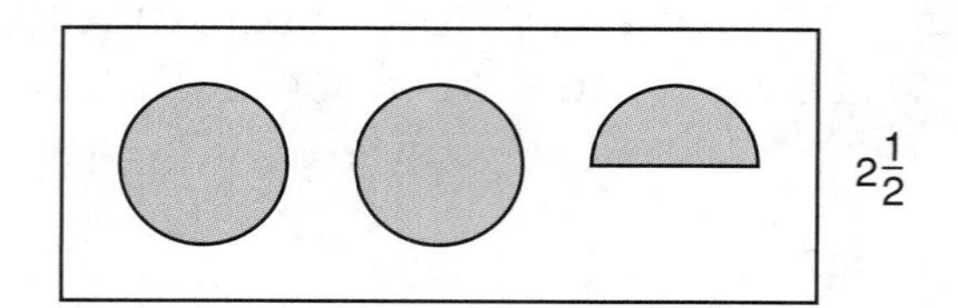

We take away $1\frac{1}{2}$. What is left is 1.

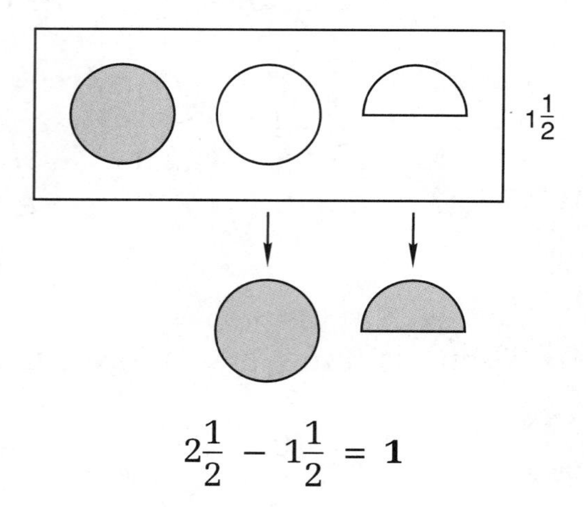

Example 5 Add: $2\frac{1}{2} + 1\frac{1}{2}$

Solution The sum is $3\frac{2}{2}$. The fraction $\frac{2}{2}$ is two halves, which is one whole. So $3\frac{2}{2}$ is 3 + 1, which is 4.

$$\begin{array}{r} 2\frac{1}{2} \\ +\ 1\frac{1}{2} \\ \hline 3\frac{2}{2} \end{array} = \mathbf{4}$$

LESSON PRACTICE

Practice set Illustrate each addition or subtraction problem. Also, write a number sentence for each problem.

a. $\frac{1}{10} + \frac{2}{10}$

b. $\frac{3}{4} - \frac{2}{4}$

c. $1\frac{1}{2} + 1\frac{1}{2}$

d. $3\frac{4}{10} - 1\frac{1}{10}$

MIXED PRACTICE

Problem set

1. (31, 32) Draw a pair of horizontal parallel line segments. Make the upper segment longer than the lower segment. Connect the ends of the segments to form a quadrilateral.

2. (30) If a birthday cake is cut into 10 equal pieces, then each piece is what fraction of the whole cake? Each piece is what percent of the whole cake?

3. (28) What year was two centuries after 1492?

For problems 4–6, write an equation and find the answer.

4. (35) The population of Colville was 340 less than the population of Sonora. The population of Colville was 4360. What was the population of Sonora?

5. (21) Jayne bought vegetable plants for her garden. She bought three flats of plants. There were six plants in each flat. How many plants did Jayne buy?

6. (16) Scotty is reading a 243-page book. If he has read through page 167, then how many pages does he have yet to read?

7. (41) $3\frac{3}{10} - 1\frac{2}{10}$

8. (41) $\frac{5}{10} + \frac{4}{10}$

9. (41) $\frac{1}{2} - \frac{1}{2}$

10. (41) $2\frac{1}{4} + 3\frac{2}{4}$

11. (38) To what mixed number is the arrow pointing?

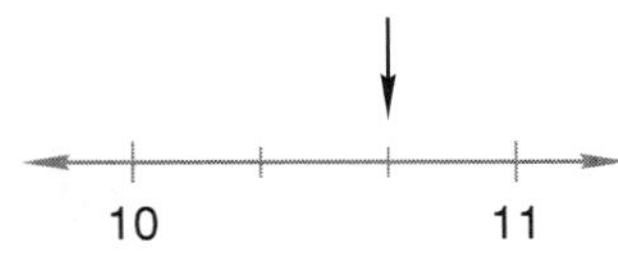

12. 3784 + 2693 + 429 + 97 + 856 + 907
(6)

13. 3106 − 528
(9)

14. \$80.00 − \$77.56
(13)

15. 804 × 700
(29)

16. 60 × 43 × 8
(18, 29)

17. $4w = 4008$
(26, 34)

18. 4228 ÷ 7
(34)

19. 9635 ÷ 8
(34)

20. \$7.98 ÷ 6
(26)

21. \$10 − (\$4.56 + \$3 + \$1.29)
(13, 24)

22. Round 98 to the nearest ten.
(33)

23. Draw an obtuse triangle.
(36)

24. One fifth of the 30 students in the class were left-handed. How many of the students were left-handed?
(Inv. 3)

25. It is evening. What time will it be in 30 minutes?
(28)

26. Four friends entered a nine-mile relay race. Each person ran one fourth of the distance. How many miles did each person run?
(40)

27. Draw two circles of the same size. Shade $\frac{1}{2}$ of one circle and $\frac{2}{3}$ of the other circle. Then compare these fractions:
(39)

$$\frac{1}{2} \bigcirc \frac{2}{3}$$

28. Which of these angles appears to be a 90° angle?
(Inv. 4)

A. B.  C.

29. Compare: $\frac{1}{4}$ of 100 ◯ 100 ÷ 4
(Inv. 2)

LESSON

42 Short Division • Divisibility by 3, 6, and 9

WARM-UP

Facts Practice: 90 Division Facts (Test D or E)

Mental Math: Name the relationship: Leticia is her father's mother's ______.
Count by 3's from 3 to 36. Count by 6's from 6 to 36.

a. Round 42 to the nearest ten.
b. 160 + 240
c. $\frac{1}{2} + \frac{1}{2}$
d. $\frac{1}{2} - \frac{1}{2}$
e. 8 × 24
f. 50% of $100.00
g. 25% of $100.00
h. 10 × 15
i. $\frac{1}{2}$ of 25
j. 8 × 8, − 1, ÷ 9, + 1, ÷ 8

Problem Solving:

One phone number is 987-6543. How many different phone numbers in the same area code could begin with 987- and end with any arrangement of all the digits 3, 4, 5, and 6?

NEW CONCEPTS

Short division

We have learned a division algorithm in which we follow four steps: divide, multiply, subtract, and bring down. This algorithm is sometimes called "long division." In this lesson we will practice a shortened form of this algorithm. The shortened form is sometimes called *short division.*

When we do short division, we follow the four steps of long division, but we do not write down every number. Instead we keep track of some numbers "in our head." We will show this by doing the same division problem both ways.

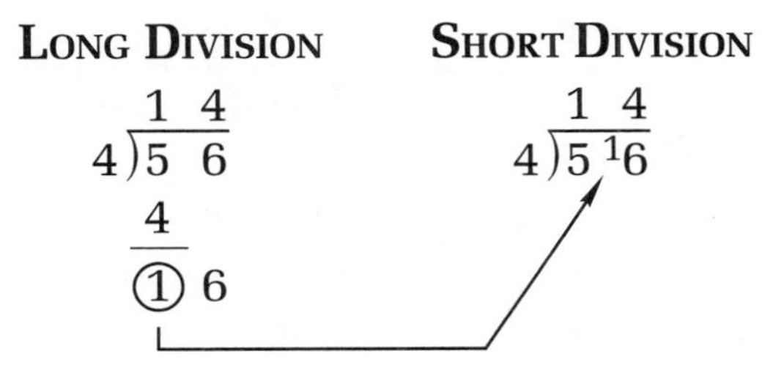

We begin both divisions by finding 4)5. We write "1" above the 5 and then we multiply. In short division we keep the multiplication answer in our head. Then we subtract. **In short**

division we write the subtraction answer in front of the next digit. Here we write a small "1" in front of the 6 to make 16. In short division we do not bring down this digit; instead we "bring up" the subtraction answer. Now we find $4\overline{)\,^{1}6}$ and write "4" above the 6. We multiply and subtract in our head and find that there is no remainder.

Example 1 Divide: $5\overline{)842}$

Solution We will use short division to find the answer. First we divide and write "1" above the 8. Then we multiply and subtract in our head to get 3. We bring up the 3 and write it in front of the next digit. Next we find $5\overline{)\,^{3}4}$. We continue to divide, multiply, subtract, and bring up. We bring up the last subtraction answer as the remainder. The answer is **168 R 2.**

$$\begin{array}{r} 1\ 6\ 8\ \text{R}\ 2 \\ 5\overline{)8\ ^{3}4\ ^{4}2} \end{array}$$

Divisibility by 3, 6, and 9

Divisibility is the "ability" of a number to be divided by another number without a remainder. In Lesson 22 we found that whole numbers ending with an even digit are divisible by 2. Whole numbers ending in zero are divisible by 2, 5, and 10, and whole numbers ending in 5 are divisible by 5.

In this lesson we will learn to identify whole numbers divisible by 3, 6, and 9. We need to look at all of the digits of a whole number to decide whether the number is divisible by 3, 6, or 9. In fact, we add the digits of the number. If the sum of the digits is divisible by 3, then the number is divisible by 3. If the sum of the digits is divisible by 9, the number is divisible by 9. Let's consider the number 438.

$$438 \longrightarrow 4 + 3 + 8 = 15$$

The sum of the digits is 15. Fifteen can be divided by 3 without a remainder, so 438 is divisible by 3. However, 15 cannot be divided by 9 without a remainder, so 438 is not divisible by 9.

A number is divisible by 6 if the number is even (divisible by 2) and divisible by 3. Since 438 is even and divisible by 3, it is also divisible by 6. Below we divide 438 by 3, 6, and 9 to show that 438 is divisible by 3 and 6 but not by 9.

$$\begin{array}{r} 146 \\ 3\overline{)438} \\ \underline{3} \\ 13 \\ \underline{12} \\ 18 \\ \underline{18} \\ 0 \end{array} \qquad \begin{array}{r} 73 \\ 6\overline{)438} \\ \underline{42} \\ 18 \\ \underline{18} \\ 0 \end{array} \qquad \begin{array}{r} 48 \text{ R } 6 \\ 9\overline{)438}\phantom{\text{ R } 6} \\ \underline{36}\phantom{0\text{ R } 6} \\ 78\phantom{\text{ R } 6} \\ \underline{72}\phantom{\text{ R } 6} \\ 6\phantom{\text{ R } 6} \end{array}$$

divisible by 3 divisible by 6 not divisible by 9

The following table summarizes the divisibility rules for 3, 6, and 9:

Divisibility Tests for 3, 6, and 9

A number is divisible by ...	
3	if the sum of its digits is divisible by 3.
6	if the number is divisible by 2 and 3.
9	if the sum of its digits is divisible by 9.

Example 2 Which of these numbers is divisible by 3 and by 6 and by 9?

A. 456 B. 567 C. 576

Solution We add the digits of each number and find whether the sums are divisible by 3 and by 9. We note that 456 and 576 are even.

456 ⟶ 4 + 5 + 6 = 15 divisible by 3 and 6
567 ⟶ 5 + 6 + 7 = 18 divisible by 3 and 9
576 ⟶ 5 + 7 + 6 = 18 divisible by 3, 6, and 9

Only choice **C. 576** is divisible by 3, by 6, and by 9.

LESSON PRACTICE

Practice set Divide using short division:

a. $3\overline{)435}$ **b.** $6\overline{)534}$ **c.** $9\overline{)567}$

d. $4\overline{)500}$ **e.** $7\overline{)800}$ **f.** $10\overline{)836}$

g. $5\overline{)600}$ **h.** $3\overline{)616}$ **i.** $6\overline{)858}$

For problems **j–o,** decide whether the number is divisible by 3, by 6, by 9, or by none of these numbers.

j. 1350 **k.** 4371 **l.** 1374

m. 436 **n.** 468 **o.** 765

MIXED PRACTICE

Problem set

1. (32) Draw a pentagon.

For problems 2–4, write an equation and find the answer.

2. (21) A rattlesnake's rattle shakes about 50 times each second. At that rate, how many times would it shake in 1 minute?

3. (35) Jim weighed 98 pounds before dinner and 101 pounds after dinner. How many pounds did Jim gain during dinner?

4. (35) Hiroshi stepped on the scale and smiled. He had lost 27 pounds since he began dieting and exercising. If Hiroshi weighs 194 pounds, how much did he weigh before starting to diet and exercise?

5. (40) Sarah cut a 21-foot long ribbon into four equal lengths. How many feet long was each length of ribbon?

6. (39) Draw two rectangles of the same size. Shade $\frac{3}{4}$ of one rectangle and $\frac{3}{5}$ of the other rectangle. Then compare these fractions:

$$\frac{3}{4} \bigcirc \frac{3}{5}$$

7. (41) $\frac{3}{10} + \frac{4}{10}$

8. (41) $1\frac{1}{3} + 2\frac{1}{3}$

9. (41) $\frac{7}{10} - \frac{4}{10}$

10. (41) $5\frac{1}{4} - 2\frac{1}{4}$

11. (40) Use a mixed number to name the number of shaded circles shown below.

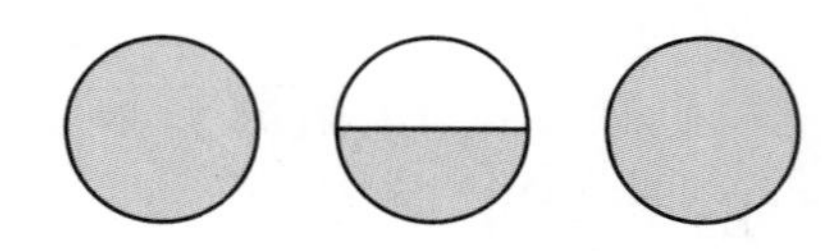

12. Round 151 to the nearest hundred.
(33)

13. To what fraction is the arrow pointing?
(38)

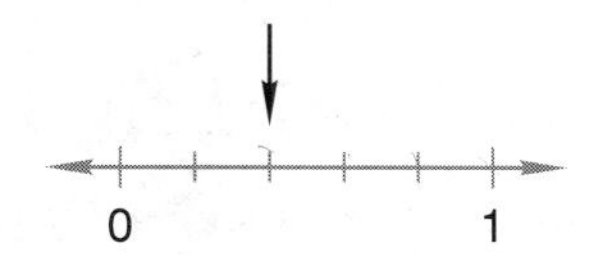

14. Compare: $\frac{1}{3}$ of 30 ◯ $\frac{1}{5}$ of 50
(Inv. 3)

15. Two fourths of a circle is what percent of a circle?
(30, 39)

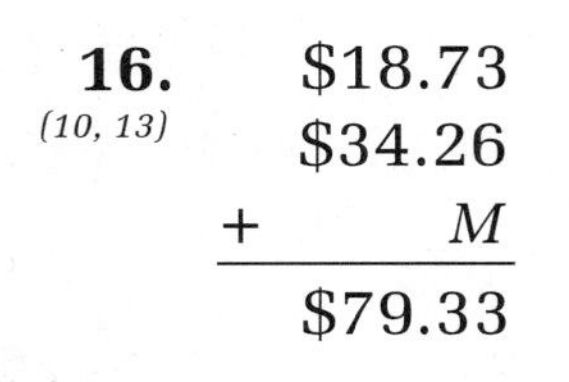

16. (10, 13)

$$\begin{array}{r} \$18.73 \\ \$34.26 \\ + \quad M \\ \hline \$79.33 \end{array}$$

17. (14)

$$\begin{array}{r} 6010 \\ - \quad R \\ \hline 543 \end{array}$$

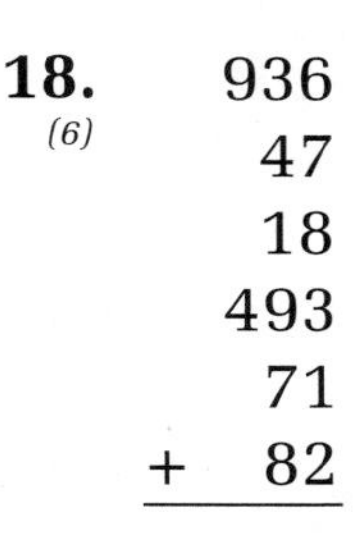

18. (6)

$$\begin{array}{r} 936 \\ 47 \\ 18 \\ 493 \\ 71 \\ + \quad 82 \\ \hline \end{array}$$

19. (29)

$$\begin{array}{r} 346 \\ \times \quad 80 \\ \hline \end{array}$$

20. (29)

$$\begin{array}{r} \$7.25 \\ \times \quad 90 \\ \hline \end{array}$$

21. (29)

$$\begin{array}{r} 670 \\ \times \quad 700 \\ \hline \end{array}$$

Divide using short division:

22. $4\overline{)1736}$ (42)

23. $8\overline{)\$17.60}$ (42)

24. $3\overline{)100}$ (42)

25. Which word names an angle that is smaller than a 90° angle?
(32, Inv. 4)

A. acute B. right C. obtuse

26. Which of these angles appears to be a 60° angle?
(Inv. 4)

A. B. C. D.

27. Which of these numbers is divisible by 3 and by 6 and by 9?
(42)

A. 369 B. 246 C. 468

28. $3 + (4 + 5) = (3 + 4) + M$
(24)

29. Two twelfths of the names of the months begin with the letter A, and three twelfths begin with the letter J. What fraction of the names of the months begin with either A or J?
(41)

LESSON

43 Writing Quotients as Mixed Numbers, Part 2 • Adding and Subtracting Whole Numbers, Fractions, and Mixed Numbers

WARM-UP

Facts Practice: 90 Division Facts (Test D or E)

Mental Math: Name the relationship: Simon's brother's son is Simon's ______.
What coin is 10% of a dollar? ... 25% of a dollar?

a. Round 162 to the nearest ten. **b.** 560 − 60
c. $\frac{2}{3} + \frac{1}{3}$ **d.** $\frac{2}{3} - \frac{1}{3}$ **e.** 8 × 35
f. 10 × 24 **g.** $\frac{1}{2}$ of 9 **h.** 25% of 16
i. 2 × 25, − 1, ÷ 7, + 1, ÷ 2, × 5, + 1, ÷ 3
j. 50% of 50

Problem Solving:

In the game tic-tac-toe, the goal is to get three X's or three O's in a row. How many ways are there to get three in a row?

X	O	X
X	O	O
X	X	O

NEW CONCEPTS

Writing quotients as mixed numbers, part 2

In Lesson 40 we studied some story problems that had mixed-number answers. We found that the remainder in a division problem can be divided to result in a fraction. In this lesson we will continue our study.

Example 1 The local pizzeria will donate 14 pizzas to the sixth-grade picnic. How many pizzas will there be for each of the three classes of sixth graders?

Solution Fourteen pizzas will need to be divided into three equal groups, so we divide 14 by 3.

$$\begin{array}{r} 4 \\ 3\overline{)14} \\ \underline{12} \\ 2 \end{array}$$

4 ← There are four whole pizzas for each class.

2 ← Two pizzas remain to be divided.

Four pizzas for each class is 12 pizzas. The remainder shows us there are still 2 pizzas to divide among the three classes. If each of the remaining pizzas is divided into 3 parts, there will be 6 parts to share. Each class can have 2 of the parts.

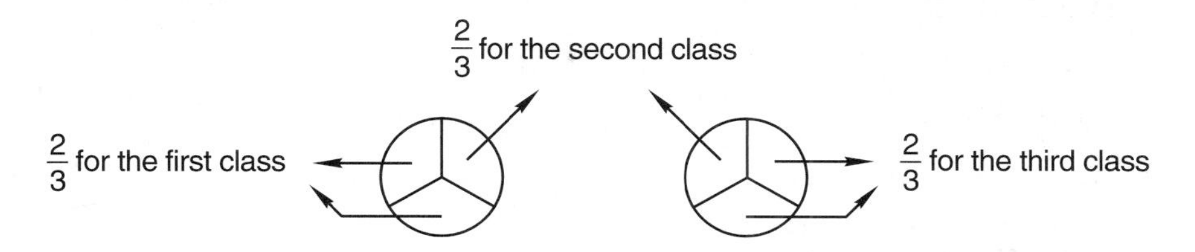

For each class there will be $\mathbf{4\frac{2}{3}}$ **pizzas.** Notice that the 2 in $\frac{2}{3}$ is the remainder, and the 3 in $\frac{2}{3}$ is the divisor.

$$\begin{array}{r} 4\frac{2}{3} \\ 3\overline{)14} \\ \underline{12} \\ 2 \end{array}$$

$4\frac{2}{3}$ ← The remainder is 2. We are dividing by 3.

We are dividing by 3. → $3\overline{)14}$

2 ← remainder

Example 2 A whole circle is 100% of the circle. If a circle is divided into thirds, then each third is what percent of the whole circle?

Solution We divide 100% by 3 to find the percent for each third.

$$\begin{array}{r} 33 \\ 3\overline{)100} \\ \underline{9} \\ 10 \\ \underline{9} \\ 1 \end{array}$$

1 ← 1% remains to be divided.

If each third of the circle were 33%, then the total would be 99%. However, the total needs to be 100%. Therefore, we need to divide the remaining 1% by 3. One divided by three is $\frac{1}{3}$. We write "$\frac{1}{3}$" after the 33. Each third of the circle is $\mathbf{33\frac{1}{3}\%}$ of the whole circle.

$$\begin{array}{r} 33\frac{1}{3} \\ 3\overline{)100} \\ \underline{9} \\ 10 \\ \underline{9} \\ 1 \end{array}$$

Adding and subtracting whole numbers, fractions, and mixed numbers

We have studied whole numbers, fractions, and mixed numbers. When adding these numbers, we add whole numbers to whole numbers and fractions to fractions. When subtracting, we subtract whole numbers from whole numbers and fractions from fractions. Remember that order matters in subtraction.

Example 3 Add: $5 + 1\frac{1}{2}$

Solution We add whole numbers to whole numbers and fractions to fractions. The sum of the whole numbers 5 and 1 is 6. There is no fraction to add to $\frac{1}{2}$. So 5 plus $1\frac{1}{2}$ is $\mathbf{6\frac{1}{2}}$.

$$\begin{array}{r} 5 \\ +\ 1\frac{1}{2} \\ \hline 6\frac{1}{2} \end{array}$$

Example 4 Add: $1 + \frac{1}{2}$

Solution We add whole numbers to whole numbers and fractions to fractions. There is no fraction to add to $\frac{1}{2}$ and no whole number to add to 1. We write the whole number and fraction together to make the mixed number $\mathbf{1\frac{1}{2}}$.

Example 5 Subtract: $3\frac{1}{2} - \frac{1}{2}$

Solution When we subtract the fractions, we find the answer is $\frac{0}{2}$, which is zero. So subtracting $\frac{1}{2}$ from $3\frac{1}{2}$ leaves **3**.

LESSON PRACTICE

Practice set*

a. Draw a diagram to illustrate this story. Then find the answer using pencil and paper. Follow example 1 in the lesson.

Anil's Pizza Shop will donate 15 pizzas to the fifth-grade picnic. How many pizzas will there be for each of the four classes of fifth graders?

b. A whole circle is divided into sevenths. Each seventh is what percent of the whole circle?

100%

Find each sum or difference:

c. $3\frac{1}{2} + 2$

d. $3\frac{2}{4} - \frac{1}{4}$

e. $6\frac{2}{3} - 3$

f. $3\frac{2}{4} + \frac{1}{4}$

g. $2\frac{1}{2} - \frac{1}{2}$

h. $\frac{3}{4} + 2$

MIXED PRACTICE

Problem set

1. (31, 32) Draw a pair of vertical parallel line segments of the same length. Connect the ends of the segments to make a quadrilateral.

For problems 2–4, write an equation and find the answer.

2. (21) Angela poured 32 ounces of juice equally into 4 cups. How many ounces of juice were in each cup?

3. (16) A stick 100 centimeters long broke into two pieces. One of the pieces was 48 centimeters long. How long was the other piece?

4. (11) Joshua has $28.75. How much more money does he need to buy a skateboard that costs $34.18?

5. (30, 37) Draw a square. Shade all but one fourth of it. What percent of the square is not shaded?

6. (33) Round 158 to the nearest ten.

7. (43) $5 + 2\frac{1}{2}$

8. (41) $12\frac{1}{2} + 12\frac{1}{2}$

9. (43) $1 + \frac{1}{3}$

10. (43) $3\frac{1}{2} - \frac{1}{2}$

11. (41) $4\frac{3}{5} - 3\frac{1}{5}$

12. (43) $5\frac{3}{8} - 1$

13. (43) A whole circle is divided into eighths. Each eighth is what percent of the whole circle?

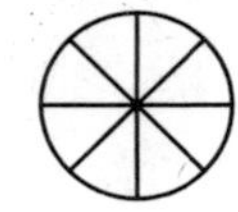

14. How many eighths equal a half?
(Inv. 3)

15. (29) $\begin{array}{r} 408 \\ \times \quad 70 \\ \hline \end{array}$

16. (29) $\begin{array}{r} \$9.67 \\ \times \quad 60 \\ \hline \end{array}$

17. (29) $\begin{array}{r} 970 \\ \times \quad 900 \\ \hline \end{array}$

18. (13) $\begin{array}{r} \$3.47 \\ \$5.23 \\ \$7.68 \\ + \ \$2.42 \\ \hline \end{array}$

19. (14) $\begin{array}{r} R \\ - \ 3977 \\ \hline 309 \end{array}$

20. (14) $\begin{array}{r} 9013 \\ - \quad W \\ \hline 3608 \end{array}$

21. (26) $7\overline{)890}$

22. (26) $6\overline{)100}$

23. (34) $4\overline{)8035}$

24. How many minutes is one tenth of an hour?
(Inv. 2, 28)

25. According to this calendar, what day of the week would February 2047 begin on?
(28)

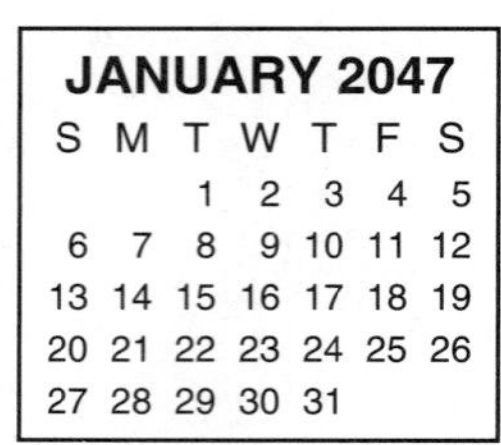

JANUARY 2047						
S	M	T	W	T	F	S
		1	2	3	4	5
6	7	8	9	10	11	12
13	14	15	16	17	18	19
20	21	22	23	24	25	26
27	28	29	30	31		

26. What mixed number names point Y on this number line?
(38)

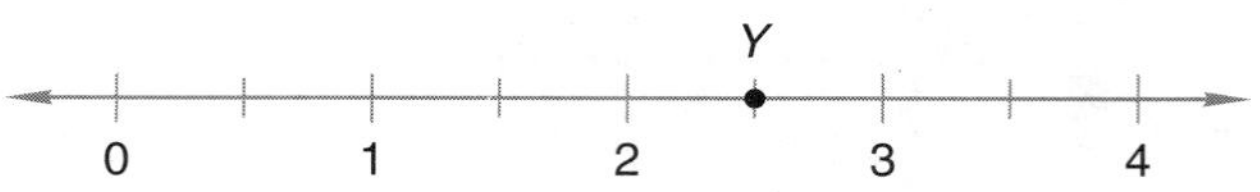

27. (a) A right angle measures how many degrees?
(23, Inv. 4)

(b) Half of a right angle is how many degrees?

(c) Use the numbers in the answers to parts (a) and (b) to write a fraction equal to $\frac{1}{2}$.

28. Which of these numbers is divisible by 6 and by 10?
(22, 42)

A. 610 B. 510 C. 410

29. What number is $\frac{1}{8}$ of 1000?
(26, Inv. 3)

LESSON

44 Measuring Lengths with a Ruler

WARM-UP

Facts Practice: 64 Multiplication Facts (Test F)

Problem Solving:

Use the information in this paragraph to complete the statements that follow. Drawing a diagram may help you with the problem.

Alejandro's mother and father are Blanca and Cesar. Alejandro's mother's parents are Dolores and Ernesto. Blanca's brother is Fidel. Fidel and his spouse Gloria have a son, Hector.

a. Hector is Alejandro's ______.
b. Fidel is Alejandro's ______.
c. Hector is Blanca's ______.
d. Cesar is Fidel's ______.
e. Blanca is Gloria's ______.
f. Dolores is Alejandro's ______.
g. Hector is Ernesto's ______.

NEW CONCEPT

You might have a ruler at your desk that has both a centimeter scale and an inch scale. We will practice using both scales in this lesson. Here we show a centimeter scale and a millimeter scale. The words *centimeter* and *millimeter* are abbreviated "cm" and "mm" respectively.

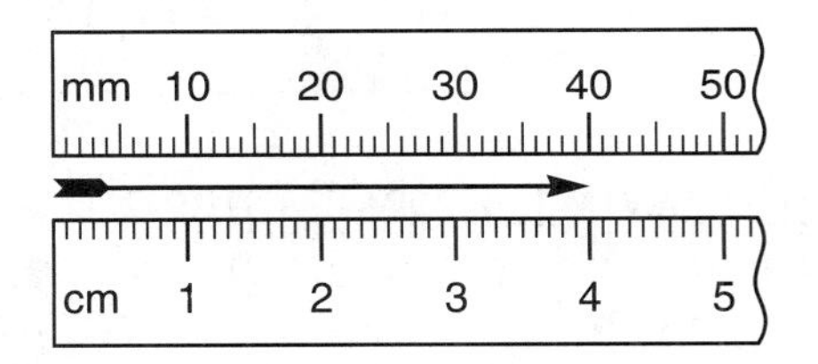

The centimeter scale is divided into segments 1 centimeter long and may be further divided into millimeters (mm). Notice that **10 millimeters equals 1 centimeter.** The arrow is 4 centimeters long. It is also 40 millimeters long.

Example 1 The distance across a nickel is about 2 centimeters. Two centimeters is how many millimeters?

Solution We remember that 1 centimeter equals 10 millimeters, so 2 centimeters equals **20 millimeters.**

Example 2 What is the length of the rectangle below?

Solution There are tick marks on the scale to mark each millimeter. Notice that the tick marks for every fifth and tenth millimeter are lengthened to make the scale easier to read. We see that the rectangle's length is 20 millimeters plus 5 more millimeters, which is **25 mm.**

Centimeters and millimeters are units of length in the **International System of Units,** sometimes called the **metric system.** The basic unit of length in the metric system is a meter. You might have a meterstick in your classroom. If you take a big step, you move about one meter. A centimeter is $\frac{1}{100}$ of a meter, and a millimeter is $\frac{1}{1000}$ of a meter. Units in the metric system are related by the number 10.

Inches, feet, yards, and miles are not units in the metric system. These units of length are part of the **U.S. Customary System.** Units in the U.S. Customary System are not related by the number 10. Instead, 12 inches equals a foot, 3 feet equals a yard, and 5280 feet (or 1760 yards) equals a mile. Inches usually are not divided into tenths; rather, they are divided into halves, fourths, eighths, sixteenths, and so on. Inches are abbreviated "in." with a period. Here we show an inch scale divided into eighths:

Below we show a magnified portion of an inch ruler. We see that $\frac{2}{8}$ of an inch equals $\frac{1}{4}$ of an inch, $\frac{4}{8}$ of an inch equals $\frac{1}{2}$ of an inch, and $\frac{6}{8}$ of an inch equals $\frac{3}{4}$ of an inch.

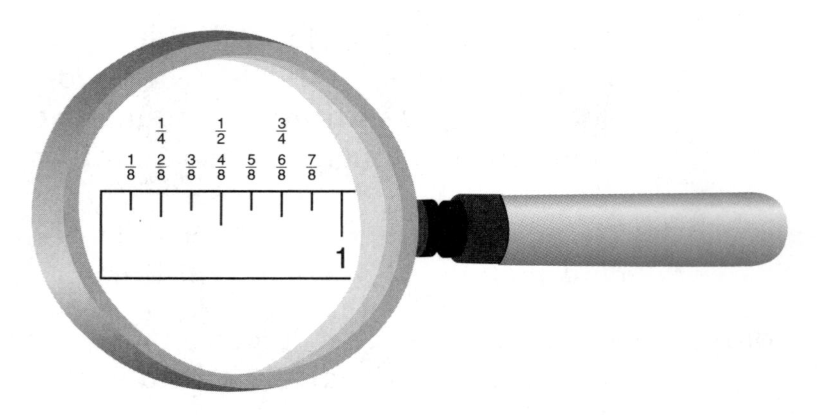

Example 3 How many inches long is this arrow?

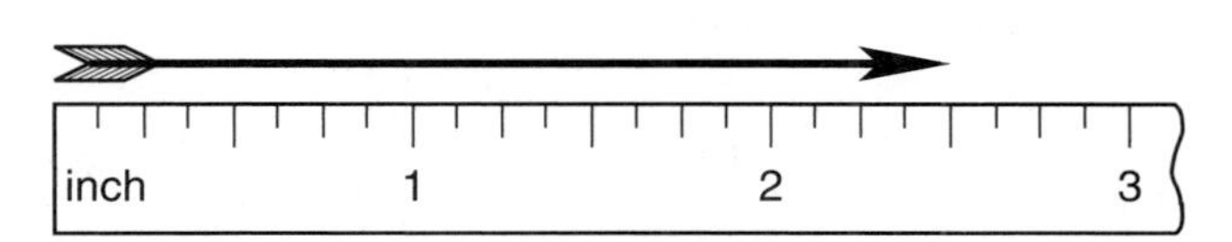

Solution The marks on the ruler divide each inch into eight smaller segments. Each small segment is one eighth of an inch long. Measuring the arrow, we see that its length is 2 full inches plus 4 small segments, or $2\frac{4}{8}$ inches. However, there are other ways to name the fraction $\frac{4}{8}$. We see that the mark at the end of the arrow is exactly halfway between 2 and 3. That mark is the two-and-one-half-inch mark. So the length of the arrow is **$2\frac{1}{2}$ inches,** or **$2\frac{1}{2}$ in.** Notice on the ruler that the half-inch marks are slightly longer than the quarter-inch marks and the eighth-inch marks.

Example 4 Is $2\frac{3}{4}$ inches closer to 2 inches or to 3 inches?

Solution Since $\frac{3}{4}$ is more than $\frac{1}{2}$, we know that $2\frac{3}{4}$ inches is closer to **3 inches.**

LESSON PRACTICE

Practice set Use a centimeter ruler to measure each segment in both centimeters and in millimeters:

a. ________________

b. ________________________________

c. ____

d. Measure the length of your math book to the nearest centimeter.

e. One centimeter is how many millimeters?

f. How many millimeters is 5 centimeters?

g. Write the abbreviations for "centimeter" and "millimeter."

h. How many millimeters long is the nail?

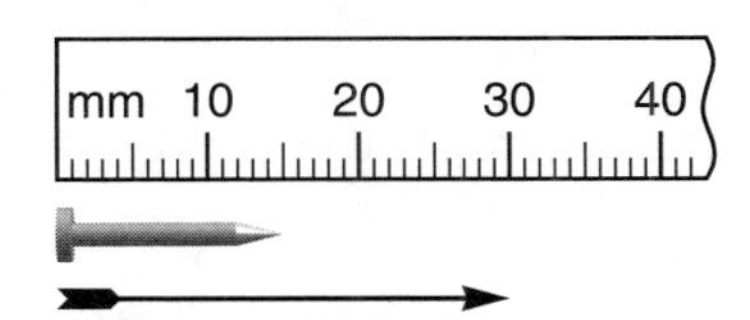

i. How many **centimeters** long is the arrow?

For problems **j–o**, name the mark on the ruler to which each arrow is pointing.

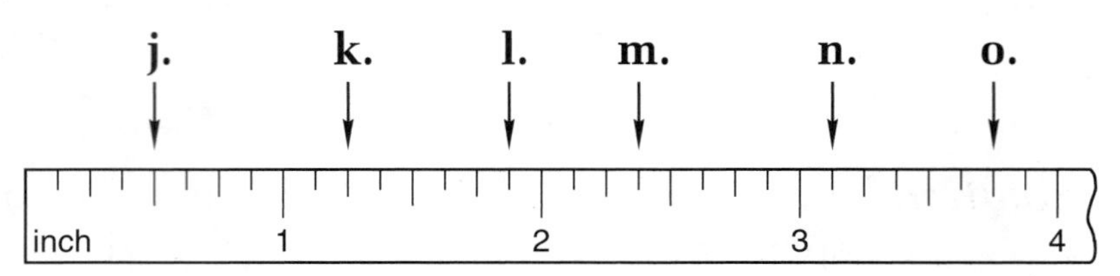

Use an inch ruler to measure each segment to the nearest eighth of an inch.

p. ________________________________

q. _________________________

r. ________________

s. Is $6\frac{1}{8}$ inches closer to 6 inches or 7 inches?

t. Round $5\frac{7}{8}$ inches to the nearest inch.

MIXED PRACTICE

Problem set

1. Draw a quadrilateral that has four right angles.
(32)

2. In her pocket Sallie has 3 pennies, 2 nickels, a dime, 3 quarters, and a half dollar. How much money is in Sallie's pocket?
(13)

For problems 3–5, write an equation and find the answer.

3. One hundred thirty-eight kindergartners climbed on three buses to go to the zoo. If the same number of children were on each bus, how many children were on each bus?
(21)

4. The distance across a nickel is about 2 centimeters. Two centimeters is how many millimeters?
(44)

5. How many years were there from 1776 to 1976?
(35)

6. Three friends want to share five oranges equally. How many oranges should each friend receive?
(40)

7. (38) What mixed number names point Z on this number line?

8. (41) $3\frac{3}{4} - 1\frac{2}{4}$

9. (43) $4\frac{1}{2} - \frac{1}{2}$

10. (43) $5\frac{1}{4} - 4$

11. (41) $33\frac{1}{3} + 33\frac{1}{3}$

12. (43) $5\frac{1}{3} + 3$

13. (43) $8\frac{3}{8} + \frac{4}{8}$

14. (44) What is the length of this rectangle?

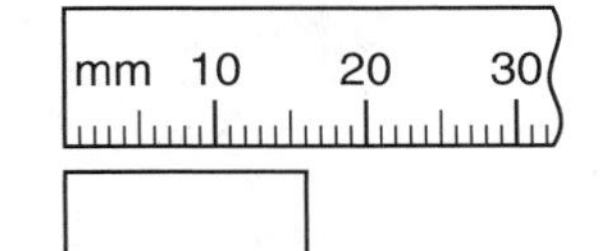

15. (6) 352 + 4287 + 593 + 7684 + 9856

16. (9) 3627 − 429

17. (24) 9104 − (2000 − 66)

18. (29) 491 × 700

19. (18, 29) 60 × 8 × 37

20. (26) $5n = 3175$

21. (26) 2964 ÷ 10

22. (Inv. 3, 37) Draw a circle. Shade all but one third of it. What percent of the circle is shaded?

23. (33) Counting by tens, the number 256 is closest to which of the following?

A. 240 B. 250 C. 260 D. 300

24. (28) It is morning. What time will it be in 30 minutes?

25. (25) List the factors of 50.

26. (Inv. 2) One foot is 12 inches. One fourth of a foot is how many inches?

27. (a) The measure of a right angle is how many degrees?
(23, Inv. 4)

(b) The measure of a straight angle is twice the measure of a right angle. A straight angle measures how many degrees?

(c) Use the numbers in the answers to parts (a) and (b) to write a fraction equal to $\frac{1}{2}$.

28. Which of these numbers is divisible by 3 and by 5?
(22, 42)

A. 305 B. 315 C. 325

29. The recipe called for $\frac{1}{3}$ cup of vegetable oil for one batch of cookies. How much vegetable oil is needed to bake two batches of cookies?
(41)

LESSON

45 Classifying Quadrilaterals

WARM-UP

Facts Practice: 90 Division Facts (Test D or E)

Mental Math: Name the relationship: Sakura is her brother's daughter's ______.
Count by 25's from 25 to 300. How many cents is 3 quarters? ... 6 quarters?

a. Round 278 to the nearest ten. **b.** $875 + 125$

c. $\frac{3}{4} + \frac{1}{4}$ **d.** $\frac{3}{4} - \frac{1}{4}$ **e.** 7×42

f. $10 \times 25¢$ **g.** 50% of 10 **h.** 10% of 50

i. 6×4, $\div 3$, $+ 2$, $\div 5$, $\times 7$, $+ 1$, $\div 3$

Problem Solving:

Copy this division problem and fill in the missing digits: $\begin{array}{r} 56 \\ 3\overline{)\_\_\_} \end{array}$

NEW CONCEPT

Recall from Lesson 32 that a quadrilateral is a polygon with four sides. Although all quadrilaterals have four sides, quadrilaterals have many different shapes. Here is an assortment of quadrilaterals:

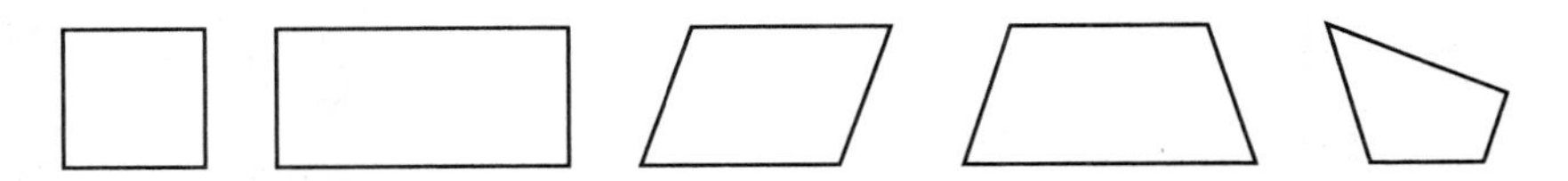

We can classify (sort) quadrilaterals into different types, such as *squares, rectangles, parallelograms,* and *trapezoids.* In this lesson we will learn ways to sort quadrilaterals, and we will practice drawing named quadrilaterals.

One way quadrilaterals are sorted is by parallel sides. Recall that parallel segments run in the same direction and remain the same distance apart. Here we show three pairs of parallel segments:

If we put the left-hand pair of segments with the center pair, we get this quadrilateral:

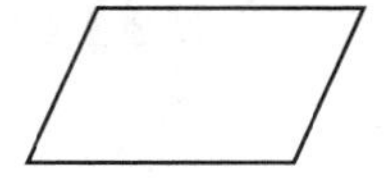

If we put the right-hand pair with the center pair, we get this quadrilateral:

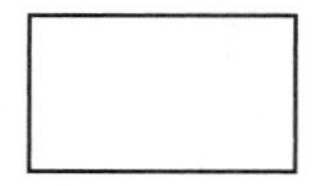

Both of these quadrilaterals have two pairs of parallel sides. Below we show some more quadrilaterals with two pairs of parallel sides. Use a finger or the eraser of your pencil to trace the pairs of parallel sides on each quadrilateral. Notice that the two segments that form a parallel pair are the same length. These quadrilaterals are called **parallelograms.**

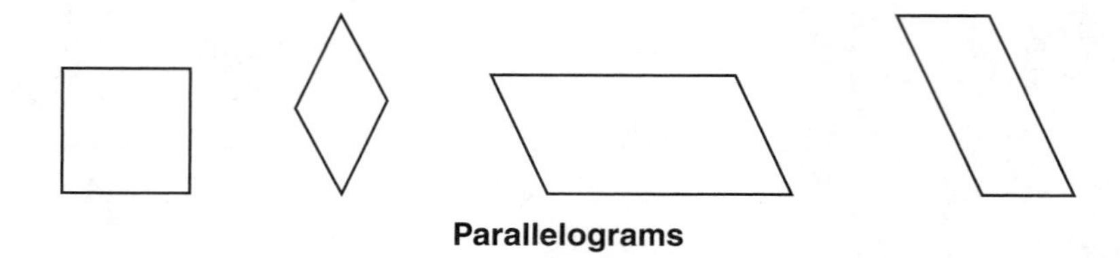

Parallelograms

Parallelograms are quadrilaterals with two pairs of parallel sides and are one classification of quadrilaterals.

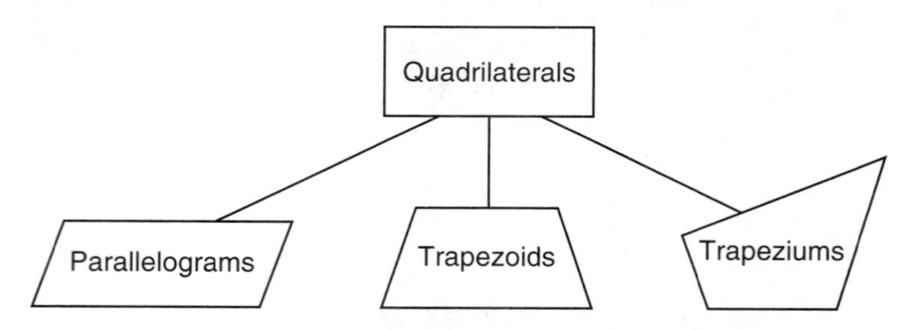

Trapezoids are another type of quadrilateral. Trapezoids have only one pair of parallel sides (the other pair of sides are not parallel). Here are some examples of trapezoids. First trace the parallel sides, and then trace the sides that are not parallel. Notice that the parallel segments in each figure are not the same length.

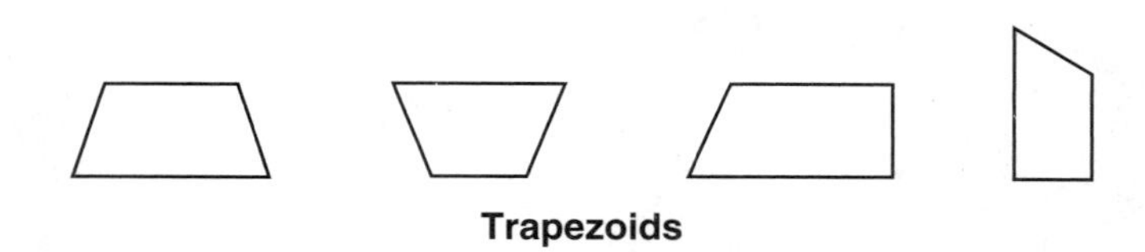

Trapezoids

Some quadrilaterals have no parallel sides. In the United States we call these shapes **trapeziums.**

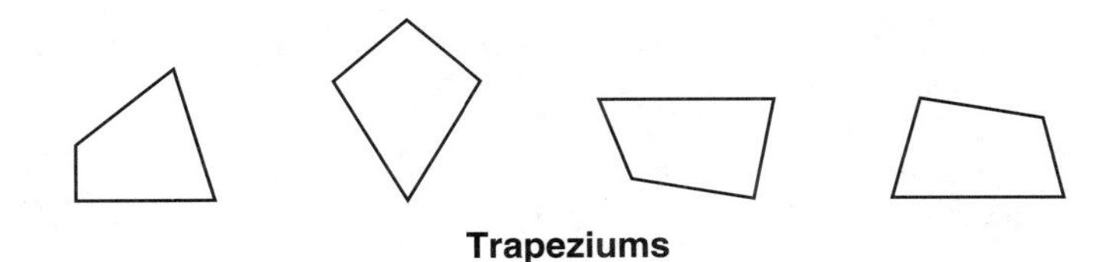

Trapeziums

Example 1 Draw an example of a parallelogram, a trapezoid, and a trapezium.

Solution To draw a parallelogram, we may begin by drawing two parallel segments of the same length.

Then we draw two more segments between the endpoints. We check these two segments to be sure they are parallel.

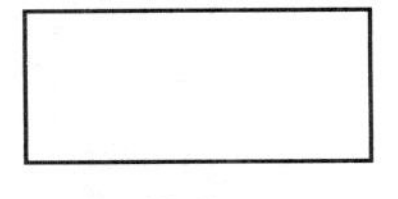

Parallelogram

This parallelogram happens to look like a rectangle. As we will see in a moment, rectangles are a special type of parallelogram.

To draw a trapezoid, we may begin by drawing two parallel segments of different lengths.

Then we draw two more segments between the endpoints.

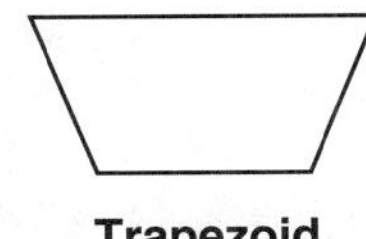

Trapezoid

To draw a trapezium, we may begin by drawing two segments that are not parallel and do not intersect.

Then we draw two segments between the endpoints. We check that these two segments are not parallel.

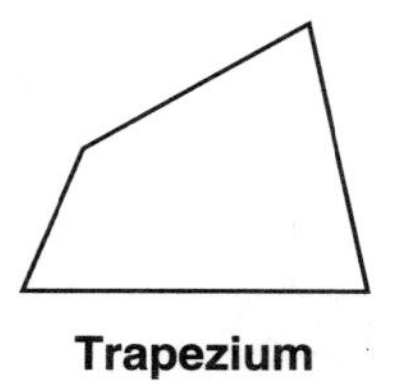

Trapezium

There are different categories of parallelograms, trapezoids, and trapeziums. In this lesson we will look at three types of parallelograms. They are *rectangles, rhombuses,* and *squares.*

Classifications of Quadrilaterals

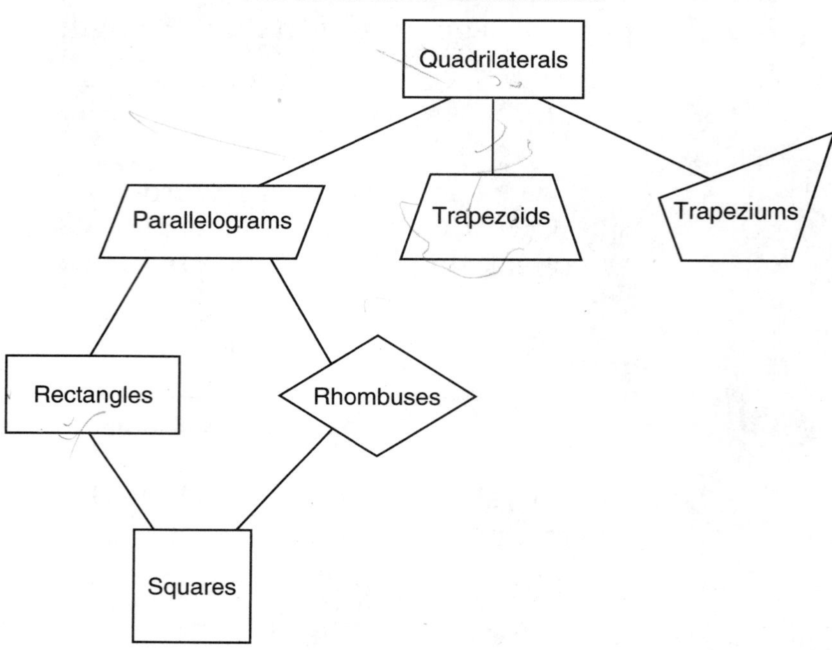

A parallelogram with four congruent angles is a **rectangle.** Each angle of a rectangle is a right angle.

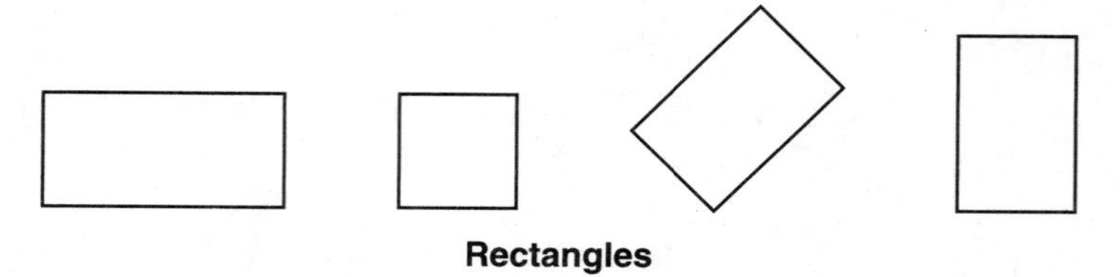

Rectangles

A parallelogram with four congruent sides (four sides of equal length) is a **rhombus.** Some people refer to a rhombus as a "diamond." A rhombus is an equilateral quadrilateral (just like a triangle with all sides of equal length is an equilateral triangle).

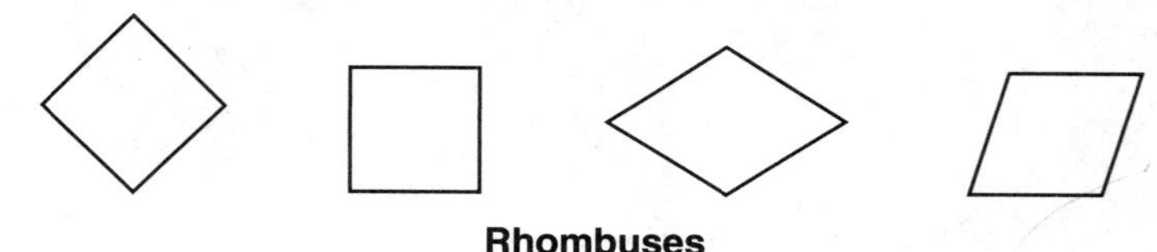

Rhombuses

Notice that a **square** is both a rectangle and a rhombus. A square has four right angles and four congruent sides.

Example 2 A square is not which of the following?

A. parallelogram B. rhombus

C. rectangle D. trapezoid

Solution A square is a quadrilateral with parallel sides of equal length that intersect at right angles. So a square is a parallelogram, a rhombus, and a rectangle. A square is not a **D. trapezoid.**

LESSON PRACTICE

Practice set The words *parallelogram, trapezoid, trapezium, rectangle, rhombus,* and *square* were used in this lesson to describe quadrilaterals. Use every word that applies to describe each quadrilateral below.

a.

b.

c. 

d.

e.

f.

g. Describe the difference between a parallelogram and a trapezoid.

h. Draw a rhombus that does not have right angles.

MIXED PRACTICE

Problem set

1. Draw a rectangle with all sides the same length.
(45)

For problems 2–4, write an equation and find the answer.

2. Julie paid $10 and got back $2.47. How much money did she spend?
(16)

3. Each of the fifty states has two U.S. senators. Altogether, how many U.S. senators are there?
(21)

4. The *Phantom of the Opera* was a hit. The theater was filled all 4 nights. If 2500 attended in all, then how many attended each night?
(21)

5. (43) There were 8 gallons of punch for the third-grade picnic. How many gallons of punch were there for each of the 3 third-grade classes?

6. (44) Ten millimeters equals how many centimeters?

7. (44) How many inches long is this arrow?

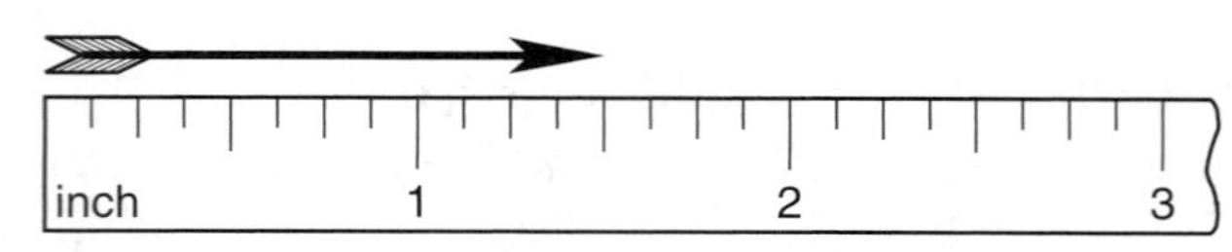

8. (41) $3\frac{1}{3} + 1\frac{1}{3}$

9. (43) $4\frac{1}{4} + 2$

10. (43) $3 + \frac{3}{4}$

11. (43) $5\frac{3}{8} - 2$

12. (41) $6\frac{3}{4} - 1\frac{2}{4}$

13. (41) $5\frac{1}{2} - 1\frac{1}{2}$

14. (13)
$$\begin{array}{r} \$87.93 \\ \$35.16 \\ \$42.97 \\ +\ \$68.74 \\ \hline \end{array}$$

15. (13)
$$\begin{array}{r} \$50.26 \\ -\ \$13.87 \\ \hline \end{array}$$

16. (14)
$$\begin{array}{r} 6109 \\ -\quad A \\ \hline 4937 \end{array}$$

17. (29)
$$\begin{array}{r} 9314 \\ \times\quad 70 \\ \hline \end{array}$$

18. (29)
$$\begin{array}{r} \$2.34 \\ \times\quad 600 \\ \hline \end{array}$$

19. (17)
$$\begin{array}{r} 4287 \\ \times\quad 5 \\ \hline \end{array}$$

20. (34) $\frac{9636}{9}$

21. (26) $8m = \$34.16$

22. (Inv. 3, 37) Draw a rectangle and shade $\frac{3}{5}$ of it. What percent of the rectangle is shaded?

23. (33) Round 256 to the nearest ten.

24. (36) Which of these triangles appears to be a right triangle?

A.

B. 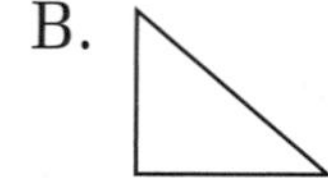

C. 

25. (27) To what number is the arrow pointing?

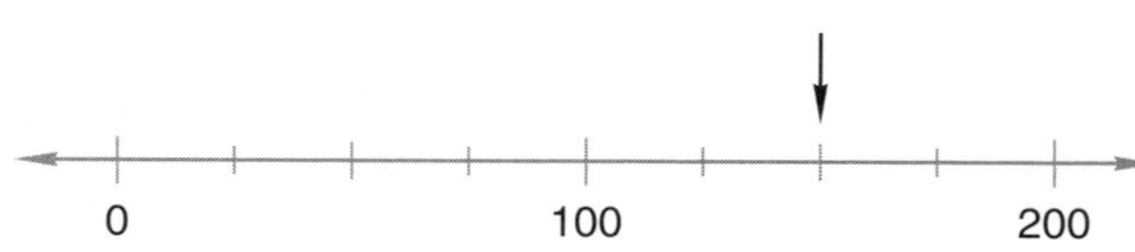

26. Use a ruler to measure the length of this segment in inches:
(44)

____

27. Show how to check this division answer. Is the answer correct?
(26)

$$8\overline{)987}\quad 123 \text{ R } 3$$

28. Draw two circles. Shade one and three fourths of them.
(40)

29. The segment from point *A* to point *B* is $1\frac{2}{8}$ inches long. The segment from point *B* to point *C* is $1\frac{5}{8}$ inches long. How long is the segment from point *A* to point *C*?
(41)

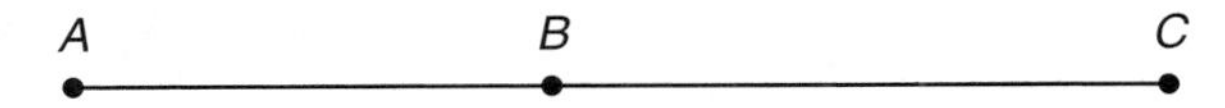

LESSON

46 Stories About a Fraction of a Group

WARM-UP

Facts Practice: 64 Multiplication Facts (Test F)

Mental Math: Name the relationship: Sal's sister's husband is Sal's ______.
What coin has a value of 50% of 50¢?

a. Round 271 to the nearest ten. **b.** 580 − 60
c. $\frac{5}{10} + \frac{2}{10}$ **d.** $\frac{5}{10} - \frac{2}{10}$ **e.** 6 × 82
f. 10 × 75¢ **g.** $\frac{1}{2}$ of 51 **h.** 25% of 24
i. 10 × 10, ÷ 2, − 1, ÷ 7, − 1, ÷ 3, − 2

Problem Solving:

In bowling, a "spare" occurs when two rolls are used to knock down all ten pins. Knocking down 3 pins on the first roll and 7 pins on the second roll is one way to bowl a spare. Make a table that lists all the possible ways to bowl a spare.

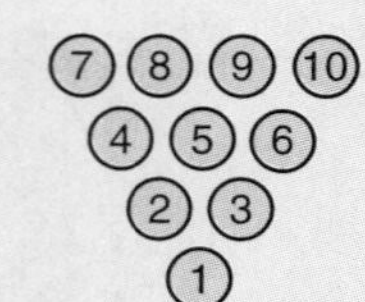

NEW CONCEPT

One type of "equal groups" story is the "fraction-of-a-group" story. The problems in these stories take two steps to answer. Here is a fraction-of-a-group story:

The teacher was pleased that $\frac{2}{5}$ of her 30 students earned an A on the test. How many students earned an A on the test?

Making a diagram for a fraction story can help us understand the problem. We draw a rectangle to stand for the whole group of 30 students. In the story the denominator of the fraction is five, so we divide the rectangle into fifths. Dividing 30 by 5, we find that there are 6 students in each fifth. We label the two fifths $\left(\frac{2}{5}\right)$ that earned an A. The rest $\left(\frac{3}{5}\right)$ did not earn an A.

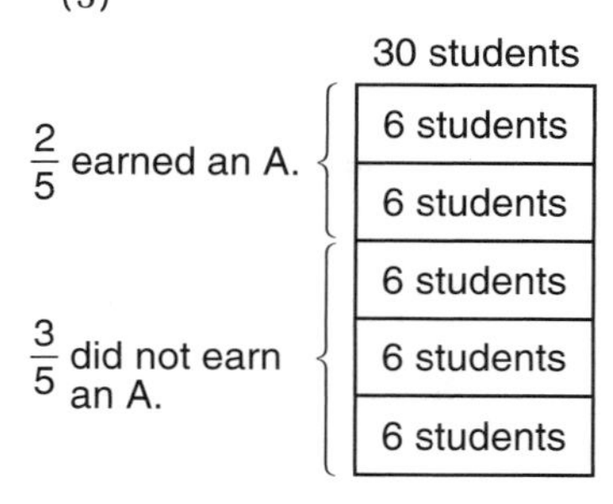

We count the number of students in two fifths of the whole and find that 12 students earned an A.

Example 1 Nia scored $\frac{2}{3}$ of her team's 36 points. How many points did she score?

Solution We draw a rectangle to stand for the team's 36 points. The denominator of the fraction in the story is 3, so we divide the rectangle into thirds. A third of 36 is 12. We write "12 points" in each third of the rectangle. Since Nia scored *two* thirds of the points, she scored 12 plus 12 points, or **24 points.**

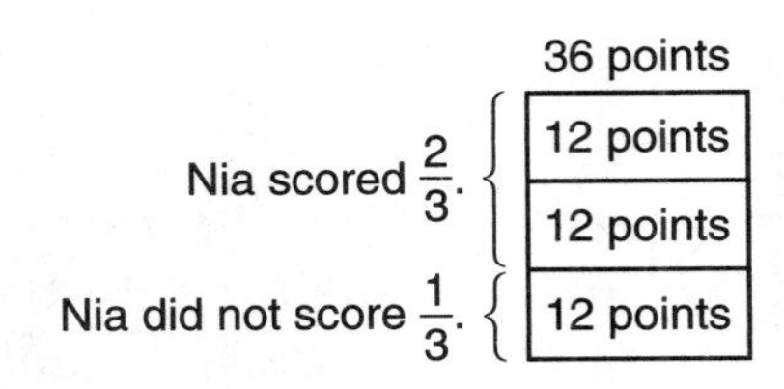

Example 2 Emily's cat ate $\frac{1}{4}$ of a dozen fish fillets. How many fish fillets were left?

Solution We draw a rectangle to stand for all 12 fish fillets. We divide the rectangle into fourths. A fourth of 12 is 3, so we write "3 fish fillets" in each of the four equal parts. Emily's cat ate 3 of the fish fillets. So **9 fish fillets** were left.

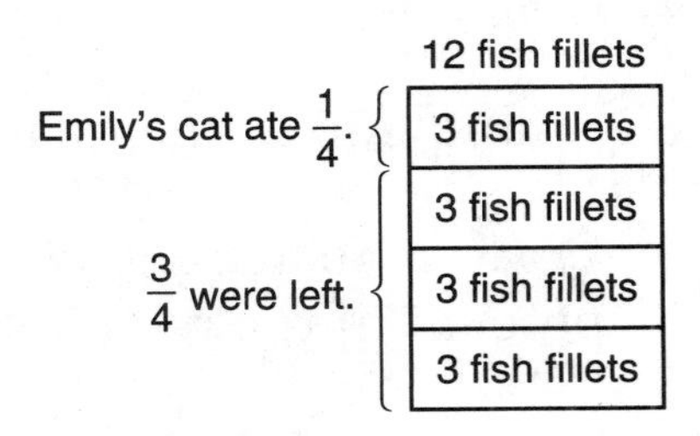

LESSON PRACTICE

Practice set Illustrate and solve each fraction story:

a. Two fifths of the 30 students in the class played in the band. How many students played in the band?

b. Susan practiced the trumpet for $\frac{3}{4}$ of an hour. For how many minutes did Susan practice the trumpet?

c. Three fifths of the 30 students were girls. How many boys were there?

MIXED PRACTICE

Problem set

1. Walking at a steady rate, Mary walked 11 miles in 3 hours. Write a mixed number that shows how many miles she walked each hour.
(43)

For problems 2–4, write an equation and find the answer.

2. The theater had 625 seats. If 139 seats were empty, how many seats were filled?
(16)

3. This line segment is 4 centimeters long. How many millimeters long is it?
(44)

____

4. Seven thousand passengers arrived on 8 ships. If each ship carried an equal number of passengers, how many passengers were on each ship?
(21)

5. What year was two centuries before 1976?
(28, 35)

6. Draw a diagram to illustrate and solve this problem:
(46)

Nia was voted "Most Valuable Player" for scoring $\frac{2}{3}$ of her team's 48 points. How many points did Nia score?

7. Compare: $\frac{1}{4}$ of 60 ◯ $\frac{1}{3}$ of 60
(Inv. 2, Inv. 3)

8. Round 256 to the nearest hundred.
(33)

9. Draw a rectangle. Shade all but two fifths of it. What percent of the rectangle is not shaded?
(Inv. 3, 37)

10. What month is 8 months after September?
(28)

11. How many inches long is this nail?
(44)

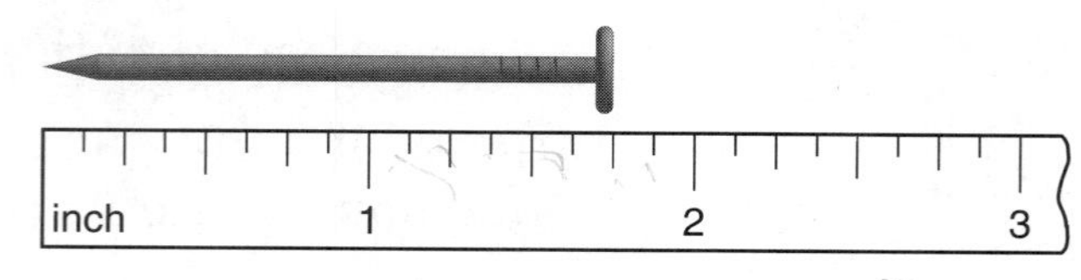

12. $3\frac{3}{7} + 2 + \frac{2}{7}$
(43)

13. $2\frac{2}{5} - 1$
(43)

14. $3\frac{2}{3} - \frac{1}{3}$
(43)

15. $6\frac{5}{12} - \left(4 + 1\frac{4}{12}\right)$
(41, 43)

16. 1396 + 727 + 854 + 4685
(6)

17. $97 + W = 512$
(10)

18. 938 × 800
(29)

19. 54 × 7 × 60
(18, 29)

20. $9n = 5445$
(26, 34)

21. 3205 ÷ 10
(34)

22. \$20 − (\$15.37 − \$12)
(13, 24)

23. 4826 + 4826 + 4826 + 4826
(13, 17)

24. A whole circle is divided into fifths. Each fifth is what percent of the whole circle?
(43)

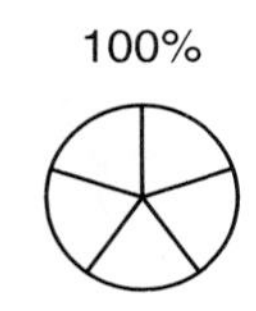

25. Which arrow could be pointing to 1375 on the number line below?
(27)

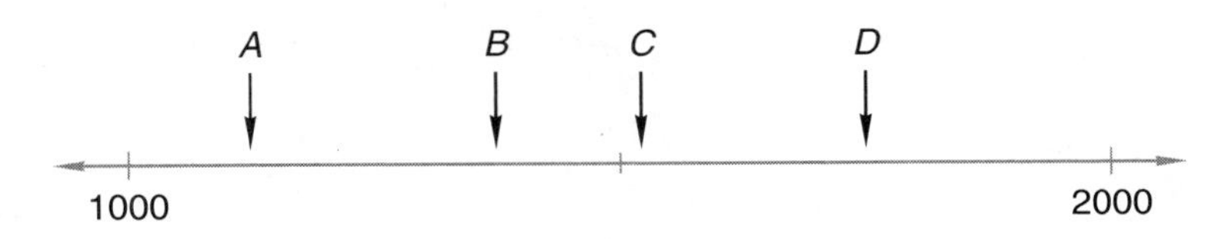

26. An angle that measures 120° is what type of angle?
(Inv. 4)

A. acute B. right C. obtuse

27. The word *rectangle* comes from Latin for "right corner." In what way is a rectangle a "right corner" polygon?
(32)

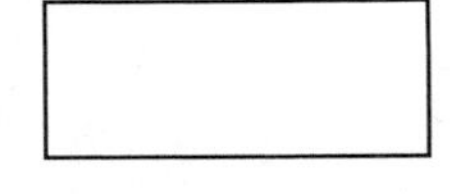

28. Which of these numbers is divisible by both 2 and 9?
(22, 42)

A. 234 B. 456 C. 567

29. Is $1\frac{3}{4}$ inches closer to 1 inch or 2 inches?
(44)

LESSON

47 Simplifying Mixed Measures

WARM-UP

Facts Practice: 64 Multiplication Facts (Test F)

Mental Math: Name the relationship: Salma is her husband's mother's ______.
A score is 20. How many is two score? ... three score? ... four score?

a. Round 381 to the nearest ten.
b. 760 − 400
c. $1\frac{1}{2} + \frac{1}{2}$
d. $1\frac{1}{2} - \frac{1}{2}$
e. Half of 3
f. 25% of 4
g. 9 × 8, − 2, ÷ 2, + 1, ÷ 4, + 1, ÷ 2
h. 3 × 34

Problem Solving:

Copy this multiplication problem and fill in the missing digits:

```
  7_
×  _
 6_5
```

NEW CONCEPT

Sometimes two units are used to name a measure. Here we show three examples:

John is 5 feet 4 inches tall.

Tomas ran a quarter mile in 1 minute 15 seconds.

The melon weighed 3 pounds 8 ounces.

In this lesson we will practice changing measures named with two units into measures named with one unit.

Example 1 John is 5 feet 4 inches tall. How many inches tall is John?

Solution Five feet 4 inches means "5 feet plus 4 inches." Before we can add, we first change 5 feet to inches. Since 1 foot equals 12 inches, we multiply 5 by 12 inches.

5 feet = 5 × 12 inches

5 feet = 60 inches

Now we add 60 inches and 4 inches.

60 inches + 4 inches = 64 inches

John is **64 inches** tall.

Example 2 Tomas ran a quarter mile in 1 minute 15 seconds. What was his time in seconds?

Solution One minute 15 seconds means "1 minute plus 15 seconds." We first change 1 minute to seconds. Then we add.

1 minute = 60 seconds

60 seconds + 15 seconds = 75 seconds

Tomas ran a quarter mile in **75 seconds.**

Example 3 The melon weighed 3 pounds 8 ounces. How many ounces did the melon weigh?

Solution Three pounds 8 ounces means "3 pounds plus 8 ounces." We change pounds to ounces first. One pound equals 16 ounces.

3 pounds = 3 × 16 ounces

3 pounds = 48 ounces

Now we add.

48 ounces + 8 ounces = 56 ounces

The melon weighed **56 ounces.**

Activity: *Simplifying Height Measurements*

Materials needed:

- yardstick

Have students use a yardstick to find their height in "feet plus inches" form and in "inches only" form. Have them show that the two forms are equal.

LESSON PRACTICE

Practice set **a.** Change 6 feet 2 inches to inches.

b. Change 3 minutes 20 seconds to seconds.

c. Change 2 hours 30 minutes to minutes.

d. Change 2 pounds 12 ounces to ounces (1 pound equals 16 ounces).

MIXED PRACTICE

Problem set For problems 1–3, write an equation and find the answer.

1. (11) When the students got on the buses to go to the picnic, there were 36 on one bus, 29 on another bus, and 73 on the third bus. Altogether, how many students were on the three buses?

2. (21, 28) Anita's grandfather has lived for seven decades. Seven decades is how many years?

3. (35) Anita is 12 years old. Her grandmother is 68 years old. Anita's grandmother is how many years older than Anita?

4. (47) When Gabriel turned 12 years old, he was 5 feet 6 inches tall. How many inches is 5 feet 6 inches?

5. (7) The 7 in 374,021 means which of the following?

   A. 7 B. 70 C. 70,000

6. (28) From March 1 of one year to May 1 of the next year is how many months?

7. (Inv. 3, 37) Draw a rectangle. Shade three eighths of it. What percent of the rectangle is shaded?

8. (44) Use a ruler to find the length of this line segment in inches:

9. (43) $4 + 3\frac{3}{4}$

10. (41) $3\frac{3}{5} + 1\frac{1}{5}$

11. (43) $2\frac{3}{8} + \frac{2}{8}$

12. (24, 43) $5\frac{1}{3} - \left(5\frac{1}{3} - \frac{1}{3}\right)$

13. (43) $2\frac{1}{2} - \frac{1}{2}$

14. (41) $3\frac{5}{9} - 1\frac{1}{9}$

15. (6) $$\begin{array}{r} \$48,748 \\ \$37,145 \\ +\ \$26,498 \\ \hline \end{array}$$

16. (9) $$\begin{array}{r} \$63,142 \\ -\ \$17,936 \\ \hline \end{array}$$

17. (29) $$\begin{array}{r} \$5.63 \\ \times\ \ 700 \\ \hline \end{array}$$

18. (17) $$\begin{array}{r} 4729 \\ \times\ \ 8 \\ \hline \end{array}$$

19. (29) $$\begin{array}{r} 9006 \\ \times\ \ 80 \\ \hline \end{array}$$

20. (26) $\dfrac{3456}{8}$

21. (34) $1836 \div 9$

22. (34) $1405 \div 7$

23. (24, 29) $(20 \times 25) + (5 \times 25)$

24. (46) Draw a diagram to illustrate and solve this problem:

In Andy's slice of watermelon, there were 60 seeds. If he swallowed $\frac{2}{5}$ of the seeds, how many seeds did he swallow?

25. (28) It is evening. What time will it be in $3\frac{1}{2}$ hours?

26. (Inv. 4) Draw an angle that has half the measure of a right angle. If drawn very carefully, the angle would measure how many degrees?

27. (22, 42) Which of these numbers is divisible by 6 and by 5?

A. 576 B. 765 C. 7650

28. (Inv. 3) Compare: $\frac{1}{5}$ of 10 ◯ $10 \div 5$

29. (44) Is $2\frac{1}{4}$ inches closer to 2 inches or to 3 inches?

LESSON

48 Reading and Writing Whole Numbers in Expanded Notation

WARM-UP

Facts Practice: 48 Uneven Divisions (Test G)

Problem Solving:

Use the information in this paragraph to complete the statements that follow. Drawing a diagram may help you with the problem.

Fred's wife is Ella. Their daughter is Gloria. Fred's sister is Helen. Helen's husband is Ishmael, and their son is Jerome. Fred's mother and father are Carlotta and Dewayne.

a. Jerome is Fred's ______.
b. Ella is Carlotta's ______.
c. Ishmael is Dewayne's ______.
d. Carlotta is Gloria's ______.
e. Jerome is Gloria's ______.
f. Ishmael is Fred's ______.
g. Jerome is Dewayne's ______.
h. Fred is Helen's ______.

NEW CONCEPT

One way to name numbers is to name the place value of each digit. The number 3256 could be named

3 thousands plus 2 hundreds plus 5 tens plus 6 ones

We could use numbers instead of words to rewrite this as

$$(3 \times 1000) + (2 \times 100) + (5 \times 10) + (6 \times 1)$$

This method of naming numbers is called **expanded notation. When we write a number in expanded notation, we write a digit times its place value,** plus the next digit times its place value, and so on.

Example 1 Write the number 5600 in expanded notation.

Solution The number 5600 is 5 thousands plus 6 hundreds plus no tens plus no ones. We write 5 times its place value, plus 6 times its place value. Since there are no tens or ones, we write only

$$\mathbf{(5 \times 1000) + (6 \times 100)}$$

Example 2 Write 750,000 in expanded notation.

Solution The number 750,000 is 7 hundred thousands plus 5 ten thousands. We write 7 times its place value, plus 5 times its place value.

$$\mathbf{(7 \times 100{,}000) + (5 \times 10{,}000)}$$

Example 3 Write the standard form for $(3 \times 100) + (2 \times 1)$.

Solution *Standard form* means "the usual way of writing numbers." We are to write the number that has a 3 in the hundreds place and a 2 in the ones place.

100's	10's	1's
3	0	2

Note that we use a zero to hold the tens place. The standard form is **302.**

LESSON PRACTICE

Practice set* Write in expanded notation:

a. 56

b. 5280

c. 250,000

Write in standard form:

d. $(6 \times 1000) + (4 \times 10)$

e. $(5 \times 100) + (7 \times 10)$

f. $(8 \times 10{,}000) + (4 \times 1000)$

g. $(9 \times 100{,}000) + (3 \times 10{,}000)$

MIXED PRACTICE

Problem set

1. Draw a square. Make each side $1\frac{1}{2}$ inches long.
(44, 45)

For problems 2–4, write an equation and find the answer.

2. Amanda is 6 years older than Jorge. If Amanda is 21, then how old is Jorge?
(35)

3. After January 25, how many days are left in a leap year?
(16, 28)

4. Eight dozen eggs is how many eggs?
(21)

5. (7) The 6 in 356,287 means which of the following?

A. 6 B. 356 C. 6000 D. 6287

6. (44) The new pencil was 19 centimeters long. How many millimeters long was the pencil?

7. (37, 43) Draw a circle. Shade one sixth of it. What percent of the circle is shaded?

8. (33) Round 287 to the nearest ten.

9. (48) Write the standard form for $(5 \times 100) + (2 \times 1)$.

10. (48) Write the number 47,000 in expanded notation.

11. (6)
$$\begin{array}{r} 98{,}572 \\ 42{,}156 \\ 37{,}428 \\ +\ 16{,}984 \\ \hline \end{array}$$

12. (14)
$$\begin{array}{r} W \\ -\ 32{,}436 \\ \hline 19{,}724 \end{array}$$

13. (14)
$$\begin{array}{r} 10{,}000 \\ -\quad Y \\ \hline 1{,}746 \end{array}$$

14. (17)
$$\begin{array}{r} \$34.78 \\ \times \quad 6 \\ \hline \end{array}$$

15. (29)
$$\begin{array}{r} 6549 \\ \times \quad 60 \\ \hline \end{array}$$

16. (29)
$$\begin{array}{r} 8037 \\ \times \quad 90 \\ \hline \end{array}$$

17. (34) $3647 \div 6$

18. (34) $5408 \div 9$

19. (26, 34) $10W = 1000$

20. (41, 43) $3\frac{1}{3} + \left(4\frac{1}{3} - 2\right)$

21. (18, 29) $6 \times 800 = 6 \times 8 \times H$

22. (13, 24) $\$10 - (\$6 + \$1.47 + \$0.93)$

23. (24, 29) $(20 \times 62) + (3 \times 62)$

24. (28, 46) How many years is one fourth of a century? You may wish to draw a diagram to help you answer the question.

25. (28) What time is 1 minute before midnight?

26. *(27)* To what number is the arrow pointing?

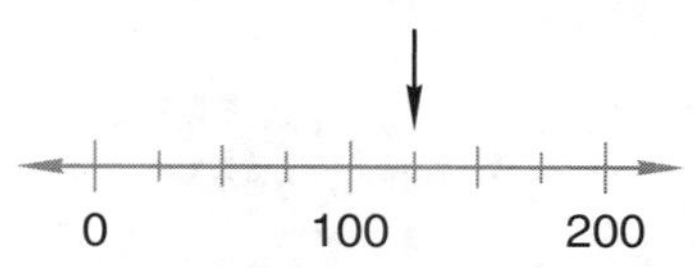

27. *(1, 42)* Here is a sequence of numbers we say when we count by nines:

9, 18, 27, 36, 45, ...

Which of the following numbers is also in this sequence?

A. 987 B. 975 C. 963

28. *(21, 30)* If a dollar's worth of dimes is divided into two equal groups, how many dimes would be in each group?

29. *(32)* A stop sign is a traffic sign that is shaped like an octagon. How many sides does a stop sign have?

LESSON

49 Solving Two-Step Word Problems

WARM-UP

Facts Practice: 48 Uneven Divisions (Test G)

Mental Math: Name the relationship: Gilberto's brother's wife is Gilberto's ______.
Count up and down by 25's between 250 and 500.

a. Round 521 to the nearest hundred.

b. 740 + 60 **c.** $\frac{4}{10} + \frac{5}{10}$ **d.** $\frac{9}{10} + \frac{1}{10}$

e. 50% of 10 **f.** 10% of 10 **g.** $\frac{1}{3}$ of 12

h. 12 ÷ 2, + 2, ÷ 2, + 2, ÷ 2, + 2

Problem Solving:

Six dots can make a triangular pattern with three dots on each side. Ten dots can make a triangular pattern with four dots on each side. How many dots are in a triangular pattern that has seven dots on each side?

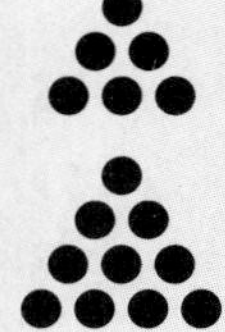

NEW CONCEPT

Many mathematical problems take more than one step to solve. We have solved several kinds of two-step problems. For example, it takes two steps to solve 10 – (6 – 3). We used two steps to convert 6 feet 2 inches to inches. We have also illustrated and solved fraction problems that take two steps.

The first step of a two-step story problem involves finding a number that does not appear in the story. We then use the number we found to help us solve the second step of the problem.

Example Huang is 5 years older than Robert. Robert is 3 years older than Sally. Sally is 15 years old. How old is Huang?

Solution This is a two-step problem. It is actually two "larger-smaller-difference" stories put together into one problem. We are asked to find Huang's age, but we cannot find his age in one step. We must first find Robert's age. Then we will be able to find Huang's age. We are given enough information to find Robert's age.

Goal: Find Huang's age.

First step: Find Robert's age.

We will draw a pair of rectangles for each comparison.

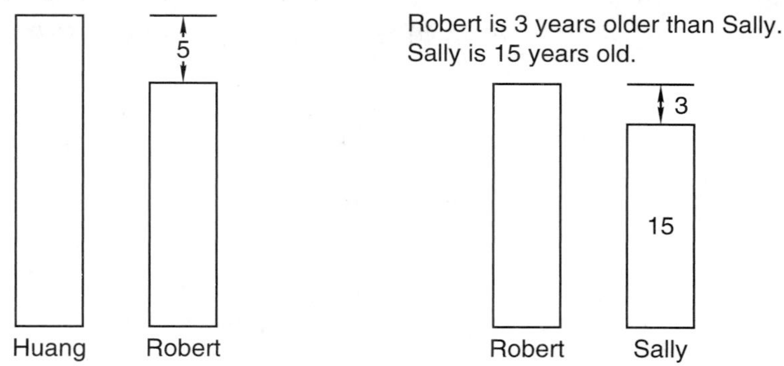

Drawing the rectangles helps us see that Robert is 18 years old. We write "18" in both rectangles that stand for Robert's age.

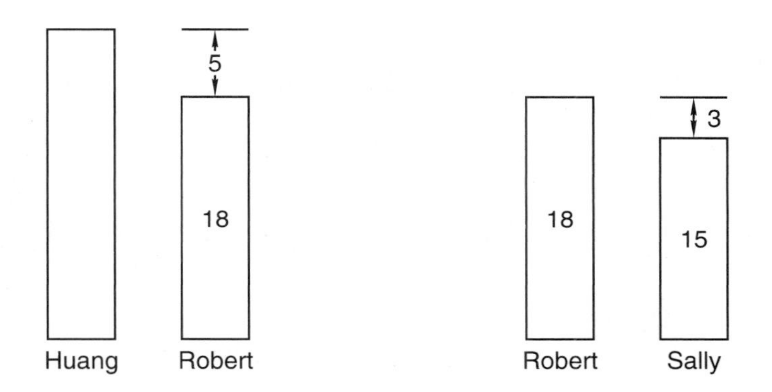

Since Huang is 5 years older than Robert, he is 18 years plus 5 years, which is **23 years old.**

LESSON PRACTICE

Practice set Solve each two-step problem:

a. The bus could carry 80 students. The 32 students from Room 8 and the 29 students from Room 12 got on the bus. How many more students can the bus carry? (Goal: Find how many more students would fit. First step: Find how many students are already on the bus.)

b. Mariabella collected 37 aluminum cans for school, and her brother collected 21. They decided to divide the cans evenly, so they put the cans into one big pile and then made two equal piles. How many cans were in each pile? (Goal: Find how many cans are in each pile. First step: Find how many cans there are in all.)

MIXED PRACTICE

Problem set

1. (31, 32) Draw a quadrilateral that has one pair of parallel sides and one pair of sides that are not parallel.

2. (49) Kyle is 10 years older than Hakim. Hakim is 5 years older than Jill. Jill is 15 years old. How old is Kyle? (Goal: Find Kyle's age. First step: Find Hakim's age.)

3. (46) Adriana's age is $\frac{1}{3}$ of her dad's age. If her dad is 36 years old, how old is Adriana?

4. (46) Draw a diagram to illustrate and solve this problem:

Jenny answered $\frac{4}{5}$ of the 20 questions correctly. How many questions did she answer correctly?

5. (7) The 7 in 754,238 means which of the following?

A. 700,000 B. 700 C. 7 D. 754

6. (48) Write the standard form for $(5 \times 100) + (6 \times 1)$.

7. (21) Kathie and Keisha will earn $5 for weeding the garden. If they share the money equally, how much money will each person receive?

8. (33) Round 234 to the nearest ten.

9. (7) Use digits to write twenty-five thousand, three hundred.

10. (37, 43) Draw a circle. Shade five sixths of it. What percent of the circle is not shaded?

11. (43) $5\frac{2}{8} + 6 + \frac{3}{8}$

12. (41, 43) $8\frac{5}{6} - \left(3\frac{5}{6} - 3\right)$

13. (6) $342 + 5874 + 63 + 285 + 8 + 96 + 87$

14. (13) $\$42.01 - \$20.14$

15. (14) $1000 - M = 1$

16. (29) 800×50

17. (18, 29) $30 \times 8 \times 25$

18. (34) $1205 \div 6$

19. (26) $\$76.32 \div 8$

20. $20 − ($12 + $4.76 + $2.89 + $0.34)
(13, 24)

21. (20 × 35) + (5 × 35)
(24, 29)

22. Use words to name the number 150,000.
(7)

23. Write 150,000 in expanded notation.
(48)

24. What are the next three terms in this counting sequence?
(1)

..., 900, 1000, 1100, _____, _____, _____, ...

25. Which place does the zero hold in 203,456?
(7)

26. To what number is the arrow pointing?
(27)

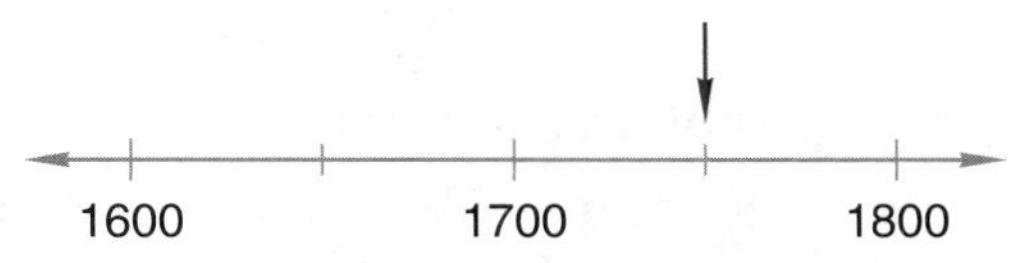

27. According to this calendar, what was the date of the first Sunday in December 1941?
(28)

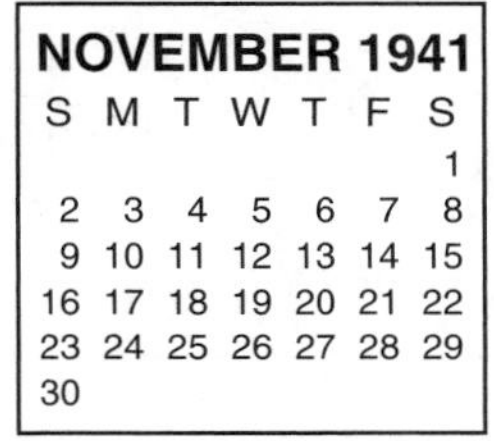

NOVEMBER 1941

S	M	T	W	T	F	S
						1
2	3	4	5	6	7	8
9	10	11	12	13	14	15
16	17	18	19	20	21	22
23	24	25	26	27	28	29
30						

28. Think of an odd number. Multiply it by 5. What is the last digit of the product?
(2, 15)

29. Which of these angles appears to measure 150°?
(Inv. 4)

A.

B. 

C.

D.

LESSON

50 Finding an Average

WARM-UP

Facts Practice: 64 Multiplication Facts (Test F)

Mental Math: Count up and down by 3's between 3 and 36. A yard is 3 feet. How many feet are in 2 yards? ... 5 yards? ... 10 yards?

a. Round 646 to the nearest hundred. **b.** $870 + 130$

c. $1\frac{1}{2} + 1\frac{1}{2}$ **d.** $3\frac{1}{3} - 2$ **e.** $\frac{1}{3}$ of 15

f. 25% of 16 **g.** 10% of 50 **h.** 8×34

i. $6 \times 5, + 3, \div 3, + 4, \div 3, + 1, \div 3$

Problem Solving:

Enrique rolled two number cubes. The total was 7. Copy this table and write all the ways Enrique could have rolled a total of 7.

First Cube	Second Cube

NEW CONCEPT

Below we show two stacks of nickels. In one stack there are 5 nickels, and in the other stack there are 9 nickels. If some nickels were moved from the taller stack to the shorter stack so that the stacks were even, how many nickels would be in each stack?

One way to answer this question is to first **find the total** number of nickels and then divide the total into two **equal groups.** Since there are 5 nickels in one stack and 9 nickels in the other stack, there are 14 nickels in all. Dividing 14 nickels into 2 equal groups, we find that there would be 7 nickels in each stack.

When we even up the number of members in groups, we are finding the **average** number of members per group. Finding an average is a two-step problem.

Step 1: Combine to find the total.

Step 2: Separate the total into equal groups.

Example 1 Refer to the picture below for this problem.

If water is poured from glass to glass until the amount of water in each glass is the same, how many ounces of water will be in each glass?

Solution The *total amount* of water will be *divided equally* among the three glasses. Finding the total amount of water is a problem about combining. It has a "some and some more" (addition) pattern. We add and find that the total amount of water is 18 ounces.

$$\begin{array}{r} 4 \text{ ounces} \\ 7 \text{ ounces} \\ +\ 7 \text{ ounces} \\ \hline \boxed{18} \text{ ounces} \end{array}$$

Finding the amount for each glass is an "equal groups" problem. "Equal groups" problems have multiplication patterns. We divide 18 ounces by 3 and find that there will be **6 ounces of water** in each glass.

$$\begin{array}{l} N \text{ ounces in each glass} \\ \times\ 3 \text{ glasses} \\ \hline 18 \text{ ounces in all 3 glasses} \end{array} \qquad 3\overline{)18} = \boxed{6}$$

Example 2 Brad's test scores are 80, 85, 90, 85, and 90. What is the average of Brad's test scores?

Solution Finding an average takes two steps. The first step is to find the total. To do this, we add Brad's scores.

$$80 + 85 + 90 + 85 + 90 = 430$$

The second step is to separate the total into equal groups. Brad took five tests, so we divide the total into five equal parts.

$$430 \div 5 = 86$$

We find that Brad's average test score is **86.** Notice that although none of Brad's scores were 86, the sum of the five scores, 430, is the same as if he had scored 86 on every test.

LESSON PRACTICE

Practice set* Solve each two-step problem by combining and then forming equal groups.

a. The number of players on the four squads was 5, 6, 9, and 8. If the squads were changed so that the same number of players were on each squad, how many players would each squad have?

5	6
9	8

b. When the class lined up, there were 11 students in one line and 17 students in the other line. If the lines were made even, how many students would be in each line?

c. This picture shows three stacks of books. If the stacks were made even, how many books would be in each stack?

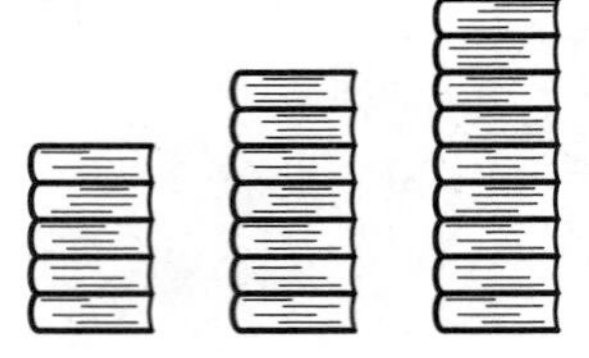

d. Here are Shauna's quiz scores:

8, 9, 7, 9, 8, 10, 6, 7

What is Shauna's average quiz score?

MIXED PRACTICE

Problem set

1. *(31, 32)* Draw a quadrilateral so that the sides that intersect are perpendicular.

2. *(49)* Kimberly is 5 years older than Loh. Miguel is 2 years older than Loh. Miguel is 13 years old. How old is Kimberly? Draw a pair of rectangles for each comparison.

3. *(50)* If water is poured from glass to glass until the amount of water in each glass is the same, how many ounces of water will be in each glass? (First use an addition pattern to find the total amount of water. Then use a multiplication pattern to divide the total equally.)

4. Draw a diagram to illustrate and solve this problem:
(46)

How many minutes is $\frac{3}{5}$ of an hour?

5. How many minutes are in 2 hours 15 minutes?
(47)

6. Four hundred years is how many centuries?
(21, 28)

7. Use digits to write fifty-four thousand, nine hundred nineteen.
(7)

8. Draw a rectangle. Shade seven eighths of it. What percent of the rectangle is not shaded?
(Inv. 3, 37)

9. There were 15 children in one line and 11 children in another line. After some children moved from the longer line to the shorter line, there were the same number of children in each line. How many children were then in each line?
(50)

10. $342 + 67 + 918 + 897 + 42$
(6)

11. $\$53.87 - \$27.59$
(13)

12. $\$34.28 \times 60$
(29)

13. $7 \times 57 \times 10$
(18, 29)

14. $(4 + 7 + 7) \div 3$
(24)

15. $(5 + 6 + 9 + 8) \div 4$
(24)

16. $4206 \div 7$
(34)

17. $\$60.24 \div 6$
(34)

18. $1000 \div 9$
(26)

19. $6D = 180$
(26, 34)

20. $1\frac{1}{7} + 2\frac{2}{7} + 3\frac{3}{7}$
(41)

21. $9\frac{9}{10} - \left(7\frac{7}{10} - 5\frac{5}{10}\right)$
(24, 41)

22. $(10 \times 43) + (2 \times 43)$
(24, 29)

23. What month is 10 months after July?
(28)

24. Counting by hundreds, 1236 is closest to which of the following numbers?
(33)

A. 1100 B. 1200 C. 1300 D. 1000

Use the information below to answer problems 25 and 26. Drawing a map may help.

> *From Sara's house, Arcadia Park is 4 miles north, Legg Lake is 5 miles south, the ocean is 32 miles west, and the mountain cabin is 98 miles east.*

25. (11) Sara's family went to the ocean one Saturday. They left home at 9 a.m. and returned home at 4 p.m. Altogether, how far did they travel going to the ocean and back home?

26. (11) How far is it from Arcadia Park to Legg Lake?

27. (50) Dexter kept a log of his bike rides. On average, how far did Dexter ride each day?

Monday	5 miles
Tuesday	8 miles
Wednesday	8 miles
Thursday	6 miles
Friday	8 miles

28. (7, 48) Write three hundred twenty thousand in expanded notation.

29. (42) Using the three digits 1, 2, and 3, what is the largest three-digit number you can make that is divisible by 6?

320,

INVESTIGATION 5

Focus on

Organizing and Analyzing Data

Your teacher collects and records information about your class, including attendance, homework, and test scores. Your school gathers information about class sizes, lunch counts, and supplies ordered. Sports teams compile information about wins and losses, points scored, and points allowed. Researchers testing new medications keep careful records of subjects who respond well, of subjects who do not respond, and of subjects who experience side effects.

These types of gathered information are called **data,** and the study of data is called **statistics.** People who work in the field of statistics collect, organize, analyze, and display data. (*Note:* The word *data* is plural; its singular form is *datum.*)

Newly gathered data are often unorganized. To be useful, these "raw" data must first be organized. In this investigation we will practice two ways of organizing data, **frequency tables** and **line plots.** Then we will practice analyzing data to solve problems.

Frequency tables

Mr. Sottong gave his 25 students a 6-question test. The number of questions that a student could have correctly answered was thus 0, 1, 2, 3, 4, 5, or 6. Here are the scores that Mr. Sottong's students received:

4, 3, 3, 4, 2, 5, 6, 1, 3, 4, 5, 2, 2,
6, 3, 3, 4, 3, 2, 4, 5, 3, 5, 5, 6

Mr. Sottong decides to organize the data in a frequency table that shows how many students earned each possible score. He lists the possible scores and then tallies the number of tests with those scores. (Since no student scored zero, that score is omitted from the table.)

Number Correct	Tally
1	|
2	||||
3	~~||||~~ ||
4	~~||||~~
5	~~||||~~
6	|||

After tallying, Mr. Sottong counts the number of marks for each score to get the **frequency** of each score. He lists these frequencies in a third column, as shown below.

Frequency Table

Number Correct	Tally	Frequency
1	|	1
2	||||	4
3	~~||||~~ ||	7
4	~~||||~~	5
5	~~||||~~	5
6	|||	3

To check whether every score was accounted for, Mr. Sottong adds the numbers in the frequency column. The total is 25, which is the number of students who took the test.

1. Twenty children were asked how many brothers and sisters each had. The responses were these numbers:

   2, 3, 0, 1, 1, 3, 0, 4, 1, 2,
   0, 1, 1, 2, 2, 3, 0, 2, 1, 1

   Make a frequency table that shows the numbers of brothers and sisters.

Sometimes we group our data in intervals of equal length.

Suppose the ABC Market offers 22 turkeys for sale the weekend before Thanksgiving. The weights of the turkeys in pounds are given below:

11, 18, 21, 23, 16, 20, 22, 14, 16, 20, 17,
19, 13, 14, 22, 19, 22, 18, 20, 12, 25, 23

Here is a frequency table for these data using intervals of 4 pounds starting with the interval 10–13 lb. (This interval includes the weights 10, 11, 12, and 13 pounds.)

Frequency Table

Weight	Tally	Frequency
10–13 lb	|||	3
14–17 lb	~~||||~~	5
18–21 lb	~~||||~~ |||	8
22–25 lb	~~||||~~ |	6

2. Use the raw data about the turkey weights at ABC Market to make a new frequency table. Instead of the weight intervals shown in the table above, use the intervals 11–15 lb, 16–20 lb, and 21–25 lb.

3. The annual rainfall in a certain Texas city for each of the last fifteen years is recorded below. Each of these data is in inches.

   17, 24, 32, 27, 18, 30, 22, 18,
   24, 31, 32, 26, 18, 19, 22

   Make a frequency table for the data, using five-inch intervals. Begin with the interval 15–19 in.

Line plots If we want to see each data point, we can create a **line plot.** We draw a number line that includes the highest and lowest values we collected. Then we place an X above each number for each data point that corresponds to that number. Above some numbers, there might be a stack of X's. Above other numbers, there might be none.

Suppose Jean-Paul recorded the ages of the first 20 people who went down a certain water slide after the water park opened at 9 a.m. Their ages were

11, 9, 8, 11, 12, 15, 13, 12, 17, 12,
12, 22, 13, 11, 21, 9, 16, 12, 13, 9

To make a line plot of this information, we first draw a number line. If we begin our number line at 5 and end it at 25, all of the data points can be included. Now we place an X on the number line for each data point. Since there are 3 data points that have value 9, we stack 3 X's above 9 on the number line.

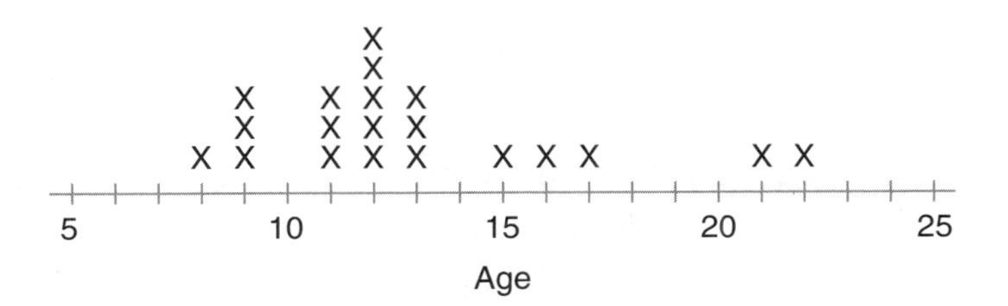

A **cluster** is a group of data points that are very close together. We see that the data points 11, 12, and 13 form a cluster. There can be more than one cluster in a data set.

An **outlier** is a data point that is distant from the majority of data points. In this set the data points 21 and 22 could be considered outliers. Refer to this line plot to answer problems 4–6.

4. Which age was recorded most frequently?

5. How many people who went down the slide were over 15?

6. Which ages were recorded exactly three times?

Below we show another line plot. Refer to this line plot to answer problems 7–11.

The Lizard Emporium pet store tracks the life span of the iguanas it sells. The data for some of the iguanas are displayed in this line plot:

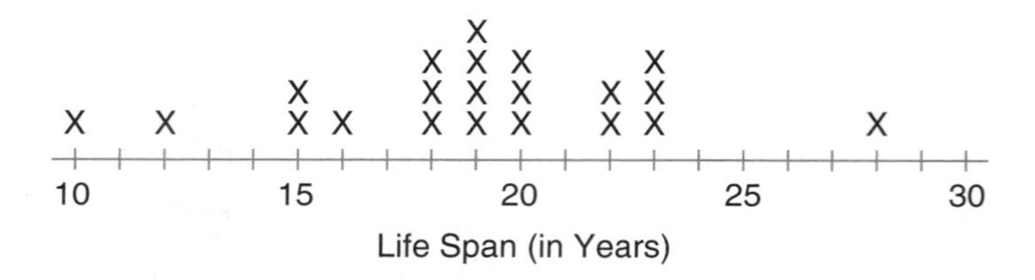

7. What was the most frequent life span of the iguanas that were tracked?

8. How many iguanas were tracked?

9. How many iguanas lived between 17 and 25 years?

10. Name 3 outliers in this plot.

11. Name 2 clusters in this plot.

12. The teacher recorded the number of correct answers that students scored on a recent test. Make a line plot for these scores. Then refer to the line plot to answer problems 13–15.

19, 18, 17, 15, 18, 20, 14, 17,
19, 11, 18, 17, 16, 18, 16

13. Which score is an outlier?

14. Which score was made most often?

15. How many students scored 16 or higher?

Analyzing data Part of the process of solving problems is finding the information needed to solve the problem. We can find information in graphs, tables, books, on the Internet, and in other places. Besides locating the information, we often need to sort through the information to find the facts needed to solve the problem.

Example 1 Read this information. Then answer the question.

In the first 6 games of the season the Rio Vista football team won 4 games and lost 2 games. They won their next game by a score of 28 to 20. The team played 10 games during the season.

In the first 7 games, how many games did the Rio Vista football team win?

Solution More information is given than is needed to answer the question. We sort through the information until we find the information necessary to answer the question. We are asked to find how many of the first 7 games were won by the team. We are told that the team won 4 of the first 6 games. We are also told that the team won the next game (that is, the seventh game). Thus, the team won **5 games** out of their first seven.

Refer to the information in example 1 to answer problems 16 and 17.

16. What is the greatest number of games the Rio Vista team could have won during the season? How do you know?

17. Is it certain that the Rio Vista team won more than half its games this season? How do you know?

Example 2 Each student in the class was asked to name his or her favorite school lunch from a choice of four menus. The results are recorded in this **bar graph.**

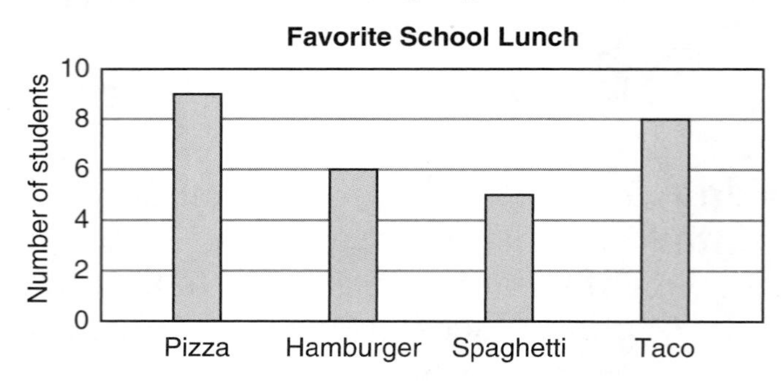

According to the bar graph, how many students chose pizza as their favorite school lunch?

Solution The bar for pizza ends halfway between the line for 8 and the line for 10. Halfway between 8 and 10 is 9. Thus, pizza was the favorite food of **9 students.**

Refer to the bar graph in example 2 to answer problems 18 and 19.

18. What is the total number of students represented by the four bars in the graph?

19. Is it true that the number of students who named pizza as their favorite was twice the number who named spaghetti? How do you know?

Example 3 Sharon recorded the number of correct answers on her math tests in the following **line graph:**

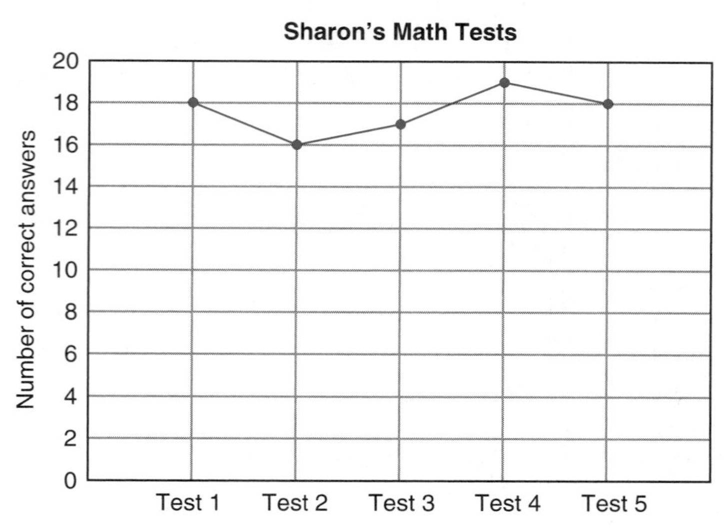

There were 20 problems on each test. How many problems did Sharon miss on Test 3?

Solution We see that on Test 3 Sharon had 17 correct answers. We subtract 17 from 20 and find that Sharon missed **3 problems** on Test 3.

Refer to the information in example 3 to answer problems 20 and 21.

20. What was the average number of correct answers on Sharon's first three tests?

21. What fraction of the problems on Test 4 did Sharon answer correctly?

LESSON

51 Multiplying by Two-Digit Numbers

WARM-UP

Facts Practice: 48 Uneven Divisions (Test G)

Mental Math: Count by 12's from 12 to 120. A foot is 12 inches. How many inches are in 2 feet? ... 3 feet? ... 4 feet?

a. 2 feet 2 inches is how many inches? **b.** 650 − 250
c. $\frac{3}{4} + \frac{1}{4}$ **d.** $3\frac{1}{4} - 1\frac{1}{4}$ **e.** Half of 25
f. 25% of 24 **g.** 10% of 20 **h.** 8 × 34
i. Find $\frac{1}{2}$ of 20, + 2, ÷ 2, + 2, ÷ 2, + 2, ÷ 2

Problem Solving:

The numbers 3, 6, and 10 may be called **triangular numbers.** This is because 3 objects, 6 objects, and 10 objects can each be arranged in a triangular pattern. What are the next three triangular numbers?

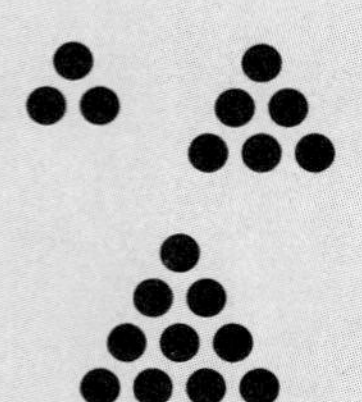

NEW CONCEPT

When we multiply by a two-digit number, we really multiply twice. We multiply by the tens, and we multiply by the ones. Here we multiply 43 by 12. Since 12 is 10 + 2, we may multiply 43 by 10 and 43 by 2. Then we add the products.

$$\begin{array}{r} 43 \\ \times\ 12 \\ \hline \end{array} \quad \text{is the same as} \quad \begin{array}{r} 43 \\ \times\ 10 \\ \hline 430 \end{array} \quad \text{plus} \quad \begin{array}{r} 43 \\ \times\ 2 \\ \hline 86 \end{array}$$

$$430 + 86 = 516$$

When we multiply by a two-digit number, we do not need to separate the problem into two problems before we start.

Example 1 Multiply: $\begin{array}{r} 43 \\ \times\ 12 \\ \hline \end{array}$

Solution First we multiply 43 by the 2 of 12. We get 86 and we write the 86 so the 6 is in the ones column under the 2.

$$\begin{array}{r} 43 \\ \times\ 12 \\ \hline 86 \end{array}$$

Next, we multiply 43 by the 10 of 12. We get 430, which we may write below the 86. Then we add 86 to 430 and find that 43 × 12 equals **516.** The numbers 86 and 430 are called **partial products.** The number 516 is the final product. Below are two ways we may show our work.

$$\begin{array}{r} 43 \\ \times\ 12 \\ \hline 86 \\ 430 \\ \hline 516 \end{array} \quad \text{or} \quad \begin{array}{r} 43 \\ \times\ 12 \\ \hline 86 \\ 43 \\ \hline 516 \end{array}$$

Some people do not write the trailing zero in the second partial product. In the method on the right, the 0 of 430 is omitted from the second partial product. After multiplying by the 2 of 12, we moved to the left one place and multiplied by the 1 of 12. This reminded us to begin writing the partial product one place to the left. The 43 means "43 tens."

Example 2 Multiply: $\begin{array}{r} \$0.35 \\ \times\quad 25 \\ \hline \end{array}$

Solution We ignore the dollar sign and the decimal point until we have a final product.

$$\begin{array}{r} \$0.35 \\ \times\quad 25 \\ \hline 1\ 75 \\ 7\ 00 \\ \hline \$8.75 \end{array} \quad \text{or} \quad \begin{array}{r} \$0.35 \\ \times\quad 25 \\ \hline 1\ 75 \\ 7\ 0 \\ \hline \$8.75 \end{array}$$

After multiplying, we place the decimal point. Since we multiplied cents, we show cents in the final product by placing the decimal point so that there are two digits to the right of the decimal point. The answer is **$8.75.**

The multiplication algorithm presented in this lesson is based on the **distributive property.** The distributive property applies to situations in which a sum is multiplied, such as

$$25 \times (10 + 2)$$

By the distributive property, we have two choices when multiplying a sum:

Choice 1: Find the sum; then multiply.

Choice 2: Multiply each addend; then add the products.

Here we illustrate these choices:

$$25 \times (10 + 2)$$

$$25 \times 12 \quad \text{or} \quad (25 \times 10) + (25 \times 2)$$

Both choices result in the same answer (which in this case is 300).

Example 3 Benito wants to multiply 35 by (20 + 4). Using the distributive property, show his two choices. Then find each answer.

Solution Here are Benito's two choices:

$$35 \times (20 + 4)$$

$$\mathbf{35 \times 24} \quad \text{or} \quad \mathbf{(35 \times 20) + (35 \times 4)}$$

Now we find each answer.

$$\begin{array}{r} 35 \\ \times\ 24 \\ \hline 140 \\ 700 \\ \hline \mathbf{840} \end{array} \qquad \begin{array}{c} (35 \times 20) + (35 \times 4) \\ \\ 700 + 140 \\ \\ \mathbf{840} \end{array}$$

Notice that 700 and 140 appear as partial products in both methods.

LESSON PRACTICE

Practice set* Multiply:

a. $\begin{array}{r} 32 \\ \times\ 12 \\ \hline \end{array}$

b. $\begin{array}{r} \$0.62 \\ \times\ \ 23 \\ \hline \end{array}$

c. $\begin{array}{r} 48 \\ \times\ 64 \\ \hline \end{array}$

d. $\begin{array}{r} 246 \\ \times\ \ 22 \\ \hline \end{array}$

e. $\begin{array}{r} \$1.47 \\ \times\ \ 34 \\ \hline \end{array}$

f. $\begin{array}{r} 87 \\ \times\ 63 \\ \hline \end{array}$

g. Molly wants to multiply 12 by (20 + 3). Show her two choices for multiplying. Find each answer.

MIXED PRACTICE

Problem set

1. (49) Jayne is reading a 320-page book. She read 47 pages the first day, 76 pages the second day, and 68 pages the third day. How many more pages does she have to read to finish the book?

2. (11) To mail the letter, Yai-Jun used one 37-cent stamp and three 23-cent stamps. How many cents did it cost to mail the letter?

3. (Inv. 2, 46) Draw a diagram to illustrate and solve this problem:

John ate $\frac{3}{4}$ of the 60 raisins. How many raisins did he eat? What percent of the raisins did he eat?

4. (48) Write $(1 \times 1000) + (1 \times 1)$ in standard form.

5. (Inv. 2, Inv. 3) Compare: $\frac{1}{2}$ of 10 ◯ $\frac{1}{3}$ of 12

6. (7) Use words to name 1760.

7. (37, 43) Draw a circle. Shade all but one sixth of it. What percent of the circle is not shaded?

8. (7) Use digits to write sixty-two thousand, four hundred ninety.

9. (33) Counting by hundreds, 2376 is closest to which number?

A. 2200 B. 2300 C. 2400 D. 2000

10. (44) How long is the line segment below?

inch 1 2 3

11. (50) Below are two stacks of coins. If some coins were taken from the taller stack and added to the shorter stack until the stacks were even, how many coins would be in each stack?

12. (51) 43×12

13. (51) $\$0.72 \times 31$

14. (51) 248×24

15. (51) $\$1.96 \times 53$

16. (6) $8762 + 3624 + 4795 + 8473$

17. (13) $\$10.00 - \$9.92$

18. (29) 600×50

19. (26) $\dfrac{\$6.00}{8}$

20. (34) $\$41.36 \div 4$

21. (26) $9x = 4275$

22. (43) $3 + \frac{1}{4} + 2\frac{2}{4}$

23. (24, 41) $\left(5\frac{5}{8} - 3\frac{3}{8}\right) - 1\frac{1}{8}$

24. (24) $(1 + 2 + 3 + 4 + 5) \div 5$

25. (47) In the running long jump Cynthia jumped 16 feet 9 inches. How many inches did she jump? (One foot equals 12 inches.)

26. (51) Kurt needs to multiply 15 by (20 + 4). Using the distributive property, show his two choices and the final product.

27. (Inv. 5) Below is a collection of scores from a class quiz. On your paper, make a frequency table, tally the scores, and count their frequency.

3, 4, 3, 5, 4, 1, 3, 4, 5, 4,
2, 4, 3, 2, 4, 1, 5, 4, 4, 3

Frequency Table

Score	Tally	Frequency
1		
2		
3		
4		
5		

28. (48) Write 205,000 in expanded notation.

29. (44) The math book was $11\frac{1}{4}$ inches long. Round $11\frac{1}{4}$ inches to the nearest inch.

LESSON

52 Naming Numbers Through Hundred Billions

WARM-UP

Facts Practice: 64 Multiplication Facts (Test F)

Mental Math: Count by 6's from 6 to 60. Count by 60's from 60 to 300. A minute is 60 seconds. How many seconds are in 2 minutes? ... 3 minutes?

a. 2 minutes 10 seconds is how many seconds?

b. \$1.00 − \$0.25

c. 50% of a minute

d. $1\frac{1}{8} + \frac{7}{8}$

e. 25% of a minute

f. 10% of a minute

g. 6 × 6, − 6, ÷ 6, + 5, ÷ 5, × 7, + 1, ÷ 3

Problem Solving:

One man said of another man, "Brothers and sisters have I none, but that man's father is my mother's son." What is the relationship between the two men?

NEW CONCEPT

The diagram below shows the values of the first twelve whole-number places.

hundred billions	ten billions	billions		hundred millions	ten millions	millions		hundred thousands	ten thousands	thousands		hundreds	tens	ones
___	___	___	,	___	___	___	,	___	___	___	,	___	___	___

Drawing the place-value diagram a different way emphasizes the repeating pattern of place values.

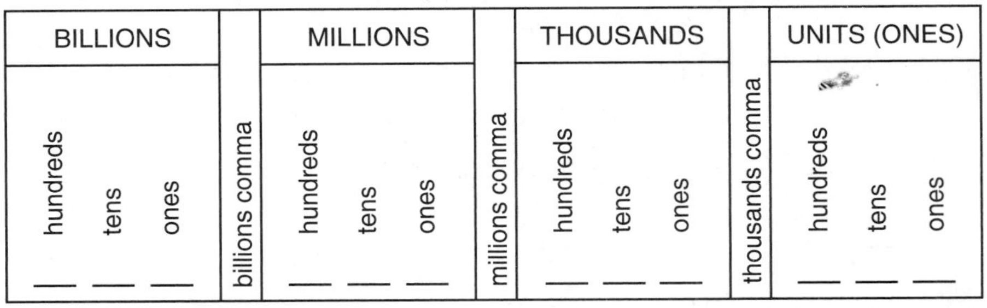

BILLIONS				MILLIONS				THOUSANDS				UNITS (ONES)		
hundreds	tens	ones	billions comma	hundreds	tens	ones	millions comma	hundreds	tens	ones	thousands comma	hundreds	tens	ones
___	___	___		___	___	___		___	___	___		___	___	___

We see that the repeating pattern of **ones, tens, hundreds** continues through the thousands, millions, and billions.

Example 1 Which digit shows the number of hundred billions in 987,654,321,000?

Solution Moving from right to left, the pattern of ones, tens, hundreds continues through the thousands, millions, and billions. The digit in the hundred-billions place is **9.**

Example 2 What is the value of the 2 in the number 12,345,678?

A. 2,000,000 B. 2000 C. 2

Solution The value of a digit depends upon its place in the number. Here the 2 means "two million." The correct choice is **A. 2,000,000.**

To name whole numbers with many digits, it is helpful to use commas. To insert commas, we count digits from the right-hand side of the whole number and put a comma after every three digits.

87,654,321

We write a comma after the millions place and after the thousands place. When reading a number with two commas, we say "million" when we come to the first comma and "thousand" when we come to the second comma.

8 7 , 6 5 4 , 3 2 1

"million" "thousand"

Using words, we name this number as follows:

eighty-seven million, six hundred fifty-four thousand, three hundred twenty-one

Example 3 Use words to name 1345200.

Solution We first put the commas in the number: 1,345,200. Then we name the number as **one million, three hundred forty-five thousand, two hundred.**

Example 4 Use digits to write one hundred thirty-four billion, six hundred fifty-two million, seven hundred thousand.

Solution **134,652,700,000**

Example 5 Write 2,500,000 in expanded notation.

Solution We write 2 times its place value, plus 5 times its place value.

$$(2 \times 1{,}000{,}000) + (5 \times 100{,}000)$$

LESSON PRACTICE

Practice set* In problems **a–d,** name the value of the place held by the zero in each number.

a. 345,052 **b.** 20,315,682

c. 1,057,628 **d.** 405,176,284

e. In 675,283,419,000, which digit is in the ten-billions place?

f. In which of the following numbers does the 7 have a value of seventy thousand?

A. 370,123,429 B. 1,372,486 C. 4,703,241

g. Use words to write the value of the 1 in 321,987,654.

Use words to name each number:

h. 21462300 **i.** 19650000000

Use digits to write each number:

j. nineteen million, two hundred twenty-five thousand, five hundred

k. seven hundred fifty billion, three hundred million

l. two hundred six million, seven hundred twelve thousand, nine hundred thirty-four

m. Write 7,500,000 in expanded notation.

MIXED PRACTICE

Problem set

1. (49) Khadija baked 5 dozen cookies and gave 24 to a friend. How many cookies did she have left?

2. (46) Marco weighs 120 pounds. His little brother weighs one half as much. How much does his brother weigh?

3. (49) Hope bought a chain for $3.60 and a lock for $4. How much should she get back in change from a $10 bill?

4. (28, 35) How many centuries were there from the year 1492 to the year 1992?

5. (48) Write (1 × 100) + (4 × 10) + (8 × 1) in standard form.

6. (Inv. 3, 37, 44) Draw a rectangle that is 2 inches long and 1 inch wide. Shade all but three eighths of it. What percent of the rectangle is not shaded?

7. (7) Use words to name the number 250,000.

8. (50) This picture shows three stacks of books. If the stacks were made even, how many books would be in each stack?

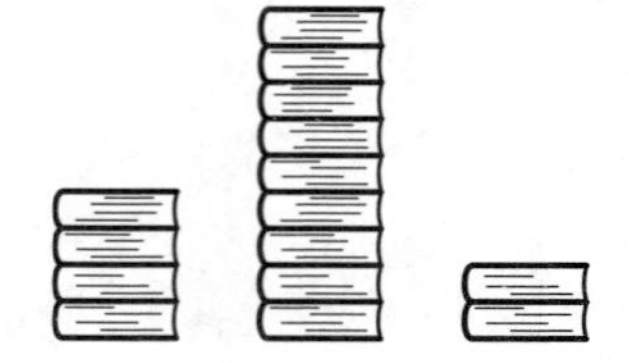

9. (52) Which digit in 789,456,321 shows the number of hundred millions?

10. (33) Round 1236 to the nearest hundred.

11. (52) Name the value of the place held by the zero in 102,345,678.

12. (51) 57×22

13. (51) $\$0.83 \times 47$

14. (51) 167×89

15. (51) $\$1.96 \times 46$

16. (6) $8437 + 3429 + 5765 + 9841$

17. (13) $\$26.38 - \$19.57$

18. (14) $3041 - W = 2975$

19. (34) $\dfrac{4328}{4}$

20. (26) $\dfrac{5670}{10}$

21. (34) $\dfrac{\$78.40}{4}$

22. (43) $\dfrac{3}{10} + 2 + 1\dfrac{4}{10}$

23. (41, 43) $5\dfrac{3}{4} - \left(2\dfrac{3}{4} - 2\right)$

24. (13, 24) $10 − ($1.43 + $2 + $2.85 + $0.79)

25. (38) Which arrow could be pointing to $3\frac{9}{10}$ on the number line below?

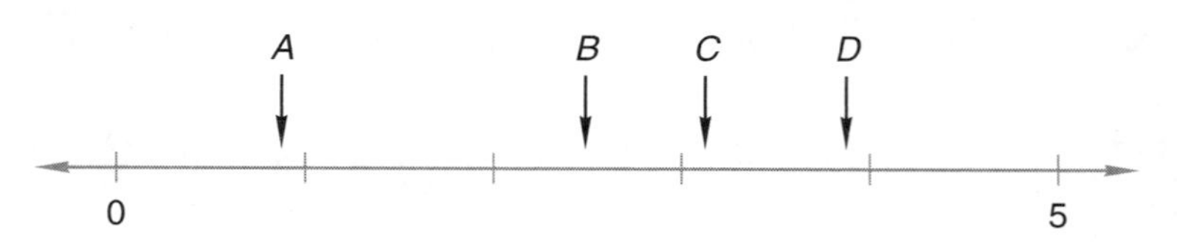

26. (51) Tucker needed to multiply 25 by 24. He thought of 24 as $20 + 4$. Using the distributive property, show two choices Tucker has for multiplying the numbers.

27. (Inv. 5) Maura counted the number of trees on each property on her block. Below are the results. Make a line plot to display these data.

2, 9, 2, 5, 4, 5, 1, 5, 4, 5, 5, 4, 12, 4

28. (Inv. 5) Refer to the plot of data in problem 27.

(a) Name two outliers.

(b) Name a data cluster.

29. (52) Write three million, two hundred thousand in expanded notation.

LESSON

53 Perimeter • Measures of a Circle

WARM-UP

Facts Practice: 64 Multiplication Facts (Test F)

Mental Math: Count by 6's from 6 to 60. Count by 60's from 60 to 360. How many minutes are 2 hours? ... 3 hours? ... 4 hours? ... 10 hours?

a. 2 hours 15 minutes is how many minutes?

b. 2000 − 500 **c.** $2\frac{1}{2} + 2\frac{1}{2}$ **d.** $2\frac{1}{2} - 2\frac{1}{2}$

e. How many minutes is $1\frac{1}{2}$ hours? ... $2\frac{1}{2}$ hours?

f. Find half of 100, ÷ 2, ÷ 5, ÷ 5, × 10, ÷ 5

Problem Solving:

The numbers 3, 6, 10, and 15 are examples of triangular numbers. The numbers 4, 9, 16, and 25 are examples of square numbers. Find a two-digit number that is both a triangular number and a square number.

NEW CONCEPTS

Perimeter

When line segments enclose an area, a polygon is formed. We can find the distance around a polygon by adding the lengths of all the segments that form the polygon. The distance around a polygon is called the **perimeter.**

We should note that the word *length* has more than one meaning. We have used length to mean the measure of a segment. But length may also mean the longer dimension of a rectangle. We use the word *width* to mean the shorter dimension of a rectangle.

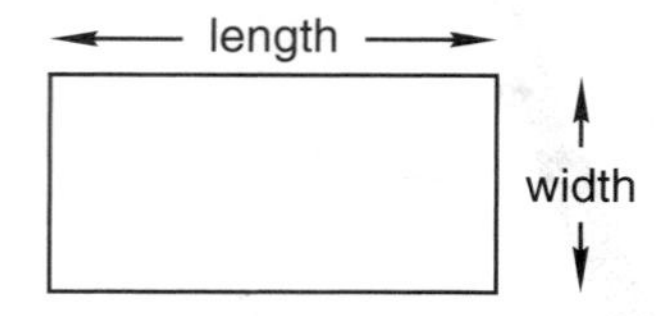

Example 1 What is the perimeter of this rectangle?

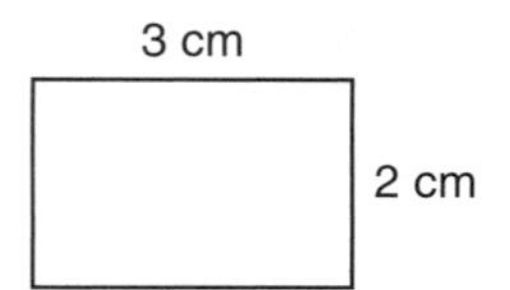

Solution The perimeter is the distance around the rectangle. This rectangle has a length of 3 cm and a width of 2 cm. The four sides measure 2 cm, 3 cm, 2 cm, and 3 cm. We add the lengths of the sides and find that the perimeter is **10 cm.**

$$2 \text{ cm} + 3 \text{ cm} + 2 \text{ cm} + 3 \text{ cm} = 10 \text{ cm}$$

A **regular polygon** has sides equal in length and angles equal in measure. For example, a square is a regular quadrilateral. Below we show some regular polygons.

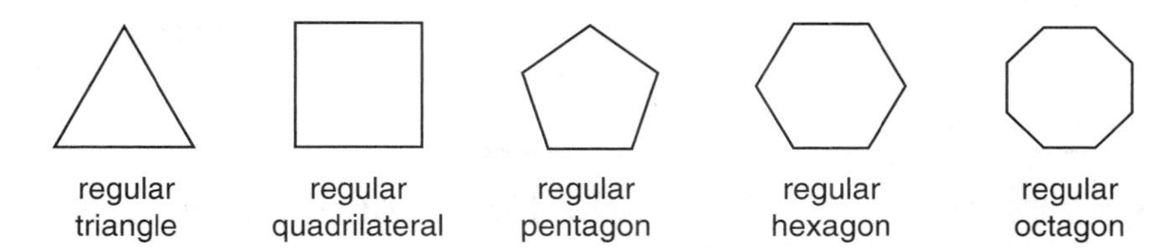

If we know the length of one side of a regular polygon, we can find the perimeter of the polygon by multiplying the length of one side by the number of sides.

Example 2 What is the perimeter of this regular triangle?

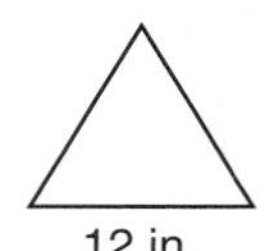

Solution The perimeter is the total of the lengths of the three sides. We can find this by multiplying the length of one side of the regular triangle by 3.

$$3 \times 12 \text{ inches} = \textbf{36 inches}$$

Measures of a circle

A **circle** is a smooth curve. The length of the curve is its **circumference.** So the circumference of a circle is the perimeter of the circle. The **center** of the circle is the "middle point" of the area enclosed by the circle. The **radius** is the distance from the center to the curve. The **diameter** is the distance across the circle through its center. Thus, the diameter of a circle is twice the radius.

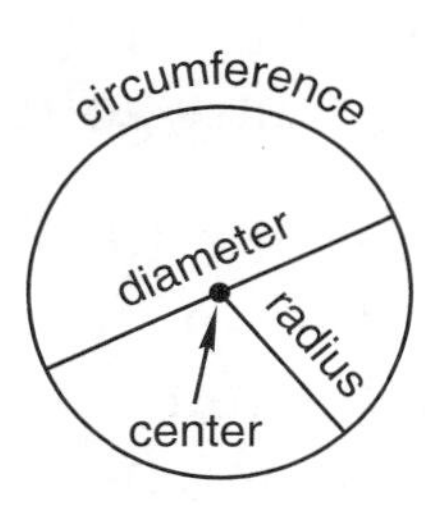

Activity: *Measuring Circles*

Materials needed:

- various circular objects such as paper plates, cups, wheels, and plastic kitchenware lids
- rulers, cloth tape measures, string, or masking tape
- photocopies of Activity Master 17 (masters available in *Saxon Math 6/5 Assessments and Classroom Masters*)

Make a list of circular objects at school and home. Measure the diameter, radius, and circumference of each object, and record the results in the table on Activity Master 17.

LESSON PRACTICE

Practice set

a. What is the length of this rectangle?

5 in.

3 in.

b. What is the width of the rectangle?

c. What is the perimeter of the rectangle?

d. What is the perimeter of this right triangle?

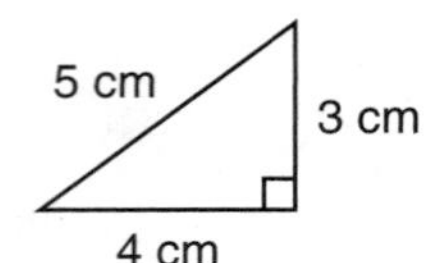

e. What is the perimeter of this square?

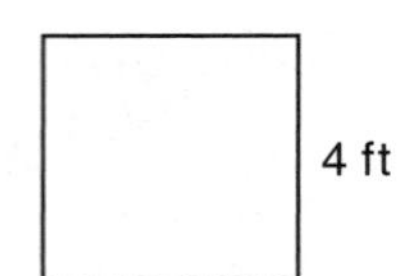

f. What do we call the perimeter of a circle?

g. What do we call the distance across a circle through its middle?

h. If the radius of a circle is 6 inches, what is the diameter of the circle?

MIXED PRACTICE

Problem set

1. (49) Atop the beanstalk Jack was excited to discover that the goose had laid 3 dozen golden eggs. Jack took 15 eggs. How many golden eggs were left?

2. (50) There are 13 players on one team and 9 players on the other team. If some of the players from one team join the other team so that the same number of players are on each team, how many players will be on each team?

3. (Inv. 3, 46) Draw a diagram to illustrate and solve this problem:

If $\frac{1}{3}$ of the 30 students earned A's on the test, how many students earned A's? What percent of the students earned A's?

4. (50) If water is poured from glass to glass until the amount of water in each glass is the same, how many ounces of water will be in each glass?

5. (52) In the number 123,456,789,000, the 2 means which of the following?

A. 2 billion B. 20 billion C. 200 billion

6. (25) Which factors of 8 are also factors of 12?

7. (28, 35) From the year 1820 to 1890 was how many decades?

8. (52) Use digits to write nineteen million, four hundred ninety thousand.

9. (24, 43) $6 + \left(4\frac{2}{3} - 2\right)$

10. (41, 43) $4\frac{2}{3} - \left(2\frac{2}{3} + 2\right)$

11. (29) 300×200

12. (29) 800×70

13. (26, 34) $5T = 500$

14. (51) $\$5.64 \times 78$

15. (51) 865×74

16. (51) 983×76

17. (13) $\$63.14 - \$42.87$

18. (9) $3106 - 875$

19. (13) $\$68.09 + \$43.56 + \$27.18 + \$14.97$

20. (26) $\dfrac{\$31.65}{5}$

21. (34) $\dfrac{4218}{6}$

22. (26) $5361 \div 10$

23. (33) Counting by tens, 1236 is closest to which number?

A. 1230 B. 1240 C. 1200 D. 1300

24. What is the length of this rectangle?
(53)

25. What is the perimeter of this rectangle?
(53)

2 cm

3 cm

26. To multiply 35 by 21, Christina thought of 21 as 20 + 1. Show two choices Christina has for multiplying the numbers.
(51)

27. Write 2,050,000 in expanded notation.
(52)

28. Draw an equilateral triangle.
(36)

29. Freddy found the circumference of the soup can to be $8\frac{5}{8}$ inches. Round $8\frac{5}{8}$ inches to the nearest inch.
(44)

LESSON

54 Dividing by Multiples of 10

WARM-UP

Facts Practice: 48 Uneven Divisions (Test G)

Mental Math:

a. $1.00 − $0.33   b. $7\frac{1}{2} + 1\frac{1}{2}$ c. 50% of 50

d. 4 ft 2 in. is how many inches?

e. 6 × 6, − 1, ÷ 5, × 2, + 1, ÷ 3, × 2

Problem Solving:

Thirty-six dots are arranged in a square array of rows and columns. How many dots are in each row?

NEW CONCEPT

In this lesson we will begin to divide by two-digit numbers that are multiples of 10. Multiples of 10 are the numbers 10, 20, 30, 40, 50, and so on. In later lessons we will practice dividing by other two-digit numbers.

We will continue to follow the four steps of the division algorithm: divide, multiply, subtract, and bring down. The divide step is more difficult when dividing by two-digit numbers because we might not quickly recall two-digit multiplication facts. To help us divide by a two-digit number, we may think of dividing by the first digit only.

To help us divide this: $30\overline{)75}$

... we may think this: $3\overline{)7}$

We use the answer to the easier division for the answer to the more difficult division. Since $3\overline{)7}$ is 2, we use 2 in the division answer. We complete the division by doing the multiplication and subtraction steps.

$$\begin{array}{r} 2\text{ R }15 \\ 30\overline{)75} \\ \underline{60} \\ 15 \end{array}$$

Notice where we placed the 2 above the box. Since we are dividing 75 by 30, we place the 2 above the 5 of 75 and not above the 7.

$$\begin{array}{r} 2 \\ 30\overline{)75} \end{array}$$

The 2 above the 5 means there are two 30's in 75. This is the correct place.

It is important to place the digits in the quotient properly.

Example 1 Divide: $30\overline{)454}$

Solution We follow the four steps: divide, multiply, subtract, and bring down. We begin by finding $30\overline{)45}$. If we are unsure of the answer, we may think "$3\overline{)4}$" to help us with the division step. We divide and write "1" above the 5 of 454. Then we multiply, subtract, and bring down. Since we brought down a digit, we divide again. This time we divide 154 by 30. To help us divide, we may mentally remove the last digit from each number and think "$3\overline{)15}$." We write "5" above the box, and then multiply and subtract. The answer to the division is **15 R 4.**

```
   15 R 4
30)454
   30
   154
   150
     4
```

Recall that we check a division answer by multiplying the quotient by the divisor and then adding any remainder. The result should equal the dividend.

```
   15
×  30
  450
+   4
  454
```

Example 2 Divide: $20\overline{)\$4.60}$

Solution When dividing money by a whole number, we place the decimal point in the quotient directly above the decimal point in the dividend. Then we ignore the decimal points and divide just as we would divide whole numbers. By adding a zero before the decimal point, we get the answer **$0.23.**

```
   $ .23
20)$4.60
     4 0
      60
      60
       0
```

LESSON PRACTICE

Practice set* Divide:

a. $30\overline{)\$4.20}$ **b.** $60\overline{)725}$ **c.** $40\overline{)\$4.80}$

d. $20\overline{)\$3.20}$ **e.** $50\overline{)610}$ **f.** $10\overline{)345}$

g. Show how to check this division answer. Is the answer correct?

```
   23 R 5
40)925
```

MIXED PRACTICE

Problem set

1. (49) Carmen went to the store with \$5.25. She bought a box of cereal for \$3.18 and a half gallon of milk for \$1.02. How much money did Carmen have left?

2. (46) A yard is 36 inches. How many inches is $\frac{2}{3}$ of a yard? Draw a diagram to illustrate the problem.

3. (33) Round 1236 to the nearest ten.

4. (52) The 7 in 987,654,321 means which of the following?

A. 700 B. 7,000,000 C. 700,000

5. (Inv. 2, 37) Draw two circles. Shade $\frac{1}{2}$ of one and $\frac{2}{4}$ of the other. What percent of a circle is $\frac{2}{4}$ of a circle?

6. (Inv. 2) (a) How many cents is $\frac{1}{4}$ of a dollar?

(b) How many cents is $\frac{2}{4}$ of a dollar?

7. (52) Use words to name the number 3,150,000,000.

8. (25) Which factors of 9 are also factors of 12?

9. (54) $30\overline{)454}$

10. (54) $40\overline{)\$5.60}$

11. (54) $50\overline{)760}$

12. (29) 500×400

13. (51) 563×46

14. (51) $68 \times \$4.32$

15. (41) $25\frac{1}{4} + 8\frac{2}{4}$

16. (41) $36\frac{2}{3} - 17\frac{2}{3}$

17. (26) $2947 \div 8$

18. (34, 54) $7564 \div (90 \div 10)$

19. (9)
$$\begin{array}{r} 12,345 \\ -\ 6,789 \\ \hline \end{array}$$

20. (13)
$$\begin{array}{r} \$3.65 \\ \$2.47 \\ \$4.83 \\ +\ \$2.79 \\ \hline \end{array}$$

21. (21) Thirty-six children were seated at tables with four children at each table. How many tables with children were there?

22. (53) What is the perimeter of this right triangle?

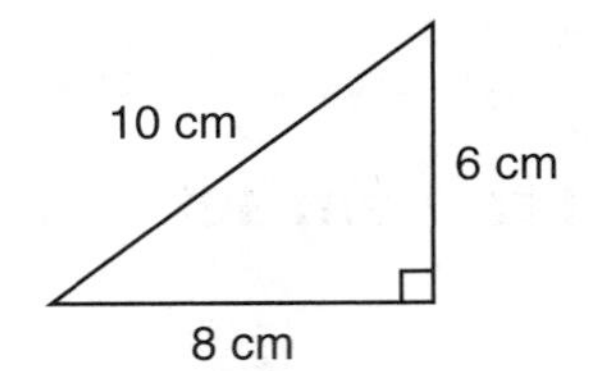

23. (44) Use a ruler to find the length of this rectangle in inches:

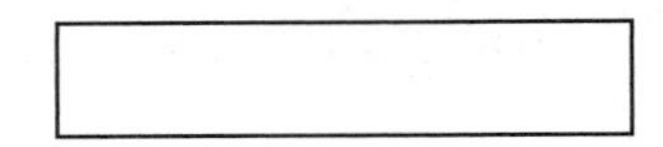

24. (28, 35) What year was five decades after 1896?

25. (53) If the diameter of this circle is 30 millimeters, then what is the radius of the circle?

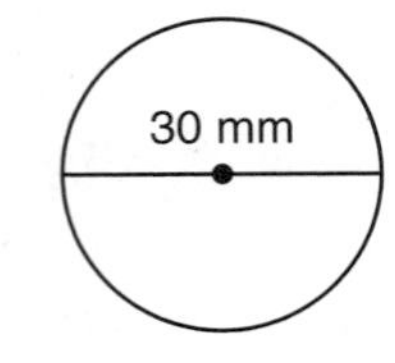

26. (51) Irene wants to multiply 150 by 12. She thinks of 12 as 10 + 2. Using the distributive property, show two ways Irene can multiply the numbers. What is the product?

27. (1, 42) Here is a sequence of numbers we say when counting by sixes:

6, 12, 18, 24, 30, ...

Which of the following numbers is not in the sequence?

A. 456 B. 654 C. 645

28. (21) If a dollar's worth of pennies is divided into four equal groups, how many pennies would be in each group?

29. (36) Could a triangle with sides 8 cm, 6 cm, and 8 cm long be a scalene triangle? Why or why not?

LESSON

55 Multiplying by Three-Digit Numbers

WARM-UP

Facts Practice: 64 Multiplication Facts (Test F)

Mental Math:

a. How many hours is 50% of a day?
b. How many hours is 25% of a day?
c. Five feet four inches is how many inches?
d. $1\frac{1}{2} + 1\frac{1}{2}$ **e.** 25% of 40
f. 6 × 8, + 1, ÷ 7, + 2, ÷ 3, + 1, ÷ 2

Problem Solving:

Two figures are congruent if they are the same shape and size. Draw a triangle that is congruent to this triangle.

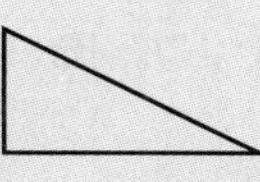

NEW CONCEPT

When we multiply by a three-digit number, we actually multiply three times. We multiply by the hundreds, we multiply by the tens, and we multiply by the ones. We demonstrate this below with the multiplications for finding 234 × 123.

$$\begin{array}{r} 234 \\ \times\ 123 \\ \hline \end{array} \text{ is the same as } \begin{array}{r} 234 \\ \times\ 100 \\ \hline 23{,}400 \end{array} \text{ plus } \begin{array}{r} 234 \\ \times\ 20 \\ \hline 4680 \end{array} \text{ plus } \begin{array}{r} 234 \\ \times\ 3 \\ \hline 702 \end{array}$$

$$23{,}400 + 4680 + 702 = 28{,}782$$

We do not need to separate a three-digit multiplication problem into three problems before we start. We may do all the multiplication within the same problem.

Example Multiply:

$$\begin{array}{r} 234 \\ \times\ 123 \\ \hline \end{array}$$

Solution

```
   234
 × 123
   702  ←  We first multiply 234 by the 3 of 123.
  4680  ←  Then we multiply by the 20 of 123.   } The zeros need not be written.
 23400  ←  Then we multiply by the 100 of 123. }
 28782  ←  We add the three partial products to find the total product.
```

We place the thousands comma and get **28,782.**

LESSON PRACTICE

Practice set Find each product:

a. 346×354 **b.** 487×634 **c.** 403×768 **d.** 705×678

MIXED PRACTICE

Problem set

1. (49) Carlos bought a hamburger for $1.65 and a drink for $0.90. He paid for the food with a $5 bill. How much should he get back in change?

2. (46) Draw a diagram to illustrate and solve this problem:

There are 276 pages in the book. If Martin has read three fourths of the book, how many pages has he read?

3. (35) Martin's 276-page book is 26 pages shorter than Jaime's book. How many pages long is Jaime's book? Write an equation and find the answer.

4. (52) Which digit in 98,765,432 is in the ten-millions place?

5. (47) Amanda can jump across a rug that is 2 yards 3 inches long. How many inches is 2 yards 3 inches? (A yard is 36 inches.)

6. (Inv. 3, 37) Draw a circle. Shade all but one third of it. What percent of the circle is shaded?

7. (52) Use digits to write six hundred seventy-nine million, five hundred forty-two thousand, five hundred.

8. $60\overline{)\$7.20}$ (54)

9. $70\overline{)850}$ (54)

10. $80\overline{)980}$ (54)

11. (55) $\begin{array}{r} 234 \\ \times\ 123 \\ \hline \end{array}$

12. (51) $\begin{array}{r} \$3.75 \\ \times\ \ \ 26 \\ \hline \end{array}$

13. (55) $\begin{array}{r} 604 \\ \times\ 789 \\ \hline \end{array}$

14. (53) Each side of this square is 10 mm long. What is the perimeter of this square?

10 mm

Use mental math to answer problems 15–20.

15. (29) 400×800

16. (29) 60×500

17. (29) 900×90

18. (6) $\begin{array}{r} 300 \\ 400 \\ +\ 500 \\ \hline \end{array}$

19. (9) $\begin{array}{r} 6000 \\ -\ 2000 \\ \hline \end{array}$

20. (54) $\dfrac{400}{20}$

21. (41) $6\frac{5}{11} + 5\frac{4}{11}$

22. (43) $3\frac{2}{3} - 3$

23. (41, 43) $7\frac{2}{3} - \left(3\frac{1}{3} - 3\right)$

Use this information to answer problems 24 and 25:

> *The Arroyo High School stadium can seat 3000 fans. Two thousand, one hundred fifty ticket-holding fans came to the first game. Arroyo won by a score of 35 to 28. Tickets to watch the game cost $2 each.*

24. (21, Inv. 5) Altogether, the fans who came to the first game paid how much money for tickets?

25. (16, Inv. 5) At the second game all but 227 seats were filled with fans. How many fans came to the second game?

26. (52) The crowd lining the parade route was estimated to be 1,200,000. Write this number in expanded notation.

27. (36) Draw an isosceles triangle.

28. (21) If a dollar's worth of dimes is divided into five equal groups, how many dimes would be in each group?

29. (44) The pencil was $5\frac{7}{8}$ inches long. Record the length of the pencil to the nearest inch.

LESSON

56 Multiplying by Three-Digit Numbers That Include Zero

WARM-UP

Facts Practice: 48 Uneven Divisions (Test G)

Mental Math: Round each length to the nearest inch:

a. $5\frac{7}{8}$ in. **b.** $12\frac{3}{8}$ in. **c.** $9\frac{3}{4}$ in.

d. How much is 600 divided by 10? ... 600 ÷ 20? ... 600 ÷ 30?

e. 7 × 42 **f.** 50% of 42

g. 6 × 8, + 6, ÷ 9, × 7, − 7, ÷ 5

Problem Solving:

Copy this multiplication problem and fill in the missing digits:

```
   3_
×   _
 ————
  333
```

NEW CONCEPT

When we multiply by a three-digit number that has a zero as one of its digits, we may find the product by doing two multiplications instead of three.

Example 1 Multiply: 243 × 120

Solution When we multiply by a number that ends with a zero, we may write the problem so that the zero "hangs out" to the right.

```
    243
×   120
———————
   4860  ←— We multiply by the 20 of 120.
  24300  ←— Then we multiply by the 100 of 120.
———————
  29160  ←— We add the two partial products to find the total product.
```

We place the thousands comma in the final product to get **29,160.**

Example 2 Multiply: 243 × 102

Solution We may write the two factors in either order. Sometimes one order is easier to multiply than the other. In the solution on the left below we multiplied three times. On the right we used a shortcut and multiplied only twice. Either way, the product is **24,786.**

```
   102              243
 × 243      or    × 102
 -----            -----
   306              486
  408             2430
 204              -----
 -----            24786
 24786
```

The shortcut on the right was to "bring down" the zero in the bottom factor rather than multiply by it. If we had not used the shortcut, then we would have written a row of zeros, as shown below.

```
   243
 × 102   ←— zero in bottom factor
 -----
   486
  000    ←— row of zeros
 243
 -----
 24786
```

In order to use this pencil-and-paper shortcut, we remember to set up multiplication problems so that factors containing zero are at the bottom.

Example 3 Multiply: $3.25 × 120

Solution We ignore the dollar sign and the decimal point until we have finished multiplying. We place the dollar sign and the decimal point in the final product to get **$390.00.**

```
    $3.25
 ×    120
 --------
    65 00
   325
 --------
  $390.00
```

LESSON PRACTICE

Practice set* Multiply:

a. 234 × 240	**b.** $1.25 × 240	**c.** 230 × 120	**d.** 304 × 120
e. 234 × 204	**f.** $1.25 × 204	**g.** 230 × 102	**h.** 304 × 102

MIXED PRACTICE

Problem set

1. (49) Diana and her sister want to buy a radio for $30. Diana has $12 and her sister has $7. How much more money do they need?

2. (46) How many seconds equal three sixths of a minute? Draw a diagram to illustrate and solve the problem.

3. (49) It is 8 blocks from Jada's house to school. How many blocks does she ride her bike traveling to and from school for 5 days?

4. (50) When the students got on the buses to go to the picnic, there were 36 on one bus, 29 on another bus, and 73 on the third bus. If students are moved so that the same number are on each bus, how many students will be on each bus?

5. (52) Which digit in 123,456,789 is in the ten-thousands place?

6. (53) The radius of this circle is 5 inches. What is the diameter of the circle?

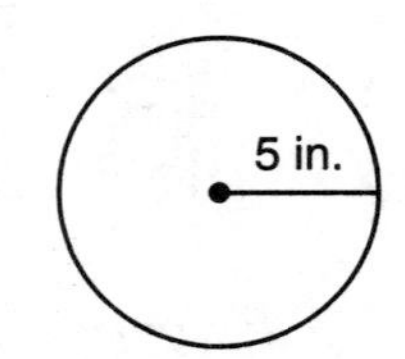

7. (52) Use digits to write the number three hundred forty-five million, six hundred fourteen thousand, seven hundred eighty-four.

8. (53) What is the perimeter of this rectangle?

20 mm

10 mm

9. 900×40 (29)

10. 700×400 (29)

11. 234×320 (56)

12. $\$3.45 \times 203$ (56)

13. 468×386 (55)

14. $\frac{w}{5} = 6$ (20)

15. $4317 \div 6$ (26)

16. $2703 \div 9$ (34)

17. $8m = \$86.08$ (26, 34)

18. (6)
$$\begin{array}{r} 79{,}089 \\ 37{,}865 \\ 29{,}453 \\ +\ 16{,}257 \\ \hline \end{array}$$

19. (9)
$$\begin{array}{r} 43{,}218 \\ -\ 32{,}461 \\ \hline \end{array}$$

20. (13)
$$\begin{array}{r} \$100.00 \\ -\ \$\quad 4.56 \\ \hline \end{array}$$

21. $3\frac{5}{6} - 1\frac{5}{6}$ (41)

22. $4\frac{1}{8} + 6$ (43)

23. Three weeks and three days is how many days? (28, 47)

24. Which arrow could be pointing to 1362? (27)

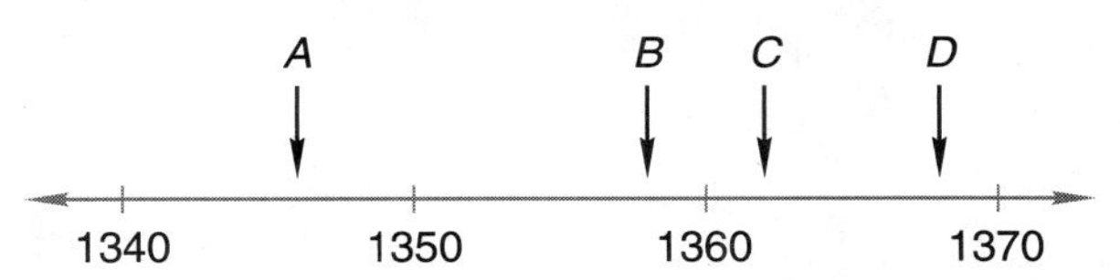

25. Use words to name the mixed number $7\frac{1}{10}$. (40)

26. Timmy needs to multiply 203 by 150. He thinks of 203 as 200 + 3. Show two ways Timmy could multiply these numbers. What is the product? (51, 56)

27. Which of these divisions has no remainder? (22, 42)

A. $543 \div 9$ B. $543 \div 5$ C. $543 \div 3$ D. $543 \div 2$

28. The large square has been divided into 100 small squares. How many small squares equal $\frac{1}{4}$ of the large square? (Inv. 2)

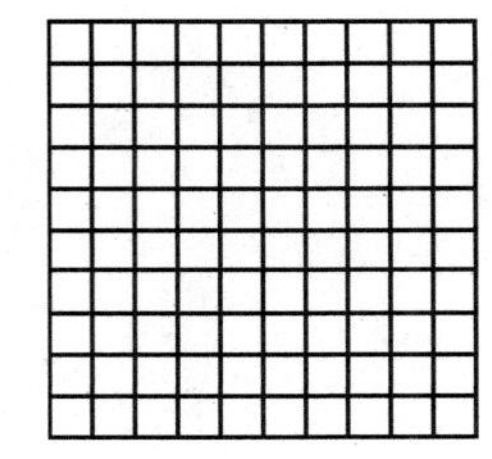

29. The circumference of the globe was $37\frac{3}{4}$ inches. Round the circumference to the nearest inch. (44)

LESSON

57 Simple Probability

WARM-UP

Facts Practice: 64 Multiplication Facts (Test F)

Mental Math: Round each length to the nearest inch:

a. $7\frac{1}{4}$ in. **b.** $4\frac{3}{4}$ in. **c.** $9\frac{1}{8}$ in.

d. What coin is 10% of a dollar?

e. $100 \div 4$ **f.** $100 \div 5$

g. $10 \times 10, \div 2, - 1, \div 7, - 1, \div 2, - 1, \div 2$

Problem Solving:

A loop of string forms this rectangle. If the same loop is used to form a square, what will be the length of each side?

8 in.

4 in.

NEW CONCEPT

There are many situations whose future outcomes are uncertain. For example, the weather forecast might say that rain is likely tomorrow, but this would only be an educated guess. It might rain or it might not rain. If we take an airplane flight, we might arrive early, we might arrive late, or we might arrive on time. We do not know for sure in advance.

Probability is a measure of **how likely** it is that an event (or combination of events) will occur. Probabilities are numbers between 0 and 1. An event that is **certain** to occur has a probability of 1. An event that is **impossible** has a probability of 0. If an event is uncertain to occur, then its probability is a fraction between 0 and 1. The more **likely** an event, the closer its probability is to 1. The more **unlikely** an event, the closer its probability is to 0. The diagram below uses words to describe the range of probabilities from 0 to 1.

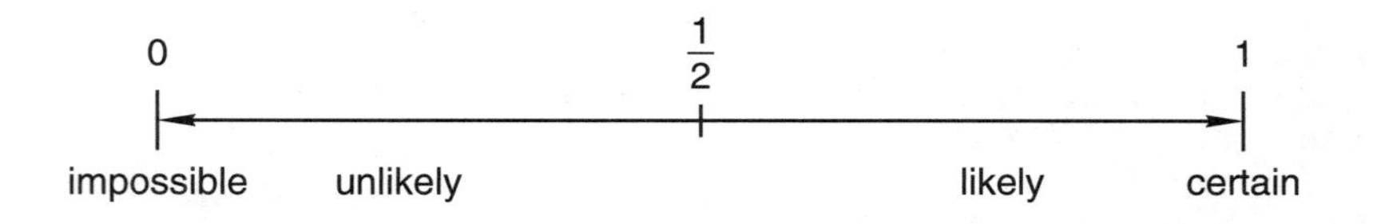

The word **chance** is also used to describe the likelihood of an event. Chance is often expressed as a percent ranging from 0% (for events that are impossible) to 100% (for events that are certain to happen). If the chance of rain is forecast as 80%, then in the meteorologist's informed judgment, it is likely to rain.

The expression "50-50 chance" means an event is equally likely to happen (50%) as it is not to happen (50%). Added together, the chances (or probability) of an event happening or not happening total 100% (or 1). So if the chance of rain is 80%, then the chance that it will not rain is 20%. If the probability of winning a drawing is $\frac{1}{1000}$, then the probability of not winning the drawing is $\frac{999}{1000}$.

Example 1 A standard dot cube is rolled once. Which word best describes each event in parts (a)–(d): *certain, likely, unlikely,* or *impossible?*

(a) The cube will stop with 3 dots on top.

(b) The cube will stop with more than 2 dots on top.

(c) The cube will stop with fewer than 7 dots on top.

(d) The cube will stop with more than 6 dots on top.

Solution (a) **Unlikely.** There are six faces and only one has 3 dots. So we would expect the cube to stop with 3 dots on top less than half the times the cube is rolled.

(b) **Likely.** Of the six faces on the dot cube, four have more than 2 dots. So we would expect that a number greater than 2 would end up on top more than half the times the cube is rolled.

(c) **Certain.** All the faces have fewer than 7 dots. So every time the cube is rolled, the upturned face will have fewer than 7 dots.

(d) **Impossible.** None of the faces have more than 6 dots. So it is not possible for an upturned face to have more than 6 dots.

Many **experiments** involve probability. Some experiments that involve probability are tossing a coin, spinning a spinner, and selecting an object from a set of objects without looking. The possible results of such experiments are called **outcomes.** The probabilities of the outcomes of any experiment always add up to 1.

Example 2 The circle below is divided into 5 equal-size **sectors.** Each sector is labeled by one of these letters: A, B, or C. Suppose the spinner is spun and stops in one of the sectors.

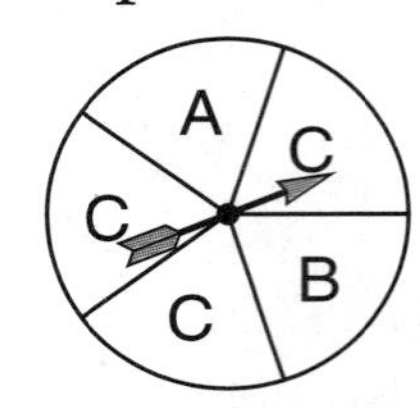

Find the probability of each of the possible outcomes A, B, and C.

Solution The probability that the spinner will stop in a certain sector is equal to that sector's fraction of the circle. Since outcome A corresponds to $\frac{1}{5}$ of the whole, the probability that the spinner will stop in sector A is $\mathbf{\frac{1}{5}}$. Outcome B also has a probability of $\mathbf{\frac{1}{5}}$. Since outcome C corresponds to $\frac{3}{5}$ of the whole, it has a probability of $\mathbf{\frac{3}{5}}$. Notice that $\frac{1}{5} + \frac{1}{5} + \frac{3}{5} = \frac{5}{5} = 1$. (The probabilities of the outcomes of an experiment always total 1.)

Example 3 A bag contains 5 red marbles, 3 blue marbles, and 2 yellow marbles. Suppose we pick one marble from the bag without looking.

(a) Find the probability that the marble is blue.

(b) Find the probability that the marble is not blue.

Solution (a) The probability that we pick a blue marble is a fraction between 0 and 1. This fraction describes the number of blue marbles as a part of the overall group of marbles. Since 3 out of 10 marbles are blue, the probability that we pick a blue marble is $\mathbf{\frac{3}{10}}$.

(b) The remaining 7 marbles are not blue, so the probability that the marble is not blue is $\mathbf{\frac{7}{10}}$.

LESSON PRACTICE

Practice set Use the circle at right to answer problems **a–d.**

a. What is the probability that the spinner will stop on 3?

b. What is the probability of spinning a number greater than three?

c. What is the probability of spinning an even number?

d. What is the probability that the spinner will stop on an odd number?

e. If the weather forecast states that the chance of rain is 40%, is it more likely to rain or not to rain?

f. If today's chance of rain is 20%, then what is the chance that it will not rain today?

g. On one toss of a standard dot cube, what is the probability of rolling a number less than 1?

h. For the experiment described in example 3, Freddy said that the probability of picking a red marble was $\frac{1}{2}$. Do you agree or disagree with Freddy? Why?

MIXED PRACTICE

Problem set

1. (47) There are 12 inches in a foot. A person 5 feet 4 inches tall is how many inches tall?

2. (21, 28) How many years is 10 centuries?

3. (53) What word is used to name the perimeter of a circle?

4. (40) Use words to name the mixed number $10\frac{7}{10}$.

5. (28, 46) How many minutes is two thirds of an hour? Draw a diagram to illustrate and solve the problem.

6. (28) Captain Hook heard the alarm go off at 6 a.m. and got up quickly. If he had fallen asleep at 11 p.m. the previous evening, how many hours of sleep did he get?

7. (20) If 4 is the divisor and 12 is the quotient, then what is the dividend?

8. (52) What is the value of the place held by the zero in 321,098,765?

9. (25) Which factors of 15 are also factors of 20?

10. (53) Assume that the sides of this regular hexagon are 3 cm long. What is the perimeter of the hexagon?

11. (24, 41) $3\frac{2}{3} - \left(2\frac{1}{3} + 1\frac{1}{3}\right)$

12. (24, 41) $3\frac{1}{3} + \left(2\frac{2}{3} - 1\frac{1}{3}\right)$

13. (54) $40\overline{)\$5.20}$

14. (26) $8\overline{)3161}$

15. (20) Which number in this problem is the divisor?

$$6 \div 3 = 2$$

16. (13)
$$\begin{array}{r} \$43.15 \\ -\ \$28.79 \\ \hline \end{array}$$

17. (56)
$$\begin{array}{r} 423 \\ \times\ 302 \\ \hline \end{array}$$

18. (6)
$$\begin{array}{r} 99 \\ 36 \\ 42 \\ 75 \\ 64 \\ 98 \\ +\ 17 \\ \hline \end{array}$$

19. (56)
$$\begin{array}{r} \$3.45 \\ \times\ \ \ 360 \\ \hline \end{array}$$

20. (56)
$$\begin{array}{r} 604 \\ \times\ 598 \\ \hline \end{array}$$

21. (41) $\frac{10}{10} - \frac{9}{10}$

22. (43) $4\frac{2}{3} - \frac{1}{3}$

23. (41) $5\frac{2}{2} - 1\frac{1}{2}$

24. (28) From May 1 of one year to August 1 of the next year is how many months?

25. (28) It is morning. What time will it be in 2 hours 20 minutes?

26. (23, 28) (a) How many years is a millennium?

(b) How many years is half of a millennium?

(c) Write a fraction equal to $\frac{1}{2}$ using the numbers in the answers to parts (a) and (b).

27. (57) If a standard dot cube is rolled once, what is the probability that it will stop with more than one dot on top?

28. (50) Elizabeth's first three test scores were 80, 80, and 95. What was the average of Elizabeth's first three test scores?

29. (57) The multiple-choice question listed four choices for the answer. Kyla figured she had a 25% chance of guessing the correct answer. What was her chance of not correctly guessing the answer?

LESSON

58 Writing Quotients as Mixed Numbers, Part 3

WARM-UP

Facts Practice: 48 Uneven Divisions (Test G)

Mental Math: Round each length to the nearest inch:

a. $18\frac{5}{8}$ in. **b.** $12\frac{3}{8}$ in. **c.** $4\frac{1}{16}$ in.

d. How much is 800 ÷ 10? ... 800 ÷ 20? ... 800 ÷ 40?

e. 50% of 800 **f.** $3\frac{1}{2} + 3\frac{1}{2}$

g. 8 × 24 **h.** $\frac{1}{3}$ of 24

Problem Solving:

Two figures are similar if they have the same shape. These two triangles are similar. Draw a triangle that is larger than but similar to these two triangles.

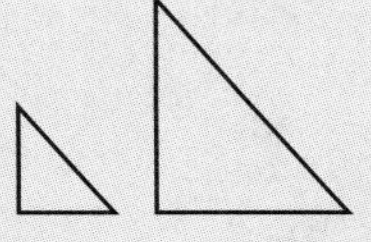

NEW CONCEPT

As we saw in Lessons 40 and 43, we sometimes need to write a division answer as a mixed number. We do this by writing the remainder as a fraction.

If two children share 5 cookies equally, how many cookies will each receive?

We divide 5 into 2 equal parts. We find that the quotient is 2 and the remainder is 1. That is, each child will receive 2 cookies, and there will be 1 extra cookie. We can take the extra cookie and divide it in half. Then each child will receive $2\frac{1}{2}$ cookies.

$$\begin{array}{r} 2\frac{1}{2} \\ 2\overline{)5} \\ \underline{4} \\ 1 \end{array}$$

To write a remainder as a fraction, we simply make the remainder the numerator of the fraction and make the divisor the denominator of the fraction.

Example 1 Divide and write the quotient with a fraction: $3\overline{)50}$

Solution We divide and find that the remainder is 2. We make the remainder the numerator of the fraction, and we make the divisor the denominator of the fraction. The quotient is $\mathbf{16\frac{2}{3}}$.

$$\begin{array}{r} 16\frac{2}{3} \\ 3\overline{)50} \\ \underline{3} \\ 20 \\ \underline{18} \\ 2 \end{array}$$

Example 2 A 15-inch string of licorice is cut into 4 equal lengths. How long is each length?

Solution We divide 15 inches by 4 and find that the quotient is not a whole number of inches. The quotient is more than 3 inches but less than 4 inches. It is 3 inches plus a fraction. To find the fraction, we write the remainder as the numerator of the fraction and write the divisor as the denominator of the fraction. We find that the length of each piece of licorice is $\mathbf{3\frac{3}{4}}$ **inches.**

$$\begin{array}{r} 3\frac{3}{4} \\ 4\overline{)15} \\ \underline{12} \\ 3 \end{array}$$

In the problem sets that follow, we will continue to write quotients with remainders, unless a problem asks that the answer be written with a fraction.

LESSON PRACTICE

Practice set* Divide. Write each quotient as a mixed number.

a. $4\overline{)17}$ **b.** $20 \div 3$ **c.** $\frac{16}{5}$

d. $5\overline{)49}$ **e.** $21 \div 4$ **f.** $\frac{49}{10}$

g. $6\overline{)77}$ **h.** $43 \div 10$ **i.** $\frac{31}{8}$

MIXED PRACTICE

Problem set

1. (49) Martin bought 8 baseball cards for 15 cents each. If he paid $2, how much should he have received in change?

2. (21, 58) Daphne bought a 21-inch string of licorice at the candy store. She cut it into 4 equal lengths to share with her friends. How long was each length of licorice? Write the answer as a mixed number.

3. Draw a diagram to illustrate and solve this problem:
(Inv. 3, 46)

Sarah used $\frac{3}{5}$ of a sheet of stamps to mail cards. If there are 100 stamps in a whole sheet, then how many stamps did Sarah use? What percent of the stamps did Sarah use?

4. Round 1776 to the nearest hundred.
(33)

5. In which of these numbers does the 5 have a value of 500,000?
(52)

A. 186,542,039 B. 347,820,516 C. 584,371,269

6. What is the perimeter of this rectangle?
(53)

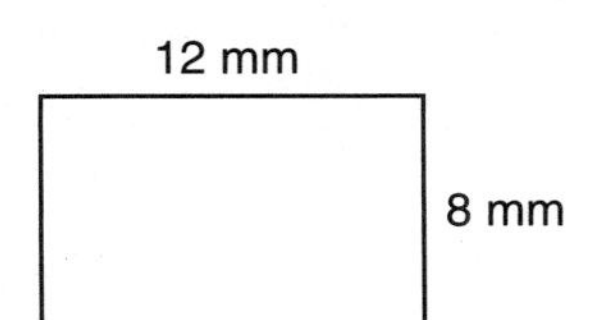

7. $30\overline{)640}$
(54)

8. $40\overline{)922}$
(54)

9. $50w = 800$
(26, 54)

10. $1400 + m = 7200$
(10)

11. $\$1.25 \times 80$
(29)

12. $700 \div 10$
(54)

13. 679×489
(55)

14. $8104 - 5647$
(9)

15. $\$2.86 + \$6.35 + \$1.78 + \$0.46 + \$0.62$
(13)

16. $\frac{4228}{7}$
(34)

17. $\frac{4635}{9}$
(26)

18. $\frac{5}{5} - \frac{1}{5}$
(41)

19. $3\frac{1}{3} - \frac{1}{3}$
(43)

20. $4\frac{6}{6} - 2\frac{5}{6}$
(41)

21. Divide and write the quotient with a fraction: $3\overline{)62}$
(58)

22. What is the denominator of the fraction in $6\frac{3}{4}$?
(Inv. 2)

23. In a division problem, if the divisor is 3 and the quotient is 9, then what is the dividend?
(20)

24. What year was five centuries before 1500?
(28, 35)

25. If the radius of this circle is 12 millimeters, then what is the diameter of the circle?
(53)

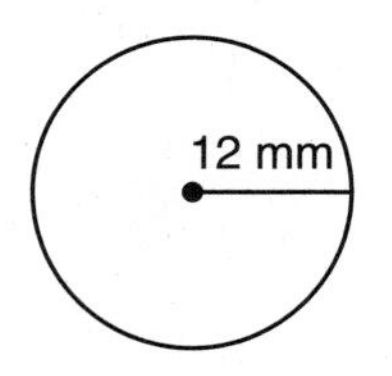

26. In a bag are 2 red marbles, 3 blue marbles, and 6 yellow marbles. If one marble is taken from the bag without looking, what is the probability that the marble will be red?
(57)

27. Which of these triangles appears to be both a right triangle and an isosceles triangle?
(36)

A. 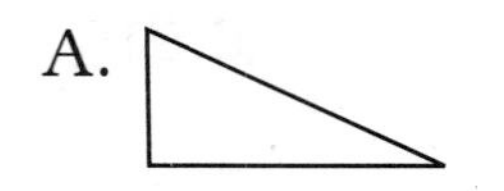 B. C.

28. The large square has been divided into 100 smaller squares. How many small squares equal $\frac{3}{4}$ of the large square?
(Inv. 2)

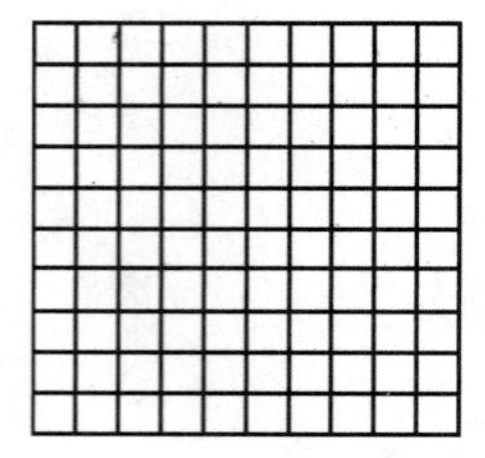

29. China has the largest population of all the countries in the world. In the year 2002 there were approximately one billion, two hundred eighty-four million, two hundred four thousand people living in China. Use digits to write the approximate number of people living in China.
(52)

LESSON

59 Fractions Equal to 1 • Subtracting a Fraction from 1

WARM-UP

Facts Practice: 64 Multiplication Facts (Test F)

Mental Math: Round each length to the nearest inch:

a. $5\frac{3}{16}$ in. **b.** $12\frac{3}{8}$ in. **c.** $8\frac{3}{4}$ in.

d. How many days are in 3 weeks 3 days?

e. Half of 101 **f.** 10% of 50

g. 6 × 6, − 1, ÷ 7, × 4, + 1, ÷ 7

Problem Solving:

One hundred dots are arranged in a square array of rows and columns. How many dots are in each column?

NEW CONCEPTS

Fractions equal to 1 We know that two halves make a whole. Similarly, it takes **three** thirds or **four** fourths or **five** fifths to make one whole.

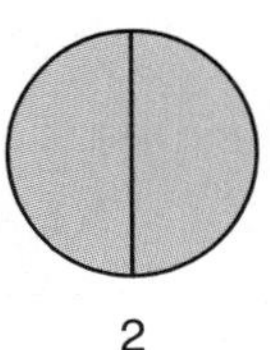

$\frac{2}{2}$

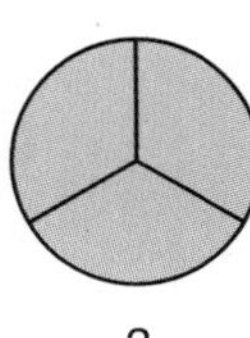

$\frac{3}{3}$

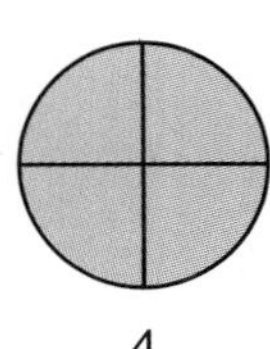

$\frac{4}{4}$

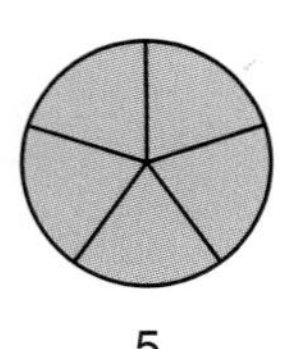

$\frac{5}{5}$

We see that each of these is a "whole pie," yet we can use different fractions to name each one. Notice that the numerator and the denominator are the same when we name a "whole pie." This is a very important idea in mathematics. Whenever the numerator and denominator of a fraction are equal (but not zero), the fraction is equal to 1.

Example 1 Write a fraction equal to 1 that has a denominator of 4.

Solution A fraction equal to 1 that has a denominator of 4 would also have a numerator of 4, so we write $\frac{4}{4}$.

Example 2 Add: $\frac{1}{4} + \frac{3}{4}$

Solution We add and find that the sum is $\frac{4}{4}$. We should always write our answers in simplest form. The simplest name for $\frac{4}{4}$ is **1.**

$$\frac{1}{4} + \frac{3}{4} = \frac{4}{4} = 1$$

Example 3 Compare: $4\frac{3}{3} \bigcirc 5$

Solution The mixed number $4\frac{3}{3}$ means $4 + \frac{3}{3}$. Since $\frac{3}{3}$ equals 1, the addition $4 + \frac{3}{3}$ is the same as $4 + 1$, which is 5. We find that $4\frac{3}{3}$ and 5 are equal.

$$\mathbf{4\frac{3}{3} = 5}$$

Example 4 Add: $1\frac{1}{2} + 1\frac{1}{2}$

Solution We add and find that the sum is $2\frac{2}{2}$. The mixed number $2\frac{2}{2}$ means $2 + \frac{2}{2}$. Since $\frac{2}{2}$ equals 1, the addition $2 + \frac{2}{2}$ is the same as $2 + 1$, which is **3.**

$$1\frac{1}{2} + 1\frac{1}{2} = 2\frac{2}{2} = 3$$

Subtracting a fraction from 1

To subtract a fraction from 1, we rewrite 1 as a fraction. There are many fractions equal to 1, such as $\frac{2}{2}$, $\frac{3}{3}$, $\frac{6}{6}$, and $\frac{10}{10}$. We look at the fraction that is subtracted to decide which name for 1 we should use.

Example 5 Subtract: $1 - \frac{1}{3}$

Solution We can show this problem with fraction manipulatives or by drawing a picture that represents a whole pie. If we remove one third of the pie, how much of the pie is still in the pan?

Before we can remove a third, we first slice the pie into three thirds. Then we can subtract one third. We see that two thirds of the pie is still in the pan. Using pencil and paper, we rewrite 1 as $\frac{3}{3}$. Then we subtract.

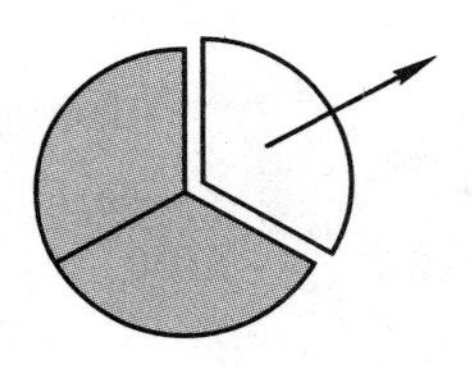

$$1 - \frac{1}{3}$$
$$\downarrow \quad \downarrow$$
$$\frac{3}{3} - \frac{1}{3} = \mathbf{\frac{2}{3}}$$

We could have chosen any name for 1, such as $\frac{2}{2}$ or $\frac{4}{4}$ or $\frac{3682}{3682}$, but we chose $\frac{3}{3}$ because it has the same denominator as the fraction that was subtracted. Remember, we can only add and subtract fractions when their denominators are the same.

LESSON PRACTICE

Practice set **a.** Write a fraction equal to 1 that has a denominator of 3.

Compare:

b. $\frac{4}{4} \bigcirc 1$

c. $5\frac{4}{4} \bigcirc 6$

Add:

d. $\frac{3}{10} + \frac{7}{10}$

e. $3\frac{3}{5} + 2\frac{2}{5}$

Subtract:

f. $1 - \frac{1}{4}$

g. $1 - \frac{2}{3}$

h. How many fraction names for 1 are there?

MIXED PRACTICE

Problem set

1. (47) Cassandra jumped rope for 3 minutes 24 seconds without stopping. How many seconds are in 3 minutes 24 seconds?

2. (49) Brady's mom baked 5 dozen cookies, and Brady ate one tenth of them. How many cookies did he eat?

3. Draw a quadrilateral that has a pair of horizontal parallel line segments of different lengths.
(31, 32)

4. Which factors of 8 are also factors of 20?
(25)

5. How many seconds is two fifths of a minute? Two fifths of a minute is what percent of a minute?
(28, Inv. 3, 46)

6. Maria stood on two scales at the same time. The scale under her right foot read 46 pounds, and the scale under her left foot read 60 pounds. If she balances her weight equally on both scales, how much will each scale read?
(50)

7. $\frac{1}{4} + \frac{3}{4}$
(59)

8. $1\frac{1}{3} + 2\frac{2}{3}$
(59)

9. $2\frac{5}{8} + \frac{3}{8}$
(59)

10. $1 - \frac{1}{4}$
(59)

11. $1 - \frac{3}{8}$
(59)

12. $2\frac{8}{8} - \frac{3}{8}$
(43)

13. $98,789 + 41,286 + 18,175$
(6)

14. $47,150 - 36,247$
(9)

15. 368×479
(55)

16. Use words to name the mixed number $8\frac{9}{10}$.
(40)

17. Divide and write the quotient as a mixed number: $\frac{15}{4}$
(58)

For problems 18 and 19, write the answer with a remainder.

18. $40\overline{)687}$
(54)

19. $60\overline{)850}$
(54)

20. $30\overline{)\$5.40}$
(54)

21. $507 \times \$3.60$
(56)

22. $(900 - 300) \div 30$
(24, 54)

23. Which of these mixed numbers is not equal to 3?
(59)

A. $2\frac{3}{3}$ B. $3\frac{2}{2}$ C. $2\frac{4}{4}$ D. $2\frac{8}{8}$

24. Write a fraction equal to 1 that has a denominator of 5.
(59)

25. What is the perimeter of this equilateral triangle?
(36, 44, 53)

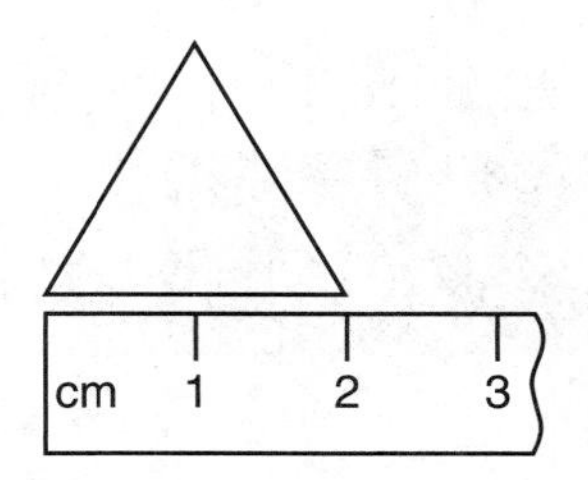

26. To multiply 35 by 21, Jim thought of 21 as 20 + 1 and performed the multiplication mentally. Show two ways Jim could multiply the numbers. Which way seems easier to perform mentally?
(51)

The face of this spinner is divided into equal sectors. Refer to the spinner to answer problems 27 and 28.

27. Which two outcomes are equally likely?
(57)

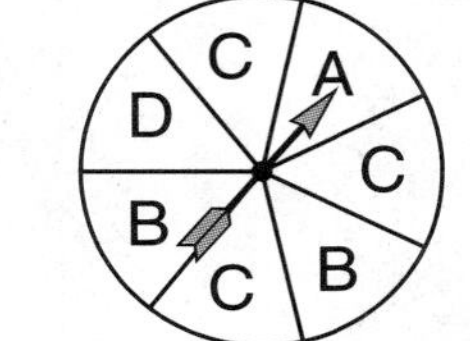

28. What is the probability that the spinner will stop on C?
(57)

29. A teacher asked 20 fifth-grade students how many televisions their families owned. Their responses made up this data set:
(Inv. 5)

1, 3, 2, 1, 4, 0, 1, 2, 3, 1, 3, 2, 2, 2, 3, 4, 3, 3, 2, 3

Copy and complete this frequency table for the number of televisions owned by the families:

Frequency Table

Televisions	Tally	Frequency
0		
1		
2		
3		
4		

LESSON

60 Finding a Fraction to Complete a Whole

WARM-UP

Facts Practice: 48 Uneven Divisions (Test G)

Mental Math: What fraction should be added to each of these fractions to get a total of 1?

a. $\frac{1}{2}$ **b.** $\frac{1}{3}$ **c.** $\frac{1}{4}$ **d.** $\frac{1}{8}$

e. How much is 900 ÷ 10? ... 900 ÷ 30? ... 900 ÷ 90?

f. 9 × 25 **g.** 25% of a dozen

h. 9 × 9, − 1, ÷ 2, − 1, ÷ 3

Problem Solving:

Two figures are congruent if they are the same shape and size. Draw a rectangle that is congruent to this rectangle.

1 in.

$\frac{1}{4}$ in.

NEW CONCEPT

Sometimes we are given one part of a whole and need to know the other part of the whole. Consider this word problem:

One third of the students are girls. What fraction of the students are boys?

We answer problems like this by thinking of the entire group as a whole. We can draw a rectangle to represent the whole group of students. The problem states that $\frac{1}{3}$ of the students are girls, so we divide the rectangle into three parts and label one of the parts "girls."

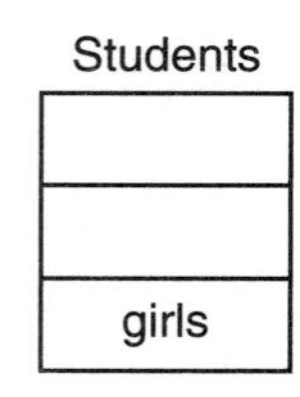

The fraction of the students that is not girls must be boys. Since the girls make up 1 of the 3 parts, the boys must make up 2 of the 3 parts. Thus, two thirds of the students are boys.

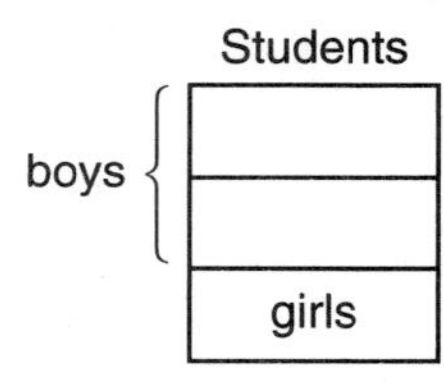

Example Bob found that commercials make up one sixth of TV airtime. What fraction of TV airtime is not commercials?

Solution We begin by thinking of TV airtime as a whole. We draw a rectangle to show this. The problem states that $\frac{1}{6}$ of the airtime is made up of commercials. So we divide the rectangle into six equal parts. We label one part "commercials." We see that the fraction of TV airtime that is not commercials is $\frac{5}{6}$.

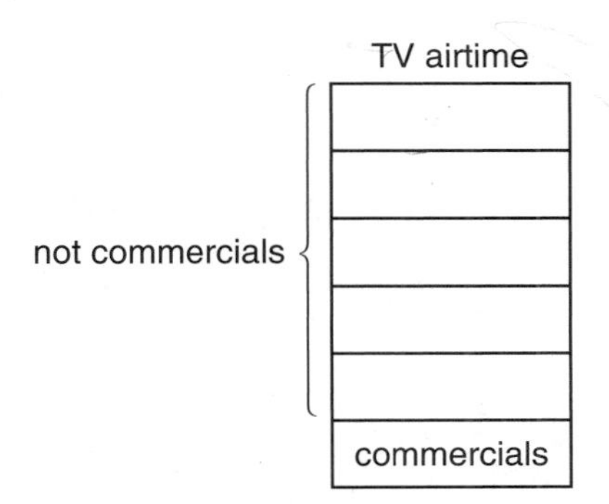

LESSON PRACTICE

Practice set

a. Lily has read one fourth of her book. What fraction of her book is left to read?

b. Five eighths of the gymnasts were able to do a back handspring. What fraction of the gymnasts were unable to do a back handspring?

c. If three fifths of the spectators were rooting for the home team, then what fraction of the spectators were not rooting for the home team?

MIXED PRACTICE

Problem set

1. (49) In one class there are three more girls than boys. There are 14 boys. How many students are in the class?

2. (49) Calvin bought two bicycle tubes for $2.39 each and a tire for $4.49. The tax was 56¢. If he paid $10, how much money should he get back in change?

3. (28, 35) From the year 1800 to the year 1900 was how many decades?

4. (53) The diameter of Kitty's bicycle wheel is 24 inches. What is the radius of the wheel?

5. Round 487 and 326 to the nearest hundred. Then add the rounded numbers. What is their sum?
(33)

6. Find each missing numerator:
(59)

(a) $\frac{\square}{7} = 1$ (b) $4 = 3\frac{\square}{4}$

7. When Mya was born, she weighed 7 pounds 12 ounces. How many ounces did Mya weigh at birth? (One pound equals 16 ounces.)
(47)

8. What is the perimeter of this square?
(53)

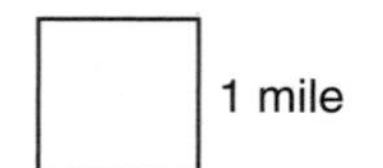

9. $\frac{1}{6} + \frac{2}{6} + \frac{3}{6}$
(59)

10. $3\frac{3}{5} + 1\frac{2}{5}$
(59)

11. $1 - \frac{1}{8}$
(59)

12. $4\frac{5}{5} - 1\frac{2}{5}$
(41)

13. $35.24 − $14.62
(13)

14. $5.78 × 467
(55)

15. $\frac{\$36.72}{9}$
(34)

16. Divide and write the quotient with a fraction: $\frac{23}{10}$
(58)

17. Selby found that commercials made up one eighth of TV airtime. What fraction of TV airtime was not commercials? What percent of TV airtime was commercials?
(Inv. 3, 60)

18. 374 × 360
(56)

19. 643 ÷ 40
(54)

20. 60 × (800 ÷ 40)
(24, 29, 54)

21. $20\overline{)1340}$
(54)

22. Compare: $\frac{4}{4} \bigcirc \frac{5}{5}$
(59)

23. Write a fraction equal to 1 that has a denominator of 8.
(59)

24. To what fraction is the arrow pointing?
(38)

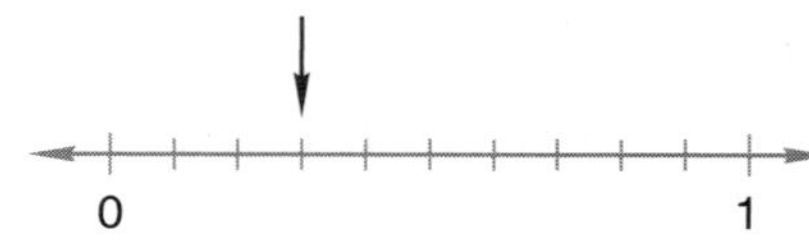

25. If the time is 11:35 a.m., how many minutes is it until noon?
(28)

26. One marble is selected from a bag containing 2 red marbles, 5 green marbles, and 6 white marbles.
(57)

(a) What fraction describes the probability that the marble is green?

(b) What fraction describes the probability that the marble is not green?

27. Which of these division problems will not result in a remainder?
(22, 42)

A. 321 ÷ 2 B. 421 ÷ 3 C. 521 ÷ 6 D. 621 ÷ 9

Refer to the information below to answer problems 28 and 29.

Christine has 30 CDs. Not every CD has the same number of songs on it. This table shows how many of Christine's CDs have 9, 10, 11, 12, 13, or 14 songs:

Frequency Table

Songs per CD	Frequency
9	1
10	4
11	7
12	13
13	3
14	2

28. Among Christine's CDs, what number of songs does a CD most frequently have?
(Inv. 5)

29. How many of Christine's CDs have more than 10 songs?
(Inv. 5)

INVESTIGATION 6

Focus on

Performing Probability Experiments

In Lesson 57 we used the word **probability** to describe how likely it is that a given event occurs in an experiment. Probabilities are fractions. If we repeat an experiment over and over, we can use probability to predict the number of times an event will occur.

A typical dot cube has six faces marked with dots representing the numbers 1, 2, 3, 4, 5, and 6.

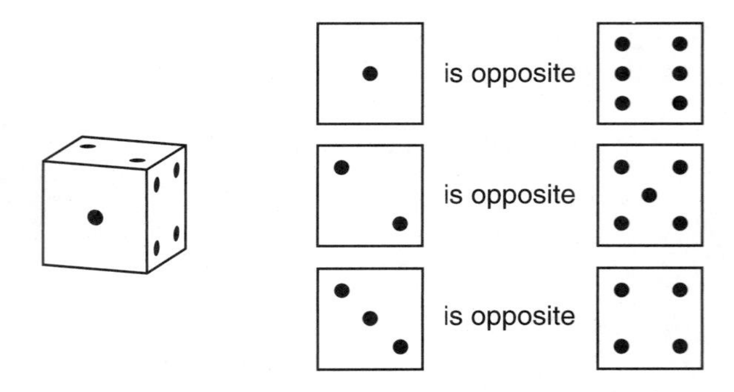

As an experiment, we can roll a dot cube and record the upturned face as an outcome. Because the 6 possible outcomes are equally likely, each outcome must have the same probability. The probabilities of all the outcomes must add up to one, so each outcome has a probability of $\frac{1}{6}$.

$$\frac{1}{6} + \frac{1}{6} + \frac{1}{6} + \frac{1}{6} + \frac{1}{6} + \frac{1}{6} = \frac{6}{6} = \mathbf{1}$$

We can add probabilities in order to determine the likelihood of one of a certain collection of outcomes. For example, the probability that the upturned face will be an even number is the sum of the probabilities of rolling a 2, a 4, or a 6.

$$\frac{1}{6} + \frac{1}{6} + \frac{1}{6} = \frac{3}{6} = \mathbf{\frac{1}{2}}$$

1. What is the probability that the upturned face is either 1 or 6?

2. What is the probability that the upturned face is less than 5?

If we roll our dot cube repeatedly, we can guess how many times certain events will occur. Our guess is based on the part-of-a-group meaning of a fraction. Suppose we rolled our dot cube 24 times. Since all the outcomes have a probability of $\frac{1}{6}$, we guess that we would roll the number 2 (or any other particular number) one-sixth of 24 times. This means we divide 24 by 6.

$$24 \div 6 = 4 \text{ times}$$

Because three faces show even numbers, we guess that we would roll an even number 3×4 times; that is, 12 times. These are only guesses; the actual number of times that an event will occur cannot be predicted with certainty.

3. If a standard dot cube is rolled 60 times, how many times would you guess that the upturned face will be 1? Explain how you arrived at your answer.

4. If a standard dot cube is rolled 60 times, how many times would you guess that the upturned face will be either 1 or 6? Explain how you arrived at your answer.

Activity: *Probability Experiment 1*

Materials needed:

- copies of Activity Master 18 (masters available in *Saxon Math 6/5 Assessments and Classroom Masters*)
- dot cubes

Distribute dot cubes and copies of Activity Master 18. (You might want to divide students into small groups for this activity.) Have each student roll a dot cube 24 times while tallying each outcome on the activity master's frequency table. Then have students use their completed tables to answer problems 5–7. A copy of the frequency table is shown below.

Upturned Number	Tally	Frequency
1		
2		
3		
4		
5		
6		

5. Use the tallies to complete the "Frequency" column on your table.

6. Which of the six outcomes occurred more frequently than you would guess?

7. How many times was the upturned face even?

We can perform probability experiments repeatedly to estimate probabilities that we do not know how to calculate. Suppose Silvia constructs a spinner with 3 regions by dividing up a circle without any definite plan. The spinner she makes is shown below.

To estimate the fraction of the whole that each region takes up, she spins the spinner 50 times. She presents the results in a **relative frequency table.** In the last column Silvia records the number of times each outcome occurred as the numerator of a fraction with denominator 50.

Outcome	Tally	Relative Frequency
1	𝍸 𝍸 𝍸 II	$\frac{17}{50}$
2	𝍸 𝍸 𝍸 𝍸 𝍸 III	$\frac{28}{50}$
3	𝍸	$\frac{5}{50}$

Since 17 of 50 spins stopped on 1, Silvia estimates the probability of outcome 1 to be $\frac{17}{50}$. In other words, Silvia guesses on the basis of her spins that region 1 takes up about $\frac{17}{50}$ of the entire circle. Similarly, she estimates the probability of outcome 2 as $\frac{28}{50}$ and the probability of outcome 3 as $\frac{5}{50}$.

8. Because $\frac{28}{50} > \frac{17}{50}$, outcome 2 seems more likely than outcome 1. Because $\frac{17}{50} > \frac{5}{50}$, outcome 1 seems more likely than outcome 3. If you just looked at the spinner and not at the table, would you make the same statements? Why?

9. Do you think $\frac{28}{50}$ overestimates the true probability of stopping on 2 or underestimates it? Give supporting reasons.

Activity: *Probability Experiment 2*

Materials needed:

- copies of Activity Master 18 (masters available in *Saxon Math 6/5 Assessments and Classroom Masters*)
- cardboard or posterboard
- scissors
- markers

For this activity, work with a partner.

Make 5 equal-size squares. While your eyes are closed, have your partner write either C, A, or T on each square. (Each letter must be used at least once.) Then have your partner mix up the squares on a table. With your eyes still closed, choose a square and have your partner tally the outcome on Activity Master 18. Repeat the process of mixing, choosing, and recording 30 times. Remember to keep your eyes closed.

10. Use the tallies to complete the "Relative Frequency" column on your table. (Remember, the denominator of each relative frequency is the number of times the experiment was performed.)

Letter	Tally	Relative Frequency
C		
A		
T		

11. From your relative frequency table, estimate the probability that the letter you choose is a T.

12. If your partner had written a letter just once, about how many times would you expect to choose it out of 30?

13. If your partner had written a letter twice, about how many times would you expect to choose it out of 30?

14. If your partner had written a letter three times, about how many times would you expect to choose it out of 30?

15. Which of the letters do you think your partner wrote once? Twice? Three times?

Extensions

a. If an experiment has N outcomes and they are equally likely, then each has probability $\frac{1}{N}$. Thus, if we flip a coin, which has two equally likely outcomes, the probability of the coin landing heads up is $\frac{1}{2}$ and the probability of the coin landing tails up is $\frac{1}{2}$. Suppose we write each letter of the alphabet on an identical tile and turn the tiles over. If we select one tile at random, what is the probability that the tile is the letter E? What is the probability that the tile is a vowel? What is the probability that the tile is a consonant?

b. Suppose we spin the spinner introduced in Lesson 57's second example 35 times. About how many times would you expect to get each of the three letters? Make an overhead spinner like the one in that example and spin it 35 times. Compare your results with your guesses.

LESSON

61 Using Letters to Identify Geometric Figures

WARM-UP

Facts Practice: 64 Multiplication Facts (Test F)

Mental Math: The following fractions are all equal to $\frac{1}{2}$. Read them aloud: $\frac{1}{2}, \frac{2}{4}, \frac{3}{6}, \frac{4}{8}, \frac{5}{10}, \frac{6}{12}, \frac{7}{14}, \frac{8}{16}, \frac{9}{18}, \frac{10}{20}$

a. $100 \div 10$

b. $100 \div 20$

c. How much is half of 5? ... half of 9? ... half of 15?

d. $1 - \frac{1}{3}$

e. $1 - \frac{1}{4}$

f. 10% of 500

Problem Solving:

If an $8\frac{1}{2}$ in. × 11 in. sheet of paper is folded from top to bottom, two congruent rectangles are formed. What are the dimensions (length and width) of each rectangle?

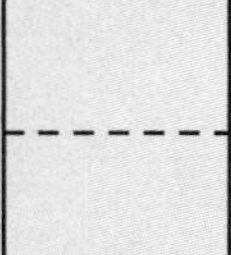

NEW CONCEPT

In geometry we often use letters to refer to points. We can identify polygons by the points at each vertex.

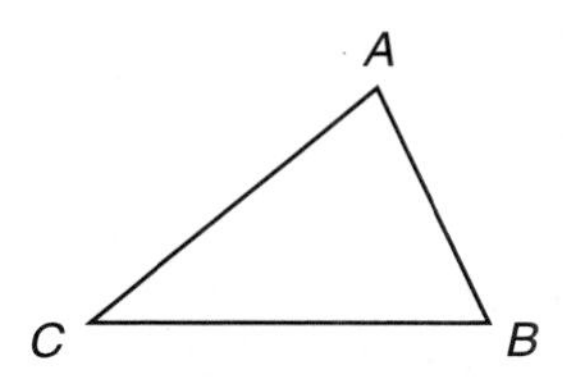

We may refer to this triangle as ΔABC ("triangle *ABC*"). We may also refer to this triangle in these ways:

$\Delta BCA \qquad \Delta CAB \qquad \Delta ACB \qquad \Delta BAC \qquad \Delta CBA$

To name a polygon, we start at one vertex and move around the perimeter, naming each vertex in order until all vertices are named. The order is important. The figure below can be named rectangle *ABCD* or rectangle *ADCB,* but not rectangle *ACBD.* Why not?

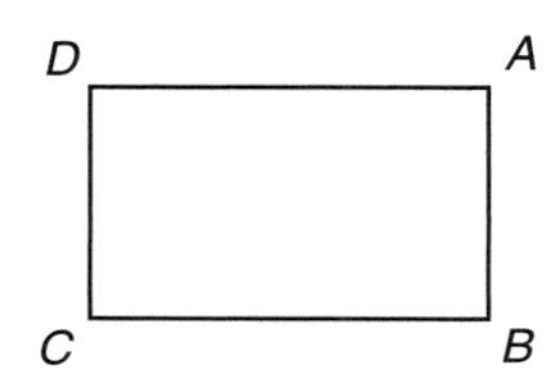

We name a line by naming any two points on the line. We refer to the line below as "line *AB*" (or "line *BA*").

We name a segment by naming the two **endpoints** of the segment. So if we wish to refer only to the portion of the line between points *A* and *B*, we would say "segment *AB*" (or "segment *BA*").

We name a ray by naming its endpoint first and then another point on the ray. This figure is "ray *AB*," but it is not "ray *BA*."

Instead of writing the word *line, segment,* or *ray,* we may draw a line, segment, or ray above the letters used to name the figure, as shown in this table:

Naming Lines, Segments, and Rays

Figure	Name	Abbreviation
A B (line)	line *AB*	$\overleftrightarrow{AB}$
A B (segment)	segment *AB*	$\overline{AB}$
A B (ray)	ray *AB*	$\overrightarrow{AB}$

When writing abbreviations for lines, segments, and rays, it is important to draw the figure above the letter pair. "Segment *AB*" and "$\overline{AB}$" both name a segment, but "*AB*" (without the word *segment* in front or the bar above) means "the distance from *A* to *B*." So "$\overline{AB}$" refers to a segment, and "*AB*" refers to the *length* of the segment.

We may name an angle using the single letter at its vertex if there is no chance for confusion. Angle *A* in the figure below is the acute angle with *A* as its vertex.

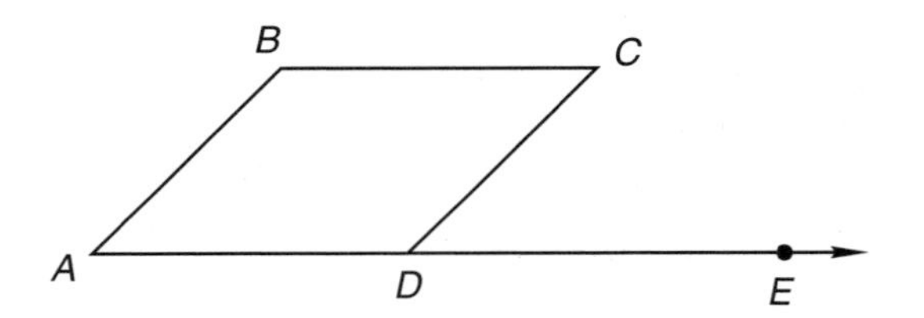

Referring to "angle *D*," however, would be unclear because there is more than one angle at *D*. In situations like this we use three letters, with the vertex letter listed second. The obtuse angle at *D* is angle *ADC* (or angle *CDA*). The acute angle at *D* is angle *CDE* (or angle *EDC*). The straight angle at *D* is angle *ADE* (or angle *EDA*). We use the symbol $\angle$ to abbreviate the word *angle*, so "angle *ADC*" may be written "$\angle ADC$."

Example 1 In rectangle *ABCD*, name the segments perpendicular to $\overline{AB}$.

Solution Each angle of a rectangle is a right angle, so both $\overline{\mathbf{AD}}$ (or $\overline{\mathbf{DA}}$) and $\overline{\mathbf{BC}}$ (or $\overline{\mathbf{CB}}$) are perpendicular to $\overline{AB}$.

Example 2 The length of segment *PQ* is 3 cm. The length of segment *PR* is 8 cm. What is the length of segment *QR*?

Solution The sum of the lengths of the two shorter segments equals the length of the longest segment.

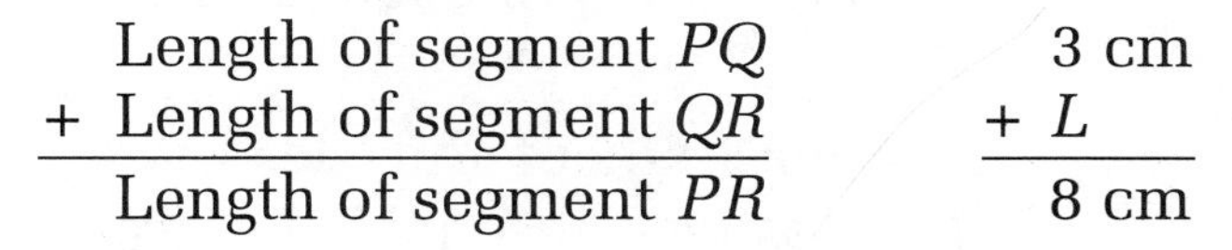

Length of segment *PQ*	3 cm
+ Length of segment *QR*	+ *L*
Length of segment *PR*	8 cm

This is a missing-addend problem. The missing addend is 5. The length of segment *QR* is **5 cm.**

Example 3 In quadrilateral *QRST*, $\angle S$ is an acute angle. Name another acute angle in the polygon.

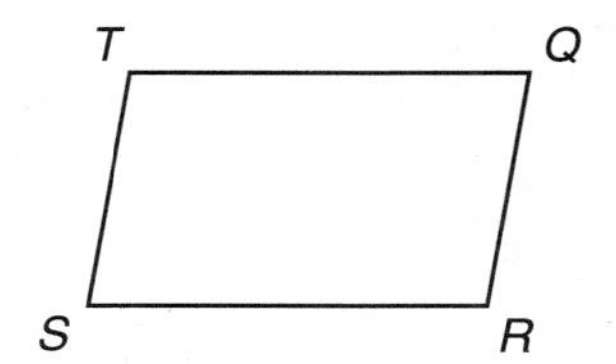

Solution The other acute angle is $\angle \mathbf{Q}$.

LESSON PRACTICE

Practice set Refer to rectangle *JKLM* to answer problems **a** and **b.**

a. Which segment is parallel to $\overline{JK}$?

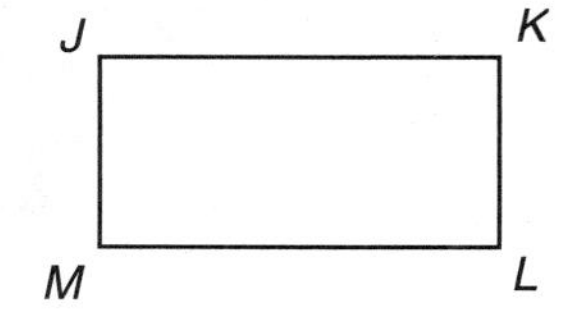

b. If $\overline{JK}$ is 10 cm long and if $\overline{JM}$ is half the length of $\overline{JK}$, then what is the perimeter of the rectangle?

Use words to show how each of these symbols is read, and draw an example of each figure:

c. $\overline{BC}$ **d.** $\overrightarrow{CD}$ **e.** $\overleftrightarrow{PQ}$

Refer to the figure at right to answer problems **f–i.**

f. Angle *AMD* is an obtuse angle. Using three letters, what is another way to name this angle?

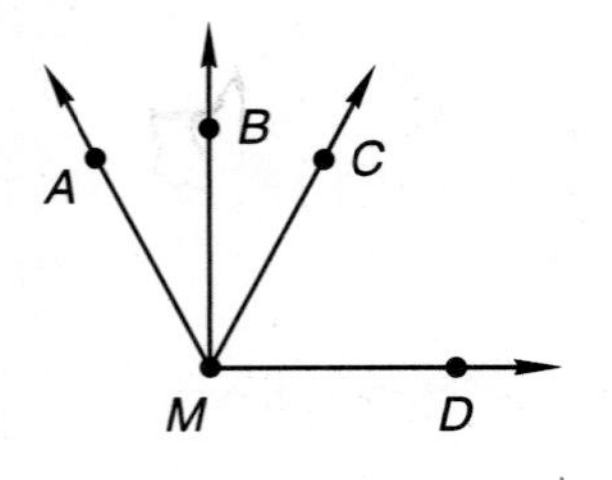

g. Which angle appears to be a right angle?

h. Which ray appears to be perpendicular to $\overrightarrow{MD}$?

i. Name one angle that appears to be acute.

MIXED PRACTICE

Problem set

1. (47) The tallest teacher at Lincoln School is 6 feet 3 inches tall. A person 6 feet 3 inches tall is how many inches tall?

2. (43, 60) One sixth of the class was absent. What percent of the class was absent? What fraction of the class was present?

3. (35) How many years were there from 1066 to 1215?

4. (48) Write the standard form for $(7 \times 1000) + (4 \times 10)$.

5. (33) Round 56 and 23 to the nearest ten. Multiply the rounded numbers. What is their product?

6. (23) Which of these fractions does not equal $\frac{1}{2}$?

A. $\frac{6}{12}$ B. $\frac{12}{24}$ C. $\frac{24}{48}$ D. $\frac{48}{98}$

7. (25) Which factors of 12 are also factors of 16?

8. (32, 53) A stop sign has the shape of a regular octagon. The sides of some stop signs are 12 inches long. What is the perimeter of a regular octagon with sides 12 inches long?

9. (59) $1 - \frac{1}{5}$

10. (59) $1 - \frac{3}{4}$

11. (41) $3\frac{3}{3} - 1\frac{2}{3}$

12. (59) $\frac{1}{10} + \frac{2}{10} + \frac{3}{10} + \frac{4}{10}$

13. (59) $5\frac{3}{4} + 4\frac{1}{4}$

14. (14)
$$\begin{array}{r} 4263 \\ -\quad Q \\ \hline 1784 \end{array}$$

15. (10, 13)
$$\begin{array}{r} \$19.34 \\ +\quad M \\ \hline \$50.00 \end{array}$$

16. (6)
$$\begin{array}{r} 58 \\ 39 \\ 24 \\ 16 \\ 52 \\ +\ 11 \\ \hline \end{array}$$

17. (56)
$$\begin{array}{r} 389 \\ \times\ 470 \\ \hline \end{array}$$

18. (34) $\frac{5445}{9}$

19. (58) Divide and write the quotient with a fraction: $\frac{25}{6}$

20. (54) $894 \div 40$

21. (54) $943 \div 30$

22. (24, 29) $(800 - 300) \times 20$

23. (38) On this number line, the arrow is pointing to what mixed number?

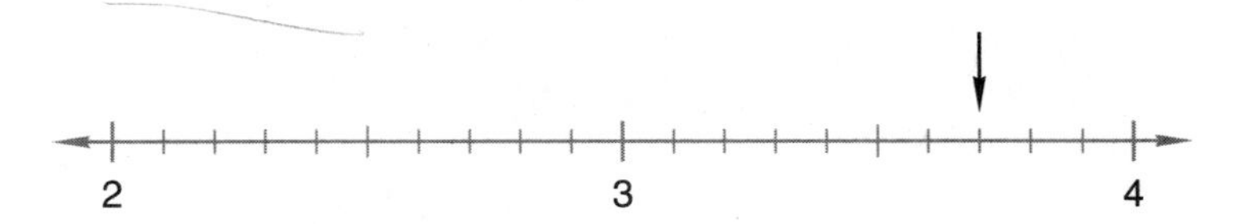

24. (23) Write two fractions equal to $\frac{1}{2}$. Make 20 the denominator of the first fraction and the numerator of the second fraction.

25. (28) What month is 15 months after November?

26. (61) The length of $\overline{RS}$ is 20 mm. The length of $\overline{RT}$ is 60 mm. What is the length of $\overline{ST}$?

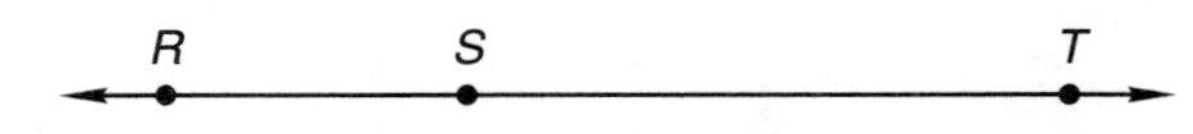

27. (61) Which angle in this figure appears to be a right angle?

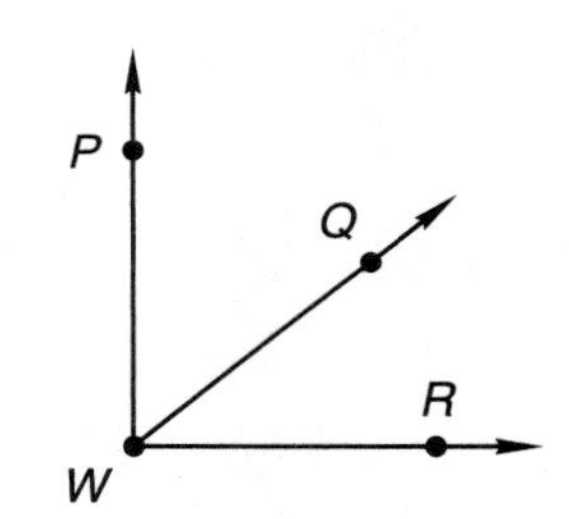

Refer to the information and line plot below to answer problems 28–30.

Rodric takes piano lessons. He records how many days he practices the piano each month and displays the data in a line plot. Here is Rodric's line plot for one full year:

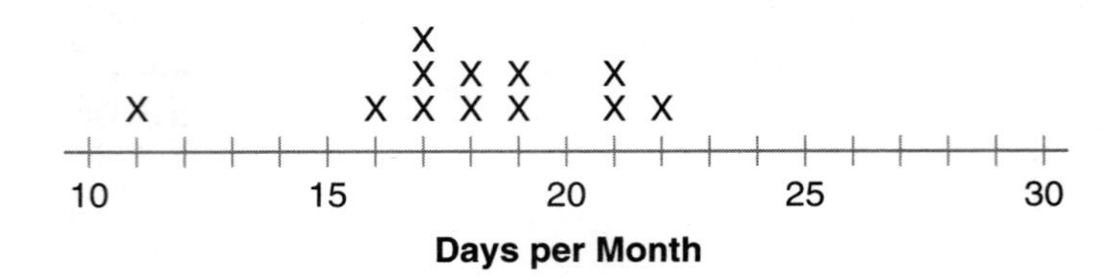

28. How many months did he practice more than 20 days?
(Inv. 5)

29. How many months did he practice between 15 and 20 days?
(Inv. 5)

30. What is the largest number of days he practiced in any month?
(Inv. 5)

LESSON

62 Estimating Arithmetic Answers

WARM-UP

Facts Practice: 48 Uneven Divisions (Test G)

Mental Math: The following fractions are equal to $\frac{1}{2}$: $\frac{1}{2}$, $\frac{2}{4}$, $\frac{3}{6}$, $\frac{4}{8}$, ... Read them aloud and continue the pattern to $\frac{10}{20}$.

a. $1000 \div 2$ **b.** $1000 \div 4$

c. One third of 7 is $2\frac{1}{3}$. How much is $\frac{1}{3}$ of 8? ... $\frac{1}{3}$ of 10?

d. $1 - \frac{1}{5}$ **e.** $1 - \frac{4}{5}$ **f.** 25% of 200

g. $100 \div 10, - 2, \div 2, - 2, \div 2$

Problem Solving:

Two figures are similar if they have the same shape. Draw a rectangle that is similar to this rectangle. Make the rectangle 2 inches long. How wide should you make the rectangle?

1 in.

$\frac{1}{2}$ in.

NEW CONCEPT

We have used arithmetic to find exact answers. For some problems, finding an exact answer may take many steps and a good deal of time. In this lesson we will practice quickly "getting close" to an exact answer. Trying to get close to an exact answer is called **estimating.** When we estimate, we use rounded numbers to make the arithmetic easier. We may even do the arithmetic mentally. An estimated answer is not an exact answer, but it is close to an exact answer.

Estimating can help us cut down on errors by showing us when a calculated answer is far from the correct answer. In other words, estimating can help us tell whether our calculated answer is *reasonable.*

Example 1 Estimate the product of 29 and 21.

Solution We estimate to quickly find *about* how much an exact answer would be. To estimate, we round the numbers *before* we do the work. The numbers 29 and 21 round to 30 and 20, which we can multiply mentally. So our estimated answer is **600.**

Example 2 Estimate the sum of \$8.95, \$7.23, \$11.42, and \$6.89 by rounding to the nearest dollar before adding.

Solution For each amount, if the number of cents is 50 or more, we round to the next dollar. If the number of cents is less than 50, we round down.

$$\$9 + \$7 + \$11 + \$7 = \mathbf{\$34}$$

Example 3 Estimate the perimeter of this rectangle by first rounding its length and width to the nearest ten millimeters.

78 mm

31 mm

Solution The length, 78 mm, rounds to 80 mm. The width, 31 mm, rounds to 30 mm.

$$80 \text{ mm} + 30 \text{ mm} + 80 \text{ mm} + 30 \text{ mm} = \mathbf{220 \text{ mm}}$$

LESSON PRACTICE

Practice set* Estimate each answer by rounding the numbers before doing the arithmetic. Often you will be able to do the work mentally, but for this practice show how you rounded the numbers. The first problem has been done for you. Refer to it as a model for showing your work.

a. $68 + 39$

Answer: $70 + 40 = 110$

b. 58×23

c. $585 + 312$

d. 38×19

e. $91 - 28$

f. 29×312

g. $685 - 391$

h. $59 \div 29$

i. $703 - 497$

j. $89 \div 31$

k. Estimate the sum of \$12.95, \$6.89, and \$8.15.

l. Estimate the perimeter of this rectangle.

57 mm

41 mm

MIXED PRACTICE

Problem set

1. (49) Mrs. Smith baked 6 dozen cookies for the party. The guests ate all but 20 cookies. How many cookies were eaten?

2. (28) A millennium is 1000 years. A millennium is how many centuries?

3. (50) If water is poured from glass to glass until the amount of water in each glass is the same, how many ounces of water will be in each glass?

4. (Inv. 3, 37) Draw a circle and shade one third of it. What fraction of the circle is not shaded? What percent of the circle is not shaded?

5. (62) Estimate the product of 39 and 41.

6. (59) $1 - \frac{1}{10}$

7. (59) $1 - \frac{3}{8}$

8. (41) $4\frac{4}{4} - 2\frac{3}{4}$

9. (59) $3\frac{1}{3} + 1\frac{2}{3}$

10. (43) $6\frac{10}{10} - \frac{1}{10}$

11. (59) $8 = 7\frac{\square}{6}$

12. (62) Estimate the sum of 586 and 317 by rounding both numbers to the nearest hundred before adding.

13. (6)

$$\begin{array}{r} 89,786 \\ 26,428 \\ 57,814 \\ +\ 91,875 \\ \hline \end{array}$$

14. (9)

$$\begin{array}{r} \$35,042 \\ -\ \$17,651 \\ \hline \end{array}$$

15. (55)

$$\begin{array}{r} 428 \\ \times\ 396 \\ \hline \end{array}$$

16. (26) $5y = 4735$

17. (18, 56) $8 \times 43 \times 602$

18. (58) Divide and write the quotient with a fraction: $\frac{15}{8}$

19. (54) $967 \div 60$

20. (54) $875 \div 40$

21. (23) (a) Which of these fractions equals $\frac{1}{2}$?

(b) Which of these fractions is less than $\frac{1}{2}$?

(c) Which of these fractions is greater than $\frac{1}{2}$?

A. $\frac{4}{7}$ B. $\frac{7}{15}$ C. $\frac{15}{30}$

22. $100 − ($24 + $43.89 + $8.67 + $0.98)
(13, 24)

23. The perimeter of this square is how many millimeters?
(44, 53)

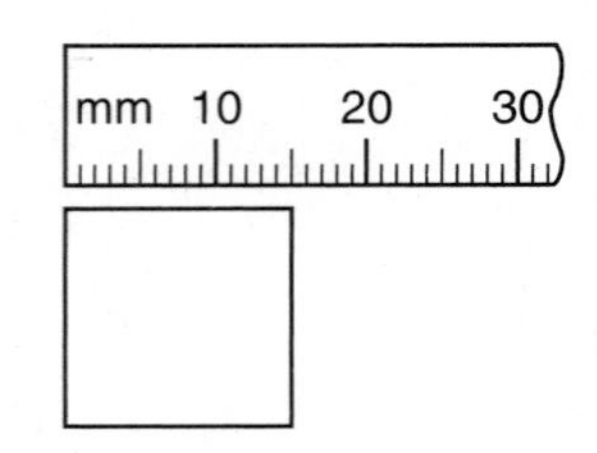

24. Think of an even number. Multiply it by 5. What number is the last digit of the product?
(2, 15)

25. It is morning. What time will be shown by this clock in 3 hours 20 minutes?
(28)

Refer to the spinner to answer problems 26 and 27.

26. What fraction names the probability that with one spin the spinner will stop on sector 3?
(57)

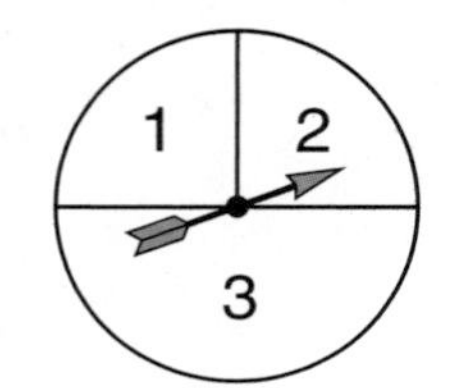

27. What fraction names the probability that with one spin the spinner will stop on sector 1?
(57)

28. Which angle in this figure appears to be an obtuse angle?
(61)

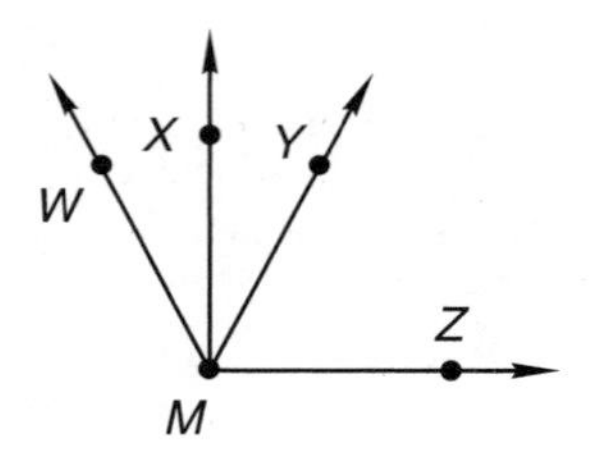

29. Estimate the perimeter of this rectangle by first rounding its length and width to the nearest inch.
(62)

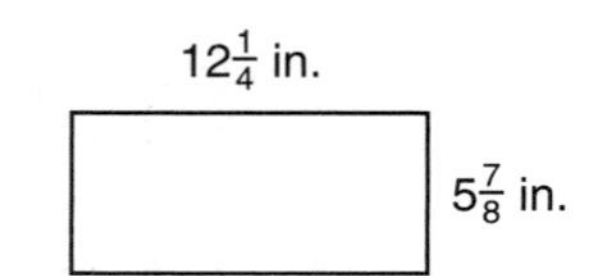

30. Half of 100 is 50, and half of 50 is 25. What number is half of 25?
(2)

LESSON

63 Subtracting a Fraction from a Whole Number Greater than 1

WARM-UP

Facts Practice: 64 Multiplication Facts (Test F)

Mental Math: The following fractions are equal to one half: $\frac{1}{2}$, $\frac{2}{4}$, $\frac{3}{6}$, $\frac{4}{8}$, ... Read the fractions aloud and continue the pattern to $\frac{12}{24}$.

a. One fifth of 6 is $1\frac{1}{5}$. How much is $\frac{1}{5}$ of 7? ... $\frac{1}{5}$ of 8?

b. $1 - \frac{1}{6}$ **c.** $1 - \frac{1}{10}$ **d.** 10% of 200

e. 500 ÷ 10, ÷ 2, + 5, ÷ 5, + 3, ÷ 3

Problem Solving:

Copy this multiplication problem and fill in the missing digits:

$$\begin{array}{r} 4\_ \\ \times \quad \_ \\ \hline 4\_4 \end{array}$$

NEW CONCEPT

Recall that when we subtract a fraction from 1, we change the 1 to a fraction name for 1. If the problem is $1 - \frac{1}{3}$, we change the 1 to $\frac{3}{3}$ so that the denominators will be the same. Then we can subtract.

We change from this form: $1 - \frac{1}{3}$

... to this form: $\frac{3}{3} - \frac{1}{3} = \frac{2}{3}$

In this lesson we will subtract fractions from whole numbers greater than 1.

Imagine we have 4 whole pies on a bakery shelf. If someone asks for half a pie, we would have to cut one of the whole pies into 2 halves. Before removing half a pie from the pan, we would have 4 pies, but we could call those pies "$3\frac{2}{2}$ pies."

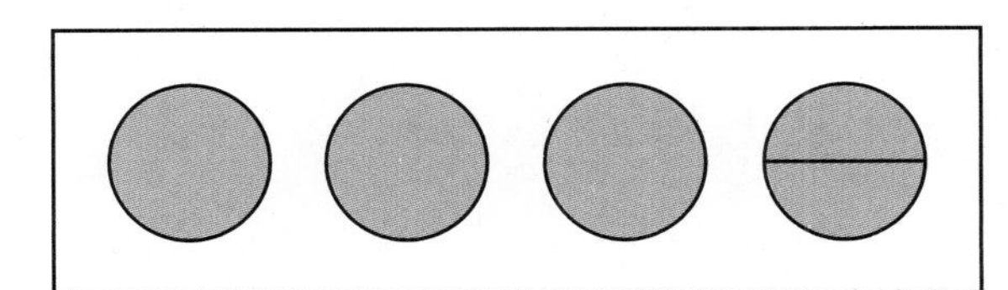

We use this idea to subtract a fraction from a whole number. We take 1 from the whole number and write it as a fraction with the same denominator as the fraction being subtracted. We will answer the problem $4 - \frac{1}{2}$ to show how we do this.

We change from this form: $4 - \frac{1}{2}$

... to this form: $3\frac{2}{2} - \frac{1}{2} = 3\frac{1}{2}$

Example 1 Name the number of shaded circles (a) as a whole number and (b) as a mixed number.

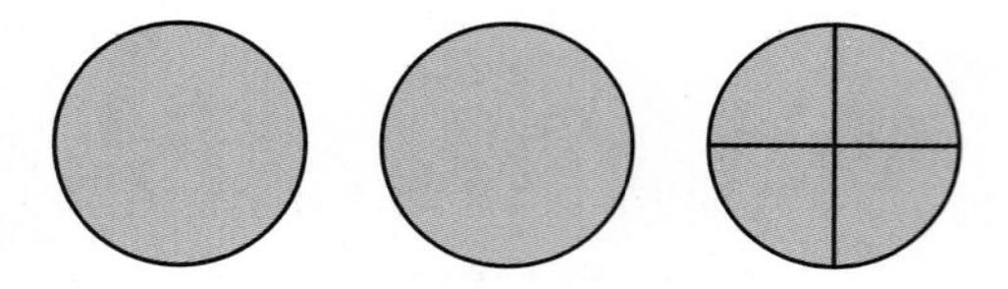

Solution (a) We see **3** whole circles.

(b) Since one of the circles is divided into fourths, we can also say that there are two whole circles and four fourths of a circle, which we write as the mixed number $\mathbf{2\frac{4}{4}}$.

Example 2 Subtract: $5 - \frac{1}{3}$

Solution We think of 5 as being $4 + 1$, which we can write as $4\frac{3}{3}$. Now we can subtract.

$$5 - \frac{1}{3}$$

$$4\frac{3}{3} - \frac{1}{3} = \mathbf{4\frac{2}{3}}$$

LESSON PRACTICE

Practice set* Subtract:

a. $4 - \frac{1}{4}$ **b.** $3 - \frac{3}{4}$ **c.** $4 - 2\frac{1}{4}$

d. $2 - \frac{1}{4}$ **e.** $4 - 1\frac{1}{2}$ **f.** $6 - 1\frac{2}{3}$

MIXED PRACTICE

Problem set

1. A 100-centimeter stick broke into 3 pieces. One piece was 7 centimeters long, and another was 34 centimeters long. How long was the third piece? (49)

2. Bill's pencil was 6 inches long. While doing his homework, Bill used up $1\frac{1}{2}$ inches of his pencil. Then how long was his pencil? (63)

3. Isabel can make 4 quarter-pound hamburgers from 1 pound of meat. How many quarter-pound hamburgers can she make from 5 pounds of meat? (21)

4. In the 4 stacks of math books, there are 18, 19, 24, and 23 books. If the number of books in each stack is made even, how many books will be in each stack? (50)

5. Estimate the sum of 398 and 487 by rounding both numbers to the nearest hundred before adding. (62)

6. Which factors of 14 are also factors of 21? (25)

7. The distance around the earth at the equator is like which measurement of a circle? (53)

   A. radius B. diameter C. circumference

8. What is the sum of five million, two hundred eighty-four thousand and six million, nine hundred eighteen thousand, five hundred? (52)

9. $7 - \frac{1}{3}$ (63)

10. $6 - 2\frac{1}{2}$ (63)

11. $8 - 3\frac{3}{4}$ (63)

12. $\frac{8}{9} + \left(\frac{2}{9} - \frac{1}{9}\right)$ (41, 59)

13. $5\frac{3}{4} - \left(3\frac{2}{4} + 1\frac{1}{4}\right)$ (24, 41)

14. $\begin{array}{r} 43{,}716 \\ -\ 19{,}537 \\ \hline \end{array}$ (9)

15. $\begin{array}{r} \$6.87 \\ \times\ 794 \\ \hline \end{array}$ (55)

16. $\frac{\$14.72}{8}$ (26)

17. Divide and write the quotient with a fraction: $\frac{20}{9}$ (58)

18. $20\overline{)951}$
(54)

19. $50\overline{)2560}$
(54)

20. 50 × (400 + 400)
(24, 29)

21. (400 + 400) ÷ 40
(24, 54)

22. 4736 + 2849 + 351 + 78
(6)

23. If three eighths of the class was absent, what fraction of the class was present? What percent of the class was present?
(Inv. 3, 60)

24. Arrange these fractions in order from least to greatest. (*Hint:* Decide whether each fraction is less than, equal to, or greater than $\frac{1}{2}$.)
(23)

$$\frac{5}{10}, \frac{5}{8}, \frac{5}{12}$$

25. What is the perimeter of this equilateral triangle?
(36, 53)

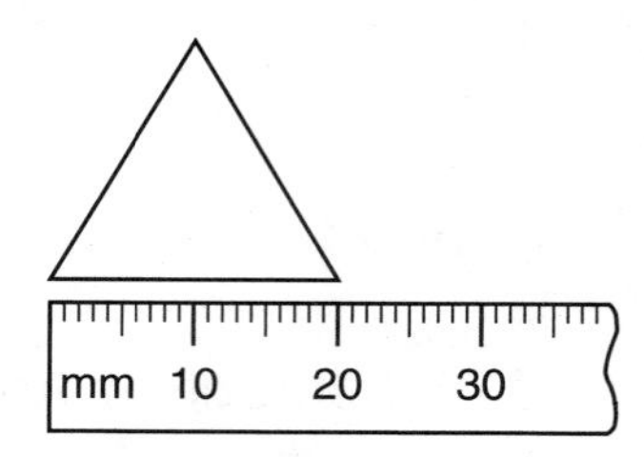

26. Mr. Chitsey gave a mathematics test to the 20 students in his class. A perfect score on the test was 100 points. Here are the students' scores:
(Inv. 5)

73, 82, 81, 70, 62, 55, 60, 90, 92, 88,
77, 92, 82, 62, 71, 83, 90, 73, 59, 69

Copy the frequency table below, and use the test-score data to complete the table.

Frequency Table

Score	Tally	Frequency
51–60		
61–70		
71–80		
81–90		
91–100		

Refer to this figure to answer problems 27–30:

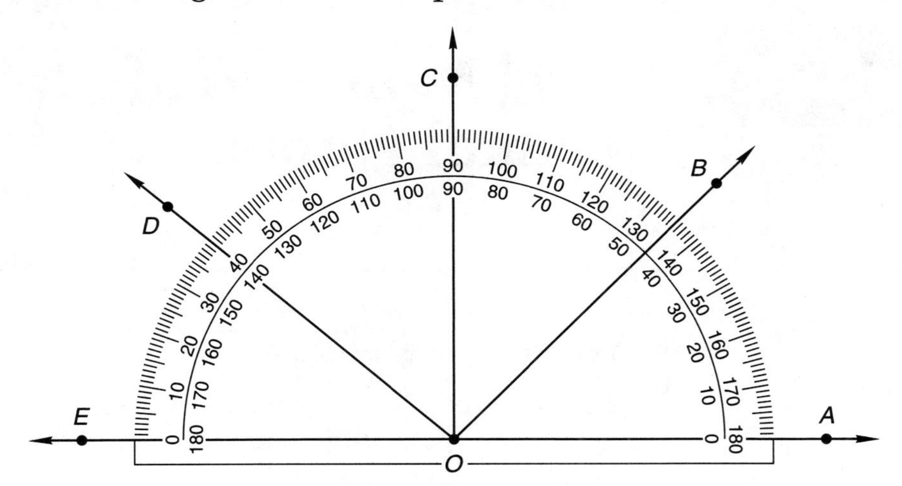

27. What is the measure of $\angle AOB$?
(Inv. 4, 61)

28. What is the measure of $\angle AOD$?
(Inv. 4, 61)

29. What type of angle is $\angle EOB$?
(32, 61)

30. Which angle is a straight angle?
(32, 61)

LESSON

64 Using Money to Model Decimal Numbers

WARM-UP

Facts Practice: 64 Multiplication Facts (Test F)

Mental Math: The following fractions are equal to $\frac{1}{2}$: $\frac{1}{2}$, $\frac{2}{4}$, $\frac{3}{6}$, ... Read them aloud and continue the pattern to $\frac{12}{24}$.

a. $1 - \frac{2}{3}$ **b.** $1 - \frac{3}{4}$ **c.** $1 - \frac{4}{5}$

d. 2 days 2 hours is how many hours?

e. $250 \div 10$ **f.** 50% of 60, + 10, ÷ 5, + 2, × 10

Problem Solving:

The symbol $\sqrt{}$ is a square root symbol. We read $\sqrt{25}$ as "the square root of 25." The expression $\sqrt{25}$ equals 5 because $5 \times 5 = 25$. What does $\sqrt{49}$ equal?

NEW CONCEPT

In this lesson we will use money to illustrate decimal numbers. Recall that in our number system the position a digit occupies in a number has a value, called **place value.**

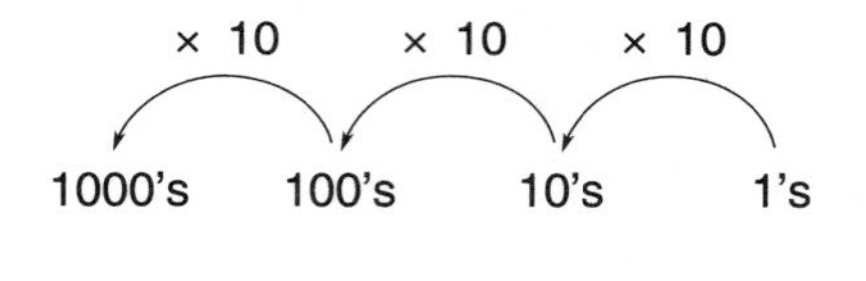

As we move to the left from the ones place, the value of each place is ten times greater than the place to its right. We have shown the value of four places, but the pattern continues without end.

Notice that as we move in the other direction (to the right), the value of each place is one tenth the value of the place to its left.

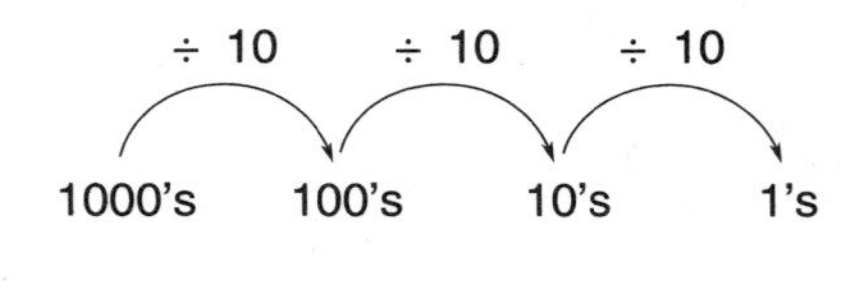

This pattern also continues without end. To the right of the ones place is the tenths place, the hundredths place, the thousandths place, and so on. These places are called **decimal places.** In this diagram we show the first three decimal places:

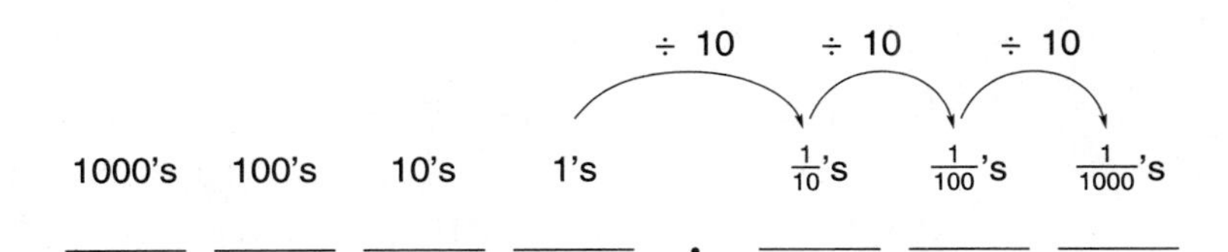

Notice the decimal point between the ones place and the tenths place. We use a decimal point as a reference point, like a landmark, so that we know where the whole-number places end and the decimal places begin. We do not need to use the decimal point to write whole numbers, because it is understood that in whole numbers the digit farthest to the right is in the ones place.

One use of decimal numbers is to write dollars and cents, such as $6.25. This collection of bills and coins totals $6.25:

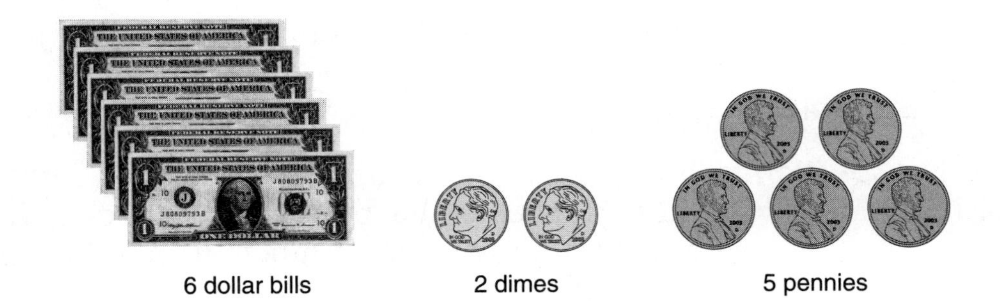

6 dollar bills 2 dimes 5 pennies

Notice that the number of bills and coins matches the digits in $6.25: 6 ones, 2 dimes, 5 pennies. We can use pennies, dimes, dollars, ten-dollar bills, and hundred-dollar bills as a model for place value.

Place-Value Chart

Place Name	hundreds	tens	ones	tenths	hundredths
Place Value	100	10	1	$\frac{1}{10}$	$\frac{1}{100}$
Place	___	___	___ .	___	___
Money Value of Place	$100 bills	$10 bills	$1 bills	dimes	pennies

The last row of the chart gives the money value of each place. The first place to the right of the decimal point is the tenths place. Since a dime is one tenth of a dollar, we may think of this as the dimes place. The second place to the right of the decimal point is the hundredths place. Since a penny is one hundredth of a dollar, we may think of this as the pennies place.

A dime demonstrates the value relationship between adjoining places. While a dime is ten times the value of a penny (thereby making it worth 10 cents), it is also one tenth the value of a dollar.

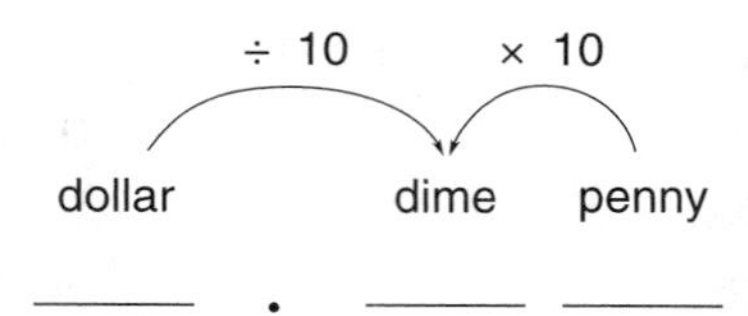

Example 1 What combination of dollars, dimes, and pennies makes $4.65 using the fewest bills and coins possible?

Solution The digits in the number $4.65 show us how many of each bill or coin to use. We need **4 dollars, 6 dimes, and 5 pennies.** (We would probably use two quarters, a dime, and a nickel to make 65 cents with actual money, but we do not use quarters and nickels to model decimal place value.)

Example 2 What is the place value of the 4 in $6.24?

Solution The 4 is in the second place to the right of the decimal point, which is the **hundredths** place. This is reasonable because 4 shows the number of pennies, and a penny is a hundredth of a dollar.

Example 3 Is $3.67 closer to $3.60 or to $3.70?

Solution To answer this question, we round $3.67 to the nearest ten cents, that is, to the tenths place. Since 7 cents is more than half a dime, $3.67 rounds up to **$3.70.**

LESSON PRACTICE

Practice set What is the place value of the 5 in each of these numbers?

a. $25.60 **b.** $54.32 **c.** $12.75 **d.** $21.50

e. What combination of dollars, dimes, and pennies makes $3.84 using the fewest bills and coins possible?

f. Is $12.63 closer to $12.60 or to $12.70?

g. Is $6.08 closer to $6.00 or to $6.10?

MIXED PRACTICE

Problem set

1. What is the sum of one hundred sixteen thousand, five hundred twenty-one and two hundred fifty-three thousand, four hundred seventy-nine?
(52)

2. At the annual clearance sale, *Shutter Shop* lowered the price of all its cameras. Terry wants to buy a new camera that costs $30.63. She has $17.85. How much more money does she need?
(11)

3. In the auditorium there were 30 rows of seats with 16 seats in each row. If there were 21 empty seats, how many seats were filled?
(49)

4. Jeremy is reading a 324-page book. If he plans to finish the book in 6 days, how many pages should he read each day?
(21)

5. Estimate the product of 68 and 52.
(62)

6. If three tenths of the bowling pins were up, what fraction of the bowling pins were down? What percent of the bowling pins were down?
(Inv. 2, 60)

7. Numbers written in dollars and cents (such as $54.63) have how many decimal places?
(64)

8. What combination of dollars, dimes, and pennies makes $3.25 using the fewest bills and coins possible?
(64)

9. Is $4.82 closer to $4.80 or to $4.90?
(64)

10. Divide 25 by 8. Write the quotient with a fraction.
(58)

11. Which factors of 20 are also factors of 30?
(25)

12. (28) What time is $1\frac{1}{2}$ hours before noon?

13. (14) $360 - a = 153$

14. (26) $5m = 875$

15. (10, 59) $\frac{3}{5} + f = 1$

16. (59) $\frac{5}{5} - z = \frac{3}{3}$

17. (34) $\$30.48 \div 6$

18. (54) $60\overline{)1586}$

19. (13)

$$\begin{array}{r} \$4.34 \\ \$0.26 \\ \$5.58 \\ \$9.47 \\ \$6.23 \\ + \ \$0.65 \\ \hline \end{array}$$

20. (15, 18) $5 \times 4 \times 3 \times 2 \times 1 \times 0$

21. (63)

$$\begin{array}{r} 7 \\ - \ 3\frac{2}{3} \\ \hline \end{array}$$

22. (59)

$$\begin{array}{r} 1\frac{1}{3} \\ + \ 2\frac{2}{3} \\ \hline \end{array}$$

23. (63)

$$\begin{array}{r} 4 \\ - \ 3\frac{3}{4} \\ \hline \end{array}$$

24. (53, 61) Figure *PQRST* is a regular pentagon. If $\overline{PQ}$ measures 12 mm, then what is the perimeter of the polygon?

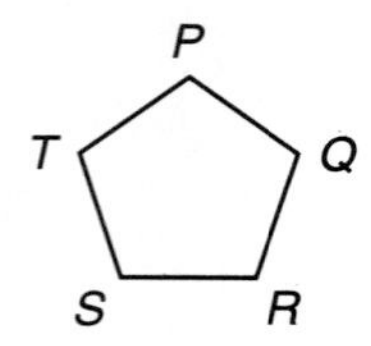

25. (22, 26) (a) When 10 is divided by 3, what is the remainder?

(b) When 100 is divided by 3, what is the remainder?

26. (44, 53) What is the perimeter of this equilateral triangle?

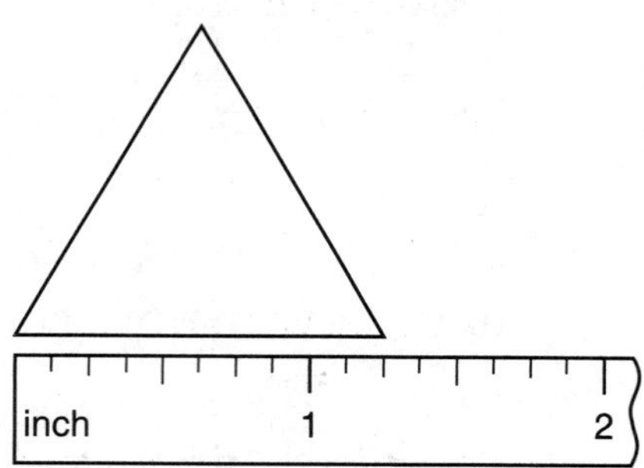

27. (57) Suppose the 7 letter tiles below are turned over and mixed up. Then one tile is selected.

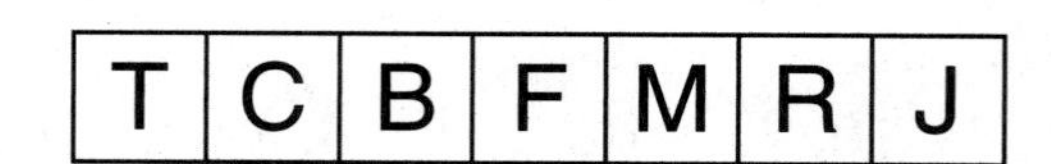

What is the probability that the letter selected will be one of the letters that follows Q in the alphabet?

Mrs. Yang's class opened bags of color-coated chocolate candy and counted how many green pieces were in each bag. The class made this frequency table to display their findings. Use the table to answer problems 28–30.

Frequency Table

Green Pieces	Frequency
1–5	3
6–10	7
11–15	9
16–20	6
21–25	2

28. How many bags did the class open?
(Inv. 5)

29. How many bags contained more than 10 green pieces?
(Inv. 5)

30. Based on the sample studied by Mrs. Yang's class, which of the following is the most likely outcome if you were to open a bag of the same kind of candy?
(Inv. 6)

A. There will be no green pieces.

B. There will be fewer than 16 green pieces.

C. There will be more than 15 green pieces.

LESSON 65 Decimal Parts of a Meter

WARM-UP

Facts Practice: 100 Multiplication Facts (Test C)

Mental Math:

a. 10% of 10 **b.** 10% of 100 **c.** 10% of 1000

d. One third of 11 is $3\frac{2}{3}$. How much is $\frac{1}{3}$ of 13? ... $\frac{1}{3}$ of 14?

e. $\sqrt{36}$ **f.** $1 - \frac{2}{5}$

g. 8 × 5, − 10, ÷ 5, × 7, − 2, ÷ 5

Problem Solving:

Two angles are congruent if they have the same measure. These two angles are congruent. They are both right angles. Draw a third angle congruent to these two angles that has a different orientation.

NEW CONCEPT

In this lesson we will use metric units of length to model decimal numbers. The basic unit of length in the metric system is a **meter.** One big step is about a meter long. Many school classrooms are about 10 meters long and 10 meters wide.

Meters are divided into tenths, hundredths, and thousandths. These smaller units are *decimeters, centimeters,* and *millimeters.*

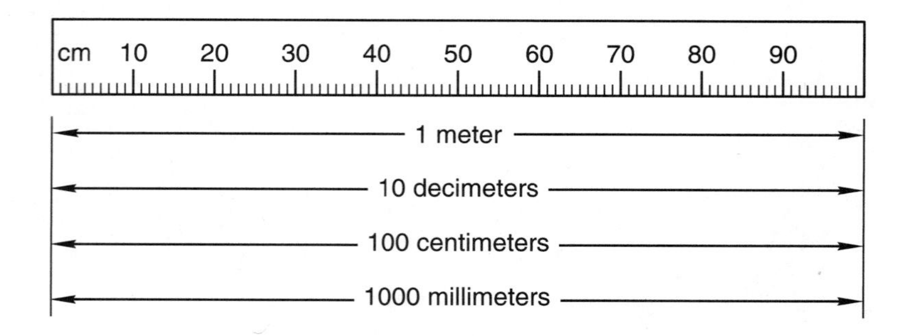

A **decimeter** is one tenth of a meter. A **centimeter** is one tenth of a decimeter and one hundredth of a meter. A **millimeter** is one tenth of a centimeter and one thousandth of a meter.

These fractional parts of a meter can represent the first three decimal places.

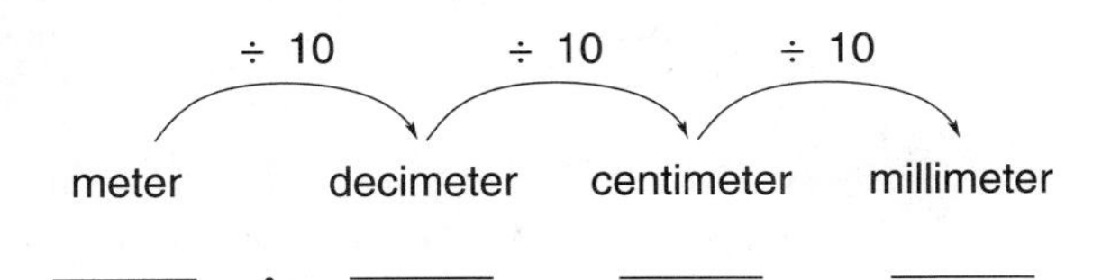

Example 1 Forty centimeters is how many decimeters?

Solution Ten centimeters equals one decimeter. So 40 centimeters equals **4 decimeters.**

Example 2 Chad measured his height with a meterstick. He was 1 meter plus 35 centimeters tall. What was Chad's height in meters?

Solution Since 10 centimeters equals 1 decimeter, we can think of 35 centimeters as 3 decimeters plus 5 centimeters. So Chad's height was 1 meter plus 3 decimeters plus 5 centimeters. We write this as **1.35 meters.**

Activity 1: *Decimal Parts of a Meter*

Materials needed:

- copy of Activity Masters 19 and 20 for each student or small group (masters available in *Saxon Math 6/5 Assessments and Classroom Masters*)
- scissors
- tape or paste

Cut and paste decimeter, centimeter, and millimeter strips from Activity Master 19 onto Activity Master 20 to show decimal parts of a meter. Use the models to compare, convert, and add the lengths specified on Activity Master 20.

Activity 2: *Measuring with a Meterstick*

Materials needed:

- meterstick
- sheet of notebook paper for each student

Use a meterstick to help you answer these questions about classroom items 1–5 listed below. Record your answers on a sheet of notebook paper.

(a) Is the measurement more than or less than a meter?

(b) What is the measurement in meters? (Use a decimal number to express the measurement.)

Item 1. height of the door
Item 2. width of the door
Item 3. height of desk
Item 4. length of bulletin board
Item 5. length of math book

LESSON PRACTICE

Practice set

a. Which of these is the most reasonable measurement for the length of an automobile?

A. 4.5 meters
B. 4.5 decimeters
C. 4.5 centimeters
D. 4.5 millimeters

b. Chuck is 1 meter plus 43 centimeters tall. Use a decimal number to write Chuck's height in meters.

c. A desk ruler is about 30 centimeters long. About how many decimeters long is a desk ruler?

MIXED PRACTICE

Problem set

1. *(31, 32)* Draw a quadrilateral with one pair of horizontal segments and one pair of vertical segments.

2. *(49)* The players are divided into 10 teams with 12 players on each team. If all the players are divided into 8 equal teams instead of 10, then how many players will be on each team?

3. *(53)* Below is a representation of a rectangular field that is 100 yards long and 40 yards wide. What is the perimeter of the field?

100 yards
40 yards

4. A yard is 36 inches. How many inches is one fourth of a
(Inv. 2, 46) yard? One fourth of a yard is what percent of a yard?

5. Pilar's school starts at 8:30 a.m. If it is now 7:45 a.m., how
(28) many minutes does she have until school starts?

6. Estimate the sum of 672 and 830 by rounding to the
(62) nearest hundred before adding.

7. (a) What fraction of the rectangle
(37) is shaded?

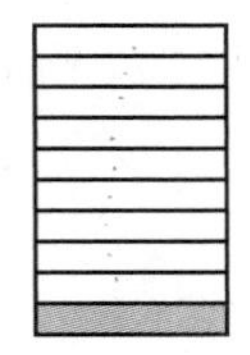

(b) What fraction of the rectangle is not shaded?

8. The refrigerator was 1 meter plus 32 centimeters in
(65) height, and it was 82 centimeters wide. Write the height of the refrigerator in meters.

9. Half a meter is how many decimeters?
(65)

10. Arrange these fractions in order from least to greatest.
(23) (*Hint:* Decide whether each fraction is less than, equal to, or greater than $\frac{1}{2}$.)

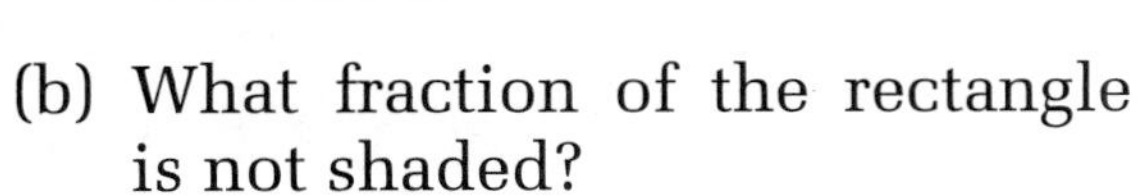

$$\frac{4}{4}, \frac{3}{8}, \frac{2}{3}, \frac{5}{10}$$

11. The number 9 has three different factors. The number 10
(25) has how many different factors?

12. Divide and write the quotient as a mixed number: $\frac{15}{4}$
(58)

13. Write the greatest odd number that uses the digits 3, 4,
(2) and 5 once each.

14. Five hundred is how much more than three hundred
(5, 35) ninety-five?

15. (6) $\begin{array}{r} 36{,}195 \\ 17{,}436 \\ +\ 42{,}374 \\ \hline \end{array}$

16. (9) $\begin{array}{r} 41{,}026 \\ -\ 39{,}543 \\ \hline \end{array}$

17. (56) $\begin{array}{r} 608 \\ \times\ 479 \\ \hline \end{array}$

18. (26) $2637 \div 4$

19. (54) $40\overline{)\$33.60}$

20. (54) $\frac{3360}{20}$

21. (59) $3\frac{3}{8} + 5\frac{5}{8}$

22. (63) $5 - 3\frac{3}{8}$

23. (43) $3\frac{3}{4} - 3$

24. (18, 56) $6 \times 42 \times 20$

25. (13, 24) $20 − ($5.63 + $12)

26. (23, 51) To find the number of eggs in $2\frac{1}{2}$ dozen, Chad thought of $2\frac{1}{2}$ as $\left(2 + \frac{1}{2}\right)$. Then he used the distributive property.

$$2\tfrac{1}{2} \text{ dozen} = 2 \text{ dozen} + \tfrac{1}{2} \text{ dozen}$$

How many eggs is $2\frac{1}{2}$ dozen?

27. (22, 42) By which of these numbers is 1080 divisible?

2, 3, 5, 6, 9, 10

Refer to the spinner to answer problems 28–30.

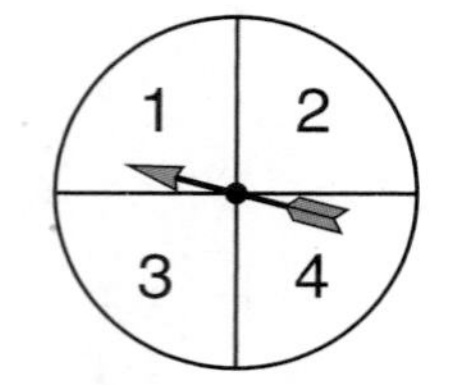

28. (57) What is the probability that with one spin the outcome will be an even number?

29. (57) What is the probability that with one spin the outcome will be a number less than 4?

30. (57) What is the probability that with one spin the outcome will be a number less than 5?

LESSON

66 Reading a Centimeter Scale

WARM-UP

Facts Practice: 64 Multiplication Facts (Test F)

Mental Math: Hold your fingers ...

a. a decimeter apart. **b.** a centimeter apart. **c.** a millimeter apart.

d. One fifth of 11 is $2\frac{1}{5}$. How much is $\frac{1}{5}$ of 16? ... $\frac{1}{5}$ of 17?

e. $\sqrt{9}$ **f.** $1 - \frac{3}{10}$ **g.** 6×23

h. $2 \times 2 \times 2 \times 2 \times 2$

Problem Solving:

Copy this multiplication problem and fill in the missing digits. Find two different solutions.

```
   2_
 ×  _
 ----
  2_2
```

NEW CONCEPT

In this lesson we will measure objects using a centimeter ruler. A desk-size ruler is usually 30 centimeters long and is further divided into millimeters. Each millimeter is one tenth of a centimeter. Here we show part of a centimeter ruler:

In Lesson 65 we learned to write lengths as decimal parts of a meter. For example, fifteen centimeters can be written as 0.15 m, which means "15 hundredths of a meter." To show the units as centimeters rather than meters, we would write fifteen centimeters without a decimal point (15 cm), changing the units from "m" to "cm." (Similarly, we write fifteen cents as 15¢ instead of $0.15 if the units are cents instead of dollars.)

How we write a particular length depends upon whether we use millimeters, centimeters, or meters as units. This segment is 15 millimeters long:

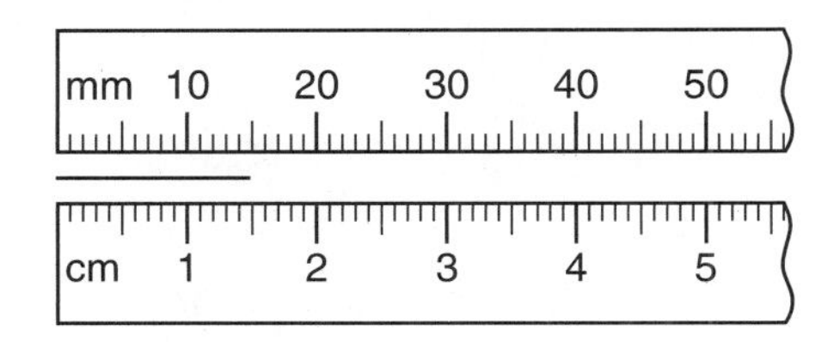

The segment is also 1.5 centimeters long. The tick marks on the centimeter scale divide each centimeter into ten equal lengths that are each $\frac{1}{10}$ of a centimeter. The end of the segment is 5 lengths past the 1 centimeter mark.

Example 1 Write the length of this segment

(a) as a number of millimeters.

(b) as a number of centimeters.

Solution (a) **24 mm**

(b) **2.4 cm**

Just as tenths of a centimeter can be written as a decimal number, so can tenths on a number line. Here we show a number line with the distance between whole numbers divided into tenths. We show the decimal numbers represented by some points on the number line.

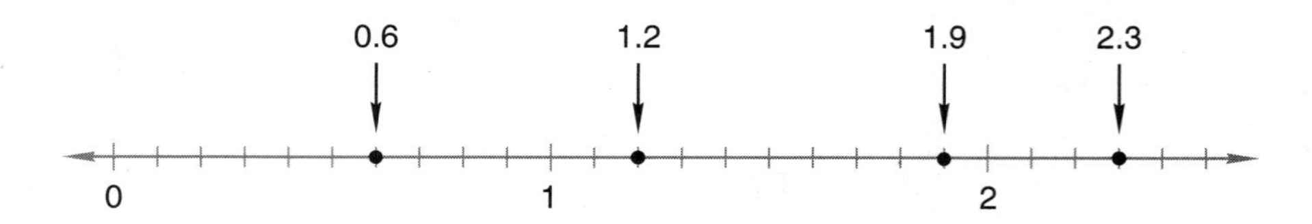

Example 2 To what decimal number is the arrow pointing?

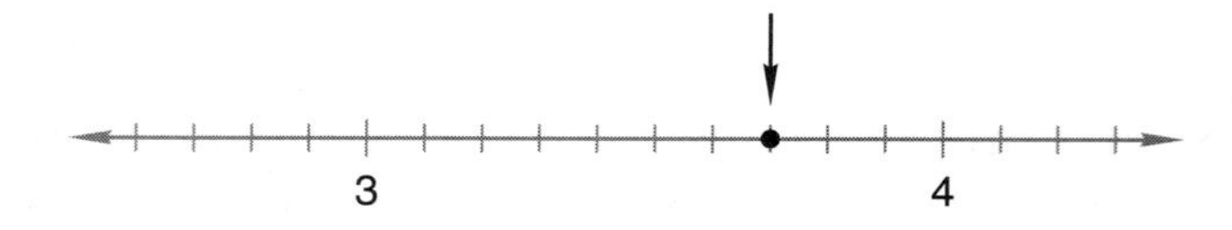

Solution The distance from 3 to 4 is divided into ten segments. The arrow indicates a point seven tenths greater than 3, which is **3.7.**

LESSON PRACTICE

Practice set Use a centimeter ruler to find the following measurements. Record each measurement twice, once as a number of millimeters and once as a number of centimeters.

a. length of your math book

b. width of your paper

c. length of this 1-inch segment: ____________

d. length of this paper clip: 

e. diameter of a dime: 

Write a decimal number to name each point marked by an arrow on the number line below:

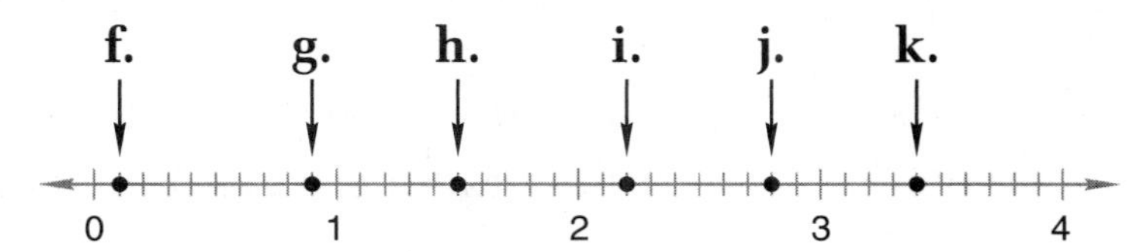

MIXED PRACTICE

Problem set

1. How many tens are in 100?
(3)

2. What is the next term in this counting sequence?
(1)

..., 2450, 2550, 2650, ______, ...

3. Estimate the difference of 794 and 312 by rounding both numbers to the nearest hundred before subtracting.
(62)

4. Fernando could carry 6 containers at one time. If 4 containers weighed 20 pounds, then how much did 6 containers weigh?
(49)

5. When one end of the seesaw is 9 inches above the ground, the other end is 21 inches above the ground. How far are the ends above the ground when the seesaw is level?
(50)

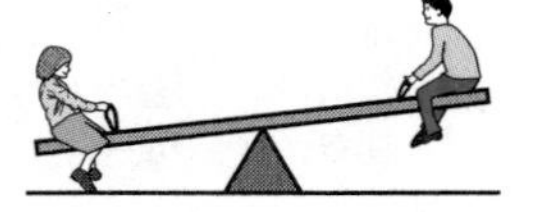

6. Compare: $\frac{3}{5} \bigcirc \frac{4}{9}$. (*Hint:* Decide whether each fraction is more than $\frac{1}{2}$ or less than $\frac{1}{2}$.)
(23)

7. Which digit in 4318 is in the same place as the 7 in 96,275?
(52)

8. (a) What fraction of this rectangle is shaded?
(37)

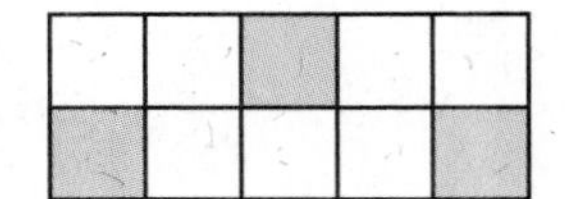

(b) What fraction of the rectangle is not shaded?

9. (66) Find the length of this tack to the nearest tenth of a centimeter.

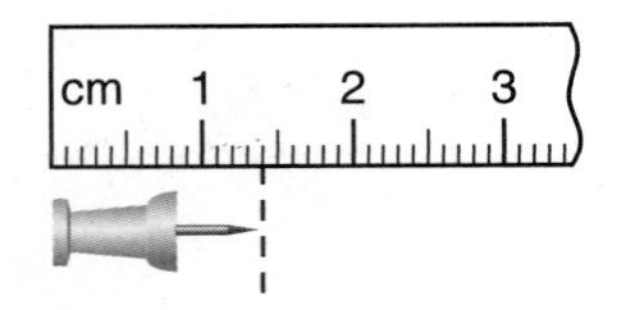

10. (65) Compare: 1 decimeter ◯ 1 centimeter

11. (58) Divide 53 by 10. Write the quotient as a mixed number.

12. (53) Rihana ran around the block. If the block is 200 yards long and 60 yards wide, how far did she run?

13. (61) Segment AB is 40 millimeters long. Segment BC is 35 millimeters long. How long is $\overline{AC}$?

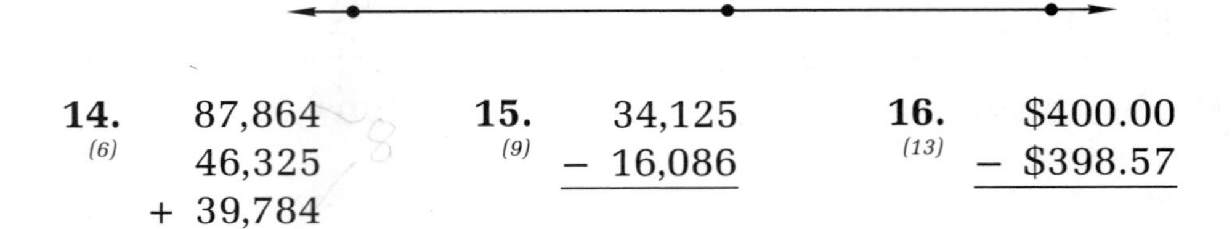

14. (6)
$$\begin{array}{r} 87{,}864 \\ 46{,}325 \\ +\ 39{,}784 \\ \hline \end{array}$$

15. (9)
$$\begin{array}{r} 34{,}125 \\ -\ 16{,}086 \\ \hline \end{array}$$

16. (13)
$$\begin{array}{r} \$400.00 \\ -\ \$398.57 \\ \hline \end{array}$$

17. (26) $\dfrac{5628}{6}$

18. (56)
$$\begin{array}{r} 807 \\ \times\ 479 \\ \hline \end{array}$$

19. (29)
$$\begin{array}{r} \$7.00 \\ \times\ \ 800 \\ \hline \end{array}$$

20. (41, 43) $3\frac{2}{3} - \left(2\frac{1}{3} + 1\right)$

21. (43, 63) $4 - \left(2 + 1\frac{1}{4}\right)$

22. (18, 29) $36 \times 60 \times 7$

23. (13, 24) $\$20 - (\$8 + \$2.07)$

Use this information to answer problems 24 and 25:

There are 16 players on the Norwood softball team. Ten players are in the game at one time. The rest of the players are substitutes. The team won 7 of its first 10 games.

24. (16, Inv. 5) The Norwood softball team has how many substitutes?

25. (Inv. 5) If the team played 12 games in all, what is the greatest number of games the team could have won?

A. 12 B. 10 C. 9 D. 7

26. Make a frequency table for the number of letters in the names of the twelve months of the year. "May" has the least (3 letters). "September" has the most (9 letters). The months of the year are listed below for reference.
(Inv. 5)

January, February, March, April, May, June, July, August, September, October, November, December

27. To what decimal number is the arrow pointing?
(66)

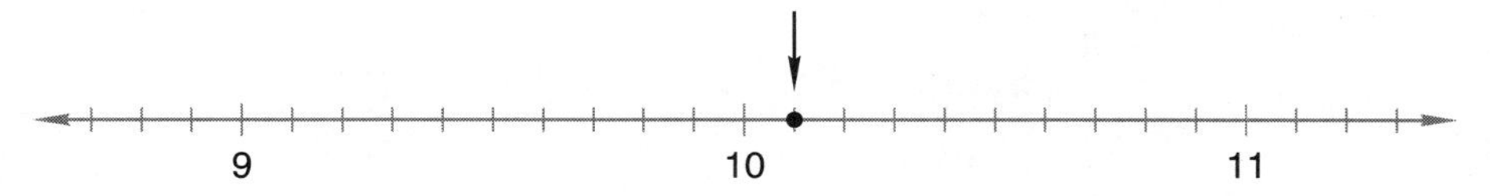

28. Draw a right triangle.
(36)

Refer to this figure to answer problems 29 and 30:

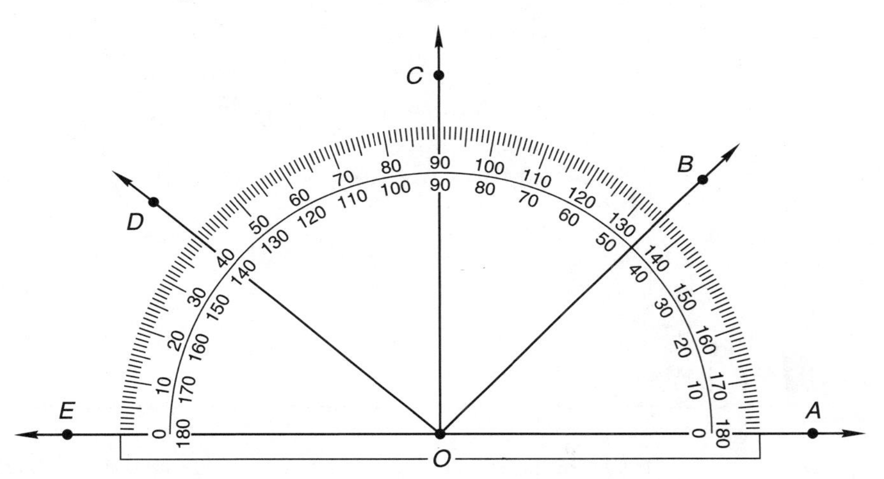

29. What is the measure of $\angle AOC$?
(Inv. 4, 61)

30. Name an acute angle that includes ray OE.
(32, 61)

LESSON

67 Writing Tenths and Hundredths as Decimal Numbers

WARM-UP

Facts Practice: 48 Uneven Divisions (Test G)

Mental Math: The following fractions are equal to $\frac{1}{2}$: $\frac{1}{2}$, $\frac{2}{4}$, $\frac{3}{6}$, ... Read them aloud and continue the pattern to $\frac{12}{24}$.

a. 10% of a decimeter is a _____. **b.** 10% of a centimeter is a _____.

c. $\sqrt{64}$ **d.** $640 \div 20$

e. 6×8, $-\ 3$, $\div\ 5$, $\times\ 3$, $+\ 1$, $\div\ 4$

Problem Solving:

Two figures are similar if they have the same shape. These two triangles are not similar. Draw a triangle similar to the top triangle. Make the sides 2 cm long.

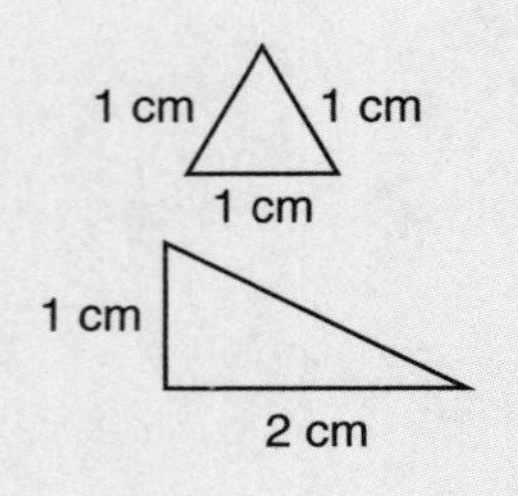

NEW CONCEPT

In this lesson we will write fractions with denominators of 10 or 100 as decimal numbers. A **common fraction** with a denominator of 10 can be written as a decimal number with one decimal place. The numerator of the common fraction is written in the tenths place of the decimal number. For example,

$$\frac{1}{10} \quad \text{can be written as} \quad 0.1$$

The common fraction $\frac{1}{10}$ and the decimal number 0.1 are both named "one tenth" and are equal in value. The zero to the left of the decimal point shows that the whole-number part of the decimal number is zero.

Example 1 Write the fraction three tenths as a common fraction. Then write it as a decimal number.

Solution Three tenths is written as a common fraction like this: $\frac{3}{10}$. A common fraction with a denominator of 10 can be written as a decimal number with one digit after the decimal point. The numerator of the fraction becomes the digit after the decimal point. We write the decimal number three tenths as **0.3.**

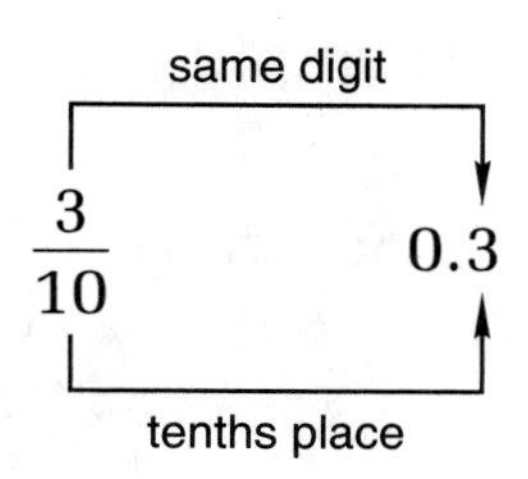

Example 2 A portion of this square is shaded. Name the shaded portion as a fraction and as a decimal number.

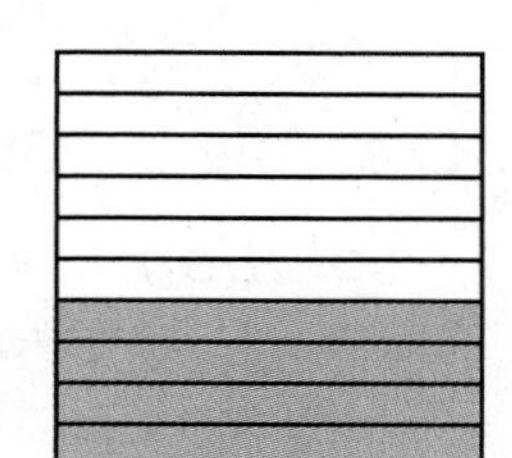

Solution The square is divided into 10 equal parts. Four of the 10 parts are shaded. We are told to name the portion that is shaded as a fraction and as a decimal number. We write $\frac{4}{10}$ and **0.4** as our answers.

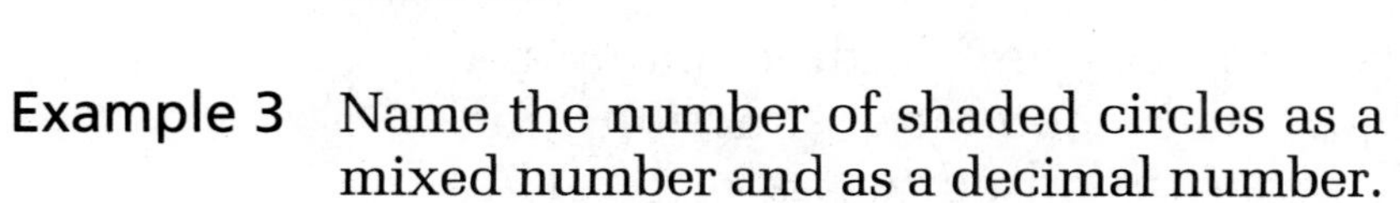

Example 3 Name the number of shaded circles as a mixed number and as a decimal number.

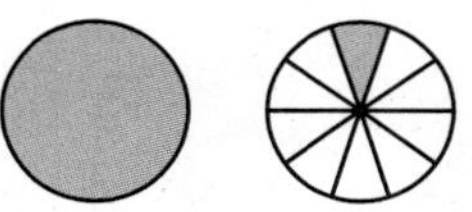

Solution One whole circle is shaded, and one tenth of another circle is shaded. We write one and one tenth as the mixed number $1\frac{1}{10}$. We write one and one tenth as a decimal number by writing the whole number and then the decimal fraction, **1.1.**

A common fraction with a denominator of 100 can be written as a decimal number with two digits after the decimal point. The digits of the numerator of the common fraction become the digits of the decimal number.

$$\frac{1}{100} \quad \text{is the same as} \quad 0.01$$

Notice that in the decimal number we placed the 1 **two places** to the right of the decimal point so that the 1 is in the **hundredths place.** Study these examples:

$$\frac{3}{100} = 0.03 \qquad \frac{30}{100} = 0.30 \qquad \frac{97}{100} = 0.97$$

Notice that when the fraction has only one digit in the numerator we still write two digits after the decimal point. In the first example above, we write the 3 in the hundredths place and a 0 in the tenths place.

Example 4 Write twelve hundredths both as a common fraction and as a decimal number.

Solution Twelve hundredths is written as a common fraction like this: $\frac{12}{100}$. A common fraction with a denominator of 100 can be written as a decimal number with two digits after the decimal point. We write the decimal number twelve hundredths as **0.12.**

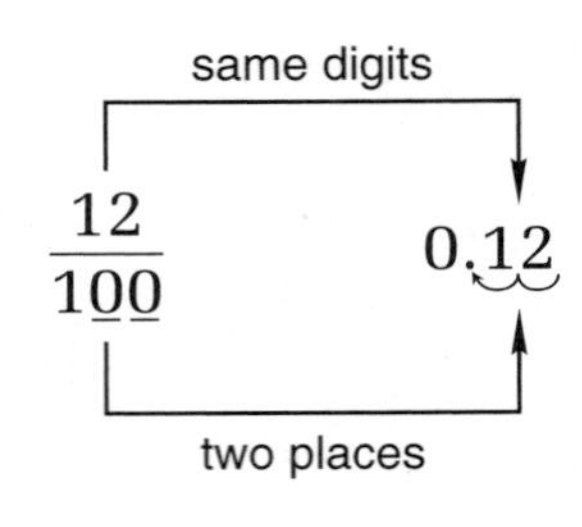

Example 5 Write $4\frac{3}{100}$ as a decimal number.

Solution We write the whole number, 4, to the left of the decimal point. To write hundredths, we use the two places to the right of the decimal point. So $4\frac{3}{100}$ is equal to **4.03.**

Example 6 Name the shaded portion of the square as a common fraction and as a decimal number.

Solution Thirty-three of the hundred parts are shaded. The common fraction for thirty-three hundredths is $\frac{33}{100}$. The decimal number is **0.33.**

Compare the shaded squares found in examples 2 and 6. Notice that more of the square is shaded to show 0.4 than to show 0.33. In the following activity you will compare decimal numbers by shading and comparing portions of squares.

Activity: *Comparing Decimal Numbers*

Materials needed:

- 1 copy of Activity Master 21 per student (masters available in *Saxon Math 6/5 Assessments and Classroom Masters*)

Distribute copies of Activity Master 21, and have students complete problems 1–5.

LESSON PRACTICE

Practice set

a. Name the shaded portion of this rectangle both as a fraction and as a decimal number.

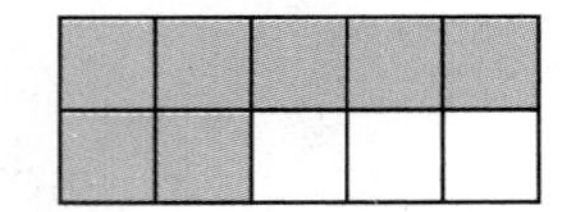

b. Name the unshaded portion of the rectangle both as a fraction and as a decimal number.

c. Name the number of shaded circles both as a mixed number and as a decimal number.

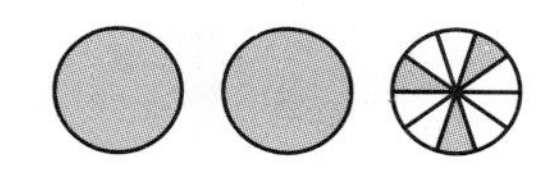

d. Name the shaded portion of the square as a fraction and as a decimal number.

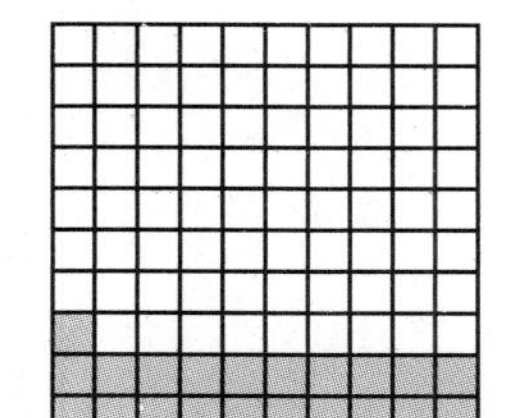

e. Name the unshaded portion of the square as a fraction and as a decimal number.

Write each fraction or mixed number as a decimal number:

f. $\frac{9}{10}$ g. $\frac{39}{100}$ h. $1\frac{7}{10}$ i. $2\frac{99}{100}$

Write each decimal number as a fraction or mixed number:

j. 0.1 k. 0.03 l. 4.9 m. 2.54

MIXED PRACTICE

Problem set

1. (49) The books are divided into 4 stacks with 15 books in each stack. If the books are divided into 5 equal stacks instead of 4, how many books will be in each stack?

2. (53) A loop of string 20 inches long is made into the shape of a square. How long is each side of the square?

3. (49) Geneviève rented 2 movies for $2.13 each. She paid for them with a $10 bill. How much money should she have gotten back?

4. (40, 67) Write the mixed number $2\frac{3}{10}$ with words and as a decimal number.

5. (67) Write the fraction twenty-one hundredths as both a common fraction and as a decimal number.

6. (67) Write the fraction $\frac{99}{100}$ as a decimal number.

7. (67) Use both a fraction and a decimal number to name the "unshaded" portion of this rectangle.

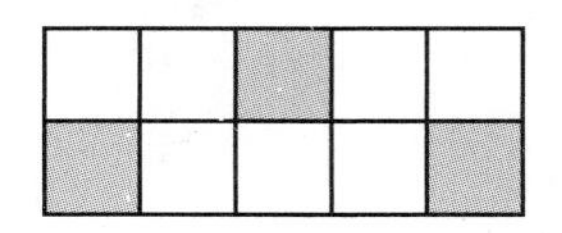

8. Find the length of this segment in centimeters and in millimeters.
(44, 66)

9. Name the shaded portion of this square as a fraction and as a decimal number.
(67)

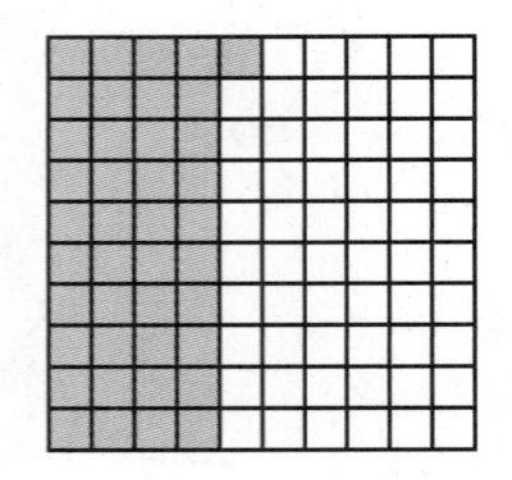

10. What is the quotient of $\frac{35}{8}$? Write your answer as a mixed number.
(58)

11. Use both a common fraction and a decimal number to name the point marked by the arrow.
(38, 66)

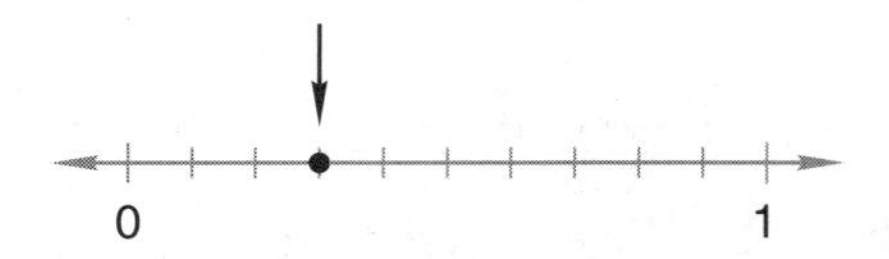

12. List the factors of 12 that are also factors of 20.
(25)

13. $\frac{12}{25} + \frac{12}{25}$
(41)

14. $3\frac{5}{8} - 1$
(43)

15. $5 - 3\frac{5}{8}$
(63)

16. \$100 − (\$90 + \$9 + \$0.01)
(13, 24)

17. $\frac{7848}{9}$
(26)

18. $\frac{3640}{70}$
(54)

19. $20{,}101 - 19{,}191$
(9)

20. $10 - \left(3 + 1\frac{1}{3}\right)$
(43, 63)

21. $3\frac{1}{4} + \left(2 - 1\frac{1}{4}\right)$
(41, 63)

22. $24 \times 8 \times 50$
(18, 29)

23. Write two fractions equal to $\frac{1}{2}$. Make 30 the denominator of the first fraction, and make 25 the numerator of the second fraction.
(23)

Use this menu to answer problems 24 and 25:

MENU	
Taco	$1.20
Nachos	$0.90
Burrito	$1.05
Drinks:	
Regular	$0.80
Small	$0.50

Prices include sales tax.

24. (11, Inv. 5) What is the total cost of one taco, two nachos, and one small drink?

25. (11, Inv. 5) Sam paid for 2 burritos with a $5 bill. How much money should he get back?

26. (32, 53) (a) What type of polygon is figure *ABCDEF*?

(b) If this polygon is regular and the perimeter is 12 inches, then how long is each side?

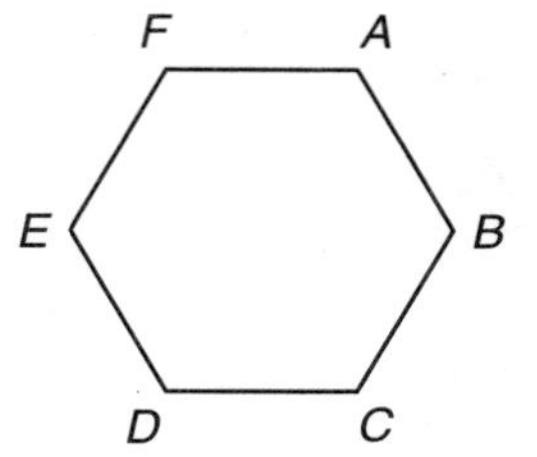

27. (57) The numbers 1, 2, 3, 4, 5, 6, 7, 8, 9, 10, and 11 are written on separate cards. The cards are then turned over and mixed up, and one card is selected.

(a) What is the probability that the number on the card is 7?

(b) What is the probability that the number on the card is odd?

28. (50) What is the average of the 11 numbers in problem 27?

29. (53, 65) A square with sides one decimeter long has a perimeter of how many centimeters?

30. (65) Which of the following choices best describes your height?

A. between 1 and 2 meters

B. between 2 and 3 meters

C. more than 3 meters

D. less than 1 meter

LESSON

68 Naming Decimal Numbers

WARM-UP

Facts Practice: 90 Division Facts (Test D or E)

Mental Math:

a. $\sqrt{81}$ **b.** $1 - \frac{5}{12}$ **c.** 10×10 **d.** $10 \times 10 \times 10$

e. One tenth of 23 is $2\frac{3}{10}$. How much is $\frac{1}{10}$ of 43? ... $\frac{1}{10}$ of 51?

f. Find 25% of 40, + 1, × 3, − 1, ÷ 4

Problem Solving:

Triangles *A* and *B* are congruent. Triangle *A* was "flipped" to the right, to form triangle *B*. Suppose triangle *B* is flipped down to form triangle *C*. Draw triangles *A, B,* and *C.*

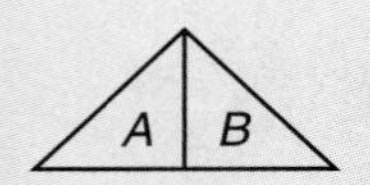

NEW CONCEPT

In this lesson we will name decimal numbers that have one, two, or three decimal places. Recall that the first three decimal places are tenths, hundredths, and thousandths.

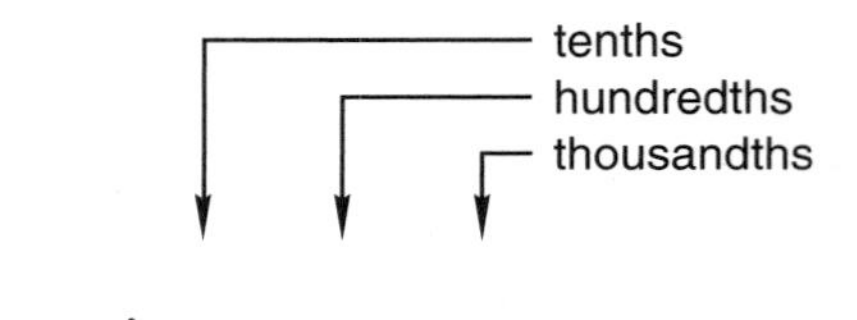

To name a decimal number that has digits on both sides of the decimal point, we mentally break the number into two parts: the whole-number part and the fraction part. The whole-number part is to the left of the decimal point. The fraction part is to the right of the decimal point.

To read this decimal number: 12.5

we mentally break it into two parts, like this: (12).(5)

We read the whole-number part first, say "and" at the decimal point, and then read the fraction part. To read the fraction part, we read the digits as though they named a whole number. Then we say the place value of the last digit. The last digit of 12.5 is 5. It is in the **tenths** place.

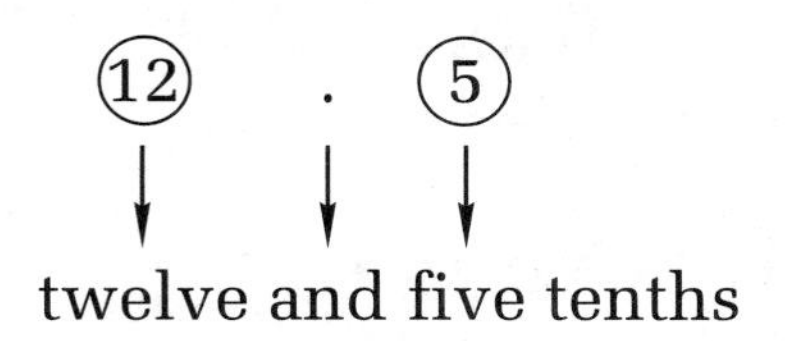

Example 1 Use words to name the decimal number 12.25.

Solution We break the number into two parts. We name the whole-number part, write "and," and then name the fraction part. Then we write the place value of the last digit, which in this case is *hundredths.* We write **twelve and twenty-five hundredths.**

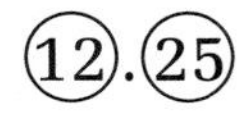

Example 2 Use digits to write the decimal number ten and twelve hundredths.

Solution The whole-number part is ten. The fraction part is twelve hundredths. The word *hundredths* means there are two places to the right of the decimal point.

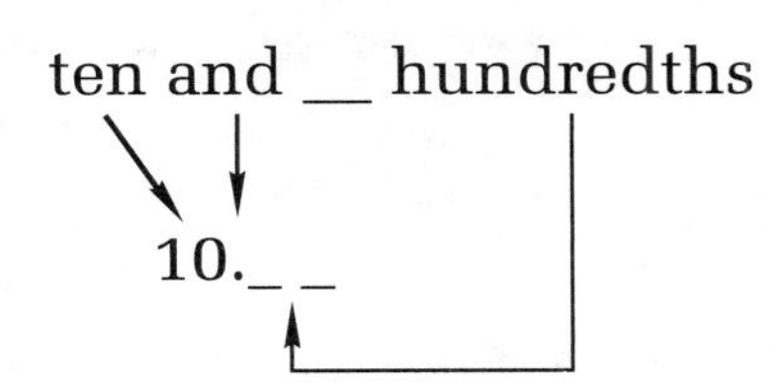

The twelve is written in the two decimal places.

10.12

Example 3 The door was 2.032 meters tall. Use words to write the height of the door.

Solution We break the number into two parts. The place value of the last digit is thousandths.

2.032

The height of the door is **two and thirty-two thousandths meters.**

LESSON PRACTICE

Practice set* Use words to name each decimal number:

a. 8.9

b. 24.42

c. 0.125

d. 10.075

Use digits to write each decimal number:

e. twenty-five and fifty-two hundredths

f. thirty and one tenth

g. seven and eighty-nine hundredths

h. two hundred thirty-four thousandths

MIXED PRACTICE

Problem set

1. *(28)* It takes Wu 20 minutes to walk to school. What time should he leave for school if he wants to arrive at 8:10 a.m.?

2. *(21)* To improve her physical condition, Donna swims, bikes, and runs. Every day Donna swims 40 lengths of a pool that is 25 meters long. How far does Donna swim each day?

3. *(Inv. 3, 46)* Cordelia has read $\frac{1}{3}$ of a 240-page book. How many pages has she read? What percent of the book has she read?

4. *(49)* If 3 tickets cost $12, how many tickets can Gus buy with $20?

5. *(23, 59)* Arrange these fractions in order from least to greatest:

$$\frac{5}{5}, \frac{3}{4}, \frac{2}{6}, \frac{1}{2}$$

6. *(22, 42)* A number is divisible by 4 if it can be divided by 4 without leaving a remainder. The numbers 8, 20, and 32 are all divisible by 4. What number between 10 and 20 is divisible by both 4 and 6?

7. (67) Use a fraction and a decimal number to name the shaded portion of this square.

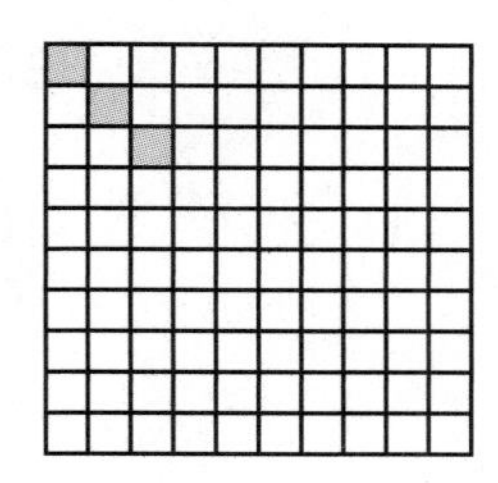

8. (68) Which digit in 16.43 is in the tenths place?

9. (68) The length of the notebook paper was 0.279 meter. Write 0.279 with words.

10. (38, 66) Use a mixed number and a decimal number to name the point on this number line marked by the arrow:

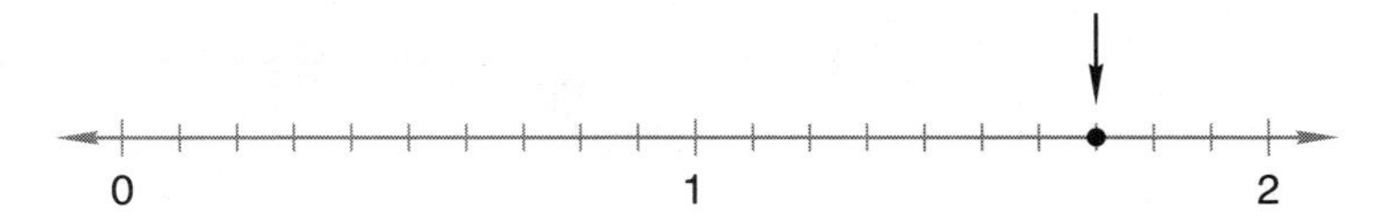

11. (67) Write the decimal number 0.03 as a fraction.

12. (58) Divide 81 by 10. Write the quotient as a mixed number.

13. (61) The length of $\overline{RT}$ is 100 millimeters. If the length of $\overline{RS}$ is 30 millimeters, then how long is $\overline{ST}$?

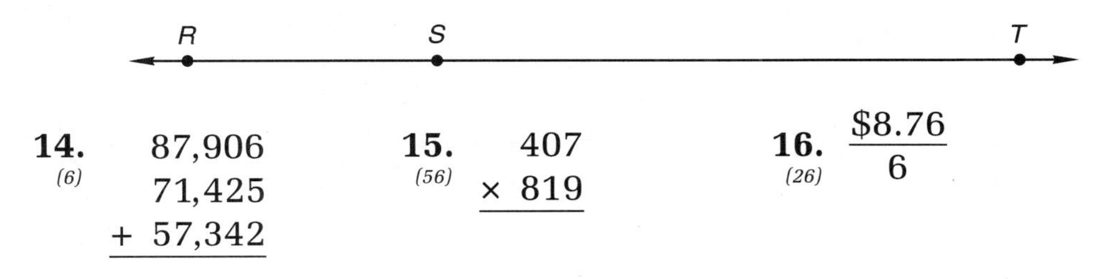

14. (6) $\begin{array}{r} 87{,}906 \\ 71{,}425 \\ +\ 57{,}342 \\ \hline \end{array}$

15. (56) $\begin{array}{r} 407 \\ \times\ 819 \\ \hline \end{array}$

16. (26) $\dfrac{\$8.76}{6}$

17. (24) $600 \div (60 \div 6)$

18. (54) $40\overline{)5860}$

19. (32, 53) If each side of a regular hexagon is 4 inches long, then what is the perimeter of the hexagon?

20. (6) $341 + 5716 + 98 + 492 + 1375$

21. (18) $7 \times 6 \times 5 \times 4$

22. (41, 63) $5\frac{1}{4} + \left(3 - 1\frac{1}{4}\right)$

23. (59) $3\frac{1}{6} + 2\frac{2}{6} + 1\frac{3}{6}$

24. (26, 54) $20w = 300$

25. Compare: $365 \times 1 \bigcirc 365 \div 1$
(4, 15)

26. William found $30,000 of misplaced money. The grateful owner gave William one tenth of the money as a reward. How much money did William receive?
(46)

27. In this figure there are three triangles. Triangle *WYZ* is a right triangle. Which triangle appears to be an obtuse triangle?
(36, 61)

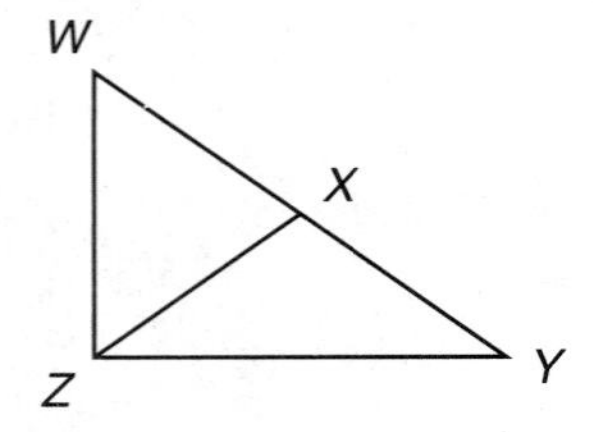

28. A coin is tossed once.
(57)

(a) List the two possible outcomes.

(b) What fraction describes the probability of each outcome?

29. Write 0.625 with words.
(68)

30. Use digits to write the decimal number twelve and seventy-five hundredths.
(68)

LESSON

69 Fractions of a Second • Comparing and Ordering Decimal Numbers

WARM-UP

Facts Practice: 64 Multiplication Facts (Test F)

Mental Math: The following fractions are equal to one half: $\frac{1}{2}, \frac{2}{4}, \frac{3}{6}, \ldots$. Read them aloud and continue the pattern to $\frac{12}{24}$.

a. 10% of \$100 **b.** 10% of \$10 **c.** $\sqrt{100}$ **d.** $3\frac{1}{2} + 3\frac{1}{2}$

e. Find $\frac{1}{3}$ of 12, × 5, − 2, ÷ 2, × 5, − 1, ÷ 4

Problem Solving:

Some small cubes were stacked together to form this larger cube. How many small cubes were used?

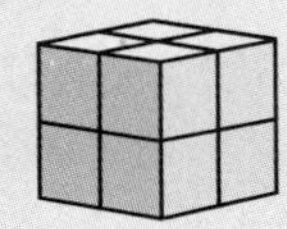

NEW CONCEPTS

Fractions of a second

Fractions of a second are usually expressed as decimals.

Cecilia ran 100 meters in 14.6 seconds.

Marlon swam 50 meters in 28.43 seconds.

Cecilia's 100-meter time was fourteen and six tenths seconds. However, athletes often state their race times in a shorter way. Cecilia might say she ran "fourteen point six" or even "fourteen six." If she runs 100 meters in 14.0 seconds, she might say she ran "fourteen flat." What is important to understand is that 14.6 seconds is more than 14 seconds but less than 15 seconds. A tenth of a second is a short period of time. It is about how long it takes to blink your eyes. A hundredth of a second is even shorter. Races timed to a hundredth of a second are timed electronically rather than by a hand-held stopwatch, because a person with a stopwatch cannot react quickly enough to get an accurate reading.

Activity: *Fractions of a Second*

A stopwatch can help us understand fractions of a second. If a stopwatch is available, try these activities:

- **Test your ability to estimate brief periods of time.** Without looking at the stopwatch display, start the watch and then try to stop it at 5 seconds. Record the time shown on the watch. Repeat the experiment once, and then calculate how close each estimate was to 5 seconds. Which estimate was closer? How close to 5 seconds did you get?
- **Test your quickness.** Start and then stop the stopwatch as quickly as you can. Repeat the experiment once and record the shorter of the two times.

Comparing and ordering decimal numbers To compare decimal numbers, we need to pay close attention to place value. The decimal point separates the whole-number part of a decimal number from the fraction part.

Example 1 Compare: 12.3 ◯ 1.23

Solution Although the same digits appear in both numbers in the same order, the numbers are not equal. The number 12.3 is a little more than 12, but it is less than 13. The number 1.23 is more than 1 but less than 2. So 12.3 is greater than 1.23.

12.3 > 1.23

Example 2 Arrange these numbers in order from least to greatest:

1.02, 1.2, 1.12

Solution Arranging the numbers vertically with the decimal points aligned can make the order easier to determine. We compare the digits column by column, beginning with the first column on the left.

1.02
1.2
1.12

The whole-number part of each number is 1, so we need to compare the fraction parts. The first digit to the right of the decimal point is in the tenths place (in money, it is the dimes place). The number 1.02 has a zero in the tenths place, the number 1.12 has a one in the tenths place, and the number 1.2 has a two in the tenths place. This is enough information to order the numbers from least to greatest.

1.02, 1.12, 1.2

LESSON PRACTICE

Practice set

a. John ran 200 meters in 38.6 seconds. Juanita ran 200 meters in 37.9 seconds. Which athlete ran faster?

b. Compare: 3.21 ◯ 32.1

c. Arrange these numbers in order from least to greatest:

2.4, 2.04, 2.21

MIXED PRACTICE

Problem set

1. (21) The ceiling was covered with square tiles. There were 30 rows of tiles with 30 tiles in each row. How many tiles covered the ceiling?

2. (49) Carlos gave the clerk $10 for a book that cost $6.95 plus $0.42 tax. How much money should he get back?

3. (21) Silvia emptied a jar of 1000 pennies and put them into rolls holding 50 pennies each. How many rolls did she fill?

4. (Inv. 2) The distance around the school track is $\frac{1}{4}$ mile. How many times must Steve run around the track to run 1 mile?

5. (42) What even number greater than 20 and less than 30 is divisible by 3?

6. (25) List the numbers that are factors of both 10 and 15.

7. (69) Compare: 44.4 ◯ 4.44

8. (52) Which digit in 56,132 is in the same place as the 8 in 489,700?

9. (67) Use both a fraction and a decimal number to name the unshaded portion of this group of circles.

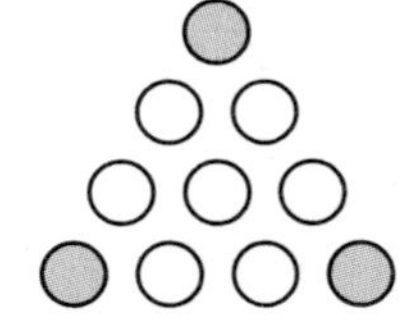

10. (66) Find the length of this segment to the nearest tenth of a centimeter.

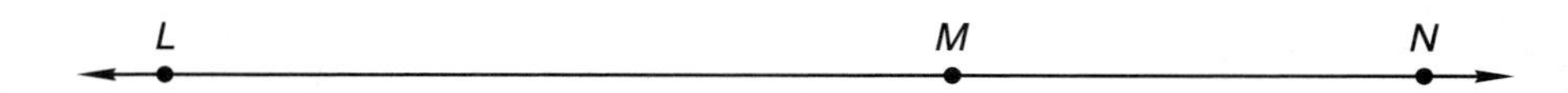

11. (64) Which digit in 67.89 is in the hundredths place?

12. (61, 63) The length of $\overline{LN}$ is 4 inches. If $\overline{MN}$ is $1\frac{1}{2}$ inches, then how long is $\overline{LM}$?

L M N

13. (68) Write 10.5 with words.

14. (68) Use digits to write the decimal number fifteen and twelve hundredths.

15. (26) $\frac{3744}{8}$

16. (9) $30{,}000 - 29{,}925$

17. (55) 973×536

18. (58) What number is half of 75?

19. (29) $\$0.65 \times 10$

20. (26) $5\overline{)\$9.60}$

21. (54) $\frac{\$54.30}{30}$

22. (43, 63) $7 - \left(3 + 1\frac{1}{3}\right)$

23. (41, 43, 59) $5\frac{2}{3} + \left(3\frac{1}{3} - 2\right)$

Use this information to answer problems 24 and 25:

> *In the school election for president, Aaron received 239 votes, Brigit received 168 votes, and Chang received 197 votes.*

24. (11, Inv. 5) One other person ran for president and received 95 votes. Altogether, how many votes were cast for president?

25. (35, Inv. 5) The winner received how many more votes than the person who came in second?

26. A number cube is rolled once. What is the probability of each of these outcomes?
(57)

(a) The number will be 6 or less.

(b) The number will be greater than 6.

(c) The number will be even.

27. What is the place value of the 7 in $6.75?
(64)

28. Name the shaded portion of this square as a fraction, as a decimal number, and with words.
(67, 68)

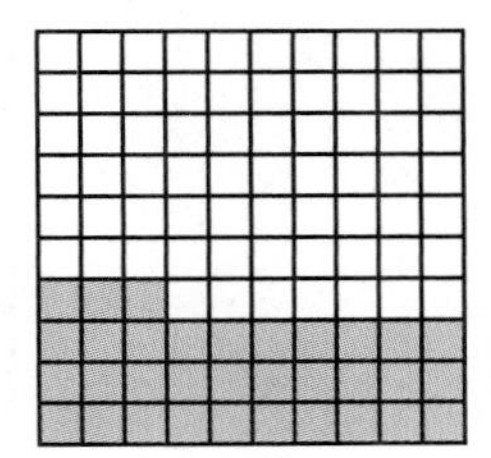

29. What mixed number is $\frac{1}{3}$ of 100?
(Inv. 3, 58)

30. The table below shows the age at which fourteen children first rode a bicycle. Copy and complete the frequency table for this data.
(Inv. 5)

Age Children First Rode a Bicycle

Name	Age	Name	Age
Phil	6	Reggie	4
Rosa	5	Scott	7
Naomi	4	Ann	5
Ali	8	Jaime	4
Ngoc	5	Russ	5
Tina	4	Ashley	4
Yuko	6	Andrew	5

Frequency Table

Age	Tally	Frequency
4		
5		
6		
7		
8		

LESSON

70 Writing Equivalent Decimal Numbers • Writing Cents Correctly

WARM-UP

Facts Practice: 48 Uneven Divisions (Test G)

Mental Math: A number is divisible by 4 if the number formed by the last two digits is a multiple of 4. So 1324 is divisible by 4, but 1342 is not. Are the numbers in problems **a–d** divisible by 4? Answer "yes" or "no" for each.

a. 1234 **b.** 3412 **c.** 2314 **d.** 4132

e. 100 ÷ 4 **f.** 200 ÷ 4 **g.** 300 ÷ 4

Problem Solving:

Copy this multiplication problem and fill in the missing digits. The digits of the product are 1, 2, and 7, though not in that order.

```
  ___
×   4
-----
  ___
```

NEW CONCEPTS

Writing equivalent decimal numbers

We may attach one or more zeros to the end of a decimal number without changing the number's value. For example, we may write 0.3 as 0.30. The zero does not change the value of the number, because it does not change the place value of the 3. In both numbers, 3 is in the tenths place. Thus, three tenths is equal to thirty hundredths.

Example 1 Write 12.6 with three decimal places.

Solution The number 12.6 is written with one decimal place. By attaching two zeros, we get **12.600,** which has three decimal places.

Example 2 Compare: 12.6 ◯ 12.600

Solution When we compare decimal numbers, we must pay close attention to place value. We use the decimal point to locate places. We see that the whole-number parts of these two numbers are the same. The fraction parts look different, but both numbers have a 6 in the tenths place. If we add two zeros to 12.6 to get 12.600, we see that the numbers are the same. So we use an equal sign in the comparison.

12.6 = 12.600

Writing cents correctly

Here are two ways to write "fifty cents":

1. As a number of cents: 50¢
2. As a number of dollars: $0.50

Sometimes we see signs with a money amount written incorrectly.

Soda
0.50¢
per can

This sign literally means that a can of soda costs $\frac{50}{100}$ of a penny, which is half a cent! The sign could be corrected by changing 0.50¢ to $0.50 or to 50¢.

Example 3 Use digits and symbols to write "five cents" both in cent form and in dollar form.

Solution **5¢ (cent form)**

$0.05 (dollar form)

Example 4 This sign is written incorrectly. Show two ways to correct the money amount shown on the sign.

Brownies
0.25¢
each

Solution **25¢ (cent form)**

$0.25 (dollar form)

Example 5 Add: $1.56 + 75¢

Solution When both forms of money are in the same problem, we first rewrite the amounts so that they are all in the same form. Then we solve the problem. Sums of money equal to a dollar or more are usually written with a dollar sign. To find $1.56 + 75¢, we can change 75¢ to dollar form and then add, as shown at right.

```
  $1.56
+ $0.75
-------
  $2.31
```

LESSON PRACTICE

Practice set Write each number with three decimal places:

a. 1.2 **b.** 4.08 **c.** 0.50000

Compare:

d. 50 ◯ 500 **e.** 0.4 ◯ 0.04

f. 0.50 ◯ 0.500 **g.** 0.2 ◯ 0.20000

Write each money amount both in cent form and in dollar form:

h. two cents **i.** fifty cents

j. twenty-five cents **k.** nine cents

Solve problems **l–o.** Write each answer in the indicated form.

l. 36¢ + 24¢ = $_____ **m.** $1.38 − 70¢ = _____¢

n. $0.25 − 5¢ = $_____ **o.** $1 − 8¢ = _____¢

Multiply. Write each product in dollar form.

p. 7 × 65¢ **q.** 20 × 18¢

MIXED PRACTICE

Problem set

1. (53) Each side of a 1-foot square is 1 foot long. What is the perimeter of a 1-foot square?

2. (35) Columbus landed in the Americas in 1492. The Pilgrims landed in 1620. The Pilgrims landed how many years after Columbus landed?

3. (62) Estimate the product of 307 and 593 by rounding both numbers to the nearest hundred before multiplying.

4. (17) Three times a number n can be written "$3n$." If n equals the number 5, then what number does $3n$ equal?

5. (Inv. 3, 60) Mike has read $\frac{1}{3}$ of his book. What fraction of his book does he still have to read? What percent of his book does he still have to read?

6. (Inv. 3, 37) Draw a circle and shade one eighth of it. What percent of the circle is shaded?

7. Divide 100 by 7. Write the quotient as a mixed number.
(58)

8. Which digit in 12.3 is in the tenths place?
(64)

9. Use a fraction and a decimal number to name the shaded portion of this square.
(67)

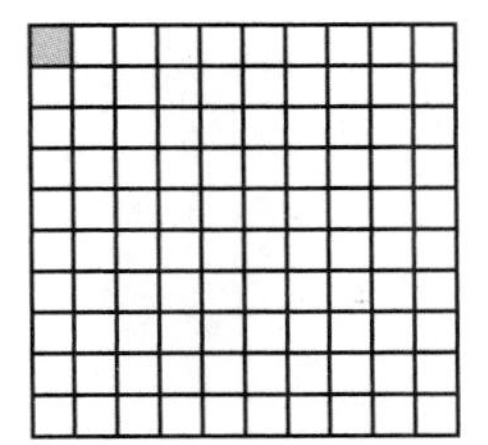

10. Which digit in 98.765 is in the thousandths place?
(68)

11. The length of $\overline{QR}$ is 3 centimeters. The length of $\overline{RS}$ is twice the length of $\overline{QR}$. How long is $\overline{QS}$?
(61)

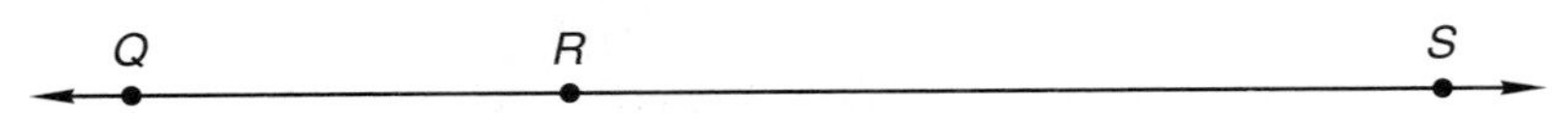

12. Use words to name the decimal number 16.21.
(68)

13. Write 1.5 with two decimal places.
(70)

14. Compare: 3.6 ◯ 3.60
(70)

15. $\begin{array}{r} 307 \\ \times\ 593 \\ \hline \end{array}$
(56)

16. $\dfrac{765}{5}$
(26)

17. $60\overline{)\$87.00}$
(54)

18. 3517 + 9636 + 48 + 921 + 8576 + 50,906
(6)

19. $2\frac{3}{10} + 1\frac{3}{10} + \frac{3}{10}$
(43)

20. $9\frac{4}{8} + \left(4 - 1\frac{7}{8}\right)$
(41, 63)

21. 40 × 50 × 60
(18, 29)

22. \$100 − (\$84.37 + \$12)
(13, 24)

23. Write "twenty-five cents"
(70)
(a) with a dollar sign.
(b) with a cent sign.

Use this graph to answer problems 24 and 25:

Favorite Sports of 100 Children

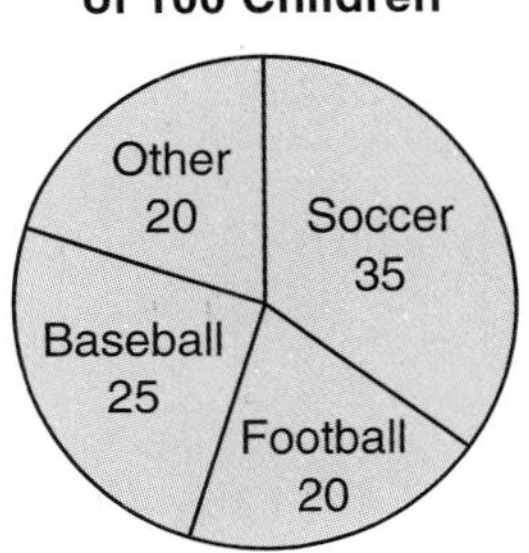

24. For how many children was either soccer or football their favorite sport?
(11, Inv. 5)

25. What was the second-most favorite sport?
(Inv. 5)

26. (57) Suppose the 8 letter tiles below are turned over and mixed up. Then one tile is selected.

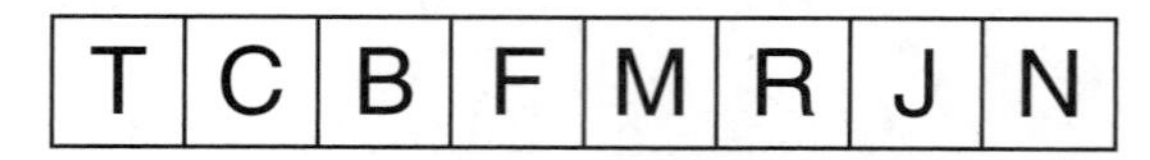

Which word best describes the following events: *likely, unlikely, certain,* or *impossible?*

(a) The letter selected is a consonant.

(b) The letter selected comes after S in the alphabet.

(c) The letter selected is either G or H.

Courtney and Lamar went fishing for trout. They caught 17 trout that were at least 7 inches long. The distribution of lengths is shown on the line plot below. Refer to this information to answer problems 27–29.

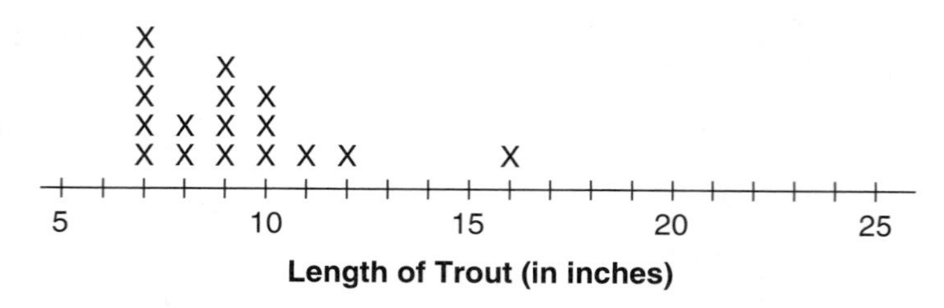

27. (Inv. 5) How many trout were less than 11 inches long?

28. (Inv. 5) Which lengths were recorded more than three times?

29. (Inv. 5) Which, if any, of the lengths are outliers?

30. (67, 68) One fourth of this square is shaded. Write the shaded portion of the square as a decimal number. Then write the decimal number with words.

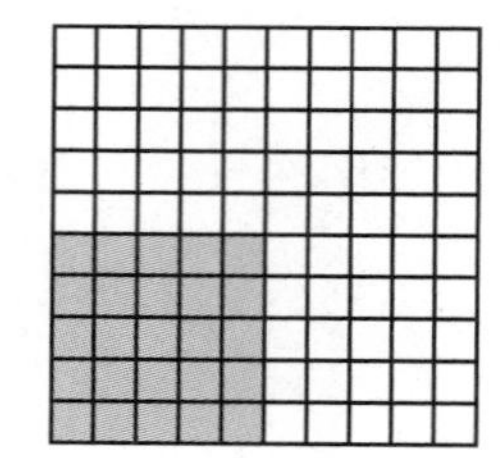

INVESTIGATION 7

Focus on

Pattern Recognition

In Lesson 1 we introduced sequences as counting patterns that continue indefinitely. In each of the examples we looked at, we either counted up or down by a fixed amount. But there are many other possible patterns that can determine the terms of a sequence. Rather than adding or subtracting a fixed number, we could also multiply by a fixed number to produce the next terms.

Example 1 What rule seems to describe this sequence? Find the next two terms in this sequence:

3, 6, 12, 24, _____, _____, ...

Solution Since $3 \times 2 = 6$ and $6 \times 2 = 12$ and $12 \times 2 = 24$, we **multiply by 2** to find the next terms. Thus, the fifth term of the sequence is $24 \times 2 =$ **48,** and the sixth term is $48 \times 2 =$ **96.**

A sequence that counts up (adds) or counts down (subtracts) by the same amount is called an **arithmetic sequence.** A sequence that increases by multiplying by the same number or decreases by dividing by the same number is called a **geometric sequence.**

In problems 1–4 below, decide whether the sequence is arithmetic or geometric. Then write the next three terms.

1. 43, 49, 55, 61, _____, _____, _____, ...

2. 2, 4, 8, 16, _____, _____, _____, ...

3. 50, 48, 46, 44, _____, _____, _____, ...

4. 2, 6, 18, 54, _____, _____, _____, ...

5. Sally has saved \$55. Each month she plans to add \$8 to her savings. If she does not spend any of the money she saves, how much will Sally have after one month? Two months? Three months? What kind of sequence are we making?

6. Without fumigation, the cockroach population in a house is expected to double each month. If there are 50 cockroaches in the house now, what will be the number of cockroaches after one month? Two months? Three months? What kind of sequence are we making?

Another kind of pattern found in sequences is *repetition.* This means that the terms of the sequence repeat themselves. The number of terms in a repeating unit is called the **period.**

Example 2 What appears to be the period of the following sequence? List the next three terms.

4, 5, 8, 4, 5, 8, 4, 5, ____, ____, ____, ...

Solution Here, the unit "4, 5, 8" appears to be repeating, so the period is **three.** Since the last term given is 5, the next three terms would be **8, 4, 5.**

Science has demonstrated that the human brain searches for patterns in events or objects. Since the sun rises every day in the east and sets in the west, we expect the same to occur tomorrow. If we see a pattern in a section of floor tiles, we assume that the pattern continues over the whole floor. We might make similar assumptions about sequences. However, if the part of the sequence we are looking at is not large enough, we might assume a pattern that is not actually there! For example, if we see that a sequence begins with

4, 6, 4, 6, ...

it is natural to assume that the sequence has period two and that the alternating 4's and 6's will continue. But those terms are also the beginning of the sequence

4, 6, 4, 6, 2, 4, 6, 4, 6, 2, ...

which seems to fit a very different pattern. Without clear information about the structure of a sequence, we must be aware that the patterns we see might not really be there.

7. Assuming the following sequence has period three, write the next three terms:

4, 5, 9, 4, 5, 9, 4, ____, ____, ____, ...

8. Assuming the following sequence has period four, write the next three terms:

5, 2, 3, 6, 5, 2, _____, _____, _____, ...

9. Assuming the following sequence has period four, write the next three terms:

B, U, L, B, _____, _____, _____, ...

10. Assuming the sequence of problem 9 has period three, write the next three terms.

There are many types of patterns that sequences can follow. In the following example, we look at another kind of pattern.

Example 3 What pattern does this sequence appear to follow?

1, 0, 1, 0, 0, 1, 0, 0, 0, 1, ...

Solution **This sequence of 0's and 1's has 1's separated by an increasing number of 0's.** First, one 0 separates 1's; then two 0's; then three 0's. It is reasonable to predict that there will be four more 0's before the next 1 that appears in the sequence.

1, 0, 1, 0, 0, 1, 0, 0, 0, 1, 0, 0, 0, 0, 1, ...

1 2 3 4 zeros

We can predict that there will then be five more 0's between 1's, six more 0's, and so on.

For problems 11–15 below, describe the pattern that the sequence appears to follow. Then write the next few terms that seem to fit the pattern.

11. 1, 1, 2, 2, 3, 3, ...

12. 0, 2, 0, 4, 0, 6, 0, ...

13. A, B, D, E, G, H, ...

14. T, ⊣, ⊥, ⊢, T, ...

15. 1, 2, 1, 2, 3, 1, 2, 3, 4, ...

Some patterns can be seen more easily by recording the increase or decrease between terms.

Example 4 What are the next three terms in this sequence?

0, 1, 3, 6, 10, ...

Solution We first find the difference between successive terms.

$$\overset{+2}{1, \; 3,} \; \overset{+3}{6,} \; \overset{+4}{10, \ldots}$$

The increasing difference from one term to the next also forms a sequence. This sequence may be continued.

+2 +3 +4 +5 +6 +7

1, 3, 6, 10, _____, _____, _____, ...

We add 5 to 10 and get 15 for the next term. We add 6 to 15 and get 21 for the following term. We add 7 to 21 and get 28. We have found the next three terms.

1, 3, 6, 10, **15**, **21**, **28**, ...

Find the next three terms in these sequences:

16. 1, 4, 9, 16, 25, _____, _____, _____, ...

17. 2, 3, 5, 8, 12, _____, _____, _____, ...

Example 5 Suppose the first two terms of a sequence are 3 and 4 and we always get the next term by adding the previous two terms together. Thus, the third term would be $3 + 4 = 7$. Find the fourth, fifth, and sixth terms of such a sequence:

3, 4, 7, _____, _____, _____, ...

Solution We find each term by adding the two preceding terms. Three and 4 were added to get the third term, 7. Now we add 4 and 7 to find the fourth term, 11.

3, 4, 7, 11, _____, _____, ...

We continue adding the two preceding terms. The sum of 7 and 11 is 18. The sum of 11 and 18 is 29.

3, 4, 7, __**11**__, __**18**__, __**29**__, ...

18. A famous sequence in mathematics is the Fibonacci sequence, which follows a pattern similar to the sequence in example 5. Many patterns found in nature fit the Fibonacci sequence. Below we show the first six terms of the Fibonacci sequence. Find the next three terms.

1, 1, 2, 3, 5, 8, ____, ____, ____, ...

We have studied some patterns in sequences. There are also patterns between pairs of numbers. Below we show a "number-changing machine." When we put a number in the machine, another number comes out of the machine. An in-number and its out-number form a pair. By studying in-out pairs we can find the pattern, or rule, the machine uses to change in-numbers to out-numbers.

In		Out
3 →	NUMBER-CHANGING MACHINE	→ 8
6 →		→ 11
9 →		→ 14

The in-out pairs of this machine are (3, 8), (6, 11), and (9, 14). Notice that the number-changing machine adds five to the in-number to produce the out-number. So if the in-number is 10, the out-number is 15.

Example 6 What is the rule of this number-changing machine?

In		Out
1 →	NUMBER-CHANGING MACHINE	→ 4
3 →		→ 12
4 →		→ 16

If 5 is the in-number, what is the out-number?

Solution **The machine multiplies the in-number by 4.** If 5 is the in-number, the out-number is 4×5, which is **20.**

Refer to this machine to answer problems 19 and 20:

In		Out
6 →	NUMBER-CHANGING MACHINE	→ 2
8 →		→ 4
10 →		→ 6

19. What does the machine do to each in-number?

A. It divides by 3. B. It adds 4.

C. It subtracts 4. D. It adds 2.

20. If 15 is used as an in-number, what will be the out-number?

Refer to this machine to answer problems 21 and 22:

In		Out
6 →	NUMBER-CHANGING MACHINE	→ 3
8 →		→ 4
10 →		→ 5

21. What does the machine do to each in-number?

A. It adds 2. B. It adds 1.

C. It subtracts 3. D. It divides by 2.

22. If the in-number is 20, what is the out-number?

LESSON

71 Fractions, Decimals, and Percents

WARM-UP

Facts Practice: 100 Multiplication Facts (Test C)

Mental Math: The following fractions are equal to one half: $\frac{1}{2}$, $\frac{2}{4}$, $\frac{3}{6}$, ... Read them aloud and continue the pattern to $\frac{15}{30}$.

a. Is 2736 divisible by 4?

b. Is 3726 divisible by 4?

c. $\frac{1}{3}$ of 10

d. $\frac{1}{3}$ of 100

e. $\sqrt{25}$, + 3, × 4, + 1, ÷ 3

Problem Solving:

All squares are similar. Each side of this square is $\frac{1}{2}$ inch long. Draw a square with sides half as long and another square with sides twice as long. Calculate the total perimeter of all three squares.

NEW CONCEPT

Fractions, decimals, and percents are three ways to name part of a whole.

$\frac{1}{2}$ of the circle is shaded.

0.5 of the circle is shaded.

50% of the circle is shaded.

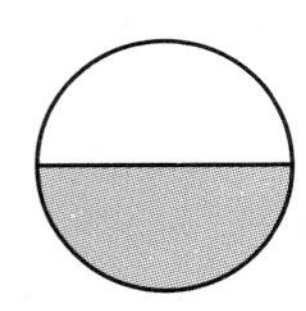

Fractions, decimals, and percents have numerators and denominators. The denominator might not be obvious.

$\frac{1}{2}$ ← The denominator of a fraction can be any number other than zero and is expressed in the fraction.

0.5 ← The denominator of a decimal number is a number from the sequence 10, 100, 1000, ... The denominator is indicated by the number of digits to the right of the decimal point.

50% ← The denominator of a percent is always 100 and is indicated by the word *percent* or by a percent sign.

To write a decimal or a percent as a fraction, we must express the denominator.

$$0.5 \text{ equals } \frac{5}{10} \qquad 50\% \text{ equals } \frac{50}{100}$$

Notice that both $\frac{50}{100}$ and $\frac{5}{10}$ equal $\frac{1}{2}$.

Example The fraction manipulative for $\frac{1}{10}$ has (a) 10% and (b) 0.1 printed on it. Change these two numbers into fractions.

Solution (a) We can write a percent as a fraction by replacing the percent sign with a denominator of 100.

$$10\% = \mathbf{\frac{10}{100}}$$

(b) A decimal number with one decimal place has a denominator of 10.

$$0.1 = \mathbf{\frac{1}{10}}$$

The above example refers to the manipulatives we used in Investigations 2 and 3 that have fractions, percents, and decimals printed on them. Here we show the numbers that are printed on the different pieces:

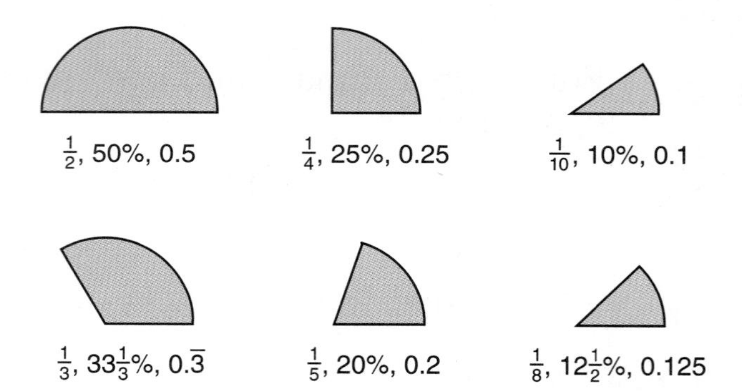

Activity: *Fractions, Decimals and Percents*

Use your fraction manipulatives or refer to the figures above to help you answer these questions:

1. If you fit three $\frac{1}{4}$ pieces together $\left(\frac{1}{4} + \frac{1}{4} + \frac{1}{4}\right)$, you have

(a) what fraction of a circle?

(b) what percent of a circle?

(c) what decimal part of a circle?

2. If you fit two $\frac{1}{5}$ pieces and one $\frac{1}{10}$ piece together $\left(\frac{1}{5} + \frac{1}{5} + \frac{1}{10}\right)$, you have
   (a) what fraction of a circle?
   (b) what percent of a circle?
   (c) what decimal part of a circle?

3. (a) If you divide 100 by 3, what mixed number is the quotient?
   (b) What percent is printed on the $\frac{1}{3}$ fraction manipulative?

4. (a) If you divide 1,000,000 by 3, what digit repeats in the quotient?
   (b) What decimal number is written on the $\frac{1}{3}$ fraction manipulative?
   (c) What is unusual about the way the number is printed?

5. (a) If you divide 1000 by 8, what is the quotient?
   (b) What decimal number is printed on the $\frac{1}{8}$ fraction manipulative?
   (c) What percent is printed on the $\frac{1}{8}$ fraction manipulative?

Compare. You may use your fraction manipulatives to help answer each problem.

6. 0.125 ◯ 0.2

7. 0.25 ◯ $0.\overline{3}$

8. 0.5 ◯ 0.25 + 0.25

9. 50% ◯ $33\frac{1}{3}$%

10. $12\frac{1}{2}$% ◯ 20%

LESSON PRACTICE

Practice set Work problems **a–d.** Use your fraction manipulatives as necessary to help answer the problems.

a. Draw a circle and shade 25% of it. What decimal part of the circle did you shade?

b. The fraction manipulative for $\frac{1}{5}$ has the numbers 20% and 0.2 printed on it. Write both 20% and 0.2 as fractions.

c. This square is divided into 100 equal parts, and 33 parts are shaded. Write the shaded portion as a fraction, as a percent, and as a decimal.

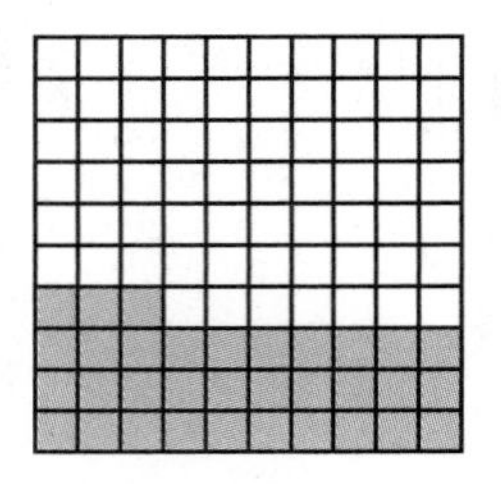

d. Refer to the figure in problem **c** to complete this comparison and to answer the question that follows.

Compare: $\frac{1}{3} \bigcirc 0.33$

How did you determine the comparison?

MIXED PRACTICE

Problem set

1. What is the total cost of a $7.98 notebook that has 49¢ tax?
(11, 70)

2. In Room 7 there are 6 rows of desks with 5 desks in each row. There are 4 books in each desk. How many books are in all the desks?
(49)

3. This year, Martin is twice as old as his sister. If Martin is 12 years old now, how old will his sister be next year?
(49)

4. Silviano saves half-dollars in a coin holder. How many half-dollars does it take to total $5?
(46)

5. Louisa put her nickel collection into rolls that hold 40 nickels each. She filled 15 rolls and had 7 nickels left over. Altogether, how many nickels did Louisa have?
(49)

6. The number 7 has how many factors?
(25)

7. Which of these fractions is not equal to $\frac{1}{2}$?
(23)

A. $\frac{6}{12}$ B. $\frac{7}{15}$ C. $\frac{8}{16}$ D. $\frac{9}{18}$

8. Allison can swim 50 meters in half a minute. Amy can swim 50 meters in 28.72 seconds. Which of the two girls can swim faster?
(69)

9. Use a mixed number and a decimal number to name the point on this number line marked by the arrow.
(66)

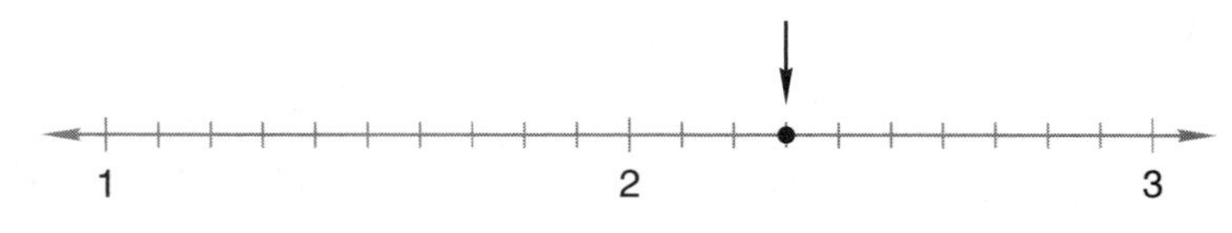

10. Which digit in 1.234 is in the thousandths place?
(68)

11. Use digits to write the decimal number ten and one tenth.
(68)

12. How many cents is $\frac{4}{5}$ of a dollar?
(Inv. 3)

13. Segment AB measures 50 millimeters. The length of $\overline{BC}$ is half the length of $\overline{AB}$. How long is $\overline{AC}$?
(61)

14. Compare: 12.3 ◯ 12.30
(69)

15. *(13)*
$$\begin{array}{r} \$5.37 \\ \$8.95 \\ \$0.71 \\ + \ \$0.39 \\ \hline \end{array}$$

16. *(13)*
$$\begin{array}{r} \$60.10 \\ - \ \$48.37 \\ \hline \end{array}$$

17. *(56)*
$$\begin{array}{r} \$9.84 \\ \times \quad 150 \\ \hline \end{array}$$

18. \$1.75 + 36¢ = \$_____
(70)

19. \$1.15 − \$0.80 = _____¢
(70)

20. 40 × 76¢
(70)

21. \$39.00 ÷ 50
(54)

22. $\frac{13}{100} + \frac{14}{100}$
(41)

23. $7 - \left(6\frac{3}{5} - 1\frac{1}{5}\right)$
(41, 63)

Use this information to answer problems 24 and 25:

> *Tyrone invented a machine to change numbers. When he puts a 7 into the machine, a 5 comes out. When he puts a 4 in the machine, a 2 comes out. When he puts a 3 in the machine, a 1 comes out.*

7 → NUMBER-CHANGING MACHINE → 5
4 → → 2
3 → → 1

24. What does the machine do to each number that Tyrone puts into it?
(Inv. 7)

A. It adds 2.
B. It subtracts 2.
C. It divides by 2.
D. It multiplies by 2.

25. If Tyrone puts in a 10, what number will come out?
(Inv. 7)

26. The sign showed that lemonade was offered for 0.20¢ per glass. Show two ways to correct the money amount shown on the sign.
(70)

27. Is the sequence below arithmetic or geometric? What are the next two terms?
(Inv. 7)

1, 3, 9, 27, _____, _____, ...

28. A bag contains 3 red marbles, 4 yellow marbles, 2 purple marbles, and 1 green marble. One marble is selected without looking.
(57)

(a) Find the probability that the marble is yellow.

(b) Find the probability that the marble is not yellow.

29. The fraction $\frac{2}{5}$ is equivalent to 0.4 and to 40%. Write 0.4 and 40% as unreduced fractions.
(71)

30. Draw a parallelogram that has a right angle.
(45)

LESSON

72 Area, Part 1

WARM-UP

Facts Practice: 90 Division Facts (Test D or E)

Mental Math:

a. How many minutes is $2\frac{1}{2}$ hours?

b. 60×70

c. Is 5172 divisible by 4?

d. 10% of 250

e. How much is $\frac{1}{2}$ of 12? ... $\frac{1}{3}$ of 12? ... $\frac{1}{4}$ of 12?

f. $\sqrt{36}$, + 1, × 7, + 1, ÷ 5, − 2, ÷ 2

Problem Solving:

Adam, Barbara, Conrad, and Debby were posing for a picture, but the photographer insisted that only three people could pose at one time. List the combinations of three people that are possible. (In this problem different arrangements of the same three people are not considered different combinations.)

NEW CONCEPT

If you look at the edges of your classroom where the floor and walls meet, you might see a strip of molding or baseboard that runs all the way around the room except at the doorways. That molding illustrates the perimeter of the floor of the room. If you were to buy molding at a store, you would buy a length of it and pay for it by the foot or yard.

The floor of your classroom might be covered by tile or carpet. That tile or carpet illustrates the **area** of the floor. Area is not a length; it is an amount of surface. If you were to buy tile or carpet at a store, you would buy a box or roll of it and pay for it by the square foot or square yard.

A square tile illustrates the units we use to measure area. Many floor tiles are squares with sides one foot long. Each of these tiles is one square foot; that is, each tile would cover one square foot of the area of a room's floor. By counting the number of one-square-foot tiles on the floor, you can determine the area of the room in square feet.

Example 1 In a classroom the floor was the shape of a rectangle and was covered with 1-foot square tiles. The room was 30 tiles long and 25 tiles wide. What was the area of the floor?

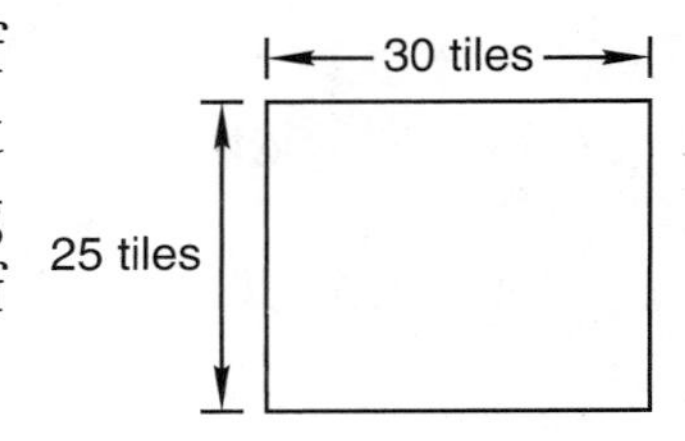

Solution By finding the number of tiles, we will find the area of the room. To find the number of tiles in 25 rows of 30 tiles, we multiply.

$$30 \times 25 = 750$$

There are 750 tiles. Each tile is one square foot. So the area of the floor is **750 square feet.**

The areas of rooms, houses, and other buildings are usually measured in square feet. Expanses of land may be measured in acres or square miles. (One square mile equals 640 acres.) Smaller areas may be measured in square inches or square centimeters.

A square that has sides 1 centimeter long is called a *square centimeter.* The square at right is the actual size of a square centimeter.

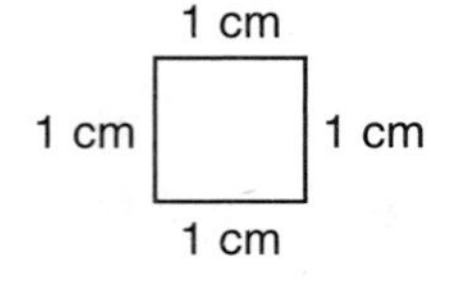

A square that has sides 1 inch long is called a *square inch.* The square below is the actual size of a square inch.

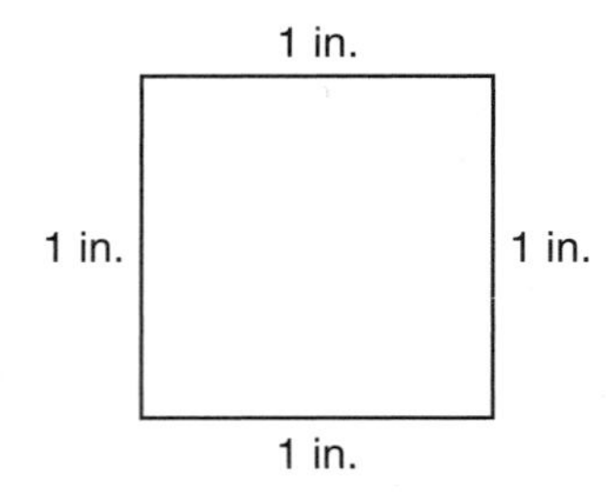

We have noted that it is sometimes necessary to draw a figure in reduced size so it will fit on a page. In the following example we have drawn the figure in reduced size.

Example 2 How many square inches are needed to cover the area of this rectangle?

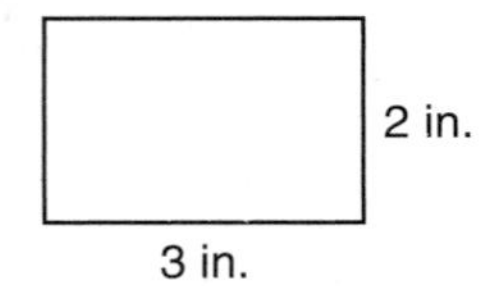

Solution The length of the rectangle is 3 inches, so we can fit 3 square inches along the length. The width is 2 inches, so we can fit 2 square inches along the width. Two rows of three means that the area can be covered with 6 square inches. We may abbreviate the answer as **6 sq. in.**

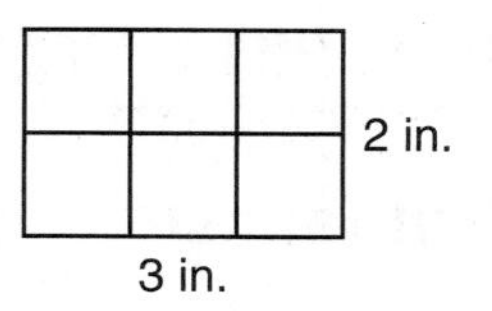

The area of a rectangle may be calculated by multiplying the length of the rectangle by its width. So a formula for finding the area of a rectangle is

$$A = L \times W$$

Example 3 Estimate the area of a room that is 14 ft 3 in. long and 12 ft 8 in. wide.

Solution To estimate the area, we round the length and width to the nearest foot and then multiply the rounded measures. If the inch part of the measure is 6 inches or more, we round up to the next foot. If the inch part is less than 6 inches, we round down. So 14 ft 3 in. rounds to 14 ft, and 12 ft 8 in. rounds up to 13 ft.

$$14 \text{ ft} \times 13 \text{ ft} = \mathbf{182 \text{ sq. ft}}$$

LESSON PRACTICE

Practice set For problems **a–d** find the area of the rectangle shown. Draw square units inside the rectangle, and then count the units.

a. 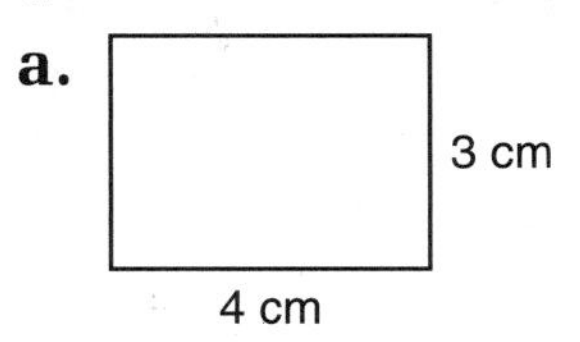

b. 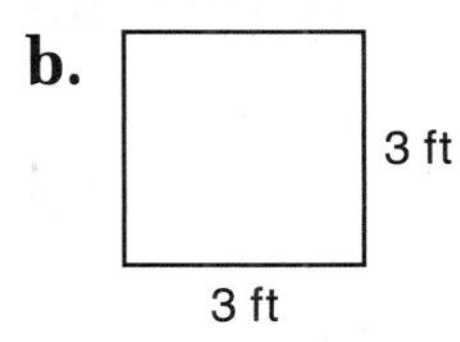

c. 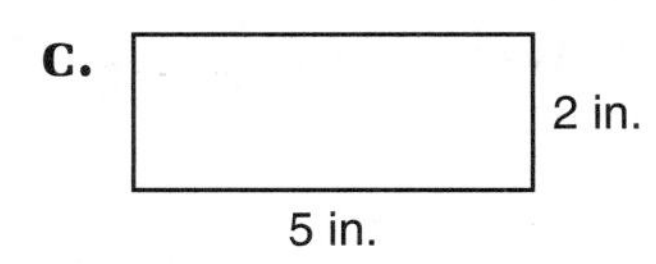

d. 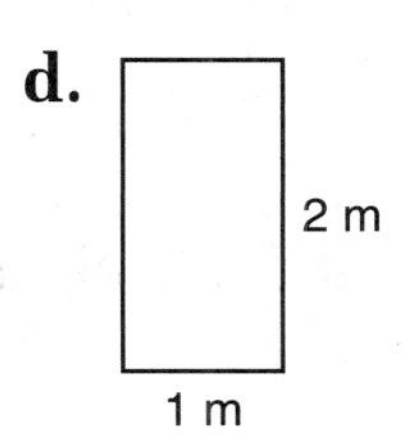

e. Lola's bedroom is 10 feet wide and 12 feet long. What is the area of the room?

f. As a class activity, calculate the area of the classroom floor. Round the length and width of the room to the nearest foot to perform the calculation.

MIXED PRACTICE

Problem set

1. (21, 70) Reggie bought a dozen candy bars for 40¢ each. What was the total cost of the candy bars? Write an equation and find the answer.

2. (21) The total cost of 4 cartons of ice cream was $10.00. If each carton had the same price, what was the price per carton? Write an equation and find the answer.

3. (Inv. 7) Write the next three terms of this sequence:

4, 5, 8, 9, 12, 13, ____, ____, ____, ...

4. (46, 60, 71) Mary has read $\frac{1}{3}$ of a 240-page book. How many pages does she still have to read to finish the book? What percent of the book does she still have to read?

5. (21) One meter equals 100 centimeters. Five meters equals how many centimeters?

6. (68) Name the decimal number 12.25 with words.

7. (23) Write a fraction that shows how many twelfths equal one half.

8. (25) List the factors of 16.

9. (68) Leroy ran 100 meters in ten and twelve hundredths seconds. Use digits to write Leroy's race time.

10. (68) Which digit in 436.2 is in the ones place?

11. (58) Write the quotient as a mixed number: $\frac{100}{3}$

12. (61) Segment *FH* measures 90 millimeters. If *GH* is 35 millimeters, then how long is $\overline{FG}$?

F G H

13. (13, 70) $10.35 + $5.18 + 8¢ + $11 + 97¢

14. (13) $\begin{array}{r} \$80.00 \\ -\ \$72.47 \\ \hline \end{array}$

15. (17) $\begin{array}{r} \$4.97 \\ \times \quad 6 \\ \hline \end{array}$

16. (55) $\begin{array}{r} 375 \\ \times\ 548 \\ \hline \end{array}$

17. (26) $7\overline{)\$40.53}$

18. (54) $60\overline{)5340}$

19. (26, 54) $30m = 6000$

20. (59) $3\frac{3}{8} + 1\frac{1}{8} + 4\frac{4}{8}$

21. (41, 63) $7\frac{3}{4} - \left(5 - 1\frac{1}{4}\right)$

22. (69) Compare: 55.5 ◯ 5.55

23. (41) $4\frac{1}{10} + 5\frac{1}{10} + 10\frac{1}{10}$

24. (43, 63) $10 - \left(4 + 1\frac{1}{8}\right)$

25. (44, 53) This rectangle is half as wide as it is long. What is the perimeter of the rectangle?

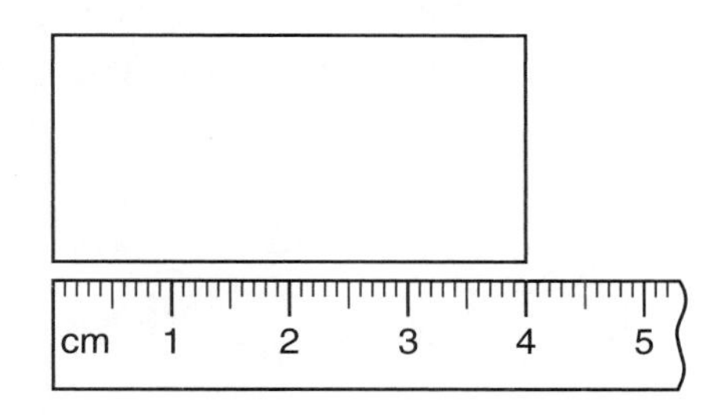

26. (72) What is the area of the rectangle in problem 25?

27. (Inv. 4) What is the measure of ∠*AOD* in the figure below?

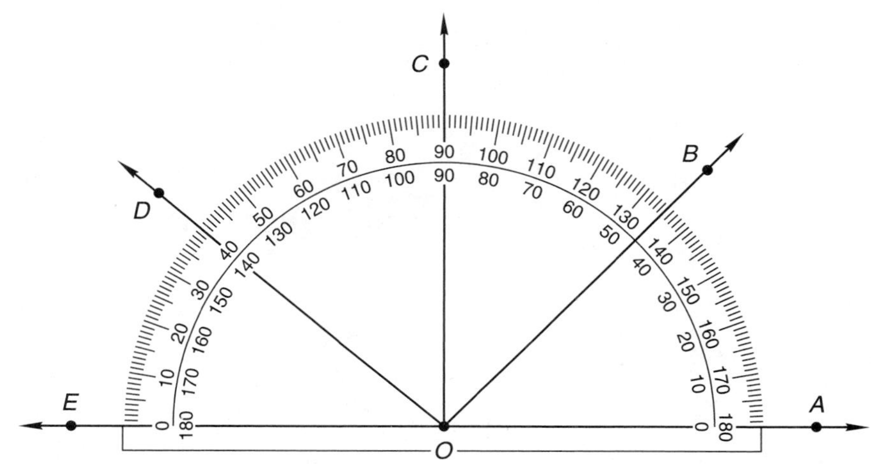

28. (57) Draw a spinner with four sectors labeled A, B, C, and D. Make the sizes of the sectors such that with one spin the probability of outcome A is $\frac{1}{2}$, the probability of outcome B is $\frac{1}{4}$, and outcomes C and D are equally likely.

29. (62) Bradley mentally kept track of his grocery purchases. As he placed each item in the cart, he rounded the item's price to the nearest dollar and then added the rounded amount to the total. Use Bradley's method to estimate the total cost of these seven items.

milk	2.59
bread	2.39
cereal	4.89
juice	2.39
vitamins	11.89
eggs	1.99
butter	2.29

30. (62) (a) Bradley does not want to spend much more than $30 on groceries. He mentally keeps a running total of his purchases. Does Bradley's calculation need to be exact, or is an estimate acceptable?

(b) At the check-out line the clerk scans Bradley's purchases and calculates the total cost of the items. Does the clerk's calculation need to be exact, or is an estimate acceptable?

LESSON

73 Adding and Subtracting Decimal Numbers

WARM-UP

Facts Practice: 64 Multiplication Facts (Test F)

Mental Math:

a. How many inches is $2\frac{1}{2}$ feet? **b.** 25% of 240
c. $124 \div 4$ **d.** $412 \div 4$ **e.** $1 - \frac{7}{10}$
f. One third of 22 is $7\frac{1}{3}$. How much is $\frac{1}{3}$ of 23? ... $\frac{1}{3}$ of 25?
g. $\sqrt{64}$, $\div$ 2, $\times$ 3, $\div$ 2, $\times$ 4, $\div$ 3

Problem Solving:

Triangles *A*, *B*, and *C* are congruent. Triangle *A* was "flipped" to the right to form triangle *B*. Then triangle *B* was flipped down to form triangle *C*. Suppose triangle *C* is flipped to the left to form triangle *D*. Draw triangles *A*, *B*, *C*, and *D*.

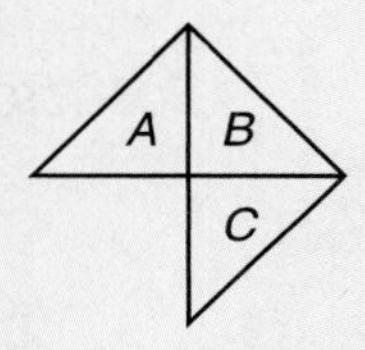

NEW CONCEPT

Recall that when we add or subtract money, we write the numbers so that the decimal points are vertically aligned. This way we are sure to add digits with the same place value. We insert the decimal point in the answer below the other decimal points, as shown here:

$$\begin{array}{r} \$3.45 \\ +\ \$1.25 \\ \hline \$4.70 \end{array} \qquad \begin{array}{r} \$3.45 \\ -\ \$1.25 \\ \hline \$2.20 \end{array}$$

We use the same procedure to add or subtract any decimal numbers. We keep the decimal points in line. This way, we add or subtract digits with the same place value. The decimal points stay in a straight line, as shown here:

$$\begin{array}{r} 2.4 \\ +\ 1.3 \\ \hline 3.7 \end{array} \qquad \begin{array}{r} 2.4 \\ -\ 1.3 \\ \hline 1.1 \end{array}$$

Example 1 Find the perimeter of the triangle at right. Units are in centimeters.

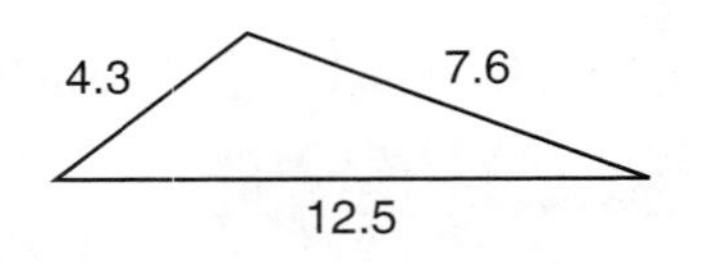

Solution We keep the decimal points aligned in the problem and answer. We add the digits column by column, just as we would add whole numbers or money.

$$\begin{array}{r} 4.3 \text{ cm} \\ 12.5 \text{ cm} \\ +\ 7.6 \text{ cm} \\ \hline \mathbf{24.4\ cm} \end{array}$$

To add or subtract decimal numbers with different numbers of decimal places, we align the decimal points, not the last digits.

Example 2 The roof was 6.37 meters above the ground. The ladder could reach only 4.2 meters. The roof was how much higher than the ladder could reach?

Solution This is a story about comparing, which we solve by subtracting. As we saw in Lesson 70, we may attach zeros to the end of a decimal number without changing the value of the number. We attach a zero to 4.2 so that there are no empty places in the problem. Then we subtract.

$$\begin{array}{r} 6.37 \text{ m} \\ -\ 4.20 \text{ m} \\ \hline \mathbf{2.17\ m} \end{array}$$

Note: Attaching zeros might make the problem easier to work. However, it is not necessary to attach zeros if we remember that an empty place has the same value as a zero in that place.

Example 3 Add: 3.45 + 6.7

Solution We line up the decimal points vertically so that we add digits with the same place value.

$$\begin{array}{r} 3.45 \\ +\ 6.7 \\ \hline \mathbf{10.15} \end{array}$$

Think about the meaning of each decimal number to be sure your answers are reasonable. In example 3, 3.45 is more than 3 but less than 4, and 6.7 is more than 6 but less than 7. So the sum should be more than 3 + 6 but less than 4 + 7.

LESSON PRACTICE

Practice set Add:

a. $\begin{array}{r} 3.4 \\ 6.7 \\ +\ 11.3 \\ \hline \end{array}$

b. $\begin{array}{r} 4.63 \\ 2.5 \\ +\ 0.46 \\ \hline \end{array}$

c. $\begin{array}{r} 9.62 \\ 12.5 \\ +\ \ \ 3.7 \\ \hline \end{array}$

Subtract:

d. $\begin{array}{r} 3.64 \\ -\ 1.46 \\ \hline \end{array}$

e. $\begin{array}{r} 5.37 \\ -\ 1.6\ \ \\ \hline \end{array}$

f. $\begin{array}{r} 0.436 \\ -\ 0.2\ \ \ \\ \hline \end{array}$

Line up the decimal points and solve. Show your work.

g. 4.2 + 2.65

h. 6.75 − 4.5

i. Find the perimeter of this square.

2.4 cm

j. The distance from Quan's house to school is 0.8 mile. How far does Quan travel going from his house to school and back again?

MIXED PRACTICE

Problem set

1. *(49, 70)* Ben bought a sheet of 37¢ stamps. The sheet had 5 rows of stamps with 8 stamps in each row. How much did the sheet of stamps cost?

2. *(49)* Ling is half the age of her brother, but she is 2 years older than her sister. If Ling's brother is 18 years old, how old is her sister?

3. *(73)* Carrie was asked to run to the fence and back. It took her 23.4 seconds to run to the fence and 50.9 seconds to run back. How many seconds did the whole trip take?

4. *(72)* The classroom floor is covered with one-foot-square tiles. There are 30 rows of tiles with 40 tiles in each row. What is the area of the floor?

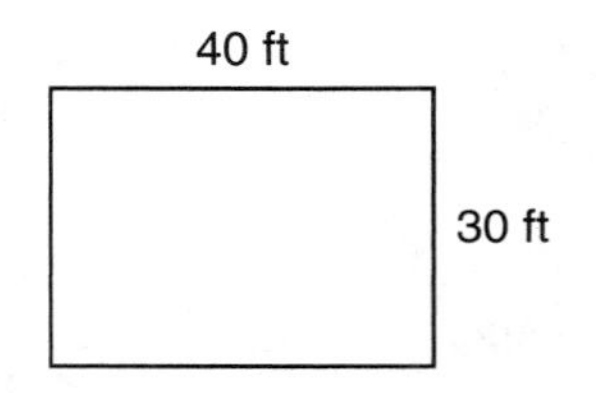

5. *(37, 71)* Draw two circles. Shade $\frac{2}{8}$ of one circle and $\frac{1}{4}$ of the other circle. What percent of each circle is shaded?

6. (Inv. 3) What fraction is equal to one half of one fourth?

7. (66, 68) To the nearest tenth of a centimeter, what is the length of this rectangle? Use words to write the answer.

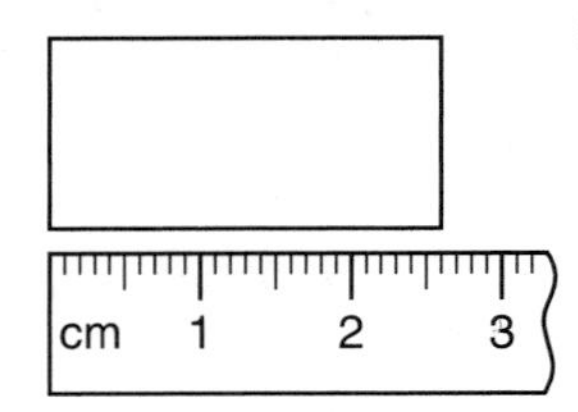

8. (25) List the numbers that are factors of both 16 and 20.

9. (18) Three times a number y can be written $3y$. If $3y = 12$, then what number does $2y$ equal?

10. (61, 73) The length of $\overline{AC}$ is 8.5 centimeters. If AB is 3.7 centimeters, then how long is $\overline{BC}$?

11. (69) Compare: 12.0 ◯ 1.20

12. (73) $\begin{array}{r} 53.46 \\ -\ 5.7 \\ \hline \end{array}$

13. (17) $\begin{array}{r} \$6.48 \\ \times\ 9 \\ \hline \end{array}$

14. (73) $4.5 + 6.75$

15. (70) $\$5 - 5¢$

16. (26) $5\overline{)\$8.60}$

17. (54) $20\overline{)\$8.60}$

18. (55) $\begin{array}{r} 378 \\ \times\ 296 \\ \hline \end{array}$

19. (29) $\begin{array}{r} 800 \\ \times\ 500 \\ \hline \end{array}$

20. (26, 54) $30w = 9870$

21. (43) $12 + 1\frac{1}{2}$

22. (63) $12 - 1\frac{1}{2}$

23. (41) $\frac{49}{99} + \frac{49}{99}$

Use this information to answer problems 24 and 25:

Kobe did yard work on Saturday. He worked for $2\frac{1}{2}$ hours in the morning and $1\frac{1}{2}$ hours in the afternoon. Kobe's parents paid him $3.50 for every hour he worked.

24. (Inv. 5, 59) How many hours did Kobe work in all?

25. (21, Inv. 5) How much money was Kobe paid in all?

Thirty-nine girls were asked to choose their favorite form of exercise. Use the frequency table below to answer problems 26 and 27.

Frequency Table

Type of Exercise	Frequency
Bicycle riding	5
Roller-skating	7
Soccer	6
Swimming	10
Walking	5
Basketball	2
Aerobic dancing	4

26. What fraction of the girls chose swimming?
(Inv. 5)

27. What fraction of the girls chose an exercise other than bicycle riding or roller-skating?
(Inv. 5)

28. Each side of a square on the playground was 10 ft 3 in. long. Estimate the area of the square.
(72)

10 ft 3 in.

29. The bill for dinner was $14.85. Janna wanted to leave a tip of about $\frac{1}{5}$ of the bill. So she rounded $14.85 to the nearest dollar and found $\frac{1}{5}$ of the rounded amount. How much did Janna leave as a tip?
(Inv. 3, 62)

30. Draw a rhombus that has a right angle.
(45)

LESSON

74 Converting Units of Length

WARM-UP

Facts Practice: 48 Uneven Divisions (Test G)

Mental Math:

a. How many cents are in two and a half dollars?

b. $128 \div 4$ c. $812 \div 4$ d. 10% of 360

e. $3\frac{1}{3} + 1\frac{2}{3}$ f. $1 - \frac{5}{8}$ g. $\frac{1}{3}$ of 360

h. $\sqrt{49}$, + 3, × 10, − 1, ÷ 9, − 1, ÷ 10

Problem Solving:

Some small cubes were stacked together to form this larger cube. How many small cubes were used?

NEW CONCEPT

The following table lists some common units of length used in the metric system and in the U.S. Customary System. Some units of length used in the metric system are millimeters (mm), centimeters (cm), meters (m), and kilometers (km). Some units of length used in the U.S. Customary System are inches (in.), feet (ft), yards (yd), and miles (mi). The table also shows equivalences between units of length.

Equivalence Table for Units of Length

U.S. Customary System	Metric System
12 in. = 1 ft	10 mm = 1 cm
3 ft = 1 yd	1000 mm = 1 m
5280 ft = 1 mi	100 cm = 1 m
1760 yd = 1 mi	1000 m = 1 km
A meter is about 3 inches longer than a yard.	

Example 1 The star player on the basketball team is 197 centimeters tall. About how many meters tall is the star player?

Solution The chart shows that 100 centimeters equals 1 meter. The prefix *cent-* can help us remember this fact because there are 100 cents in 1 dollar. Since 197 centimeters is nearly 200 centimeters, the height of the basketball player is about **2 meters.**

Example 2 Two yards is the same length as how many inches?

Solution The equivalence table shows that 1 yard equals 3 feet and that each foot equals 12 inches.

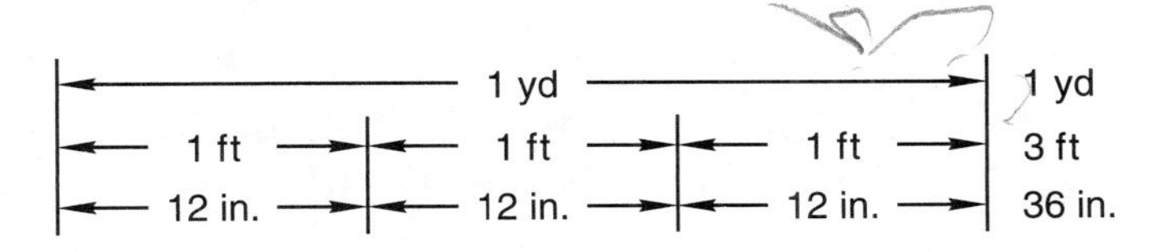

Thus, 1 yard equals 36 inches. Two yards is twice that amount. So two yards equals **72 inches.**

LESSON PRACTICE

Practice set

a. How many yards are in one fourth of a mile?

b. Fifty millimeters is how many centimeters?

c. Tom's height is 5 feet 1 inch. How many inches tall is he?

d. A 10K race is a 10-kilometer race. How many meters is 10 kilometers?

MIXED PRACTICE

Problem set

1. *(49)* Gizmos come in a carton. A carton holds 6 packages. Each package holds 10 small boxes. Each small box holds 12 gizmos. How many gizmos come in a carton?

2. *(68, 73)* When the decimal number two and three tenths is added to three and five tenths, what is the sum?

3. *(21)* Bacchus bought 7 pounds of grapes for $3.43. What was the price for 1 pound of grapes? Write an equation and find the answer.

4. (23) Compare: $\frac{3}{6} \bigcirc \frac{6}{12}$

5. (74) One of the players on the basketball team is 2 meters tall. Two meters is how many centimeters?

6. (66) Use a fraction and a decimal number to name the point marked by the arrow on this number line.

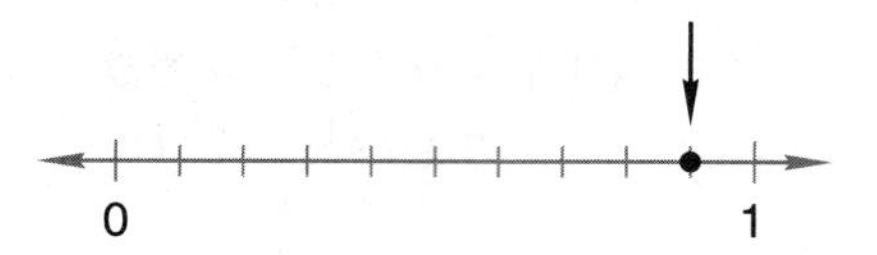

7. (68) Joanne ran the 100-meter dash in 11.02 seconds. Use words to name the decimal number 11.02.

8. (74) Three yards is the same length as how many inches?

9. (61, 63) Segment RT measures 4 inches. If $\overline{RS}$ is $2\frac{1}{4}$ inches long, then how long is $\overline{ST}$?

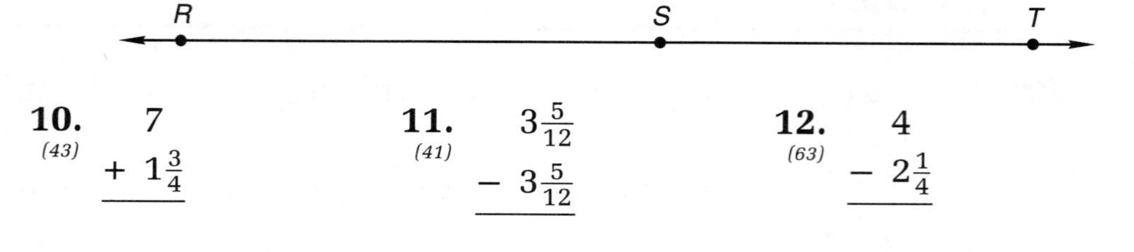

10. (43) $7 + 1\frac{3}{4}$

11. (41) $3\frac{5}{12} - 3\frac{5}{12}$

12. (63) $4 - 2\frac{1}{4}$

13. (73) $16.2 + 1.25$

14. (73) $30.1 - 14.2$

15. (29) $\$12.98 \times 40$

16. (26) $6\overline{)\$45.54}$

17. (26) $\frac{4384}{8}$

18. (51) 12×12

19. (70) $12 + 84¢ + $6.85 + 9¢ + $8 + $98.42 + $55.26

20. (58) Write the quotient as a mixed number: $\frac{18}{5}$

21. (70) Write a decimal number equal to 2.5 that has three decimal places.

22. (53) The perimeter of a certain square is 24 inches. How long is each side of the square?

23. (72) What is the area of the square described in problem 22?

24. Show two ways to correct the money amount shown on this sign.
(70)

Use the map below to answer problems 25–27.

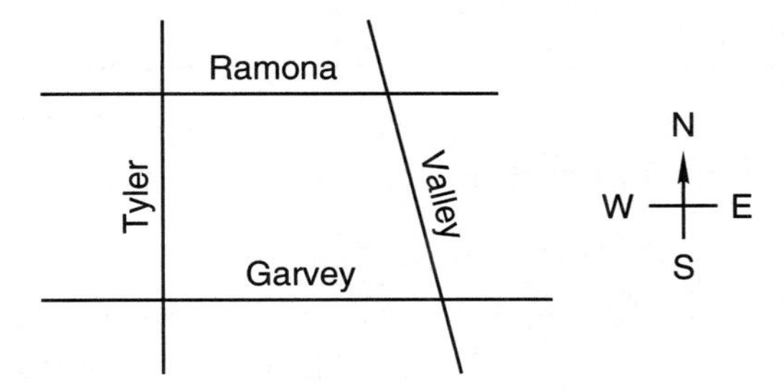

25. Which street runs straight north and south?
(Inv. 5)

26. Which street is parallel to Ramona?
(31)

27. Which street is neither perpendicular nor parallel to Garvey?
(31)

28. Write the next two terms in this sequence:
(Inv. 7)

Z, X, V, T, _____, _____, ...

29. (a) One foot is what fraction of a yard?
(71, 74)

(b) One foot is what percent of a yard?

30. Draw a circle and shade $\frac{1}{2}$ of it. What percent of the circle is shaded?
(71)

LESSON

75 Changing Improper Fractions to Whole or Mixed Numbers

WARM-UP

Facts Practice: 100 Multiplication Facts (Test C)

Mental Math:

a. How many inches are in a foot? How many feet are in a yard? Hold your fingers one inch apart. Hold your hands one yard apart.

b. $\frac{1}{4}$ of 36 **c.** $\frac{1}{4}$ of 360 **d.** $\frac{1}{3}$ of 36

e. 360 ÷ 30 **f.** $\sqrt{81}$, − 1, × 10, + 1, ÷ 9, − 9

Problem Solving:

Sam takes about 600 steps when he walks around the block. In 6 steps Sam travels about 15 feet. About how many feet does Sam travel when he walks around the block?

NEW CONCEPT

A fraction may be less than 1, equal to 1, or greater than 1. A fraction that is less than 1 is called a **proper fraction.** A fraction that is equal to 1 or greater than 1 is called an **improper fraction.** An improper fraction has a numerator equal to or greater than its denominator.

Less than 1	Equal to 1	Greater than 1
$\frac{3}{4}$	$\frac{4}{4}$	$\frac{5}{4}$
Proper fraction	Improper fractions	

Every improper fraction can be changed either to a whole number or to a mixed number. Consider the fractions above. The fraction $\frac{4}{4}$ is equal to 1, and the fraction $\frac{5}{4}$ is equal to $\frac{4}{4} + \frac{1}{4}$, which is $1\frac{1}{4}$.

$$\frac{5}{4} = \frac{4}{4} + \frac{1}{4} = 1\frac{1}{4}$$

$\frac{1}{4}$ $\frac{1}{4}$ $\frac{1}{4}$ $\frac{1}{4}$ $\frac{1}{4}$

Example 1 Separate $\frac{8}{3}$ into fractions equal to 1 plus a proper fraction. Then write the result as a mixed number.

Solution The denominator is 3, so we separate eight thirds into groups of three thirds. We make two whole groups and two thirds remain.

$$\frac{8}{3} = \frac{3}{3} + \frac{3}{3} + \frac{2}{3} = \mathbf{2\frac{2}{3}}$$

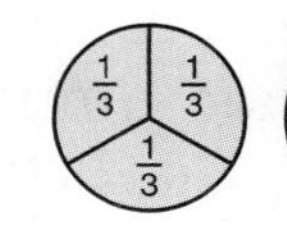

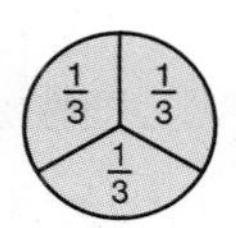

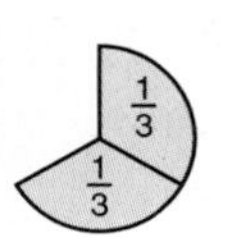

When the answer to an arithmetic problem is an improper fraction, we usually convert the answer to a whole number or a mixed number.

Example 2 Add: $\frac{3}{5} + \frac{4}{5}$

Solution We add and find that the sum is the improper fraction $\frac{7}{5}$.

$$\frac{3}{5} + \frac{4}{5} = \frac{7}{5}$$

Then we convert the improper fraction to a mixed number.

$$\frac{7}{5} = \frac{5}{5} + \frac{2}{5} = \mathbf{1\frac{2}{5}}$$

When adding mixed numbers, the fraction part of the answer may be an improper fraction.

$$1\frac{2}{3} + 2\frac{2}{3} = 3\frac{4}{3} \longleftarrow \text{Improper fraction}$$

We convert the improper fraction to a whole number or mixed number and add it to the whole-number part of the answer.

$$\begin{aligned} 3\frac{4}{3} &= 3 + \frac{3}{3} + \frac{1}{3} \\ &= 3 + 1\frac{1}{3} \\ &= 4\frac{1}{3} \end{aligned}$$

Example 3 Add: $2\frac{1}{2} + 2\frac{1}{2} + 2\frac{1}{2}$

Solution We add and get the sum $6\frac{3}{2}$. The fraction part of $6\frac{3}{2}$ is an improper fraction. We find that $\frac{3}{2}$ equals $1\frac{1}{2}$. We add $1\frac{1}{2}$ to 6 and get $7\frac{1}{2}$.

$$\begin{aligned} 6\frac{3}{2} &= 6 + \frac{2}{2} + \frac{1}{2} \\ &= 6 + 1\frac{1}{2} \\ &= \mathbf{7\frac{1}{2}} \end{aligned}$$

LESSON PRACTICE

Practice set* Convert each improper fraction into a whole number or a mixed number:

a. $\frac{2}{2}$ **b.** $\frac{5}{2}$ **c.** $\frac{5}{3}$ **d.** $\frac{9}{4}$

e. $\frac{3}{2}$ **f.** $\frac{3}{3}$ **g.** $\frac{6}{3}$ **h.** $\frac{10}{3}$

i. $\frac{4}{2}$ **j.** $\frac{4}{3}$ **k.** $\frac{7}{3}$ **l.** $\frac{15}{4}$

Add. Simplify each answer.

m. $\frac{4}{5} + \frac{4}{5}$ **n.** $8\frac{1}{3} + 8\frac{1}{3} + 8\frac{1}{3}$

o. $\frac{5}{8} + \frac{3}{8}$ **p.** $7\frac{4}{8} + 8\frac{7}{8}$

q. What is the perimeter of a square with sides $2\frac{1}{2}$ inches long?

MIXED PRACTICE

Problem set

1. (49, 70) Robin bought 10 arrows for 49¢ each and a package of bow wax for $2.39. How much did she spend in all?

2. (50) On the shelf there are three stacks of books. In the three stacks there are 12, 13, and 17 books. If the number of books in each stack were made the same, how many books would be in each stack?

3. (69, 73) Arrange these numbers in order from least to greatest. Then find the difference between the least and greatest numbers.

32.16 32.61 31.26 31.62

4. (2) What is the largest four-digit even number that has the digits 1, 2, 3, and 4 used only once each?

5. (67) Name the total number of shaded circles as a mixed number and as a decimal number.

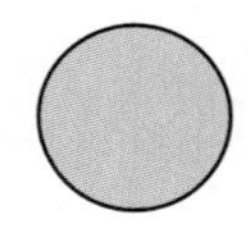
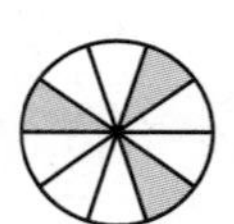

6. Compare: $\frac{4}{3} \bigcirc \frac{3}{4}$
(75)

7. Write 4.5 with the same number of decimal places as 6.25.
(70)

8. Use a mixed number and a decimal number to name the point marked by the arrow on this number line.
(66)

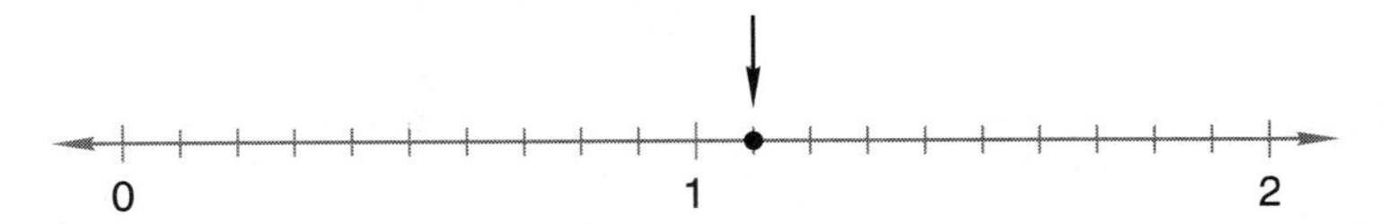

9. Daniel ran a 5-kilometer race in 15 minutes 45 seconds. How many meters did he run?
(74)

10. The length of $\overline{PQ}$ is $1\frac{1}{4}$ inches. The length of $\overline{QR}$ is $1\frac{3}{4}$ inches. How long is $\overline{PR}$?
(59, 61)

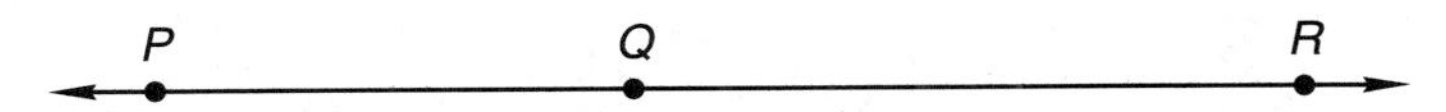

11. Seven twelfths of the months have 31 days, and the rest have fewer than 31 days. What fraction of the months have fewer than 31 days?
(60)

12. $60.45 - 6.7$
(73)

13. $4.8 + 2.65$
(73)

14. $3d = \$20.01$
(26)

15. $36 \times 9 \times 80$
(18, 29)

16. $\begin{array}{r} 506 \\ \times\ 478 \\ \hline \end{array}$
(56)

17. $\dfrac{4690}{70}$
(54)

18. $\begin{array}{r} \$30.75 \\ \times\ \quad 8 \\ \hline \end{array}$
(17)

19. $10 + $8.16 + 49¢ + $2 + 5¢
(70)

20. $\frac{4}{5} + \frac{4}{5}$
(75)

21. $\frac{5}{9} + \frac{5}{9}$
(75)

22. $16\frac{2}{3} + 16\frac{2}{3}$
(75)

23. If each side of a square is 1 foot, then the perimeter of the square is how many inches? Each side of a square is what percent of the square's perimeter?
(53, 71, 74)

24. (a) What is the area of the square in problem 23 in square feet?
(72, 74)

(b) What is the area in square inches?

Use the graph below to answer problems 25 and 26.

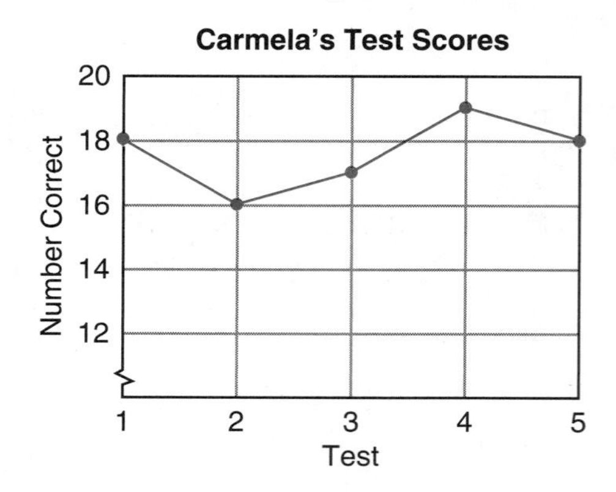

25. How many answers did Carmela get correct on her best test?
(Inv. 5)

26. There are 20 questions on each test. How many did Carmela miss on Test 2?
(Inv. 5)

27. Name the coin that is equal to half of a half-dollar.
(Inv. 2)

Use a centimeter ruler to measure this rectangle. Then answer problems 28 and 29.

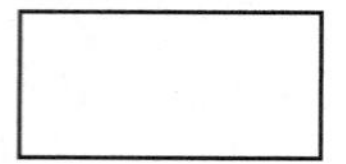

28. What is the perimeter of the rectangle?
(44, 53)

29. What is the area of the rectangle?
(44, 72)

30. What is the measure of $\angle EOD$ in the figure below?
(Inv. 4)

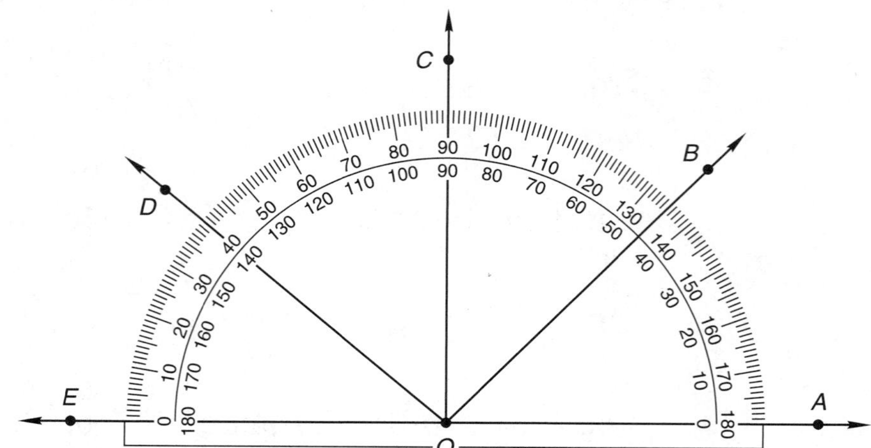

LESSON

76 Multiplying Fractions

WARM-UP

Facts Practice: 60 Improper Fractions to Simplify (Test H)

Mental Math:

a. How many millimeters are in a centimeter? How many centimeters are in a meter? Hold two fingers one centimeter apart. Hold your hands one meter apart.

b. Is 3828 divisible by 4?

c. Is 2838 divisible by 4?

d. 50% of $5.00

e. $1 - \frac{7}{12}$

f. $\sqrt{9}$, × 9, + 1, ÷ 4, + 3, × 8, + 1, ÷ 9

Problem Solving:

Copy this multiplication problem and fill in the missing digits:

```
   _7
×   _
 5_6
```

NEW CONCEPT

We have added and subtracted fractions. Adding and subtracting fractions involves counting same-size parts. In this lesson we will multiply fractions. When we multiply fractions, the sizes of the parts change. Consider this multiplication problem: How much is one half of one half?

Using our fraction manipulatives, we show one half of a circle. To find one half of one half, we divide the half circle in half. We see that the answer is one fourth.

$\frac{1}{2}$ → $\frac{1}{4}$

$$\frac{1}{2} \text{ of } \frac{1}{2} \text{ is } \frac{1}{4}$$

Using pencil and paper, the problem looks like this:

$$\frac{1}{2} \times \frac{1}{2} = \frac{1}{4}$$

Notice that the word *of* is another way to say "times." Also notice that we find the answer to a fraction multiplication problem by multiplying the numerators to get the numerator of the product and multiplying the denominators to get the denominator of the product.

Example 1 Bruce found $\frac{1}{4}$ of a pizza in the refrigerator and ate half of it. What fraction of the whole pizza did Bruce eat?

Solution One half of one fourth is one eighth.

$\frac{1}{2}$ of $\frac{1}{4}$

$$\frac{1}{2} \times \frac{1}{4} = \frac{1}{8}$$

Bruce ate $\frac{1}{8}$ of the whole pizza.

Example 2 What fraction is one half of three fourths?

Solution First we show three fourths.

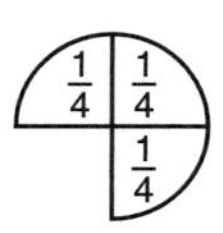

To find one half of three fourths, we may either divide each fourth in half or divide three fourths in half.

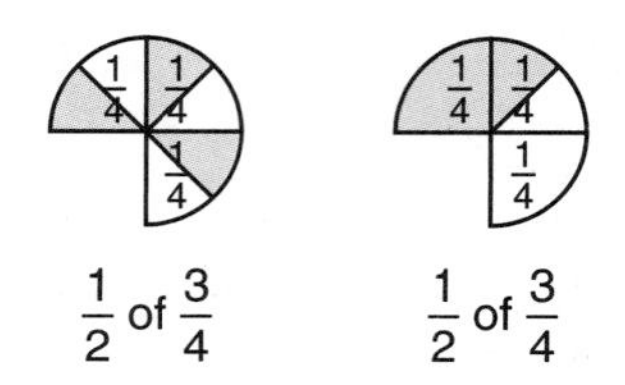

Since one half of one fourth is one eighth, one half of three fourths is three eighths. We may also find one half of three fourths by multiplying.

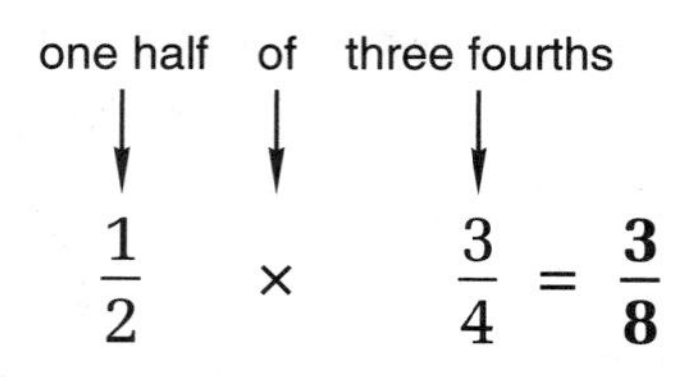

$$\frac{1}{2} \times \frac{3}{4} = \mathbf{\frac{3}{8}}$$

We multiplied the numerators to find the numerator of the product, and we multiplied the denominators to find the denominator of the product.

Example 3 (a) A nickel is what fraction of a dime?

(b) A dime is what fraction of a dollar?

(c) A nickel is what fraction of a dollar?

(d) The answers to parts (a)–(c) show that one half of one tenth is what fraction?

Solution (a) $\mathbf{\frac{1}{2}}$ (b) $\mathbf{\frac{1}{10}}$

(c) $\mathbf{\frac{1}{20}}$ (d) $\frac{1}{2} \times \frac{1}{10} = \mathbf{\frac{1}{20}}$

Example 4 Multiply: $\frac{2}{3} \times \frac{4}{5}$

Solution We find two thirds of four fifths by multiplying.

$$\frac{2}{3} \times \frac{4}{5} = \mathbf{\frac{8}{15}}$$

Example 5 (a) What fraction of the square is shaded?

(b) What is the area of the shaded rectangle?

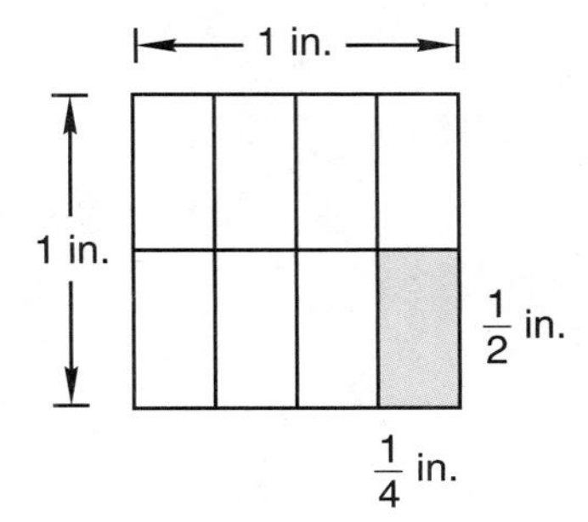

Solution (a) One of eight equal parts is shaded, so $\mathbf{\frac{1}{8}}$ of the square is shaded.

(b) To find the area of the shaded part, we multiply the length by the width.

$$\frac{1}{2} \text{ in.} \times \frac{1}{4} \text{ in.} = \mathbf{\frac{1}{8} \textbf{ sq. in.}}$$

LESSON PRACTICE

Practice set*

a. Draw a semicircle (one half of a circle). Shade one half of the semicircle. The shaded part of the semicircle shows that $\frac{1}{2}$ of $\frac{1}{2}$ is what fraction?

b. A penny is what fraction of a dime? A dime is what fraction of a dollar? A penny is what fraction of a dollar? The answers to these questions show that $\frac{1}{10}$ of $\frac{1}{10}$ is what fraction?

c. What fraction is three fourths of one half?

d. What fraction is one half of one third?

e. What fraction is two fifths of two thirds?

Multiply:

f. $\frac{1}{3} \times \frac{2}{3}$ **g.** $\frac{3}{5} \times \frac{1}{2}$ **h.** $\frac{2}{3} \times \frac{2}{3}$ **i.** $\frac{1}{2} \times \frac{2}{2}$

j. Half of the students were girls, and one third of the girls had blond hair. What fraction of the students were blond-haired girls?

k. What is the area of a square with sides $\frac{1}{2}$ inch long?

MIXED PRACTICE

Problem set

1. (11) After two days the troop had hiked 36 miles. If the troop hiked 17 miles the first day, how many miles did the troop hike the second day? Write an equation and find the answer.

2. (21, 50) The troop hiked 57 miles in 3 days. The troop averaged how many miles per day? Write an equation and find the answer.

3. (68, 73) When the decimal number six and thirty-four hundredths is subtracted from nine and twenty-six hundredths, what is the difference?

4. (25) Which factors of 6 are also factors of 12?

5. (18) If $3n = 18$, then what number does $2n$ equal?

6. (72) What is the area of a square with sides 10 cm long?

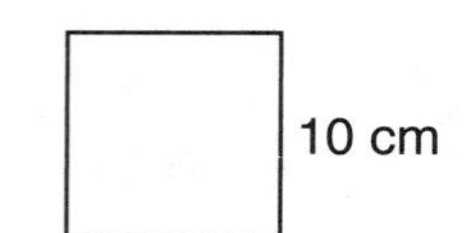

7. (70) Compare: 4.5 ◯ 4.500

8. (23, 59, 75) Arrange these fractions in order from least to greatest:

$$\frac{2}{3}, \frac{1}{2}, \frac{4}{3}, \frac{3}{8}, \frac{5}{5}$$

9. (46, 71, 76) One half of the 64 squares on the board were black. The other half were red. One half of the black squares had checkers on them. None of the red squares had checkers on them.

(a) How many squares on the board were black?

(b) How many squares had checkers on them?

(c) What fraction of the squares had checkers on them?

(d) What percent of the squares had checkers on them?

10. (61) The length of segment AC is 78 millimeters. If BC is 29 millimeters, then what is the length of segment AB?

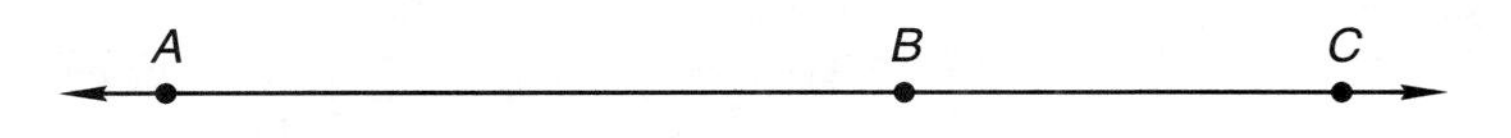

11. (73) 24.86 − 9.7

12. (73) 9.06 − 3.9

13. (26, 34) $8m = \$36.00$

14. (26, 54) $50w = 7600$

15. (17) $\$16.08 \times 9$

16. (56) 638×570

17. (59) $3\frac{1}{3} + 1\frac{2}{3}$

18. (75) $1\frac{2}{3} + 1\frac{2}{3}$

19. (63) $4 - 1\frac{2}{5}$

20. (76) $\frac{1}{2}$ of $\frac{3}{5}$

21. (76) $\frac{1}{3} \times \frac{2}{3}$

22. (76) $\frac{1}{2} \times \frac{6}{6}$

Use this information to answer problems 23–25:

Tyrone has a machine that changes numbers. He fixed the machine so that when he puts in a 3, a 6 comes out. When he puts in a 4, an 8 comes out. When he puts in a 5, a 10 comes out.

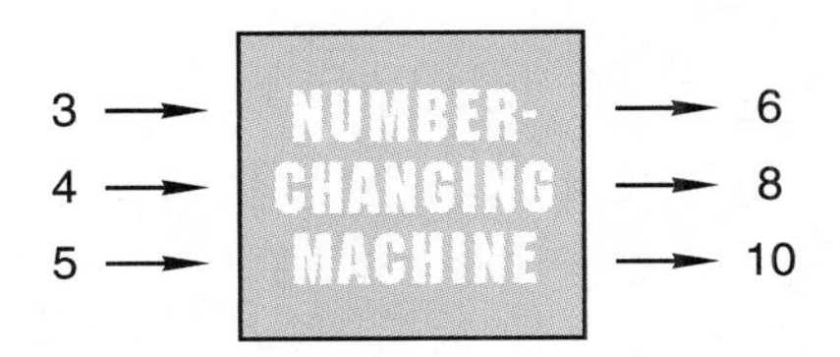

23. (Inv. 7) What does the machine do to each number that Tyrone puts into it?

A. It adds 3.

B. It doubles the number.

C. It divides by 2.

D. It multiplies by 3.

24. (Inv. 7) If Tyrone puts in a 12, what number will come out?

25. (Inv. 7) Tyrone put in a number, and a 20 came out. What number did Tyrone put into the machine?

Refer to this rectangle to answer problems 26 and 27:

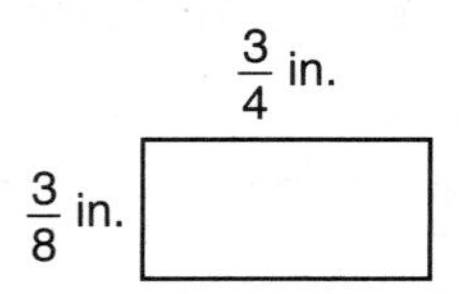

26. What is the area of the rectangle?
(76)

27. Draw a rectangle that is similar to the rectangle but has sides twice as long.
(32, 44)

28. (a) Which number on the spinner is the most unlikely outcome of a spin?
(57)

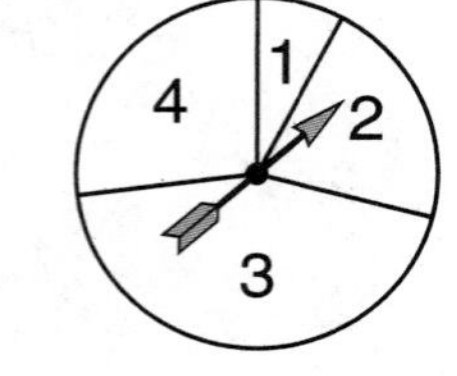

(b) Which outcomes have probabilities that exceed $\frac{1}{4}$ with one spin of the spinner?

29. (a) A nickel is what fraction of a quarter?
(76)

(b) A quarter is what fraction of a dollar?

(c) A nickel is what fraction of a dollar?

(d) The answers to parts (a)–(c) show that one fifth of one fourth is what fraction?

30. List the factors of 100.
(25)

LESSON

77 Converting Units of Weight and Mass

WARM-UP

Facts Practice: 60 Improper Fractions to Simplify (Test H)

Mental Math:

a. How many centimeters are in one meter? How many millimeters are in one meter? Hold your fingers one centimeter apart. Hold your hands one meter apart.

b. 100 ÷ 4 **c.** 1000 ÷ 4

Problem Solving:

If rectangle I is rotated a quarter of a turn clockwise around point *A*, it will be in the position of rectangle II. If it is rotated again, it will be in the position of rectangle III. If it is rotated again, it will be in the position of rectangle IV. Draw the congruent rectangles I, II, III, and IV.

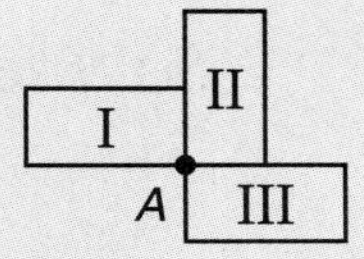

NEW CONCEPT

When you go to the doctor for a checkup, the doctor measures many things about you. The doctor might measure your height and your temperature. The doctor might also measure your blood pressure and heart rate. To measure your **weight** or mass†, the doctor has you step onto a scale. Measurements of weight and mass are used to compare "how much stuff" is in objects. Heavier objects contain more stuff than lighter objects.

To measure weight in the U.S. Customary System, we use units such as ounces (oz), pounds (lb), and tons (tn). One slice of bread weighs about 1 ounce. A shoe weighs about 1 pound. The weight of a small car is about 1 ton. To measure the mass of an object in the metric system, we use units such as milligrams (mg), grams (g), kilograms (kg), and metric tons (t). The wing of a housefly is about 1 milligram. A paper clip is about 1 gram. A pair of shoes is about 1 kilogram, and a small

†There is a technical difference between the terms *weight* and *mass* that will be clarified in other coursework. In this book we will use the word *weight* to include the meanings of both terms.

car is about a metric ton. The table below lists some common units of weight in the U.S. Customary System and units of mass in the metric system. The chart also gives equivalences between different units.

Equivalence Table for Units of Weight

U.S. Customary System	Metric System
16 oz = 1 lb 2000 lb = 1 tn	1000 mg = 1 g 1000 g = 1 kg 1000 kg = 1 t
On Earth a kilogram is a little more than 2 pounds, and a metric ton is about 2200 pounds.	

Example 1 The large elephant weighs about 4 tons. About how many pounds does the large elephant weigh?

Solution One ton is 2000 pounds. Four tons is 4 times 2000 pounds. The large elephant weighs about **8000 pounds.**

Example 2 The watermelon's mass was 6 kilograms. The mass of the watermelon was how many grams?

Solution One kilogram is 1000 grams. Six kilograms is 6 times 1000 grams. The watermelon's mass was **6000 grams.**

Example 3 Which is heavier, a car that weighs a ton or a car that weighs a metric ton?

Solution One ton is 2000 pounds. One metric ton is 1000 kilograms, which on Earth is about 2200 pounds. **So a car that weighs a metric ton is heavier** than a car that weighs a ton.

LESSON PRACTICE

Practice set

a. One half of a pound is how many ounces?

b. If a pair of tennis shoes is about 1 kilogram, then one tennis shoe is about how many grams?

c. Ten pounds of potatoes weighs how many ounces?

d. Sixteen tons is how many pounds?

e. How many 500 mg vitamin tablets equal 1 gram?

MIXED PRACTICE

Problem set

1. (35) Samuel Clemens wrote *Huckleberry Finn* using the pen name Mark Twain. Clemens turned 74 in 1909. In what year was he born?

2. (68, 73) Add the decimal number sixteen and nine tenths to twenty-three and seven tenths. What is the sum?

3. (69) Arrange these decimal numbers in order from least to greatest:

2.13, 1.32, 13.2, 1.23

4. (46) One fourth of the 36 students earned A's on the test. One third of the students who earned A's scored 100%.

(a) How many students earned A's?

(b) How many students scored 100%?

(c) What fraction of the students scored 100%?

5. (77) A small car weighs about one ton. How many pounds is 1 ton?

6. (71) Use a fraction, a decimal number, and a percent to name the shaded portion of this square.

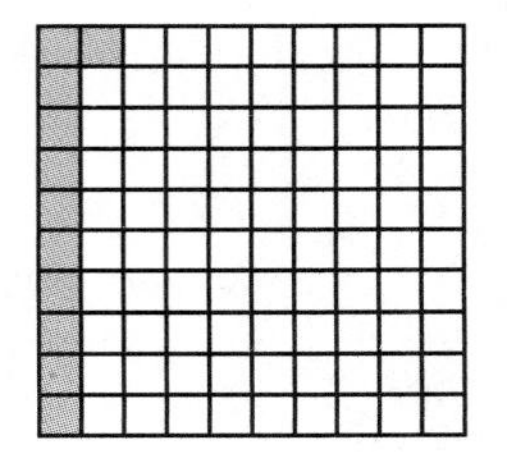

7. (77) A 2-pound box of cereal weighs how many ounces?

8. (77) Three kilograms is how many grams?

9. (61, 73) *AB* is 3.5 centimeters. *BC* is 4.6 centimeters. Find *AC*.

A *B* *C*

10. (75) $\frac{3}{4} + \frac{3}{4} + \frac{3}{4}$

11. (75) $\frac{3}{3} + \frac{2}{2}$

12. (75) $3\frac{5}{8} + 4\frac{6}{8}$

13. (6) $463 + 2875 + 2489 + 8897 + 7963$

14. (76) $\frac{1}{2} \times \frac{5}{6}$

15. (76) $\frac{2}{3} \times \frac{3}{4}$

16. (76) $\frac{1}{2} \times \frac{2}{2}$

17. (73) 401.3 − 264.7

18. (29) \$5.67 × 80

19. (55) 347 × 249

20. (29) 50 × 50

21. (24, 70) (\$5 + 4¢) ÷ 6

22. (34) 64,275 ÷ 8

23. (26, 54) $60w = 3780$

Use the drawing below to answer problems 24–26.

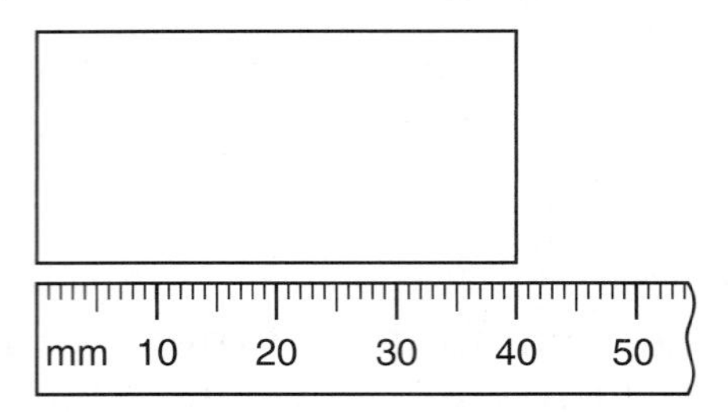

24. (44) How long is the rectangle?

25. (53) If the rectangle is half as wide as it is long, then what is the perimeter of the rectangle?

26. (72) What is the area of the rectangle in square millimeters?

27. (Inv. 7) Assume that this sequence repeats with period three. What are the next four terms of the sequence?

7, 3, 5, 7, _____, _____, _____, _____, ...

28. (74, 76) (a) An inch is what fraction of a foot?

(b) A foot is what fraction of a yard?

(c) An inch is what fraction of a yard?

(d) The answers to parts (a)–(c) show that $\frac{1}{12}$ of $\frac{1}{3}$ is what fraction?

29. (77) The mass of a dollar bill is about

A. 1 milligram B. 1 gram C. 1 kilogram

30. (76) One square inch is divided into quarter-inch squares, as shown at right.

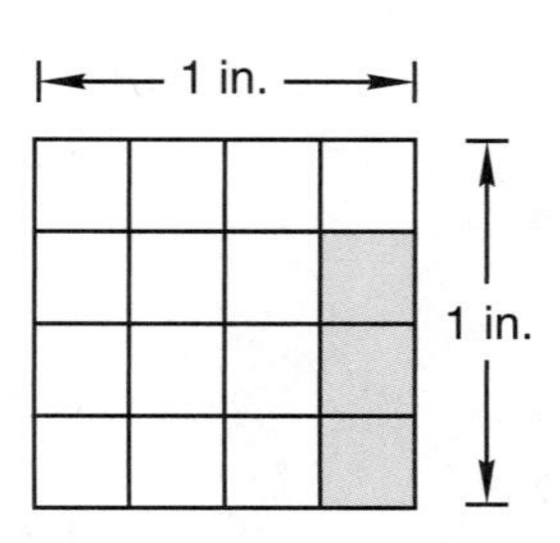

(a) What fraction of the square inch is shaded?

(b) What is the area of the shaded region?

LESSON

78 Exponents and Powers

WARM-UP

Facts Practice: 60 Improper Fractions to Simplify (Test H)

Mental Math:

a. How many ounces are in a pound? How many pounds are in a ton? How many pounds are in 2 tons? ... 3 tons? ... 4 tons?

b. Is 4218 divisible by 4?

c. Is 8124 divisible by 4?

d. What number is 50% of 5?

e. $\sqrt{16}$, × 2, + 2, ÷ 10, − 1, × 5

Problem Solving:

Some 1-inch cubes were stacked together to build this rectangular prism. How many 1-inch cubes were used?

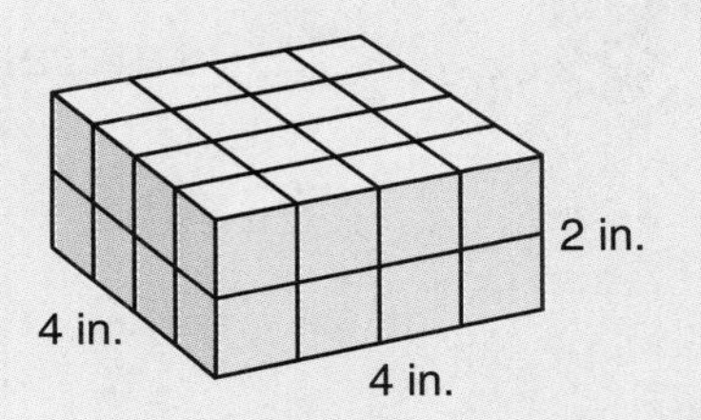

NEW CONCEPT

To show repeated addition, we may use multiplication.

$$5 + 5 + 5 = 3 \times 5$$

To show repeated multiplication, we may use an **exponent.**

$$5 \times 5 \times 5 = 5^3$$

In the expression 5^3, the exponent is 3 and the **base** is 5. The exponent shows how many times the base is used as a factor.

$$5^3 = 5 \times 5 \times 5 = 125$$

Together, the base and exponent are called a **power.** Below are some examples of how expressions with exponents are read. The examples are "powers of three."

3^2 "three squared"

3^3 "three cubed"

3^4 "three to the fourth power"

3^5 "three to the fifth power"

We could read 3^2 as "three to the second power," but we usually say "squared" when the exponent is 2. The word *squared* is a geometric reference to a square. Here we illustrate "three squared":

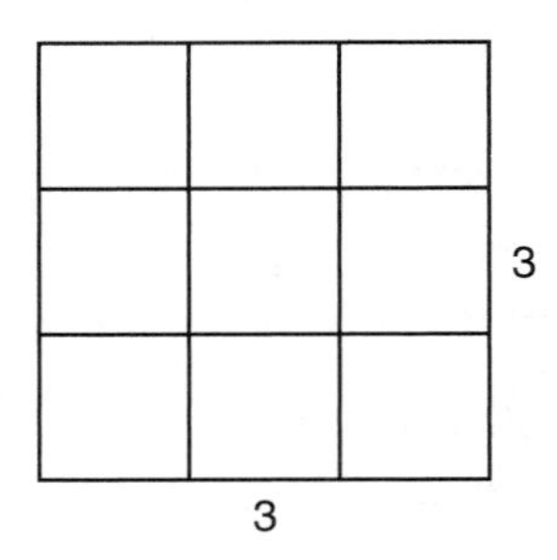

Each side is 3 units long, and the area of the square is 3^2 square units.

When the exponent is 3, we usually say "cubed" instead of "to the third power." The word *cubed* is also a geometric reference. Here we illustrate "three cubed":

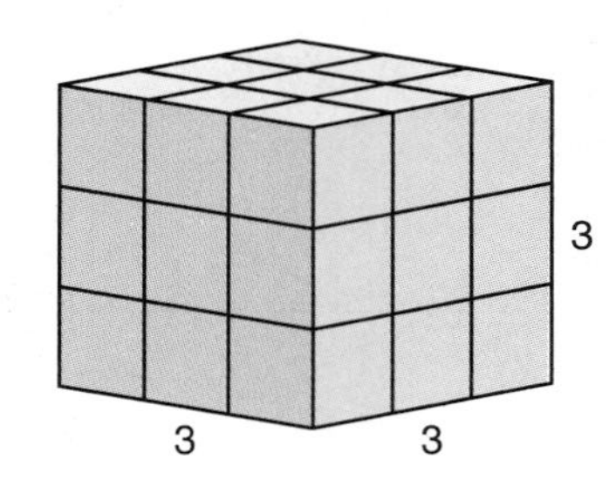

Each edge is three units long, and the number of blocks in the cube is 3^3.

Example 1 Write 3^3 as a whole number.

Solution We find the value of 3^3 by multiplying three 3's.

$$3^3 = 3 \times 3 \times 3 = \mathbf{27}$$

Example 2 If $2n = 6$, then what does n^2 equal?

Solution The expression $2n$ means "2 times n" (or "$n + n$"). If $2n = 6$, then $n = 3$. The expression n^2 means "n times n." To find n^2 when n is 3, we multiply 3 by 3. So n^2 equals **9.**

Example 3 Here we show four powers of 10:

$$10^1, 10^2, 10^3, 10^4$$

Write each power as a whole number.

Solution $10^1 = \mathbf{10}$

$10^2 = 10 \times 10 = \mathbf{100}$

$10^3 = 10 \times 10 \times 10 = \mathbf{1000}$

$10^4 = 10 \times 10 \times 10 \times 10 = \mathbf{10{,}000}$

Powers of 10 can be used to show place value, as we show in the following diagram:

hundred millions	ten millions	millions		hundred thousands	ten thousands	thousands		hundreds	tens	ones
10^8	10^7	10^6		10^5	10^4	10^3		10^2	10^1	10^0
—	—	—	,	—	—	—	,	—	—	—

Notice that the power of 10 in the ones place is 10^0, which equals 1.

Example 4 Write 4,500,000 in expanded notation using powers of 10.

Solution In expanded notation, 4,500,000 is expressed like this:

$$(4 \times 1{,}000{,}000) + (5 \times 100{,}000)$$

Using powers of 10, we replace 1,000,000 with 10^6, and we replace 100,000 with 10^5.

$$\mathbf{(4 \times 10^6) + (5 \times 10^5)}$$

LESSON PRACTICE

Practice set

a. This figure illustrates "five squared," which we can write as 5^2. There are five rows of five small squares. Draw a similar picture to illustrate 4^2.

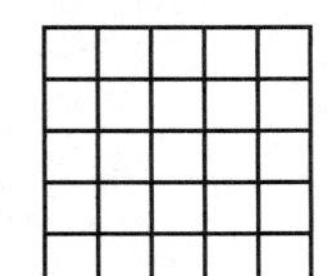

b. This picture illustrates "two cubed," which we can write as 2^3. Two cubed equals what whole number?

Write each power as a whole number. Show your work.

c. 3^4 **d.** 2^5 **e.** 11^2

f. If $2m = 10$, then what does m^2 equal?

Write each number in expanded notation using powers of 10:

g. 250,000 **h.** 3,600,000 **i.** 60,500

MIXED PRACTICE

Problem set

1. (76) One half of the students are girls. One third of the girls have long hair. What fraction of the students are girls with long hair? What percent of the students are girls with long hair?

2. (16) Friendly Fred bought a car for \$860 and sold it for \$1300. How much profit did he make?

3. (50) Heather read a 316-page book in 4 days. She averaged how many pages per day?

4. (77) The pickup truck could carry $\frac{1}{2}$ ton. How many pounds is $\frac{1}{2}$ ton?

5. (77) The baby kitten weighed one half of a pound. How many ounces did it weigh?

6. (Inv. 3, 39) Which shaded circle below is equivalent to this shaded circle?

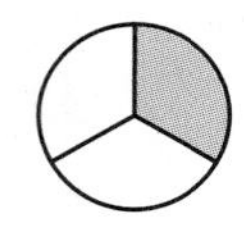

A. B. 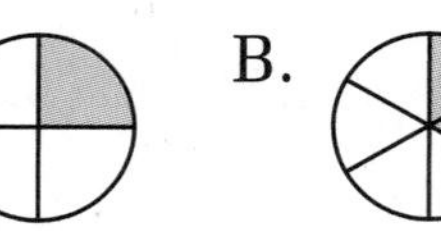 C. 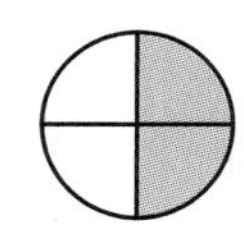

7. (23) Which of these fractions does not equal one half?

A. $\frac{50}{100}$ B. $\frac{1000}{2000}$ C. $\frac{16}{30}$ D. $\frac{6}{12}$

8. (66) Find the length of this segment twice, first in millimeters and then in centimeters.

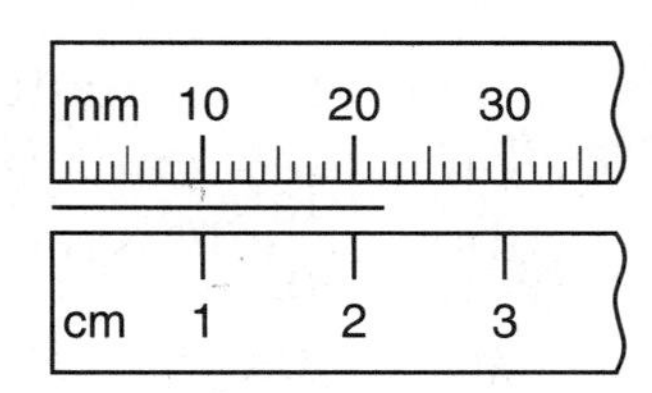

9. List the numbers that are factors of both 6 and 8.
(25)

10. *LN* is 6.4 centimeters. *LM* is 3.9 centimeters. Find *MN*.
(61, 73)

11. $\frac{2}{3} + \frac{2}{3} + \frac{2}{3}$
(75)

12. $\frac{3}{3} - \frac{2}{2}$
(59)

13. $9\frac{4}{10} + 4\frac{9}{10}$
(75)

14. 4.6 + 3.27
(73)

15. $40.00 − $13.48
(13)

16. $20.50 × 8
(17)

17. 9)$56.70
(34)

18. 13^2
(78)

19. 80)4650
(54)

20. Divide and write the quotient as a mixed number: $\frac{98}{5}$
(58)

21. $\frac{3}{4}$ of $\frac{1}{2}$
(76)

22. $\frac{3}{2} \times \frac{3}{4}$
(76)

23. $\frac{1}{3} \times \frac{2}{2}$
(76)

Use this information to answer problems 24 and 25:

It is 1.5 miles from Kiyoko's house to school. It takes Kiyoko 30 minutes to walk to school and 12 minutes to ride her bike to school.

24. How far does Kiyoko travel going to school and back in 1 day?
(Inv. 5, 73)

25. If Kiyoko leaves her house at 7:55 a.m. and rides her bike, at what time will she get to school?
(28, Inv. 5)

26. Assume that this sequence repeats with period four. Write the next four terms of the sequence.
(Inv. 7)

7, 3, 5, 7, _____, _____, _____, _____, ...

27. (57) Suppose the 7 letter tiles below are turned over and mixed up. Then one tile is selected.

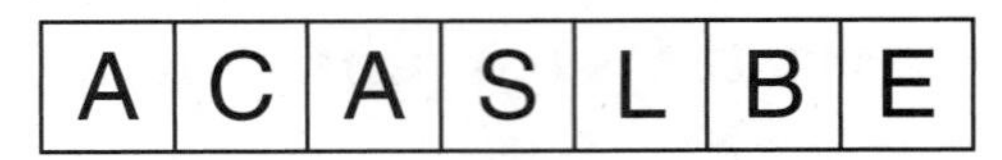

(a) What is the probability that the letter selected is A?

(b) What is the probability that the letter selected is a vowel?

(c) What is the probability that the letter selected comes before G in the alphabet?

28. (61, 72) Each angle of quadrilateral *ABCD* is a right angle. If *AB* is 10 cm and *BC* is 5 cm, what is the area of the quadrilateral?

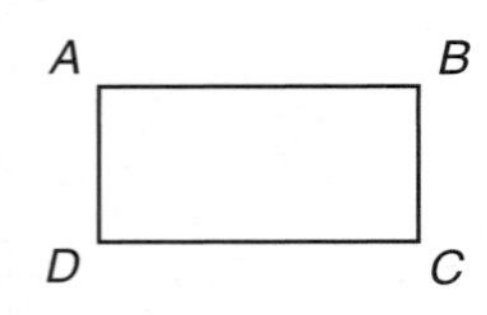

29. (45) Which of these terms does not apply to quadrilateral *ABCD* in problem 28?

A. rectangle B. parallelogram C. rhombus

30. (78) Write 25,000,000 in expanded notation using powers of 10.

LESSON

79 Finding Equivalent Fractions by Multiplying by 1

WARM-UP

Facts Practice: 60 Improper Fractions to Simplify (Test H)

Mental Math:

a. How many centimeters are in one meter? How many meters are in one kilometer? Hold two fingers one centimeter apart. Hold your hands one yard apart.

b. $\frac{1}{4}$ of 20 **c.** $\frac{1}{4}$ of 200 **d.** $\frac{1}{5}$ of 16

e. 10% of $5.00 **f.** $\sqrt{49}$, − 2, ÷ 2, − 2

Problem Solving:

Draw a triangle that is similar to this triangle with sides that are twice as long.

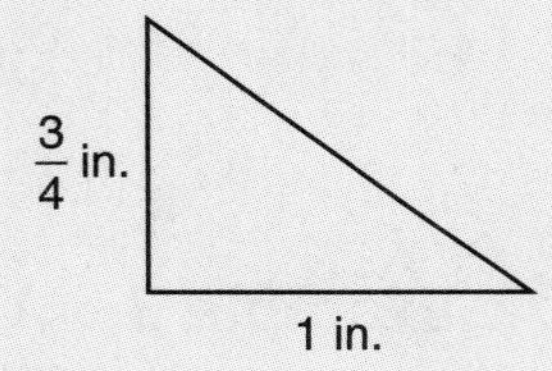

NEW CONCEPT

In Lesson 15 we learned that when a number is multiplied by 1, the value of the number does not change. This property is called the **identity property of multiplication.** We can use this property to find **equivalent fractions.** Equivalent fractions are different names for the same number. For example, $\frac{1}{2}$, $\frac{2}{4}$, $\frac{3}{6}$, and $\frac{4}{8}$ are equivalent fractions. To find equivalent fractions, we multiply a number by different fraction names for 1.

$\frac{1}{2} \times \frac{2}{2} = \frac{2}{4}$	$\frac{1}{2} \times \frac{3}{3} = \frac{3}{6}$	$\frac{1}{2} \times \frac{4}{4} = \frac{4}{8}$

As we see above, we can find fractions equivalent to $\frac{1}{2}$ by multiplying by $\frac{2}{2}$, $\frac{3}{3}$, and $\frac{4}{4}$. By multiplying $\frac{1}{2}$ by $\frac{5}{5}$, $\frac{6}{6}$, $\frac{7}{7}$, and so on, we find more fractions equivalent to $\frac{1}{2}$:

$$\frac{1}{2} \times \frac{n}{n} = \frac{5}{10}, \frac{6}{12}, \frac{7}{14}, \frac{8}{16}, \frac{9}{18}, \frac{20}{20}, \ldots$$

Example 1 By what name for 1 should $\frac{3}{4}$ be multiplied to make $\frac{6}{8}$?

$$\frac{3}{4} \times \frac{?}{?} = \frac{6}{8}$$

Solution To change $\frac{3}{4}$ to $\frac{6}{8}$, we multiply by $\mathbf{\frac{2}{2}}$. The fraction $\frac{2}{2}$ is equal to 1, and when we multiply by 1 we do not change the value of the number. Therefore, $\frac{3}{4}$ equals $\frac{6}{8}$.

Example 2 Write a fraction equal to $\frac{2}{3}$ that has a denominator of 12.

$$\frac{2}{3} = \frac{?}{12}$$

Solution We can change the name of a fraction by multiplying by a fraction name for 1. To make the 3 a 12, we must multiply by 4. So the fraction name for 1 that we will use is $\frac{4}{4}$. We multiply $\frac{2}{3} \times \frac{4}{4}$ to form the equivalent fraction $\mathbf{\frac{8}{12}}$.

$$\frac{2}{3} \times \frac{}{4} = \frac{}{12}$$

$$\frac{2}{3} \times \frac{4}{4} = \frac{8}{12}$$

Example 3 Write a fraction equal to $\frac{1}{3}$ that has a denominator of 12. Then write a fraction equal to $\frac{1}{4}$ that has a denominator of 12. What is the sum of the two fractions you made?

Solution We multiply $\frac{1}{3}$ by $\frac{4}{4}$ and $\frac{1}{4}$ by $\frac{3}{3}$.

$$\frac{1}{3} \times \frac{4}{4} = \mathbf{\frac{4}{12}} \qquad \frac{1}{4} \times \frac{3}{3} = \mathbf{\frac{3}{12}}$$

Then we add $\frac{4}{12}$ and $\frac{3}{12}$ to find their sum.

$$\frac{4}{12} + \frac{3}{12} = \mathbf{\frac{7}{12}}$$

Example 4 Write $\frac{3}{4}$ as a fraction with a denominator of 100. Then write that fraction as a percent.

Solution To change fourths to hundredths, we multiply by $\frac{25}{25}$.

$$\frac{3}{4} \times \frac{25}{25} = \mathbf{\frac{75}{100}}$$

The fraction $\frac{75}{100}$ is equivalent to **75%**.

LESSON PRACTICE

Practice set* Find the fraction name for 1 used to make each equivalent fraction:

a. $\frac{3}{4} \times \frac{?}{?} = \frac{9}{12}$

b. $\frac{2}{3} \times \frac{?}{?} = \frac{4}{6}$

c. $\frac{1}{3} \times \frac{?}{?} = \frac{4}{12}$

d. $\frac{1}{4} \times \frac{?}{?} = \frac{25}{100}$

Find the numerator that completes each equivalent fraction:

e. $\frac{1}{3} = \frac{?}{9}$

f. $\frac{2}{3} = \frac{?}{15}$

g. $\frac{3}{5} = \frac{?}{10}$

h. Write a fraction equal to $\frac{1}{2}$ that has a denominator of 6. Then write a fraction equal to $\frac{1}{3}$ that has a denominator of 6. What is the sum of the two fractions you made?

i. Write $\frac{3}{5}$ as a fraction with a denominator of 100. Then write that fraction as a percent.

MIXED PRACTICE

Problem set

1. Mr. MacDonald bought 1 ton of hay for his cow, Geraldine. Every day Geraldine eats 50 pounds of hay. At this rate 1 ton of hay will last how many days?
(21, 77)

2. A platypus is a mammal with a duck-like bill and webbed feet. A platypus is about $1\frac{1}{2}$ feet long. One and one half feet is how many inches?
(23, 74)

3. Sam bought 3 shovels for his hardware store for \$6.30 each. He sold them for \$10.95 each. How much profit did Sam make on all 3 shovels? (Sam's profit for each shovel can be found by subtracting how much Sam paid from the selling price.)
(49)

4. Add the decimal number ten and fifteen hundredths to twenty-nine and eighty-nine hundredths. Use words to name the sum.
(68, 73)

5. By what fraction name for 1 should $\frac{2}{3}$ be multiplied to make $\frac{6}{9}$?
(79)

$$\frac{2}{3} \times \frac{?}{?} = \frac{6}{9}$$

6. Draw a rectangle whose sides are all 1 inch long. What is the area of the rectangle?
(44, 45, 72)

7. List the numbers that are factors of both 9 and 12.
(25)

8. Write a fraction equal to $\frac{3}{4}$ that has a denominator of 12. Then write a fraction equal to $\frac{2}{3}$ that has a denominator of 12. What is the sum of the fractions you wrote?
(79)

9. *AC* is 9.1 centimeters. *BC* is 4.2 centimeters. Find *AB*.
(61, 73)

A — B — C

10. $1\frac{1}{5} + 2\frac{2}{5} + 3\frac{3}{5}$
(75)

11. $5 - \left(3\frac{5}{8} - 3\right)$
(43, 63)

12. \$10 − 10¢
(70)

13. \$10 ÷ 4
(34)

14. 9 × 64¢
(70)

15. $24.6 + M = 30.4$
(10, 73)

16. $W - 6.35 = 2.4$
(14, 73)

17. $9n = 6552$
(26)

18. $7\overline{)43{,}859}$
(26)

19. 15^2
(78)

20. $80\overline{)4137}$
(54)

21. $\frac{1}{2}$ of $\frac{1}{5}$
(76)

22. $\frac{3}{4} \times \frac{2}{2}$
(76)

23. $\frac{3}{5} \times \frac{5}{4}$
(76)

The graph below shows the number of ice cream cones sold at the snack bar from June through August. Use the information in the graph to answer problems 24 and 25.

Ice Cream Cone Sales

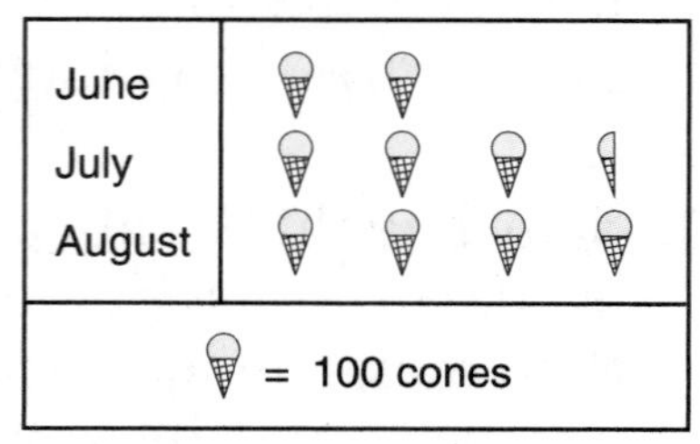

24. How many ice cream cones were sold in July?
(Inv. 5)

A. $3\frac{1}{2}$ B. 300 C. 305 D. 350

25. Altogether, how many ice cream cones were sold during June, July, and August?
(Inv. 5)

26. A standard number cube is rolled once. What is the probability that the upturned face is not 4?
(57)

27. To multiply 12 by 21, Tom thought of 21 as 20 + 1. Then he mentally calculated this problem:
(51)

$$(20 \times 12) + (1 \times 12)$$

Try mentally calculating the answer. What is the product of 12 and 21?

28. Fourteen books were packed in a box. The mass of the packed box could reasonably be which of the following?
(77)

A. 15 milligrams B. 15 grams C. 15 kilograms

29. What is the perimeter of this equilateral triangle?
(53, 73)

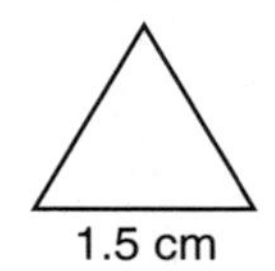

30. Compare: 500 mg ◯ 1.0 g
(77)

LESSON

80 Prime and Composite Numbers

WARM-UP

Facts Practice: 60 Improper Fractions to Simplify (Test H)

Mental Math:

a. How many grams equal one kilogram? A pair of shoes weighs about one kilogram. One shoe weighs about how many grams?

b. 25% of 16 **c.** 25% of 160 **d.** $\frac{1}{3}$ of 16

e. 25% of $20.00 **f.** $\sqrt{81}$, − 2, ÷ 2, − 1, × 2, − 5

Problem Solving:

Find the next three numbers in this Fibonacci sequence:

1, 1, 2, 3, 5, 8, 13, __, __, __, ...

NEW CONCEPT

We have practiced listing the factors of whole numbers. Some whole numbers have many factors. Other whole numbers have only a few factors. In one special group of whole numbers, each number has exactly two factors.

Below, we list the first ten counting numbers and their factors. Numbers with exactly two factors are **prime numbers.** Numbers with more than two factors are **composite numbers.** The number 1 has only one factor and is neither prime nor composite.

Number	Factors	Type
1	1	
2	**1, 2**	**prime**
3	**1, 3**	**prime**
4	1, 2, 4	composite
5	**1, 5**	**prime**
6	1, 2, 3, 6	composite
7	**1, 7**	**prime**
8	1, 2, 4, 8	composite
9	1, 3, 9	composite
10	1, 2, 5, 10	composite

We often think of a prime number as a number that is not divisible by any other number except 1 and itself. Listing prime numbers will quickly give us a feel for which numbers are prime numbers.

Example 1 The first three prime numbers are 2, 3, and 5. What are the next three prime numbers?

Solution We list the next several whole numbers after 5. A prime number is not divisible by any number except 1 and itself, so we mark through numbers that are divisible by some other number.

~~6~~, 7, ~~8~~, ~~9~~, ~~10~~, 11, ~~12~~, 13, ~~14~~, ~~15~~, ~~16~~, 17, ~~18~~

The numbers that are not marked through are prime numbers. The next three prime numbers after 5 are **7, 11,** and **13.**

Every number in the shaded part of this multiplication table has more than two factors. So every number in the shaded part is a composite number.

	1	2	3	4	5	6	7	8	9	10	11
1	1	(2)	(3)	4	(5)	6	(7)	8	9	10	(11)
2	(2)	4	6	8	10	12	14	16	18	20	22
3	(3)	6	9	12	15	18	21	24	27	30	33
4	4	8	12	16	20	24	28	32	36	40	44
5	(5)	10	15	20	25	30	35	40	45	50	55
6	6	12	18	24	30	36	42	48	54	60	66
7	(7)	14	21	28	35	42	49	56	63	70	77
8	8	16	24	32	40	48	56	64	72	80	88

In this multiplication table prime numbers appear *only* in the row and column beginning with 1. We have circled the prime numbers that appear in the table. Even if the table were extended, prime numbers would appear only in the row and column beginning with 1.

We can use arrays to illustrate the difference between prime and composite numbers. An **array** is a rectangular arrangement of numbers or objects in rows and columns. Here we show three different arrays for the number 12:

```
XXXX     XXXXXX     XXXXXXXXXXXX
XXXX     XXXXXX        1 by 12
XXXX      2 by 6
3 by 4
```

Twelve is a composite number, which is demonstrated by the fact that we can use **different pairs** of factors to form arrays for 12. By turning the book sideways, we can actually form three more arrays for 12 (4 by 3, 6 by 2, and 12 by 1), but these arrays use the same factor pairs as the arrays shown above.

For the prime number 11, however, there is only **one pair** of factors that forms arrays: 1 and 11.

XXXXXXXXXXX

1 by 11

Example 2 Draw three arrays for the number 16. Use different factor pairs for each array.

Solution The multiplication table can guide us. We see 16 at 4 × 4 and at 2 × 8. So we can draw a 4-by-4 array and a 2-by-8 array. Of course, we can also draw a 1-by-16 array.

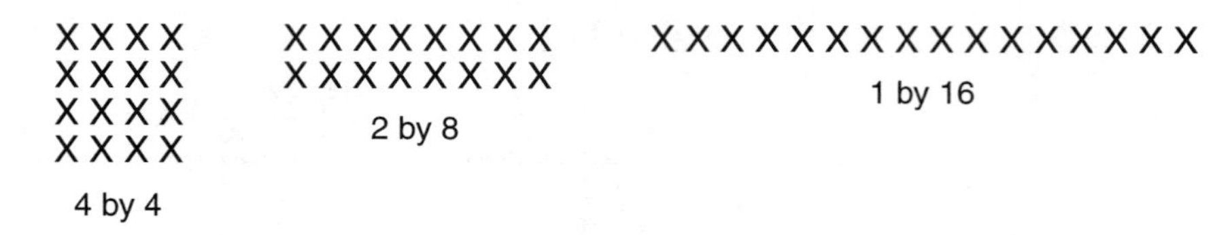

LESSON PRACTICE

Practice set

a. The first four prime numbers are 2, 3, 5, and 7. What are the next four prime numbers?

b. List all the factors of 21. Is the number 21 prime or composite? Why?

c. Which counting number is neither prime nor composite?

d. Draw two arrays of X's for the composite number 9. Use different factor pairs for each array.

MIXED PRACTICE

Problem set

1. *(49, 70)* The student store buys one dozen pencils for 96¢ and sells them for 20¢ each. How much profit does the store make on a dozen pencils?

2. *(21, 77)* A small car weighs about 1 ton. If its 4 wheels carry the weight evenly, then each wheel carries about how many pounds?

3. *(25)* List the numbers that are factors of both 8 and 12.

4. *(80)* The first five prime numbers are 2, 3, 5, 7, and 11. What are the next three prime numbers?

5. By what fraction name for 1 should $\frac{3}{4}$ be multiplied to make $\frac{9}{12}$?
(79)

$$\frac{3}{4} \times \frac{?}{?} = \frac{9}{12}$$

6. Write a fraction equal to $\frac{1}{2}$ that has a denominator of 6. Then write a fraction equal to $\frac{2}{3}$ that has a denominator of 6. What is the sum of the fractions you wrote?
(79)

7. Think of a prime number. How many different factors does it have?
(80)

8. Arrange these numbers in order from least to greatest:
(23, 59)

$$\frac{3}{8}, \frac{4}{6}, \frac{5}{6}, \frac{6}{12}, \frac{7}{7}$$

9. One mile is 1760 yards. How many yards is $\frac{1}{8}$ mile?
(46, 74)

10. *XZ* is 84 millimeters. *XY* equals *YZ*. Find *XY*.
(61)

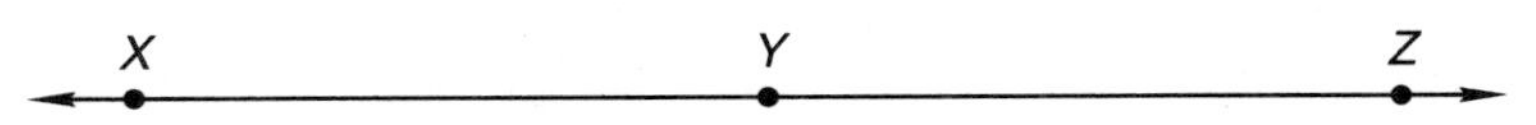

11. \$8.43 + 68¢ + \$15 + 5¢
(70)

12. 6.505 − 1.4
(73)

13. \$12 − 12¢
(70)

14. \$18.07 × 6
(17)

15. $6w$ = \$76.32
(26)

16. 2^6
(78)

17. $70\overline{)4791}$
(54)

18. Divide 365 by 7. Write the quotient as a mixed number.
(58)

19. $\frac{3}{4}$ of $\frac{3}{4}$
(76)

20. $\frac{3}{2} \times \frac{3}{2}$
(76)

21. $\frac{3}{10} = \frac{?}{100}$
(79)

22. $3\frac{2}{3} + 1\frac{2}{3}$
(75)

23. $5 - \frac{1}{5}$
(63)

24. $\frac{7}{10} - \frac{7}{10}$
(41)

25. It is evening. What time will be shown by this clock in $6\frac{1}{2}$ hours?
(28)

26. The Sun is about 92,956,000 miles from Earth. Which digit in 92,956,000 is in the millions place?
(52)

27. The Sun is about 150,000,000 kilometers from Earth. Write that distance in expanded notation using powers of 10.
(78)

28. Is the sequence below arithmetic or geometric? Find the next two terms in the sequence.
(Inv. 7)

2, 4, 8, 16, _____, _____, ...

29. As the coin was tossed, the team captain called, "Heads!" What is the probability that the captain's guess was correct?
(57)

30. The fraction $\frac{4}{5}$ is equivalent to 0.8 and 80%. Write 0.8 and 80% as unreduced fractions.
(71)

INVESTIGATION 8

Focus on

Displaying Data

Data that are gathered and organized may be displayed in various types of charts and graphs. One type of graph is a **bar graph.** A bar graph uses rectangles, or bars, to display data. Below we show the test scores and frequency table from Investigation 5 and a bar graph that displays the data.

Test scores: 4, 3, 3, 4, 2, 5, 6, 1, 3, 4, 5, 2, 2,
6, 3, 3, 4, 3, 2, 4, 5, 3, 5, 5, 6

Frequency Table

Number Correct	Frequency
1	1
2	4
3	7
4	5
5	5
6	3

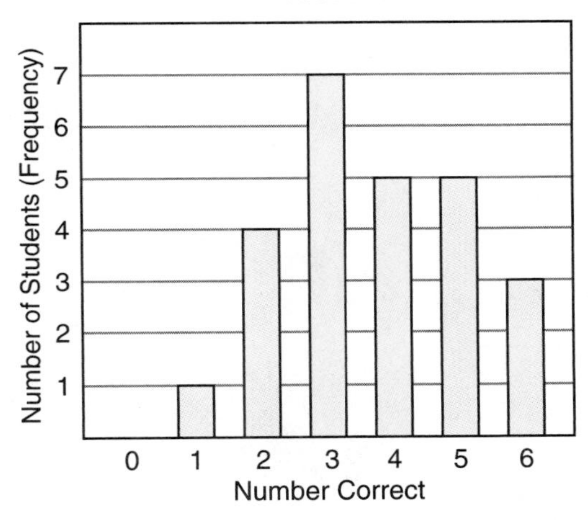

Notice how the information from the frequency table is presented in the bar graph. The scale across the bottom of the bar graph, the horizontal axis, lists all possible scores on the tests. It shows the same values as the first column of the frequency table. The scale along the left side of the bar graph, the vertical axis, lists the number of students. The height of a bar tells how often the score shown below the bar was achieved. In other words, it tells the frequency of the score.

Now we will practice making bar graphs using a new situation. Twenty children in a class were asked how many siblings (brothers *and* sisters) each had. The data from their responses, as well as a frequency table to organize the data, are shown below.

Number of siblings: 2, 3, 0, 1, 1, 3, 0, 4, 1, 2,
0, 1, 1, 2, 2, 3, 0, 2, 1, 1

Frequency Table

Number of Siblings	Tally	Frequency
0	||||	4
1	~~||||~~ ||	7
2	~~||||~~	5
3	|||	3
4	|	1

1. Copy and complete this bar graph to display the data.

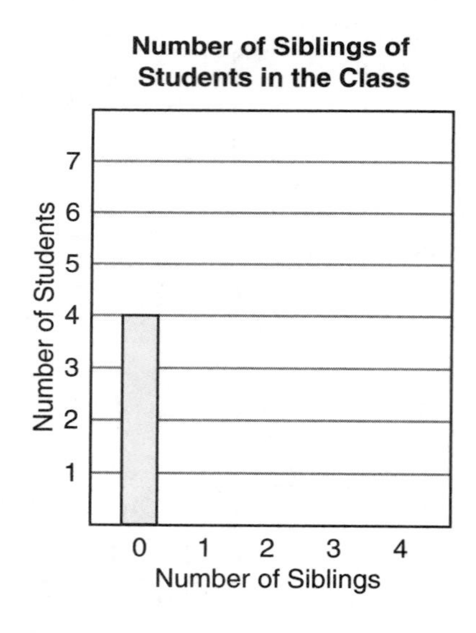

In Investigation 5 we made frequency tables with data grouped in intervals of equal size. From that investigation, recall that ABC Market offered turkeys with these weights (in pounds):

11, 18, 21, 23, 16, 20, 22, 14, 16, 20, 17,
19, 13, 14, 22, 19, 22, 18, 20, 12, 25, 23

Here is the frequency table for these data using intervals of 4 pounds, starting with the interval 10–13 lb:

Frequency Table

Weight	Tally	Frequency
10–13 lb	|||	3
14–17 lb	𝍸	5
18–21 lb	𝍸 |||	8
22–25 lb	𝍸 |	6

To graph data grouped in intervals, we can make a **histogram.** A histogram is a type of bar graph. In a histogram the widths of the bars represent the selected intervals, and there are no spaces between the bars. Below is a histogram for the turkey-weight data. The intervals in the histogram match the intervals in the frequency table above.

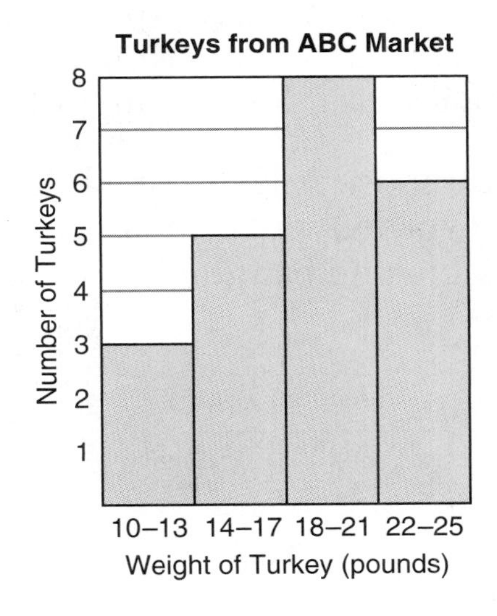

2. Create a frequency table and histogram for the turkey-weight data using these weight intervals:

11–13 lb, 14–16 lb, 17–19 lb, 20–22 lb, 23–25 lb

Another way to display these turkey weights is in a **stem-and-leaf plot.** The "stems" are the tens digits of the weights. The "leaves" for each stem are the ones digits of the weights that begin with that tens digit. Here is the stem-and-leaf plot for the first row of weights in the list. Notice that the leaves are listed in increasing order.

Stem	Leaf
1	1 4 6 6 7 8
2	0 0 1 2 3

3. Make a stem-and-leaf plot for the second row of weights in the list.

4. Use the information in the stem-and-leaf plots for the first and second rows of the list of weights to make a stem-and-leaf plot for the weights of all 22 turkeys.

Numerical data represent such quantities as ages, heights, weights, temperatures, and points scored. But data also come in **categories,** or **classes.** People, concepts, and objects belong to categories. Examples of categories include occupations, days of the week, after-school activities, foods, and colors.

Suppose Angela asked the students in her class to name their favorite type of soda and displayed the data in this frequency table and bar graph:

Frequency Table

Soda	Frequency
Cola	9
Root beer	5
Lemon-lime	6
Orange	4

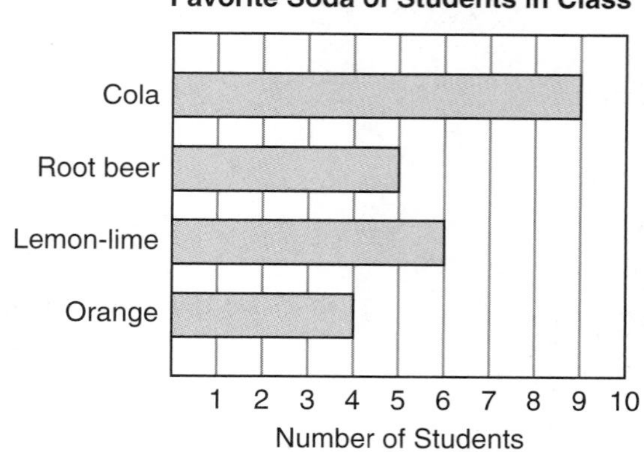

The bar graph Angela made is called a *horizontal bar graph* because the bars run horizontally. The categories that Angela used for her data are types of soda: cola, root beer, lemon-lime, and orange.

Sixty students were asked to give their position among the children in their family. Their responses were put into the four categories shown in this frequency table:

Frequency Table

Category	Frequency
Only child	22
Youngest child	16
Oldest child	14
Middle child	8

5. Make a horizontal bar graph for the data in the table above. Be sure to label each bar along the vertical side of the graph with one of the four categories. Along the bottom of the graph, use even numbers to label the number of students.

A **pictograph** uses symbols, or **icons,** to compare data that come from categories. An icon can represent one data point or a group of data points. In pictographs we include a **legend** to show what each icon represents.

Suppose 96 children were asked to choose their food preference among hamburgers, hot dogs, and pizza. The data that was collected is displayed in this pictograph:

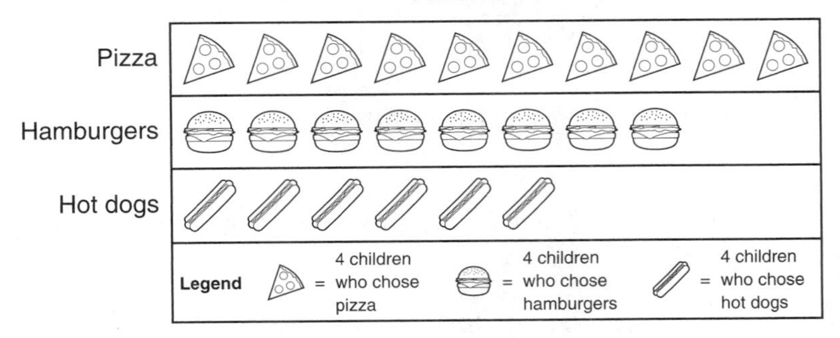

We count 8 hamburger icons in the pictograph. To find how many children 8 icons represent, we read the legend and find that we should multiply the number of icons by 4. So 8×4, or 32, children prefer hamburgers.

6. How many children prefer pizza?

7. How many children prefer hot dogs?

8. Draw a second pictograph for the food preferences in which each icon represents 8 children.

Sometimes we are interested in seeing how categories break down into parts of a whole group. The best kind of graph for this is a **circle graph.** A circle graph is sometimes called a *pie chart.*

The following circle graph shows how Greg spends a typical 24-hour day during the summer:

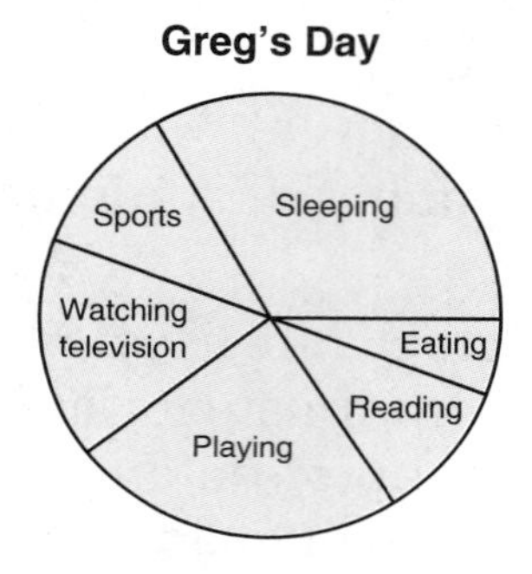

9. Which activity consumes most of Greg's time? Which activity consumes the least amount of Greg's time?

10. Which activity consumes about the same amount of time as sports?

11. Which activities consume more time than sports?

12. List activities that together consume about 12 hours.

In the examples we have presented, the categories do not overlap. This means that each data point fits in only one category. We say that such categories are **mutually exclusive.** But sometimes we have data points that fit into more than one category. To display such data, we can use a **Venn diagram.** The categories in a Venn diagram are represented by overlapping circles.

Suppose Karim surveys his 21 classmates about family pets. Karim asks whether they have a dog at home, and he asks whether they have a cat at home. There are four possibilities. A family could have both kinds of pets. A family could have a dog only. A family could have a cat only. Finally, a family could have neither kind of pet. Karim finds that 12 of his classmates have dogs and 8 have cats. Of these, 5 have both. We will make a Venn diagram to display this information.

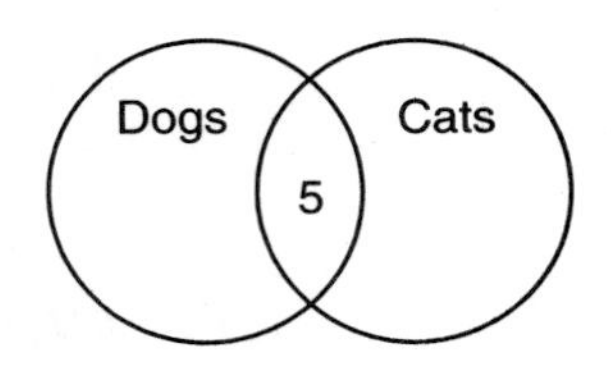

We begin by drawing two overlapping circles. One circle represents the families with dogs. The other circle represents the families with cats. We must make the circles overlap, because some families have both dogs and cats. We are told that 5 families have both kinds of pets, so we write a 5 in the overlapping region.

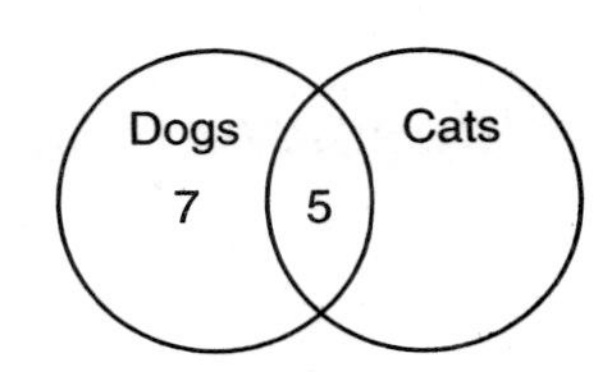

We have accounted for 5 of the 12 families who have dogs. So there are 7 families that have dogs but not cats. We write "7" in the region of the diagram for families with dogs only.

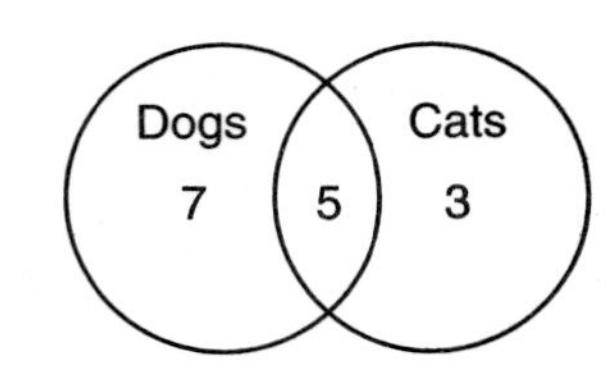

We have also accounted for 5 of the 8 families that have cats. So there are 3 more families that have cats. We write "3" in the region of the diagram for families with cats only.

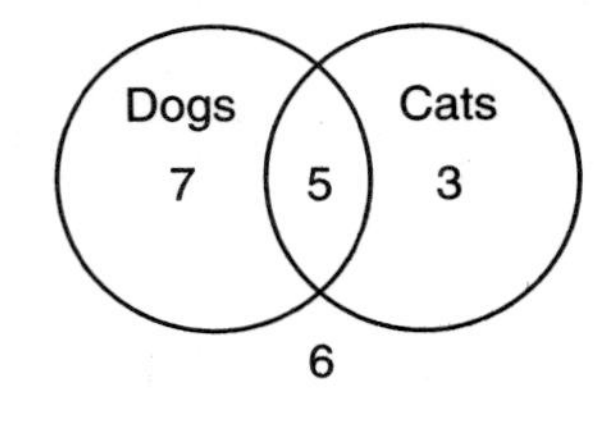

We have now accounted for 7 families with dogs only, 5 families with both dogs and cats, and 3 families with cats only. That is a total of 15 families. However, 21 families were part of the survey, so 6 families have neither animal. We can write a 6 outside of the circles to represent these families.

13. How many families had just one of the two types of animals?

Suppose Tito asked each of his classmates whether he or she plays soccer. Refer to this Venn diagram he created to answer problems 14–19.

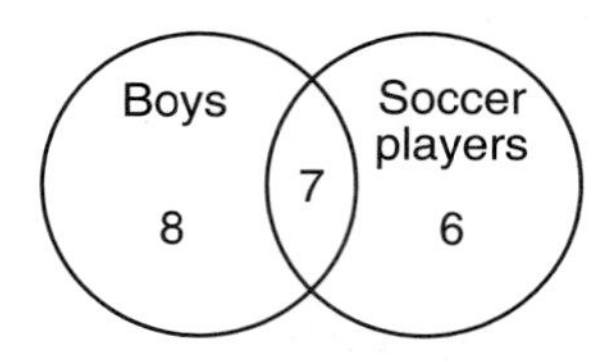

14. How many boys are in Tito's class?

15. How many soccer players are in Tito's class?

16. How many of the boys play soccer?

17. How many soccer players are girls?

18. If there are 30 students in his class, how many girls are there?

19. If there are 30 students in his class, how many girls do not play soccer?

Extensions

a. Ask the students in the class to name the month in which they were born. Display the results in a horizontal bar graph.

b. Ask the students in the class to name the season of the year in which they were born: winter, spring, summer, or fall. (Use December 22, March 22, June 22, and September 22 as the first day of each season.) Display the results in a histogram.

c. Ask each student in the class these two questions:

- Do you like to watch cartoons on television?
- Do you like to watch sports on television?

Copy and complete the Venn diagram below to display the number of students who answer "yes" to either or both questions.

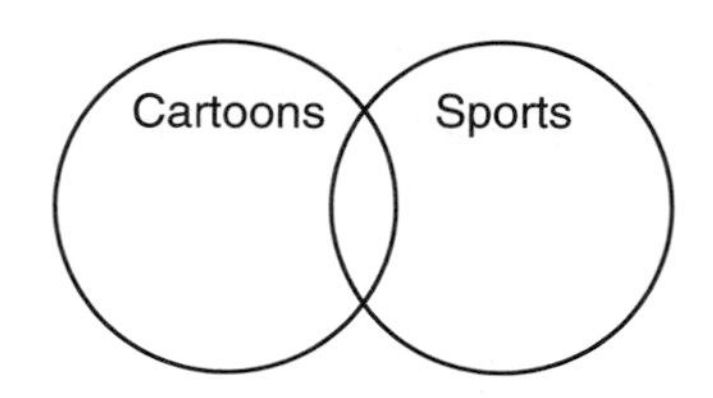

LESSON

81 Reducing Fractions, Part 1

WARM-UP

Facts Practice: 60 Improper Fractions to Simplify (Test H)

Mental Math:

a. How many feet are in one yard? How many feet are in one mile? Hold your hands about one foot apart. Hold your hands about one yard apart.

b. $\frac{1}{4}$ of 30 c. $\frac{1}{4}$ of 300 d. 10% of \$300

e. $\frac{1}{2}$ of 30 minutes f. 30 × 30, + 100, ÷ 2, − 100, ÷ 4

Problem Solving:

Triangles I, II, and III are congruent. If triangle I is rotated a quarter turn in a clockwise direction around point *A*, it will be in the position of triangle II. If it is rotated again, it will be in the position of triangle III. If it is rotated again, it will be in the position of triangle IV. Draw triangles I, II, III, and IV.

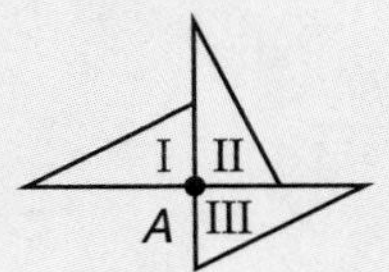

NEW CONCEPT

In Lesson 79 we practiced making equivalent fractions by multiplying by a fraction name for 1. We changed the fraction $\frac{1}{2}$ to the equivalent fraction $\frac{3}{6}$ by multiplying by $\frac{3}{3}$.

$$\frac{1}{2} \times \frac{3}{3} = \frac{3}{6}$$

Multiplying by $\frac{3}{3}$ made the **terms** of the fraction greater. The terms of a fraction are the numerator and the denominator. The terms of $\frac{1}{2}$ are 1 and 2. The terms of $\frac{3}{6}$ are 3 and 6.

Sometimes we can make the terms of a fraction smaller by dividing by a fraction name for 1. Here we change $\frac{3}{6}$ to $\frac{1}{2}$ by dividing both terms of $\frac{3}{6}$ by 3:

$$\frac{3}{6} \div \frac{3}{3} = \frac{1}{2} \quad \begin{matrix}(3 \div 3 = 1) \\ (6 \div 3 = 2)\end{matrix}$$

Changing a fraction to an equivalent fraction with smaller terms is called **reducing.** We reduce a fraction by dividing both terms of the fraction by the same number.

Example 1 Reduce the fraction $\frac{6}{8}$ by dividing both the numerator and the denominator by 2.

Solution We show the reducing process below.

$$\frac{6 \div 2}{8 \div 2} = \mathbf{\frac{3}{4}}$$

The reduced fraction $\frac{3}{4}$ has smaller terms than $\frac{6}{8}$. We can see from the picture below, however, that $\frac{3}{4}$ and $\frac{6}{8}$ are equivalent fractions.

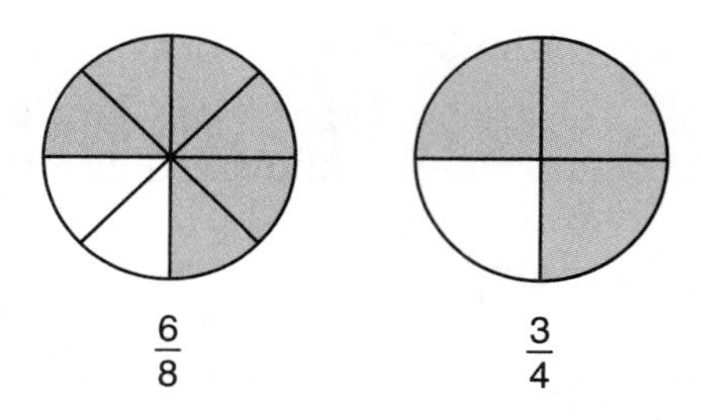

Not all fractions can be reduced. Only fractions whose terms can be divided by the same number can be reduced.

Example 2 Which of these fractions cannot be reduced?

A. $\frac{2}{6}$ B. $\frac{3}{6}$ C. $\frac{4}{6}$ D. $\frac{5}{6}$

Solution We will consider each fraction:

A. The terms of $\frac{2}{6}$ are 2 and 6. Both 2 and 6 are even numbers, so they can be divided by 2. The fraction $\frac{2}{6}$ can be reduced to $\frac{1}{3}$.

B. The terms of $\frac{3}{6}$ are 3 and 6. Both 3 and 6 can be divided by 3, so $\frac{3}{6}$ can be reduced to $\frac{1}{2}$.

C. The terms of $\frac{4}{6}$ are 4 and 6. Both 4 and 6 are even numbers, so they can be divided by 2. The fraction $\frac{4}{6}$ can be reduced to $\frac{2}{3}$.

D. The terms of $\frac{5}{6}$ are 5 and 6. The only whole number that divides both 5 and 6 is 1. Since dividing by 1 does not make the terms smaller, the fraction $\frac{5}{6}$ cannot be reduced. So the answer to the question is **D.** $\mathbf{\frac{5}{6}}$.

Example 3 Add and reduce the answer: $\frac{1}{8} + \frac{5}{8}$

Solution We add $\frac{1}{8}$ and $\frac{5}{8}$.

$$\frac{1}{8} + \frac{5}{8} = \frac{6}{8}$$

The terms of $\frac{6}{8}$ are 6 and 8. We can reduce $\frac{6}{8}$ by dividing each term by 2.

$$\frac{6 \div 2}{8 \div 2} = \frac{3}{4}$$

We find that the sum of $\frac{1}{8}$ and $\frac{5}{8}$ is $\mathbf{\frac{3}{4}}$.

$$\frac{1}{8} + \frac{5}{8} = \frac{3}{4}$$

Example 4 Subtract and reduce the answer: $5\frac{5}{6} - 2\frac{1}{6}$

Solution First we subtract.

$$5\frac{5}{6} - 2\frac{1}{6} = 3\frac{4}{6}$$

Then we reduce $3\frac{4}{6}$. We reduce a mixed number by reducing its fraction. Since the fraction $\frac{4}{6}$ reduces to $\frac{2}{3}$, the mixed number $3\frac{4}{6}$ reduces to $\mathbf{3\frac{2}{3}}$.

If an answer contains a fraction that can be reduced, we should reduce the fraction. Be alert for this as you work the problems in the problem sets.

LESSON PRACTICE

Practice set **a.** Reduce $\frac{8}{12}$ by dividing both 8 and 12 by 4.

b. Which of these fractions cannot be reduced?

A. $\frac{2}{8}$ B. $\frac{3}{8}$ C. $\frac{4}{8}$

Add, subtract, or multiply as indicated. Remember to reduce your answers.

c. $\frac{2}{3} \times \frac{1}{2}$ d. $\frac{3}{10} + \frac{3}{10}$ e. $\frac{3}{8} - \frac{1}{8}$

Rewrite each mixed number with a reduced fraction:

f. $1\frac{3}{9}$ g. $2\frac{6}{9}$ h. $2\frac{5}{10}$

Find each sum or difference. Remember to reduce your answers.

i. $1\frac{1}{4} + 2\frac{1}{4}$ j. $1\frac{1}{8} + 5\frac{5}{8}$ k. $5\frac{5}{12} - 1\frac{1}{12}$

MIXED PRACTICE

Problem set

1. (35) In 3 games Sherry's bowling scores were 109, 98, and 135. Her highest score was how much more than her lowest score?

2. (50) Find the average of the three bowling scores listed in problem 1.

3. (74) Liam is 5 feet 4 inches tall. How many inches is 5 feet 4 inches?

4. (68, 73) When twenty-six and five tenths is subtracted from thirty-two and six tenths, what is the difference?

5. (79) Write a fraction equal to $\frac{2}{3}$ that has a denominator of 12. Then write a fraction equal to $\frac{1}{4}$ that has a denominator of 12. What is the sum of the two fractions you made?

6. (80) List the prime numbers between 20 and 30.

7. (81) Reduce the fraction $\frac{10}{12}$ by dividing both 10 and 12 by 2.

8. (46, 81) One fourth of the 24 students earned A's on the test. One half of the students who earned A's on the test were girls.

(a) How many students earned A's on the test?

(b) How many girls earned A's on the test?

(c) What fraction of the 24 students were girls who earned A's on the test?

9. (44, 53) If the width of this rectangle is half its length, then what is the perimeter of the rectangle?

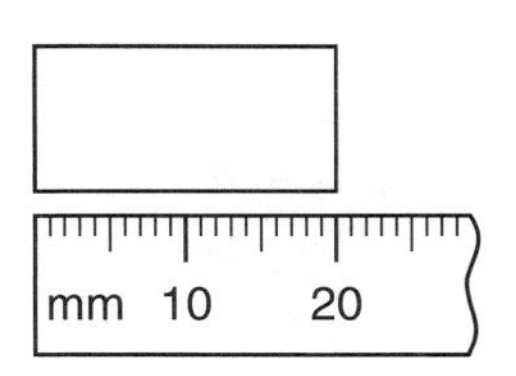

10. (44, 72) What is the area of the rectangle in problem 9?

11. (61) $\overline{QR}$ is 48 millimeters. Segment RS is one half as long as $\overline{QR}$. Find QS.

12. (73) $3.4 + 6.25$

13. (73) $6.25 - 3.4$

14. (78) The figure at right illustrates four squared (4^2). Using this model, draw a figure that illustrates three squared (3^2).

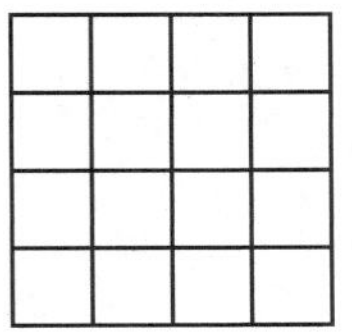

15. (34) $6\overline{)\$87.00}$

16. (54) $40\overline{)2438}$

17. (58) Divide 5280 by 9. Write the quotient as a mixed number with a reduced fraction.

18. (24, 70) $\$10 - (\$5.80 + 28¢)$

19. (59, 63) $5\frac{3}{5} + \left(4 - 1\frac{3}{5}\right)$

20. (81) Reduce: $\frac{3}{6}$

21. (76) $\frac{4}{3} \times \frac{1}{2}$

22. (76) $\frac{10}{7} \times \frac{7}{10}$

23. (28) From September 1 of one year to June 1 of the next year is how many months?

Use this information to answer problems 24 and 25:

> *Rosario has a paper route. She delivers papers to 30 customers. At the end of the month she gets $6.50 from each customer. She pays the newspaper company $135 each month for the newspapers.*

24. (21, Inv. 5) How much money does Rosario get each month from all her customers?

25. (16, Inv. 5) How much profit does she make each month for her work?

26. A standard number cube is rolled once.
(57, Inv. 6)

(a) What is the probability that the upturned face is an even number?

(b) Describe a different event that has the same probability.

27. The histogram below shows how many books some students read during the last year.
(Inv. 8)

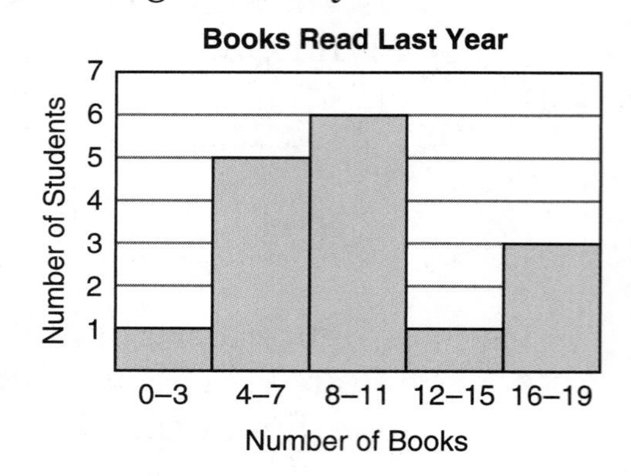

(a) How many students read 12 books or more?

(b) How many students read 15 books or fewer?

28. Which of these Venn diagrams illustrates the relationship between rectangles (R) and squares (S)? Explain your answer.
(45, Inv. 8)

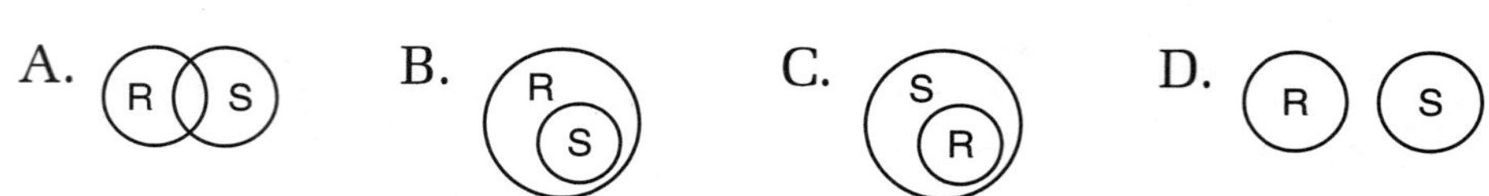

29. Write 15% as a fraction. Then reduce the fraction by dividing both terms by 5.
(71, 81)

30. Compare: $\frac{1}{2} \times \frac{1}{2} \bigcirc \frac{1}{2}$
(Inv. 2, 76)

LESSON

82 Greatest Common Factor (GCF)

WARM-UP

Facts Practice: 60 Improper Fractions to Simplify (Test H)

Mental Math:

In the expression 3(40 + 6), the sum of 40 and 6 is multiplied by 3. By using the *distributive property,* we can first multiply each addend and then add the partial products.

$$3(40 + 6)$$

$$120 + 18 = 138$$

Use the distributive property to solve problems **a** and **b.**

a. 3(20 + 7)

b. 4(30 + 6)

c. How many ounces are in one pound? ... two pounds?

d. Reduce: $\frac{2}{4}$, $\frac{2}{6}$, $\frac{2}{8}$, $\frac{2}{10}$

e. $\frac{1}{3}$ of 100

f. 25% of 60 minutes

g. $\sqrt{81}$, + 1, × 5, − 2, ÷ 4

Problem Solving:

A pizza was cut into 10 equal slices. Aretha ate four of the slices. What fraction of the pizza did Aretha eat?

NEW CONCEPT

We have practiced finding the factors of whole numbers. In this lesson we will practice finding the **greatest common factor** of two numbers. The greatest common factor of two numbers is the largest whole number that is a factor of both numbers. The letters **GCF** are sometimes used to stand for the term *greatest common factor.*

To find the greatest common factor of 12 and 18, we first list the factors of each. We have circled the common factors, that is, the numbers that are factors of both 12 and 18.

Factors of 12: ①, ②, ③, 4, ⑥, 12

Factors of 18: ①, ②, ③, ⑥, 9, 18

The greatest of these common factors is 6.

Example 1 Find the greatest common factor (GCF) of 8 and 20.

Solution We will first find the factors and identify the common factors. The factors of 8 and 20 are listed below with the common factors circled.

Factors of 8: ①, ②, ④, 8

Factors of 20: ①, ②, ④, 5, 10, 20

We see that there are three common factors. The greatest of the three common factors is **4.**

We may use greatest common factors to help us reduce fractions.

Example 2 Use the GCF of 8 and 20 to reduce $\frac{8}{20}$.

Solution In example 1 we found that the GCF of 8 and 20 is 4. So we can reduce $\frac{8}{20}$ by dividing both 8 and 20 by 4.

$$\frac{8 \div 4}{20 \div 4} = \mathbf{\frac{2}{5}}$$

LESSON PRACTICE

Practice set* Find the greatest common factor (GCF) of each pair of numbers:

a. 6 and 9 **b.** 6 and 12 **c.** 15 and 100

d. 6 and 10 **e.** 12 and 15 **f.** 7 and 10

Reduce each fraction by dividing the terms of the fraction by their GCF:

g. $\frac{6}{9}$ **h.** $\frac{6}{12}$ **i.** $\frac{15}{100}$

MIXED PRACTICE

Problem set

1. (49) Javier was paid $34.50 for working on Saturday. He worked from 8 a.m. to 2 p.m. He earned how much money per hour?

2. (62) Estimate the product of 396 and 507 by rounding to the nearest hundred before multiplying.

3. What is the next number in this counting sequence?
(1)

..., 3452, 3552, 3652, _____, ...

4. Most adults are between 5 and 6 feet tall. The height of most cars is about:
(74)

A. 4 to 5 feet   B. 8 to 10 feet   C. 40 to 50 feet

5. When sixty-five and fourteen hundredths is subtracted from eighty and forty-eight hundredths, what is the difference?
(68, 73)

6. If one side of a regular octagon is 12 inches long, then what is the perimeter of the octagon?
(32, 53)

7. Which of these numbers is not a prime number?
(80)

A. 11   B. 21   C. 31   D. 41

8. (a) Find the greatest common factor (GCF) of 20 and 30.
(82)

(b) Use the GCF of 20 and 30 to reduce $\frac{20}{30}$.

9. How many inches is $\frac{3}{4}$ of a foot?
(46, 74)

10. *AC* is 4 inches. *BC* is $\frac{3}{4}$ inch. Find *AB*.
(61, 63)

A B C

11. (a) What number is $\frac{1}{3}$ of 12?
(Inv. 3)

(b) What number is $\frac{2}{3}$ of 12?

12. Reduce: $\frac{6}{12}$
(81)

13. Compare: $2^3 \bigcirc 3^2$
(78)

14. $\frac{5}{7} + \frac{3}{7}$
(75)

15. $\frac{4}{4} - \frac{2}{2}$
(59)

16. $\frac{2}{3} \times \square = \frac{6}{9}$
(79)

17. (73)
$$\begin{array}{r} 976.5 \\ 470.4 \\ 436.7 \\ +\ 98.6 \\ \hline \end{array}$$

18. (13)
$$\begin{array}{r} \$40.00 \\ -\ \$32.85 \\ \hline \end{array}$$

19. (29)
$$\begin{array}{r} \$8.47 \\ \times\ 70 \\ \hline \end{array}$$

20. $6\overline{)43,715}$
(26)

21. $\frac{2640}{30}$
(54)

22. (55)
$$\begin{array}{r} 367 \\ \times\ 418 \\ \hline \end{array}$$

23. $3\frac{1}{4} + 3\frac{1}{4}$
(81)

24. $18.64 ÷ 4
(26)

Use the graph below to answer problems 25–27.

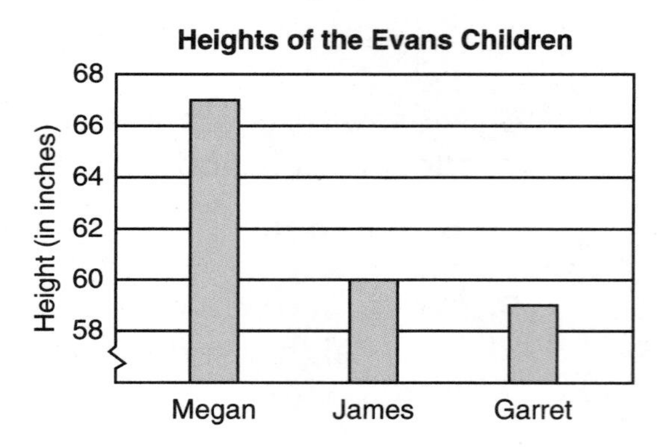

25. (Inv. 8) How many inches does Garret need to grow to be as tall as Megan?

26. (Inv. 8) Which child is exactly 5 feet tall?

27. (50, Inv. 8) What is the average height of the three children?

28. (57, 81) Find the probability that with one spin the spinner will not stop on A. Write the answer as a reduced fraction.

29. (45, Inv. 8) Which of these Venn diagrams illustrates the relationship between rectangles (R) and parallelograms (P)? Explain your answer.

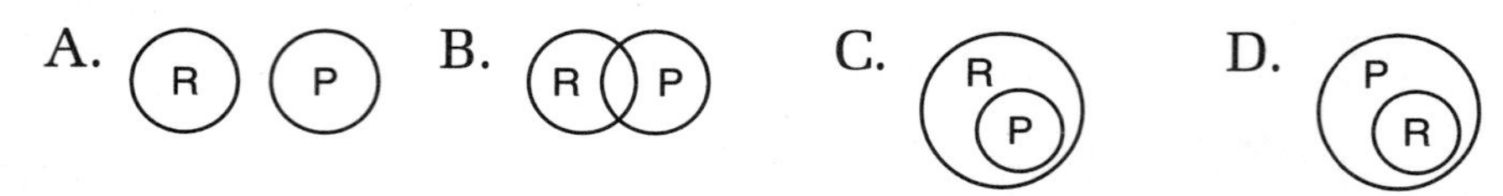

30. (71, 81) Write 22% as a fraction. Then reduce the fraction by dividing both terms by 2.

LESSON

83 Properties of Geometric Solids

WARM-UP

Facts Practice: 60 Improper Fractions to Simplify (Test H)

Mental Math:

Use the distributive property to solve problems **a** and **b.**

a. $6(20 + 3)$ **b.** $7(30 + 5)$ **c.** Reduce: $\frac{2}{8}$, $\frac{4}{8}$, $\frac{6}{8}$

d. How many inches are in one yard? How many feet are in one mile? Hold your hands about one yard apart. Hold your hands about one inch apart.

e. $33\frac{1}{3} + 33\frac{1}{3}$ **f.** 50% of $\sqrt{36}$, $\times$ 4, $\div$ 2, $\times$ 6

Problem Solving:

Carlos, Bao, and Sherry drew straws. Carlos's $3\frac{3}{4}$-inch straw was a quarter inch longer than Bao's straw and half an inch shorter than Sherry's straw. How long were Bao's and Sherry's straws?

NEW CONCEPT

We have practiced identifying geometric shapes such as triangles, rectangles, and circles. These are "flat" shapes and are called **plane figures.** They take up a certain amount of area, but they do not take up space. Objects that take up space are things like baseballs, houses, dogs, and people.

Geometric shapes that take up space are called **solids,** even though real-world objects that are similar to these shapes may not be "solid." We can make three-dimensional models of geometric solids, but they are difficult to draw on paper because paper is flat (it does not have depth). To give a sense of depth when drawing solids, we can include "hidden" edges and create optical illusions with carefully chosen angles. We draw and name some geometric solids in the table at right.

Geometric Solids

Shape	Name
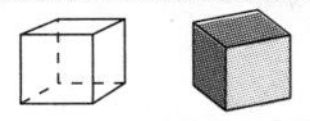	Cube
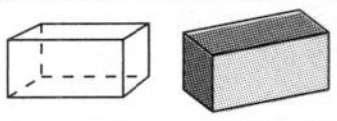	Rectangular solid
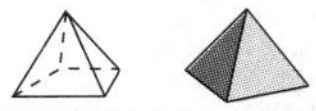	Pyramid
	Cylinder
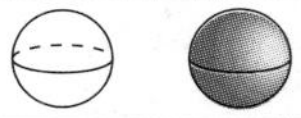	Sphere
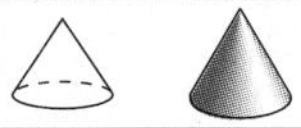	Cone

The flat surfaces of solids are called **faces.** A cube has six faces. The six faces of a standard dot cube, for example, each show a number of dots from 1 to 6. Two faces meet at an **edge.** A cube has 12 edges. Three edges meet at a **vertex.** The plural of vertex is *vertices.* A cube has eight vertices.

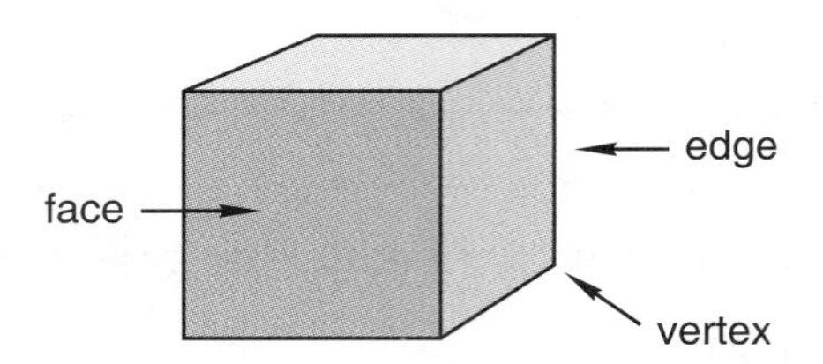

Example 1 (a) Name this shape.

(b) How many faces does it have?

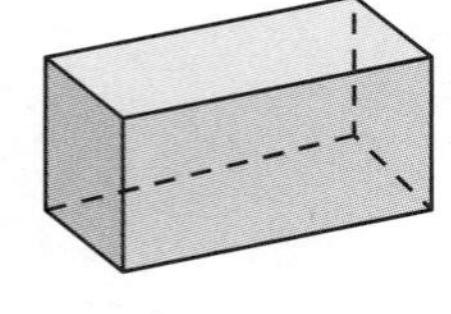

Solution (a) This shape is a **rectangular solid.**

(b) The solid has **6 faces.**

Example 2 What is the shape of a basketball?

Solution A basketball is not a circle. A circle is a "flat" shape (a plane figure), but a basketball takes up space. A basketball is a **sphere.**

LESSON PRACTICE

Practice set Name the geometric shape of each of these real-world objects:

a. brick

b. soup can

c. ice cream cone

d. shoebox

Refer to the pyramid to answer problems **e–h.**

e. The pyramid has how many triangular faces?

f. The pyramid has how many rectangular faces?

g. The pyramid has how many edges?

h. The pyramid has how many vertices?

MIXED PRACTICE

Problem set

1. *(28)* Alycia left for school at a quarter to eight in the morning and arrived home $7\frac{1}{2}$ hours later. What time was it when Alycia arrived home?

2. *(64, 70)* Mark has 5 coins in his pocket that total 47¢. How many dimes are in his pocket?

3. Use digits to write the number twenty-three million, two hundred eighty-seven thousand, four hundred twenty.
(52)

4. (a) What number is $\frac{1}{3}$ of 24?
(Inv. 3)
(b) What number is $\frac{2}{3}$ of 24?

5. List the prime numbers between 10 and 20.
(80)

6. (a) What is the greatest common factor (GCF) of 4 and 8?
(82)
(b) Use the GCF of 4 and 8 to reduce $\frac{4}{8}$.

7. (a) Name this shape.
(83)
(b) How many faces does it have?

8. Which geometric figure best describes the shape of the earth?
(83)

A. circle B. cylinder C. sphere D. plane

9. Write a decimal number equal to the mixed number $1\frac{7}{10}$.
(71)

10. Which word names the distance across a circle?
(53)

A. center B. circumference

C. radius D. diameter

11. $3.62 + 4.5$
(73)

12. $3.704 - 2.918$
(73)

13. 21^2
(78)

14. $\$6.25 \times 4$
(17)

15. $6w = \$14.58$
(26)

16. Write a fraction equal to $\frac{1}{3}$ that has a denominator of 12. Then write a fraction equal to $\frac{3}{4}$ that has a denominator of 12. What is the sum of the two fractions you wrote?
(79)

17. Reduce: $\frac{6}{8}$
(81)

18. $\frac{3}{4} = \frac{\square}{12}$
(79)

19. $4\frac{1}{6} + 2\frac{1}{6}$
(43, 71)

20. $3\frac{3}{4} + 1\frac{1}{4}$
(59)

21. $5 - 1\frac{1}{4}$
(63)

22. Compare: 0.1 ◯ 0.01
(69)

23. (75) The multiplication $3 \times \frac{1}{2}$ means $\frac{1}{2} + \frac{1}{2} + \frac{1}{2}$. So $3 \times \frac{1}{2}$ equals what mixed number?

Use the information and the table below to answer problems 24 and 25.

Mr. and Mrs. Minick took their children, Samantha and Douglas, to a movie. Ticket prices are shown in the table.

Movie Ticket Prices

Adults	$9.00
Ages 9–12	$4.50
Under 9	$3.75

24. (11, Inv. 5) Samantha is 12 years old and Douglas is 8 years old. What is the total cost of all four tickets?

25. (23, Inv. 5) Before 5 p.m., adult tickets are half price. How much money would the Minicks save by going to the movie before 5 p.m. instead of after 5 p.m.?

26. (83) Which of these figures is an illustration of an object that "takes up space"?

A. 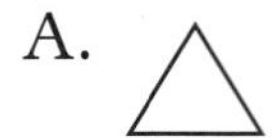 B. 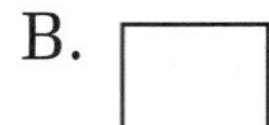 C. 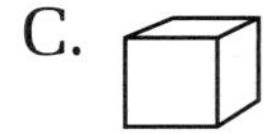 D. 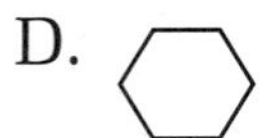

27. (62, 72) Estimate the area of a room that is 14 ft 2 in. long and 10 ft 3 in. wide.

28. (Inv. 8) The pie chart at right shows how a family's monthly expenses are divided.

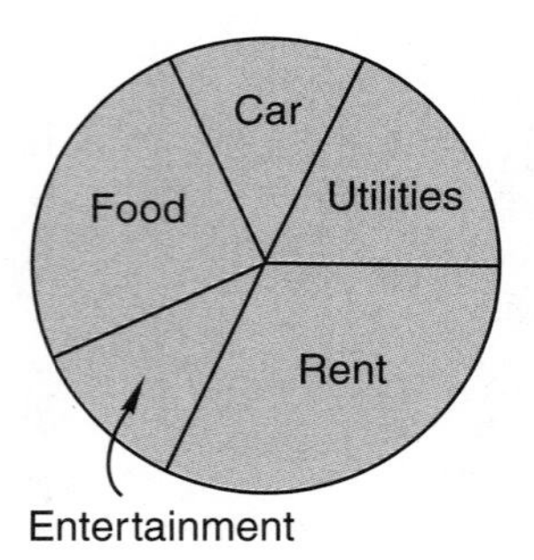

(a) Which expense consumes about one third of the budget?

(b) About what fraction of the budget does food consume?

29. (45, 73) What is the perimeter of a rhombus with sides 2.4 centimeters long?

30. (78) Light travels about 186,000 miles in one second. Write that number in expanded notation using powers of 10.

LESSON

84 Mean, Median, Mode, and Range

WARM-UP

Facts Practice: 60 Improper Fractions to Simplify (Test H)

Mental Math:

Use the distributive property to solve problems **a** and **b.**

a. 9(30 + 2) **b.** 8(30 + 4) **c.** Reduce: $\frac{2}{6}, \frac{3}{6}, \frac{4}{6}$

d. How many pounds are in one ton? How many pounds are in half a ton?

e. 10% of $500 **f.** $12\frac{1}{2} + 12\frac{1}{2}$

g. $\frac{1}{3}$ of 60, + 1, ÷ 3, × 5, + 1, ÷ 4

Problem Solving:

Is the total weight of the students in your class at least a ton?

NEW CONCEPT

In Lesson 50 we found the **average** of a set of numbers. The average is also called the **mean.** To find a mean, we add and then divide. For example, suppose we wanted to know the mean number of letters in the following names: Andrei, Raj, Althea, Mary, Bedros, Ann, and Yolanda. We would first add the seven numbers 6, 3, 6, 4, 6, 3, and 7. Then we would divide the resulting sum by 7.

Add: 6 + 3 + 6 + 4 + 6 + 3 + 7 = 35

Divide: 35 ÷ 7 = 5

The mean number of letters is 5. Notice that no name contains 5 letters. So the mean of a set of numbers does not have to be one of the numbers. In fact, the mean of a set of whole numbers can even be a mixed number.

Example 1 Find the mean of this data set: 2, 7, 3, 4, 3

Solution We divide the sum of the data points (19) by the number of data points (5). We write the remainder as a fraction and find that the mean of the data set is $\mathbf{3\frac{4}{5}}$.

$$\begin{array}{r} 3\frac{4}{5} \\ 5\overline{)19} \\ \underline{15} \\ 4 \end{array}$$

The middle number of a data set, when the data are arranged in numerical order, is called the **median.**

Example 2 Find the median of this data set: 3, 5, 8, 2, 5, 7, 7, 6, 1

Solution We first put the data in numerical order: 1, 2, 3, 5, 5, 6, 7, 7, 8. The middle object in a row of objects has the same number of objects on its left as it has on its right.

1 2 3 5 (5) 6 7 7 8

4 objects to the left — 4 objects to the right

We see that the median is **5.** If a data set has an even number of data points, there are two middle numbers. In these cases the median is the average of the two middle numbers.

Example 3 Find the median of this data set: 2, 5, 1, 6, 9, 8, 3, 10

Solution We arrange the numbers in numerical order to get the list 1, 2, 3, 5, 6, 8, 9, 10. The two middle numbers are 5 and 6. The median is the average of 5 and 6. We add 5 and 6 and then divide the resulting sum by 2.

1 2 3 (5 6) 8 9 10

$$\frac{5 + 6}{2} = \frac{11}{2} = 5\frac{1}{2}$$

The median is $\mathbf{5\frac{1}{2}}$.

Returning to our list of names at the beginning of this lesson, we find that the most common number of letters in a name is 6. There are three names with 6 letters: Andrei, Althea, and Bedros. If some data points occur more than once, then the one that occurs most often is called the **mode.** There can be more than one mode for a data set.

Example 4 Find the mode of each data set:

(a) 3, 5, 8, 2, 5, 7, 7, 6, 1 (b) 2, 5, 1, 6, 9, 8, 3, 10

Solution (a) The numbers 5 and 7 both appear twice. No other numbers appear more than once. So there are two modes, **5** and **7.**

(b) No data point appears more than once. This data set has **no mode.**

Mean, median, and mode are different ways to describe the *center* of a data set. They are called **measures of central tendency.** We might also be interested in the **spread** of a data set. Spread refers to how the data are stretched out. The simplest measure of spread is the **range,** which is the difference between the largest and smallest data points. For example, in the data set 3, 5, 8, 2, 5, 7, 7, 6, 1, the largest number is 8 and the smallest is 1. So the range for the data set is 7, because $8 - 1 = 7$.

LESSON PRACTICE

Practice set Find the mean, median, mode, and range of each data set in problems **a–c.**

a. 3, 7, 9, 9, 4

b. 16, 2, 5, 7, 11, 13

c. 3, 10, 2, 10, 10, 1, 3, 10

d. Find the mean, median, mode, and range for the ages of the students in this table:

Name	Andrei	Raj	Althea	Mary	Bedros	Ann	Yolanda
Age	13	10	10	11	11	10	11

MIXED PRACTICE

Problem set

1. (21) Tom and his two friends found $2418 of treasure buried in the cave. If they share the treasure equally, how much will each receive?

2. (Inv. 3, 37) Draw a circle and shade $\frac{1}{3}$ of it. What percent of the circle is shaded?

3. (77) This math book weighs about 1 kilogram. A kilogram is how many grams?

4. (62) Estimate the product of 732 and 480 by rounding the numbers to the nearest hundred before multiplying.

5. (28, 32) At which of these times do the hands of a clock form an acute angle?

A. 3:00 B. 6:15 C. 9:00 D. 12:10

6. Arrange these decimal numbers in order from least to greatest:
(69)

0.1, 0.01, 1.0, 1.01

7. (a) Find the greatest common factor (GCF) of 8 and 12.
(82)

(b) Use the GCF of 8 and 12 to reduce $\frac{8}{12}$.

8. (a) What number is $\frac{1}{4}$ of 80?
(Inv. 2)

(b) What number is $\frac{3}{4}$ of 80?

9. $\frac{1}{2} \times \square = \frac{3}{6}$
(79)

10. Reduce: $\frac{4}{6}$
(81)

11. Name the total number of shaded circles as a mixed number and as a decimal number.
(71)

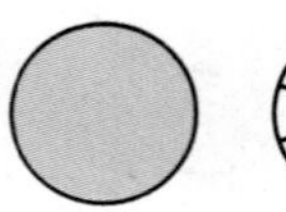
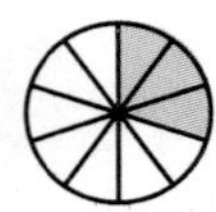

12. 9.9 + 6.14 + 7.5 + 8.31
(73)

13. \$10 − 59¢
(70)

14. $30\overline{)672}$
(54)

15. 5 × 68¢ = \$_____
(70)

16. \$3.40 ÷ 5
(26)

17. $10 - 3\frac{1}{3}$
(63)

18. $\frac{3}{4} \times \frac{5}{4}$
(76)

19. What is the name of this geometric solid?
(83)

20. In rectangle *MNOP*, which segment is parallel to $\overline{MN}$?
(31, 61)

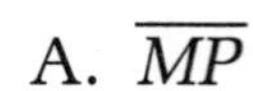

A. $\overline{MP}$ B. $\overline{PO}$

C. $\overline{NO}$ D. $\overline{MO}$

21. Which angle in this figure appears to be a right angle?
(32, 61)

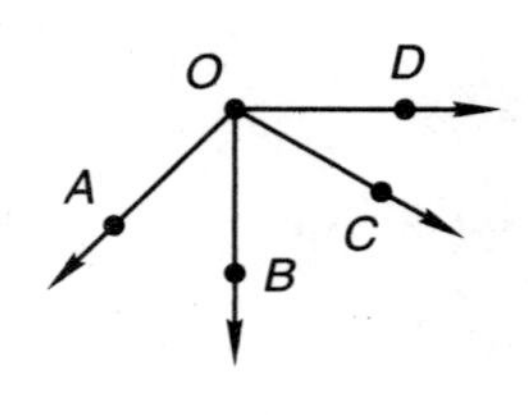

A. ∠*AOB* B. ∠*BOC*

C. ∠*BOD* D. ∠*AOD*

Use the grocery receipt to answer problems 22–26.

Milk	0.97
Milk	0.97
Milk	0.97
Milk	0.97
Apple juice	0.69
Apple juice	0.69
Eggs	1.55
Cereal	1.99
TOTAL	8.80

22. (73, Inv. 5) How much money was spent on eggs, juice, and cereal?

23. (50, Inv. 5) What was the average (mean) price of the eight items?

24. (Inv. 5, 84) What is the median price of the eight items?

25. (84) What is the mode of the prices?

26. (84) What is the range of the eight prices? Show your work.

27. (Inv. 7) The first three *triangular numbers* are 1, 3, and 6, as illustrated by the number of dots that make up each of the figures below. Find the next triangular number by drawing the next figure in the pattern.

28. (53, 59) Find the perimeter of this right triangle. Units are in inches.

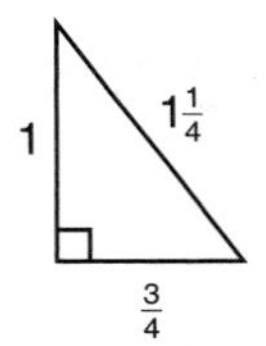

29. (71, 81) Write 90% as a fraction. Then reduce the fraction by dividing both terms by 10. What decimal number does the fraction equal?

30. (27) On the Fahrenheit scale, how many degrees are between the temperature at which water freezes and the temperature at which water boils?

LESSON

85 Converting Units of Capacity

WARM-UP

Facts Practice: 60 Improper Fractions to Simplify (Test H)

Mental Math:

Use the distributive property to solve problems **a** and **b**.

a. 3(2 pounds 4 ounces) **b.** 3(4 pounds 6 ounces)

c. How many ounces are in one pound? How many ounces are in half a pound?

d. Reduce: $\frac{2}{10}$, $\frac{4}{10}$, $\frac{6}{10}$, $\frac{8}{10}$ **e.** 10% of $500

f. $33\frac{1}{3} + 66\frac{2}{3}$ **g.** $\frac{1}{2}$ of 5, × 2, × 5, × 4

Problem Solving:

Draw and shade circles to show that $\frac{2}{3} < \frac{3}{4}$.

NEW CONCEPT

When we buy milk, soda pop, or fruit juice at the store, we are buying a quantity of liquid. In the U.S. Customary System, liquid quantities are measured in ounces (oz), pints (pt), quarts (qt), and gallons (gal). In the metric system, liquid quantities are measured in liters (L) and milliliters (mL). Here we show some common containers of liquids:

$\frac{1}{2}$ gallon

2 liters

1 quart

These cartons and bottles are said to have **capacity.** A container's capacity refers to the amount of liquid it can hold. Many containers in the U.S. Customary System are related by a factor of 2. One gallon is 2 half gallons. A half gallon is 2 quarts.

A quart is 2 pints. A pint is 2 cups. We show these relationships in the following diagram:

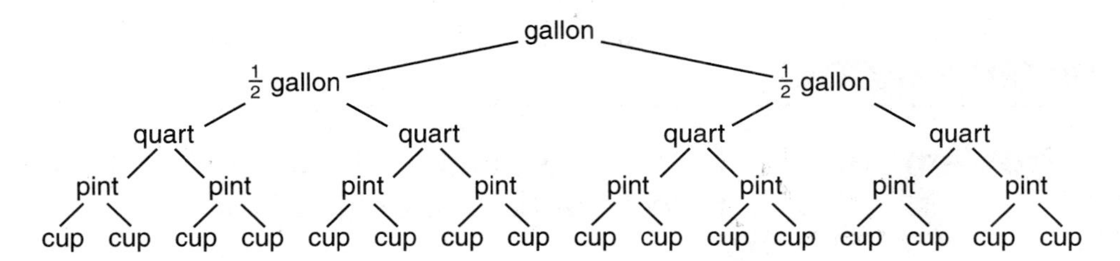

The table below shows some common units of liquid measure. The table also shows equivalences between the units.

Equivalence Table for Units of Liquid Measure

U.S. Customary System	Metric System
16 oz = 1 pt 2 pt = 1 qt 4 qt = 1 gal	1000 mL = 1 L
A liter is about 2 ounces more than a quart.	

Example 1 One quart of juice is how many ounces of juice?

Solution The table tells us that a quart is 2 pints and that each pint is 16 ounces. Since 2 times 16 is 32, 1 quart is the same as **32 ounces.**

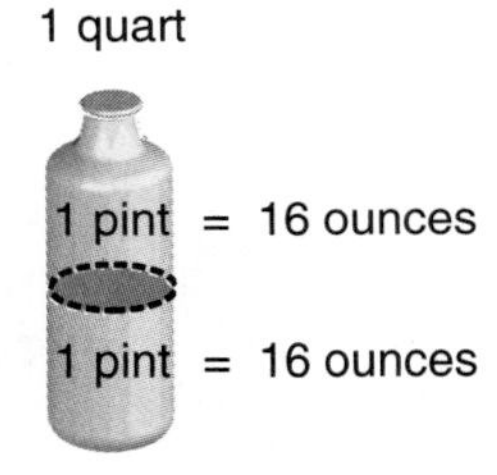

Note: The word *ounce* is used to describe a weight as well as an amount of liquid. The liquid measurement *ounce* is often called a **fluid ounce.** Although *ounce* has two meanings, a fluid ounce of water does weigh about 1 ounce.

Example 2 A half gallon of milk is how many quarts of milk?

Solution A whole gallon is equal to 4 quarts. A half gallon is equal to half as many quarts. A half gallon equals **2 quarts.**

LESSON PRACTICE

Practice set

a. One fourth of a dollar is a quarter. What is the name for one fourth of a gallon?

b. How many pints equal 1 gallon?

c. How many milliliters equal 2 liters?

d. A cup is one half of a pint. A cup is how many ounces?

MIXED PRACTICE

Problem set

1. Draw a rectangle. Shade all but two fifths of it. What percent of the rectangle is shaded?
(71)

2. Write a three-digit prime number using the digits 4, 1, and 0 once each.
(80)

3. Find the length of this segment in centimeters and in millimeters.
(66)

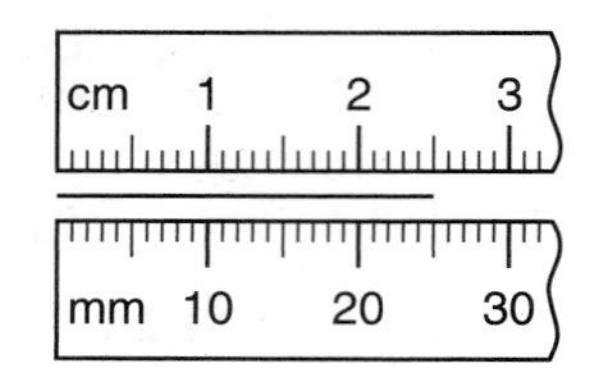

4. Tisha counted her heartbeats. Her heart beat 20 times in 15 seconds. At that rate, how many times would it beat in 1 minute?
(49)

5. In this quadrilateral, which segment appears to be perpendicular to $\overline{AB}$?
(31, 61)

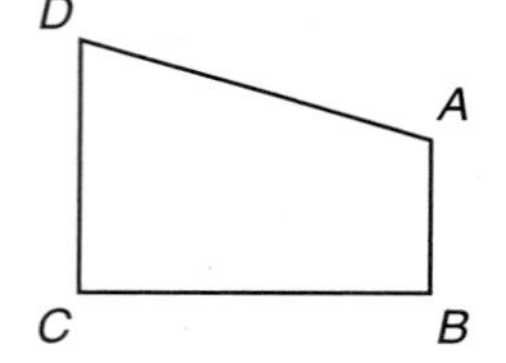

A. $\overline{BC}$ B. $\overline{CD}$ C. $\overline{DA}$

6. (a) Find the greatest common factor (GCF) of 6 and 9.
(82)
(b) Use the GCF of 6 and 9 to reduce $\frac{6}{9}$.

7. (a) What number is $\frac{1}{5}$ of 60?
(Inv. 3)
(b) What number is $\frac{2}{5}$ of 60?

8. AB is $1\frac{1}{4}$ inches. BC is $2\frac{1}{4}$ inches. Find AC.
(61, 81)

9. Arrange these numbers in order from least to greatest:
(69)

0.1, 0, 0.01, 1.0

10. Four quarts of water is how many pints of water?
(85)

11. Three liters equals how many milliliters?
(85)

12. Divide 100 by 6 and write the quotient as a mixed number. Then rewrite the quotient by reducing the fraction part of the mixed number.
(58, 81)

13. \$17.56 + \$12 + 95¢
(70)

14. 4.324 − 1.91
(73)

15. $\begin{array}{r} 396 \\ \times\ 405 \\ \hline \end{array}$
(56)

16. \$1.25 × 20
(29)

17. $9\overline{)3605}$
(34)

18. \$2.50 ÷ 10
(54)

19. Reduce: $\frac{15}{20}$
(81)

20. $3 - \left(2\frac{2}{3} - 1\right)$
(43, 63)

21. Write a fraction equal to $\frac{3}{5}$ that has a denominator of 10. Then write a fraction equal to $\frac{1}{2}$ that has a denominator of 10. What is the sum of the two fractions you wrote?
(75, 79)

22. Find the sum when five and twelve hundredths is added to six and fifteen hundredths. Write your answer as a decimal number.
(68, 73)

23. Since $\frac{1}{4} + \frac{1}{4} + \frac{1}{4} = \frac{3}{4}$, how many $\frac{1}{4}$'s are in $\frac{3}{4}$?
(Inv. 2)

Use this information to answer problems 24 and 25:

Stan is 6 inches taller than Roberta. Roberta is 4 inches shorter than Pedro. Pedro is 5 feet 3 inches tall.

24. How tall is Roberta?
(49)

25. How tall is Stan?
(49)

26. The first term of a certain sequence is 4. Each term that follows is found by doubling the number before it. Write the first four terms of the sequence.
(Inv. 7)

27. If you toss a coin 50 times, about how many times would you expect it to land heads up?
(Inv. 6)

The scores of Katie's first seven tests are listed below. Use this information to answer problems 28–30.

90, 85, 80, 90, 95, 90, 100

28. What is the range of the scores?
(84)

29. What is the mode of the scores? Why?
(84)

30. What is the median score? Why?
(84)

LESSON

86 Multiplying Fractions and Whole Numbers

WARM-UP

Facts Practice: 60 Improper Fractions to Simplify (Test H)

Mental Math:

Use the distributive property to solve problems **a** and **b**.

a. 3($6 and 25¢) **b.** 5($3 and 25¢)

c. How many quarters are in a dollar? How many quarts are in a gallon?

d. Reduce: $\frac{3}{6}$, $\frac{3}{9}$, $\frac{3}{12}$, $\frac{3}{15}$ **e.** 25% of $60

f. $\frac{1}{3}$ of 90, + 3, ÷ 3, × 9

Problem Solving:

Marissa is covering a 5-by-3-foot bulletin board with blue and gold construction paper squares, making a checkerboard pattern. Each square is 1 foot by 1 foot. Copy this diagram on your paper, and complete the checkerboard pattern. How many squares of each color does Marissa need?

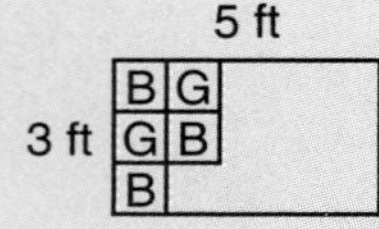

NEW CONCEPT

We have found a fraction of a whole number by dividing the whole number by the denominator of the fraction.

$$\frac{1}{3} \text{ of } 6 \text{ is } 2. \quad (6 \div 3 = 2)$$

In this lesson we will find a fraction of a whole number by multiplying.

$$\text{What number is } \frac{1}{3} \text{ of } 2?$$

We know that $\frac{1}{2}$ of 2 is 1. Since $\frac{1}{3}$ is less than $\frac{1}{2}$, $\frac{1}{3}$ of 2 must be less than 1. We can find the answer by multiplying.

$$\frac{1}{3} \text{ of } 2$$

$$\downarrow \quad \downarrow \quad \downarrow$$

$$\frac{1}{3} \times \frac{2}{1}$$

Notice that we wrote the whole number 2 as a fraction, $\frac{2}{1}$. Since 2 divided by 1 is 2, the fraction $\frac{2}{1}$ equals 2. Writing the whole number as a fraction gives us a numerator and a denominator to multiply. The product is $\frac{2}{3}$.

$$\frac{1}{3} \times \frac{2}{1} = \frac{2}{3}$$

Recall the commutative property of multiplication. This property tells us that changing the order of factors does not affect the product. So another way to approach this problem is to switch the positions of $\frac{1}{3}$ and 2.

$$\frac{1}{3} \times 2$$

$$2 \times \frac{1}{3}$$

We may reverse the order of factors when we multiply.

Since $2 \times \frac{1}{3}$ means $\frac{1}{3} + \frac{1}{3}$, we again find that the product is $\frac{2}{3}$.

Example What number is $\frac{2}{3}$ of 4?

Solution We know that $\frac{2}{3}$ of 4 is greater than 2 because $\frac{1}{2}$ of 4 is 2, and $\frac{2}{3}$ is greater than $\frac{1}{2}$. We also know that $\frac{2}{3}$ of 4 is less than 4. We multiply to find the answer.

$$\frac{2}{3} \text{ of } 4$$

$$\downarrow \quad \downarrow \quad \downarrow$$

$$\frac{2}{3} \times \frac{4}{1} = \frac{8}{3} = \mathbf{2\frac{2}{3}}$$

We converted the improper fraction to a mixed number. Since $2\frac{2}{3}$ is greater than 2 but less than 4, the answer is reasonable. We can check the answer by reversing the order of factors.

$$4 \times \frac{2}{3} \quad \text{means} \quad \frac{2}{3} + \frac{2}{3} + \frac{2}{3} + \frac{2}{3}$$

Again we get $\frac{8}{3}$, which equals $2\frac{2}{3}$.

LESSON PRACTICE

Practice set* Multiply. Simplify answers when possible. Reverse the order of factors to check your answer.

a. $\frac{1}{3} \times 4$ **b.** $\frac{3}{5} \times 2$ **c.** $\frac{2}{3} \times 2$

d. What number is $\frac{1}{5}$ of 4?

e. What number is $\frac{1}{6}$ of 5?

f. What number is $\frac{2}{3}$ of 5?

MIXED PRACTICE

Problem set

1. (31, 32) Draw a pair of horizontal parallel segments. Make the lower segment longer than the upper segment. Make a quadrilateral by connecting the endpoints.

2. (62) Estimate the difference of 6970 and 3047 by rounding the numbers to the nearest thousand and then subtracting.

3. (6) Write the following sentence using digits and symbols:

The sum of six and four is ten.

4. (85) A 2-liter bottle of soft drink contains how many milliliters of liquid?

5. (71) Name the shaded portion of this square as a fraction, as a decimal number, and as a percent.

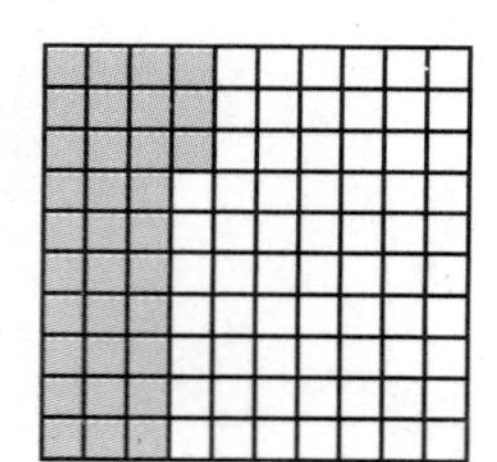

6. (86) (a) What number is $\frac{1}{3}$ of 120?

(b) What number is $\frac{2}{3}$ of 120?

7. (53, 61) Which segment names a diameter of this circle?

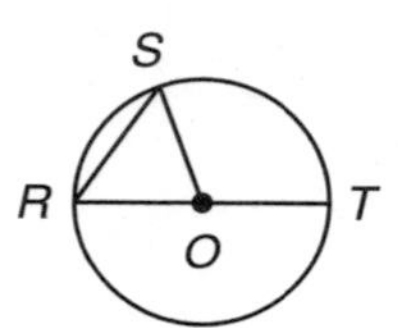

A. $\overline{RS}$ B. $\overline{RT}$

C. $\overline{OS}$ D. $\overline{OT}$

8. List these fractions in order from least to greatest:
(23, 59, 75)

$$\frac{9}{18}, \frac{8}{7}, \frac{7}{16}, \frac{6}{6}, \frac{5}{8}$$

9. To what mixed number is the arrow pointing?
(38)

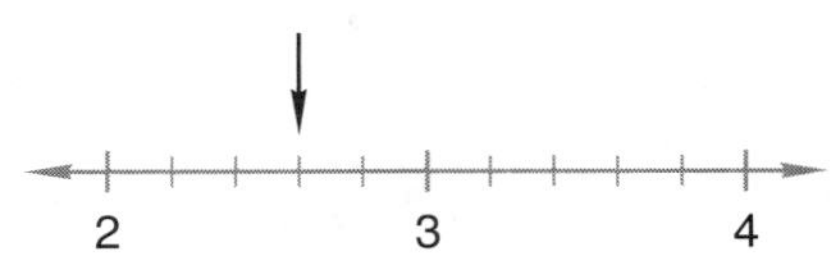

Multiply to find each product in problems 10 and 11. Then reverse the order of factors to check your answers.

10. $\frac{2}{3} \times 2$
(86)

11. $\frac{3}{4}$ of 4
(86)

12. $3 - \left(2\frac{3}{5} - 1\frac{1}{5}\right)$
(41, 63)

13. $4.7 + 3.63 + 2.0$
(73)

14. $301.4 - 143.5$
(73)

15. 476×890
(56)

16. $4\overline{)348}$
(26)

17. $40\overline{)3480}$
(54)

18. $\$42.36 \div 6$
(34)

19. 22^2
(78)

20. (a) What is the greatest common factor (GCF) of 60 and 100?
(82)

(b) Use the GCF of 60 and 100 to reduce $\frac{60}{100}$.

21. Write a fraction equal to $\frac{3}{4}$ that has a denominator of 12. Then write a fraction equal to $\frac{2}{3}$ that has a denominator of 12. Subtract the second fraction from the first fraction.
(79)

22. Since $\frac{3}{4} + \frac{3}{4} + \frac{3}{4} = \frac{9}{4}$, how many $\frac{3}{4}$'s are in $\frac{9}{4}$?
(Inv. 2)

23. (a) What is the name of this solid?
(83)

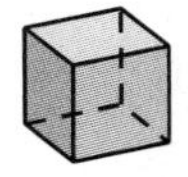

(b) How many vertices does it have?

Use the graph below to answer problems 24–26.

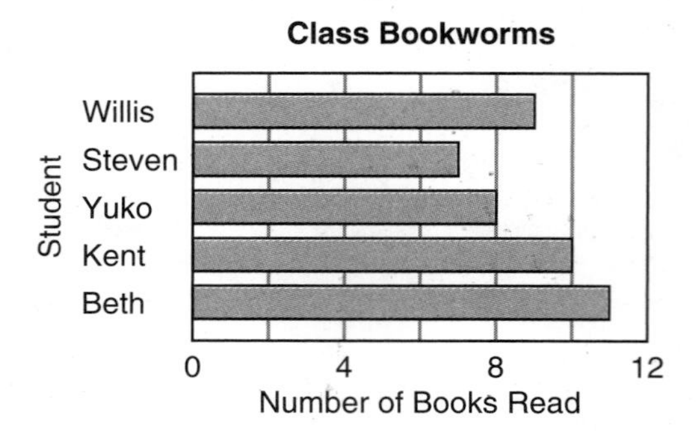

24. (Inv. 8) How many more books must Steven read to reach the goal of 12 books?

25. (Inv. 8) Each book must have 180 pages or more. Kent has read at least how many pages so far?

26. (Inv. 8, 84) What is the median number of books read by the five students?

27. (57, 81) What is the probability of rolling a number less than five with one toss of a standard number cube? Write the probability as a reduced fraction.

28. (71, 85) A quart is called a quart because it is a quarter of a gallon. What percent of a gallon is a quart?

29. (85) Compare: 1 quart ◯ 1 liter

30. (27, Inv. 7) A difference of 100° on the Celsius scale is a difference of 180° on the Fahrenheit scale. So a 10° change on the Celsius scale is an 18° change on the Fahrenheit scale. Copy this thermometer on your paper, and label the remaining tick marks on the Fahrenheit scale.

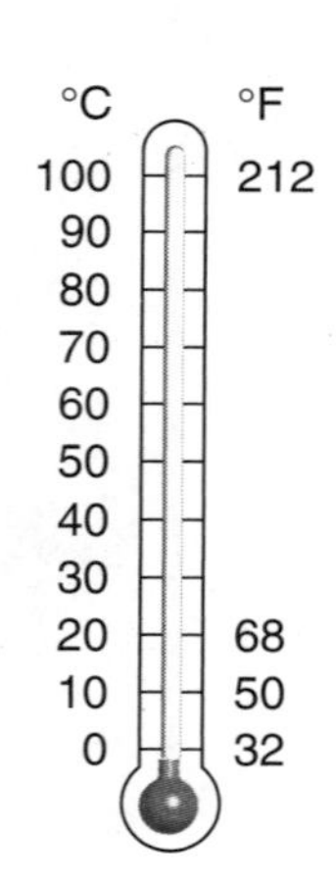

LESSON

87 Using Manipulatives and Sketches to Divide Fractions

WARM-UP

Facts Practice: 60 Improper Fractions to Simplify (Test H)

Mental Math:

Use the distributive property to solve problems **a** and **b.**

a. 3(2 ft 4 in.) **b.** 4(3 ft 4 in.)

c. How many quarts are in one gallon? ... two gallons? ... three gallons?

d. Reduce: $\frac{2}{10}$, $\frac{2}{12}$, $\frac{2}{14}$, $\frac{2}{16}$

e. $\frac{1}{4}$ of 400, ÷ 2, − 5, ÷ 5, × 4, ÷ 6

Problem Solving:

A **permutation** is an arrangement of numbers or objects in a particular order. For example, if we take the combination (1, 2, 3) from the set of counting numbers, we can form six permutations. Four of the permutations are (1, 2, 3), (1, 3, 2), (2, 1, 3), and (2, 3, 1). What are the remaining two permutations for these three numbers?

NEW CONCEPT

In this lesson we will use our fraction manipulatives and make sketches to help us divide fractions. First, let us think about what dividing fractions means. The expression

$$\frac{3}{4} \div \frac{1}{8}$$

means, "How many one eighths are in three fourths?" For example, how many one-eighth slices of pizza are in three fourths of a pizza?

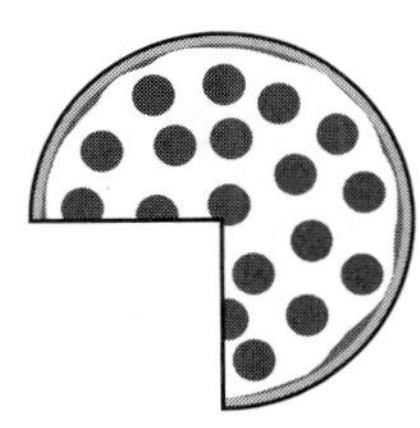

Using manipulatives, we place three fourths on our desk.

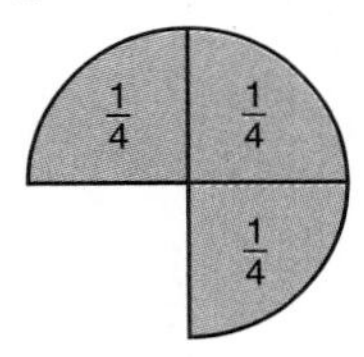

If we cover the three fourths with eighths, we can see that there are 6 one eighths in three fourths.

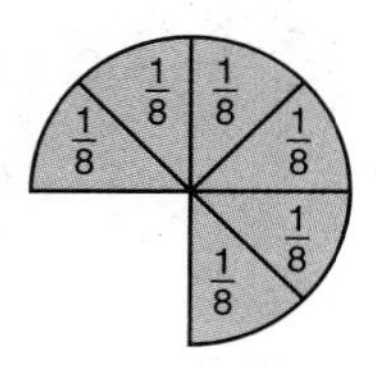

$$\frac{3}{4} \div \frac{1}{8} = 6$$

Example 1 How many one eighths are in one half?

Solution This is a division question. It could also be written

$$\frac{1}{2} \div \frac{1}{8}$$

Using our fraction manipulatives, we place one half on our desk.

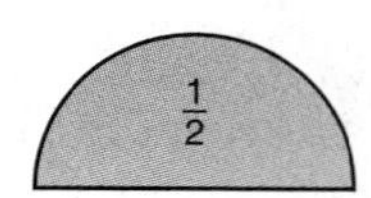

To find how many one eighths are in one half, we cover the half with eighths and then count the eighths.

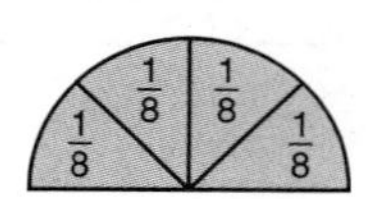

The answer is **4.** There are 4 one eighths in one half.

Example 2 Divide: $\frac{3}{4} \div \frac{1}{4}$

Solution This problem means, "How many one fourths are in three fourths?" We make three fourths of a circle using fourths. Then we count the fourths.

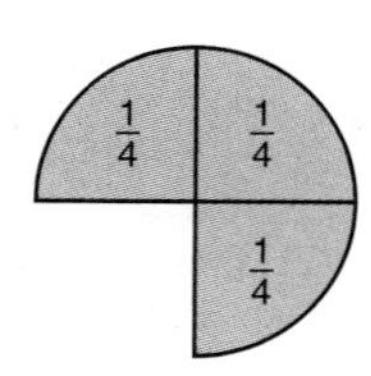

There are 3 one fourths in three fourths.

$$\frac{3}{4} \div \frac{1}{4} = \mathbf{3}$$

Example 3 Divide: $1 \div \frac{1}{3}$

Solution This problem means, "How many one thirds are in one?" Using our manipulatives, we want to find the number of one-third pieces needed to make one whole circle.

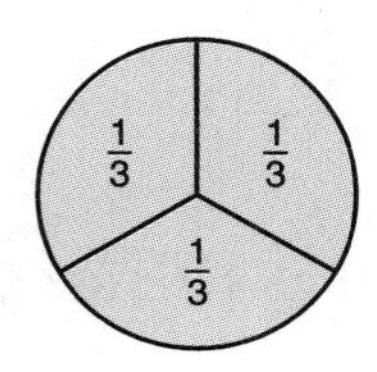

There are 3 one thirds in one.

$$1 \div \frac{1}{3} = \mathbf{3}$$

We can use the image of a clock face to help us sketch models for twelfths and sixths. We draw a circle and make twelve tick marks where the numbers 1 through 12 are positioned. To show twelfths, we draw segments from the center of the circle to each of the tick marks. To show sixths, we draw segments only to the tick marks for 2, 4, 6, 8, 10, and 12.

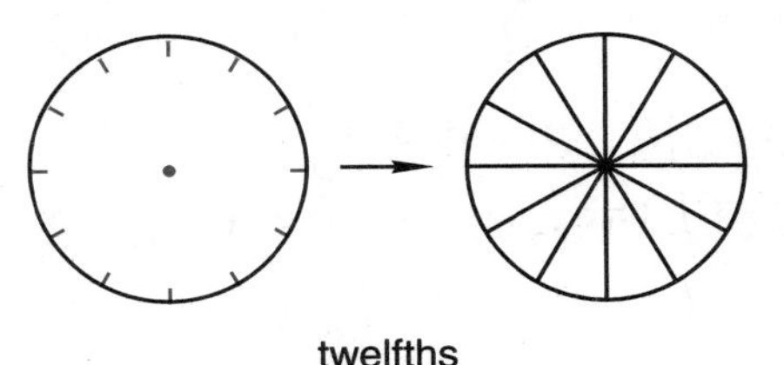

twelfths sixths

Example 4 Make a sketch to show the division $\frac{1}{4} \div \frac{1}{12}$. What is the quotient?

Solution We draw a circle divided into twelfths. Then we lightly shade $\frac{1}{4}$ of the circle. To find how many $\frac{1}{12}$'s are in $\frac{1}{4}$, we count the number of $\frac{1}{12}$'s in the shaded portion of the circle. We find that the quotient is **3**.

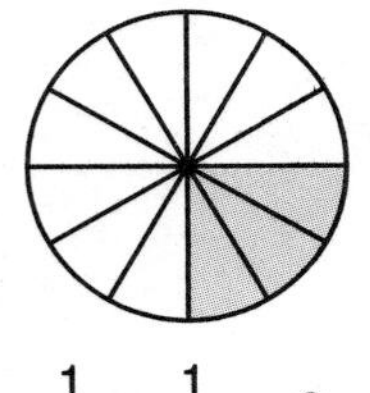

$\frac{1}{4} \div \frac{1}{12} = 3$

LESSON PRACTICE

Practice set Make sketches to help you answer problems **a** and **b**.

a. How many one sixths are in one half?

b. How many one twelfths are in one third?

Find each quotient. Try answering the problems mentally.

c. $\frac{2}{3} \div \frac{2}{3}$ **d.** $1 \div \frac{1}{4}$ **e.** $\frac{2}{3} \div \frac{1}{3}$ **f.** $1 \div \frac{1}{2}$

MIXED PRACTICE

Problem set

1. (53) Mary's rectangular garden is twice as long as it is wide. Her garden is 10 feet wide. What is the perimeter of her garden?

2. (72) What is the area of the garden described in problem 1?

3. (68) In which of these numbers does the 1 mean $\frac{1}{10}$? Use words to name the number.

A. 12.34 B. 21.43 C. 34.12 D. 43.21

4. (69, 71) Arrange these numbers in order from least to greatest:

$$1, 0, \frac{1}{2}, 0.3$$

5. (85) Two quarts of juice is how many ounces of juice?

6. (21, 30) (a) A quarter is what fraction of a dollar?

(b) How many quarters equal 1 dollar?

(c) How many quarters equal 3 dollars?

7. (71) Name the shaded portion of this rectangle as a fraction, as a decimal number, and as a percent.

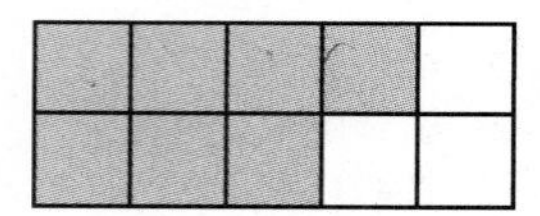

8. (17) If $a = 3$, then $2a + 5$ equals which of the following?

A. 10 B. 11 C. 16 D. 28

9. (61) AC is 84 millimeters. AB is one fourth of AC. Find BC.

A B C

10. (79) Write a fraction equal to $\frac{1}{2}$ that has a denominator of 6. Then write a fraction equal to $\frac{1}{3}$ that has a denominator of 6. Subtract the second fraction from the first fraction.

11. (82) (a) Find the greatest common factor (GCF) of 20 and 50.

(b) Use the GCF of 20 and 50 to reduce $\frac{20}{50}$.

12. (86) $\frac{3}{5}$ of 4

13. (87) $\frac{1}{2} \div \frac{1}{12}$

14. (81) $3\frac{7}{8} - 1\frac{1}{8}$

15. $2250 \div 50$
(54)

16. $5\overline{)225}$
(26)

17.
(6)
$$\begin{array}{r} 5365 \\ 428 \\ 3997 \\ 659 \\ 7073 \\ +\ 342 \\ \hline \end{array}$$

18. $4\overline{)\$8.20}$
(34)

19. 20^2
(78)

20.
(29)
$$\begin{array}{r} \$12.75 \\ \times\ \ \ \ 80 \\ \hline \end{array}$$

21. Divide 100 by 8 and write the quotient as a mixed number. Then rewrite the quotient by reducing the fraction part of the mixed number.
(58, 81)

22. How many one eighths are in one fourth?
(87)

23. Since $\frac{2}{3} + \frac{2}{3} + \frac{2}{3} = 2$, how many $\frac{2}{3}$'s are in 2?
(87)

The map below shows the number of miles between towns. Use this map to answer problems 24 and 25.

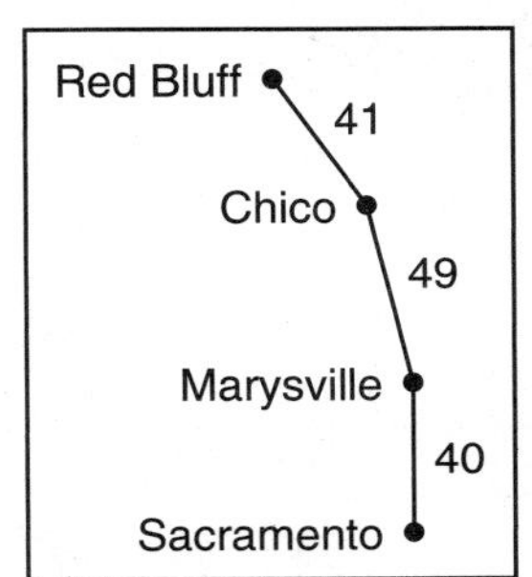

24. The distance from Marysville to Red Bluff is how many miles greater than the distance from Marysville to Sacramento?
(35, Inv. 5)

25. Allen was traveling from Sacramento to Chico. When he was halfway to Marysville, how far did he still have to go to get to Chico?
(11, 23, Inv. 5)

Kenji asked 20 classmates whether they ate eggs or cereal (or neither) for breakfast. He displayed the answers in the Venn diagram at right. Use this information to answer problems 26–30.

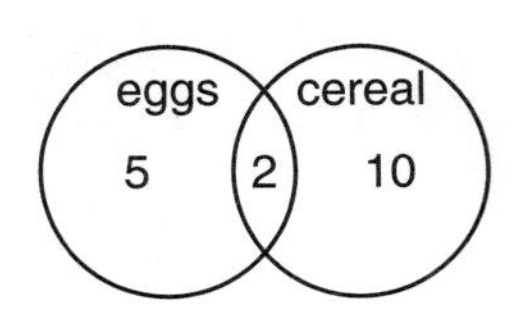

26. How many students ate cereal for breakfast?
(Inv. 8)

27. How many students ate eggs for breakfast?
(Inv. 8)

28. How many students ate both cereal and eggs for breakfast?
(Inv. 8)

29. Altogether, how many students ate eggs or cereal or both?
(Inv. 8)

30. How many students ate neither eggs nor cereal for breakfast?
(Inv. 8)

LESSON

88 Transformations

WARM-UP

Facts Practice: 60 Improper Fractions to Simplify (Test H)

Mental Math:

Use the distributive property to solve problems **a** and **b**.

a. 5(30 + 4) **b.** 5(34) **c.** Reduce: $\frac{3}{12}$, $\frac{10}{12}$, $\frac{9}{12}$

d. How many ounces are in a pound? How many ounces are in a pint? What is the meaning of the rhyme, "A pint's a pound the world around"?

e. 50% of 50, − 1, ÷ 3, + 2, × 10

Problem Solving:

Some 1-inch cubes were used to build this rectangular solid. How many 1-inch cubes were used?

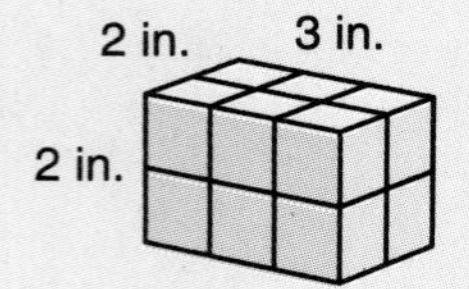

NEW CONCEPT

Recall that two figures are **congruent** if one figure has the same shape and size as the other figure. One way to decide whether two figures are congruent is to position one figure "on top of" the other. The two figures below are congruent.

To position the left-hand figure on the right-hand figure, we make three different kinds of moves. First, we **rotate** (turn) the left-hand figure a quarter turn.

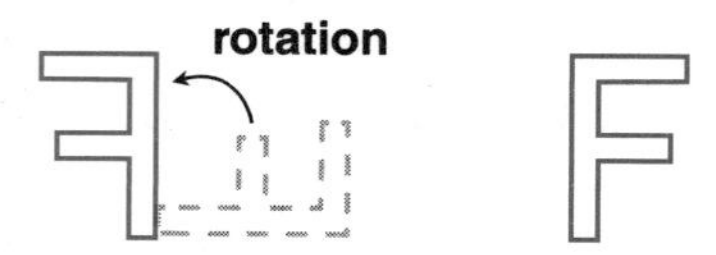

Second, we **translate** (slide) the left-hand figure so that the two figures are back-to-back.

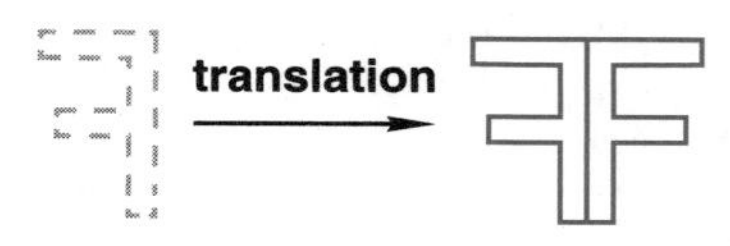

Third, we **reflect** (flip) the left-hand figure so that it is positioned on top of the right-hand figure.

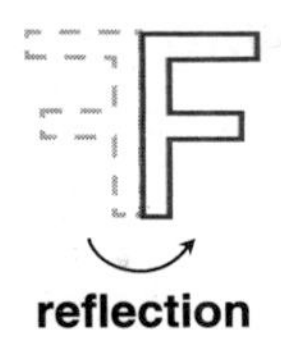

reflection

The three different kinds of moves we made are called **transformations.** We list them in the following table:

Transformations

Name	Movement
Translation	sliding a figure in one direction without turning the figure
Reflection	reflecting a figure as in a mirror or "flipping" a figure over a certain line
Rotation	turning a figure about a certain point

Rotations can be described by their degree and direction.

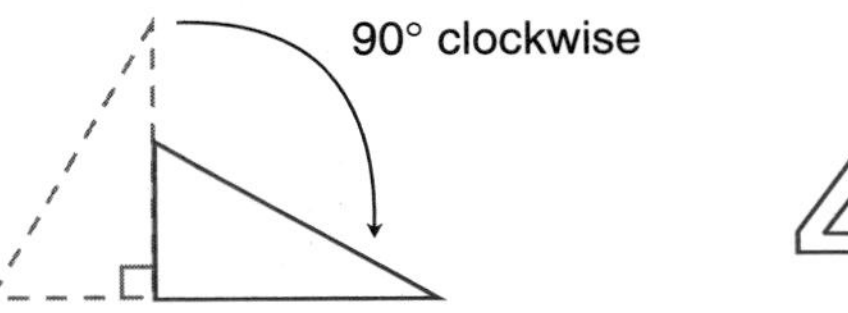

A full turn is 360°. A half turn is 180°. If you make a 180° turn, you will face in the opposite direction. A quarter turn is 90°. Clockwise is the direction the hands of a clock turn (to the right). Counterclockwise is the opposite direction (to the left).

Example 1 The figure at right is a number 3. Sketch the figure after a 90° clockwise rotation.

3

Solution A 90° clockwise rotation is a quarter turn to the right.

Example 2 Triangles *ABC* and *XYZ* are congruent. Name the transformations that would move triangle *ABC* to the position of triangle *XYZ*.

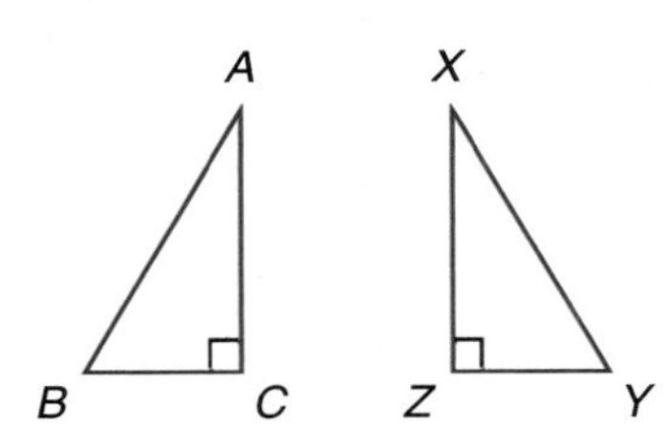

Solution If we **reflect** triangle ABC in line AC, then triangle ABC will have the same orientation as triangle XYZ. Then we **translate** triangle ABC to the position of triangle XYZ.

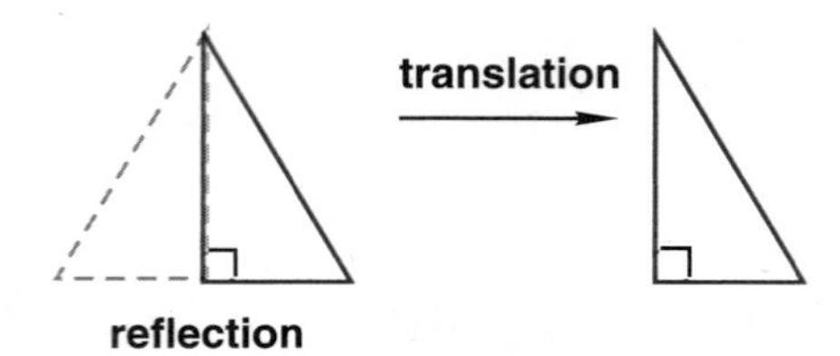

LESSON PRACTICE

Practice set

a. Sketch an uppercase letter R after a reflection in its vertical segment.

R

b. Sketch an uppercase letter R after a 180° counterclockwise rotation.

Name the transformations that could be used to position triangle A on triangle B:

c.

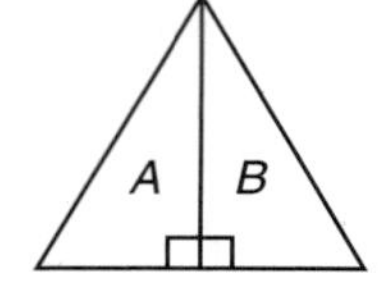

d.

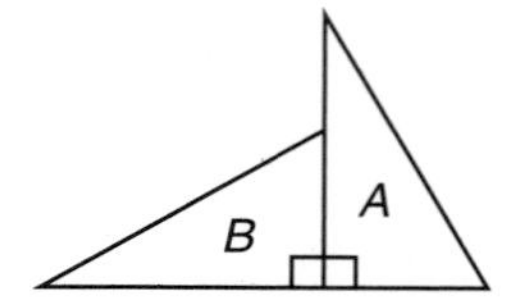

e.

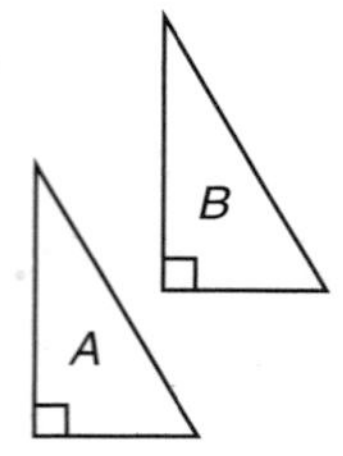

f.

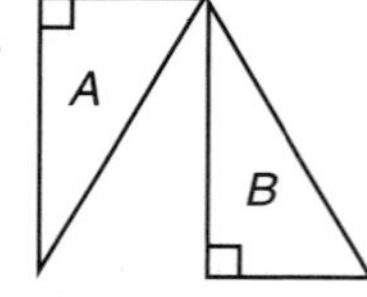

MIXED PRACTICE

Problem set

1. (81) Pam lives $\frac{1}{4}$ mile from school. How far does she travel each day going to school and back?

2. (28) According to this calendar, what is the date of the first Friday in April 2070?

MARCH 2070

S	M	T	W	T	F	S
						1
2	3	4	5	6	7	8
9	10	11	12	13	14	15
16	17	18	19	20	21	22
23	24	25	26	27	28	29
30	31					

3. (73) When the decimal number three and twelve hundredths is subtracted from four and twenty-five hundredths, what is the difference?

4. (a) How many dimes equal \$1?
(21, 30)
(b) How many dimes equal \$5?

5. What number is $\frac{2}{3}$ of 150?
(86)

6. A half gallon of milk is how many quarts of milk?
(85)

7. Which part of a bicycle wheel is most like a radius?
(53)

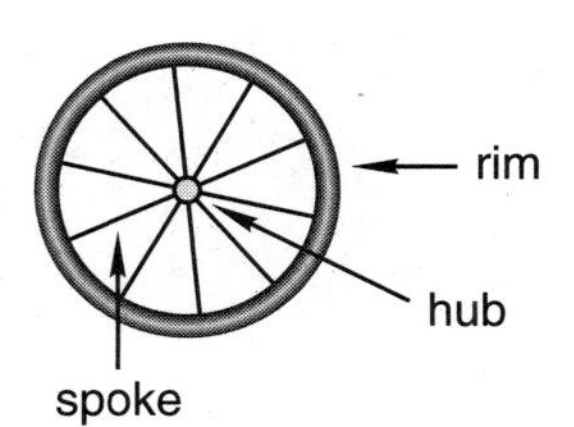

A. rim B. spoke C. hub

8. Write a fraction equal to one third that has a denominator of six. Then subtract that fraction from five sixths. Remember to reduce the answer.
(79)

9. (a) What fraction of this rectangle is shaded? Reduce the answer.
(71, 81)

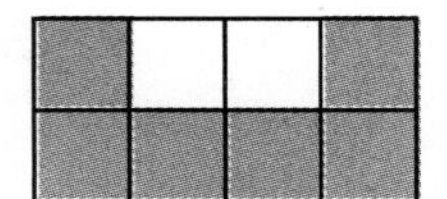

(b) What percent of this rectangle is shaded?

10. *RT* is 84 millimeters. *RS* is one third of *RT*. Find *ST*.
(61)

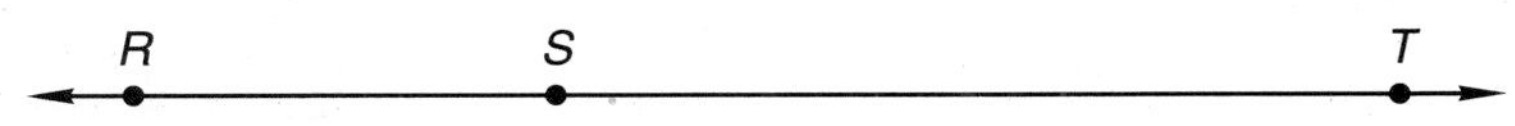

11. Compare: $\frac{3}{5} + \frac{3}{5} + \frac{3}{5} \bigcirc 3 \times \frac{3}{5}$
(86)

12. In this drawing, which angle appears to be obtuse?
(32, 61)

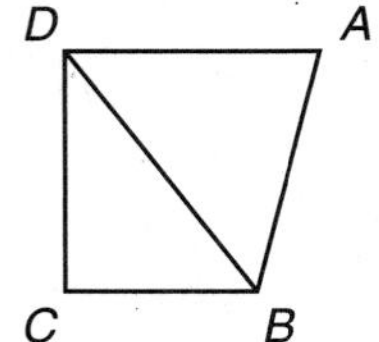

A. $\angle ABC$ B. $\angle ABD$

C. $\angle BDC$ D. $\angle DAB$

13. $\frac{1}{8} \times 3$
(86)

14. $\frac{3}{8} \div \frac{1}{8}$
(87)

15. (a) How many one fourths are in one?
(86, 87)

(b) $\frac{1}{6} \times 4$

16. $\frac{1}{4} + \frac{1}{4}$
(81)

17. $\frac{7}{8} - \frac{1}{8}$
(81)

18. $5 - 1\frac{3}{10}$
(63)

19. $6.57 + 38¢ + $16
(70)

20. 421.05 − 125.7
(73)

21. 30^2
(78)

22. 340 × 607
(56)

23. 9)$7.65
(26)

Use the school schedule below to answer problems 24 and 25.

School Schedule

Reading	8:00–8:50
Math	8:50–9:40
Recess	9:40–10:10
Language	10:10–10:50
Science	10:50–11:30
Lunch	11:30–12:30

24. How many total minutes are spent each morning in reading and language?
(28, Inv. 5)

25. If students come back for 2 hours 10 minutes after lunch, then at what time does school end?
(28, Inv. 5)

26. Room 16 is 30 feet long and 30 feet wide. What is the floor area of the room?
(72)

27. Write the next three terms of this arithmetic sequence:
(Inv. 7)

$\frac{1}{2}$, 1, $1\frac{1}{2}$, 2, $2\frac{1}{2}$, 3, ____, ____, ____, ...

28. Without looking, Sam chose one marble from a bag containing 2 red marbles, 3 white marbles, and 10 black marbles. Find the probability that the marble Sam chose was black. Write the answer as a reduced fraction.
(57, 81)

29. Which of these Venn diagrams illustrates the relationship between parallelograms (P) and trapezoids (T)? Explain your answer.
(45, Inv. 8)

A. (P)(T) overlapping B. (P) (T) separate C. P containing T D. T containing P

30. A portion of this square inch is shaded. What is the area of the shaded rectangle?
(72, 76)

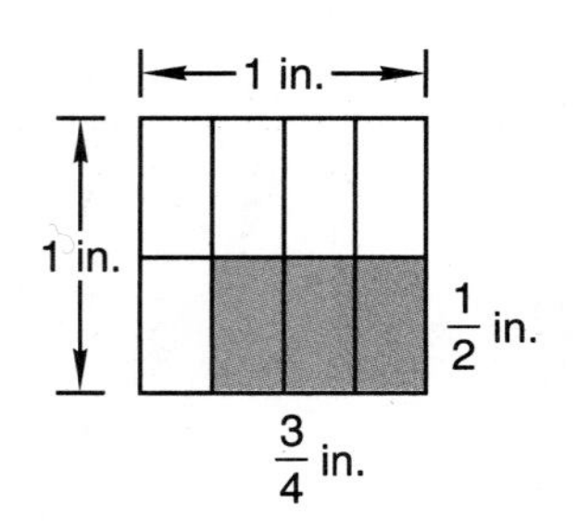

LESSON

89 Finding a Square Root

WARM-UP

Facts Practice: 64 Multiplication Facts (Test F)

Mental Math:

Use the distributive property to solve problems **a** and **b**.

a. $4(30 + 4)$ **b.** $4(34)$ **c.** Reduce: $\frac{4}{12}$, $\frac{6}{12}$, $\frac{8}{12}$

d. How many ounces are in a pint? How many pints are in a quart? How many ounces are in a quart?

e. $\frac{1}{10}$ of 1000, $-$ 1, $\div$ 9, $+$ 1, $\times$ 4, $+$ 1, $\div$ 7

Problem Solving:

Two cups equal a pint. Two pints equal a quart. Two quarts equal a half gallon. Two half gallons equal a gallon. One pint and one quart is a total of how many cups?

NEW CONCEPT

If we know the area of a square, then we can find the length of each side. The area of this square is 25 square units. Each side must be 5 units long, because $5 \times 5 = 25$.

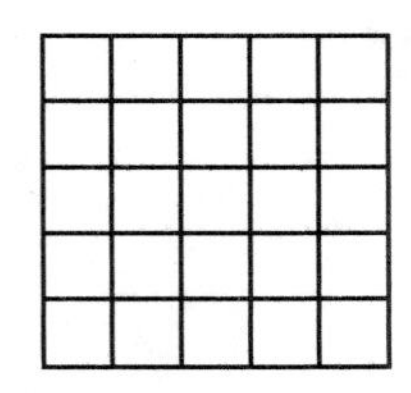

When we find the length of the side of a square from the area of the square, we are finding a **square root.**

Example 1 The area of this square is 36 square centimeters. How long is each side?

Solution The sides of a square have equal lengths. So we need to find a number that we can multiply by itself to equal 36.

$$\_\_\_\_\_ \times \_\_\_\_\_ = 36$$

We recall that $6 \times 6 = 36$, so each side of the square has a length of **6 centimeters.**

We use the symbol $\sqrt{\ }$ to indicate the positive square root of a number.

$$\sqrt{36} = 6$$

We say, "The square root of thirty-six equals six."

Example 2 Find $\sqrt{100}$.

Solution The square root of 100 is **10** because $10 \times 10 = 100$.

A **perfect square** has a whole-number square root. Here we shade the perfect squares on a multiplication table:

	1	2	3	4	5
1	1	2	3	4	5
2	2	4	6	8	10
3	3	6	9	12	15
4	4	8	12	16	20
5	5	10	15	20	25

The perfect squares appear diagonally on the multiplication table.

Example 3 Compare: $\sqrt{9 + 16} \bigcirc \sqrt{9} + \sqrt{16}$

Solution On the left, 9 and 16 are under the same square root symbol. We add the numbers and get $\sqrt{25}$. On the right, 9 and 16 are under different square root symbols. We do not add until we have found their square roots.

$$\sqrt{9 + 16} \bigcirc \sqrt{9} + \sqrt{16}$$

$$\sqrt{25} \bigcirc \sqrt{9} + \sqrt{16}$$

$$5 \bigcirc 3 + 4$$

$$\mathbf{5 < 7}$$

LESSON PRACTICE

Practice set Find each square root in problems **a–d.**

a. $\sqrt{1}$ **b.** $\sqrt{4}$ **c.** $\sqrt{16}$ **d.** $\sqrt{49}$

e. Compare: $\sqrt{36} \bigcirc 3^2$

f. Find the square roots; then subtract: $\sqrt{25} - \sqrt{16}$

MIXED PRACTICE

Problem set

1. *(35)* Thomas Jefferson wrote the Declaration of Independence in 1776. He died exactly 50 years later. In what year did he die?

2. *(21, 54)* Shannon won \$10,000. She will be paid \$20 a day until she receives the entire \$10,000. How many days will she be paid \$20?

3. Which of these numbers is divisible by both 4 and 5?
(20, 22)

A. 15 B. 16 C. 20 D. 25

4. Arrange these numbers in order from least to greatest:
(69, 71)

$$0.5, \frac{3}{2}, 1, 0, 1.1$$

5. (a) How many half-gallon cartons of milk equal 1 gallon?
(85)

(b) How many half-gallon cartons of milk equal 3 gallons?

6. Use digits to write the number one million, three hundred fifty-four thousand, seven hundred sixty.
(52)

7. Write a fraction equal to $\frac{1}{2}$ that has a denominator of 6. Then subtract that fraction from $\frac{5}{6}$. Remember to reduce the answer.
(79, 81)

8. (a) What fraction of the circles is shaded? Reduce the fraction.
(71, 81)

(b) What percent of the circles is shaded?

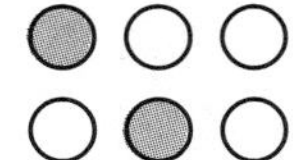

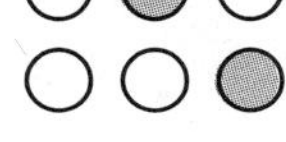

9. (a) Name this shape.
(83)

(b) How many edges does it have?

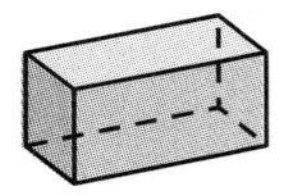

10. Write the length of the segment below as a number of centimeters and as a number of millimeters.
(66)

11. $\frac{2}{5}$ of 3
(86)

12. $\frac{2}{5} + \frac{2}{5} + \frac{2}{5}$
(75)

13. $1\frac{1}{4} + 1\frac{1}{4}$
(81)

14. $3\frac{5}{6} - 1\frac{1}{6}$
(81)

15. 42.6 + 49.76 + 28.7 + 53.18
(73)

16. \$10 − (57¢ + \$2.48)
(24, 70)

17. 42 × 5 × 36
(18, 56)

18. \$6.15 × 10
(29)

19. $40\overline{)2760}$
(54)

20. $4W = 276$
(26)

21. $\frac{1}{2} \div \frac{1}{10}$
(87)

22. $\frac{1}{2} \times \frac{6}{8}$
(81)

23. Divide 371 by 10 and write the answer with a remainder.
(54)

Use this information to answer problems 24 and 25.

> *When Jenny was born, her dad was 29 years old. Her brothers are Tom and Monty. Tom is 2 years older than Jenny and 2 years younger than Monty. Monty is 10 years old.*

24. How old is Jenny?
(49, Inv. 5)

25. How old is Jenny's dad?
(49, Inv. 5)

26. $\sqrt{25} - \sqrt{9}$
(89)

27. $3^2 + 4^2$
(78)

28. A dime is tossed and then a quarter is tossed. One possible outcome is dime : heads, quarter : tails. List the three other possible outcomes.
(Inv. 6)

29. (a) A pint is what fraction of a quart?
(85, 76)

(b) What fraction of a gallon is a quart?

(c) What fraction of a gallon is a pint?

(d) The answer to parts (a)–(c) show that one half of one fourth equals what fraction?

30. Draw a circle and shade $\frac{1}{3}$ of it. What percent of the circle is shaded?
(71)

LESSON

90 Reducing Fractions, Part 2

WARM-UP

Facts Practice: 64 Multiplication Facts (Test F)

Mental Math:

Use the distributive property to solve problems **a** and **b**.

a. 6(40 + 6) **b.** 6(46) **c.** Reduce: $\frac{4}{8}$, $\frac{4}{12}$, $\frac{4}{16}$

d. How many milliliters are in a liter? Which U.S. Customary unit is nearly equal to a liter?

e. $\frac{1}{4}$ of 36, + 1, × 2, + 1, ÷ 3, × 7, + 1, ÷ 2

Problem Solving:

A **line of symmetry** divides a figure into mirror images. If a rectangle is longer than it is wide, then it has exactly two lines of symmetry: one lengthwise and one widthwise. The lines of symmetry for this rectangle are shown with dashes. On your paper, draw a rectangle that is longer than it is wide, and show its lines of symmetry.

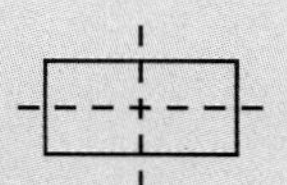

NEW CONCEPT

The equivalent fractions pictured below name the same amount. We see that $\frac{4}{8}$ is equivalent to $\frac{1}{2}$.

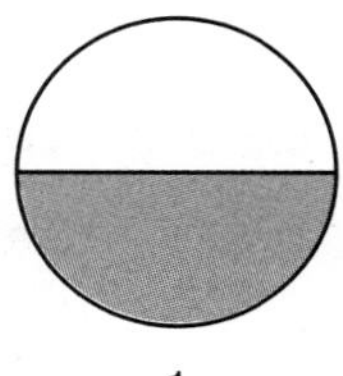

$\frac{1}{2}$

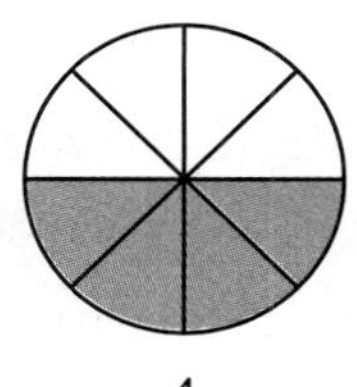

$\frac{4}{8}$

We can reduce $\frac{4}{8}$ by dividing 4 and 8 by 2.

$$\frac{4 \div 2}{8 \div 2} = \frac{2}{4}$$

If we reduce $\frac{4}{8}$ by dividing both terms by 2, we find that $\frac{4}{8}$ is equal to $\frac{2}{4}$. However, fractions should be reduced to **lowest terms.** The fraction $\frac{2}{4}$ can also be reduced, so we reduce again.

$$\frac{2 \div 2}{4 \div 2} = \frac{1}{2}$$

The fraction $\frac{4}{8}$ reduces to $\frac{2}{4}$, which reduces to $\frac{1}{2}$. We reduce twice to find that $\frac{4}{8}$ equals $\frac{1}{2}$.

We can avoid the need to reduce more than once if we divide by the greatest common factor (GCF) of the terms. The GCF of 4 and 8 is 4. If we reduce $\frac{4}{8}$ by dividing both terms by 4, we reduce only once.

$$\frac{4 \div 4}{8 \div 4} = \frac{1}{2}$$

Example 1 Reduce: $\frac{8}{12}$

Solution The terms of $\frac{8}{12}$ are 8 and 12. Since 8 and 12 are divisible by 2, we may reduce $\frac{8}{12}$ by dividing both terms by 2. This gives us $\frac{4}{6}$, which also can be reduced.

REDUCE TWICE

$$\frac{8 \div 2}{12 \div 2} = \frac{4}{6}$$

$$\frac{4 \div 2}{6 \div 2} = \mathbf{\frac{2}{3}}$$

We save a step if we reduce by the GCF of the terms. The GCF of 8 and 12 is 4. If we divide 8 and 12 by 4, then we reduce only once.

REDUCE ONCE

$$\frac{8 \div 4}{12 \div 4} = \mathbf{\frac{2}{3}}$$

Example 2 Write 40% as a reduced fraction.

Solution We first write 40% as the fraction $\frac{40}{100}$. Since the numerator and denominator both end in zero, we know they are divisible by 10.

$$\frac{40 \div 10}{100 \div 10} = \frac{4}{10}$$

Since the terms of $\frac{4}{10}$ are both even, we can continue to reduce by dividing both terms by 2.

$$\frac{4 \div 2}{10 \div 2} = \mathbf{\frac{2}{5}}$$

The GCF of 40 and 100 is 20. So we could have reduced $\frac{40}{100}$ in one step by dividing both terms by 20.

$$\frac{40 \div 20}{100 \div 20} = \mathbf{\frac{2}{5}}$$

LESSON PRACTICE

Practice set* Reduce each fraction to lowest terms:

a. $\frac{4}{12}$ **b.** $\frac{6}{18}$ **c.** $\frac{16}{24}$

d. $\frac{4}{16}$ **e.** $\frac{12}{16}$ **f.** $\frac{60}{100}$

Solve. Reduce each answer to lowest terms:

g. $\frac{7}{16} + \frac{1}{16}$ **h.** $\frac{3}{4} \times \frac{4}{5}$ **i.** $\frac{19}{24} - \frac{1}{24}$

Write each percent as a reduced fraction:

j. 25% **k.** 60% **l.** 90%

MIXED PRACTICE

Problem set **1.** This little poem is about what number?
(6)

I am a number, not 1, 2, or 3.
Whenever I'm added, no difference you'll see.

2. Write fractions equal to $\frac{1}{2}$ and $\frac{3}{5}$ with denominators of 10. Then add the fractions. Remember to convert the answer to a mixed number.
(75, 79)

3. (a) How many quarts of milk equal a gallon?
(85)
(b) How many quarts of milk equal 6 gallons?

4. Find the sum when the decimal number fourteen and seven tenths is added to the decimal number four and four tenths.
(68, 73)

5. Name the shaded portion of this rectangle as a decimal number, as a reduced fraction, and as a percent.
(71)

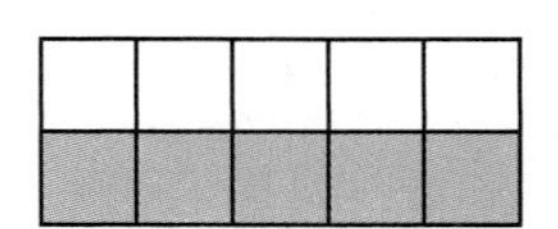

6. What is the shape of a basketball?
(83)

Refer to rectangle *ABCD* to answer problems 7–9.

7. In this rectangle, which segment is parallel to $\overline{AB}$?
(31, 61)

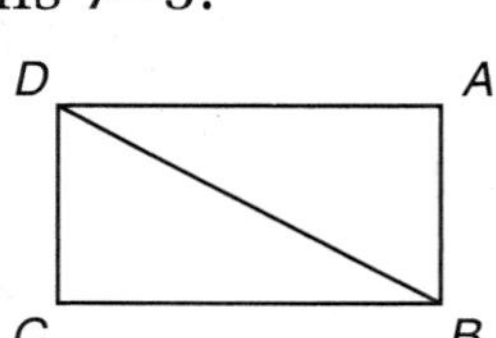

A. $\overline{BC}$ B. $\overline{CD}$

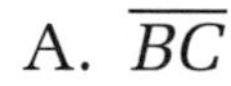

C. $\overline{BD}$ D. $\overline{DA}$

8. Classified by angles, what type of triangle is triangle *BCD*?
(36, 61)

9. What transformation would move triangle *DAB* to the position of triangle *BCD*?
(88)

10. $\frac{5}{6} + \frac{5}{6}$
(75)

11. $\frac{5}{6} \times 2$
(86)

12. $\frac{2}{5} \div \frac{1}{10}$
(87)

13. $\frac{1}{12} + \frac{7}{12}$
(41, 90)

14. $6\frac{2}{3} - \left(4 - \frac{1}{3}\right)$
(41, 63)

15. $\frac{2}{3} \times \frac{3}{4}$
(76, 90)

16. $26.4 + 2.64$
(73)

17. $8.36 - 4.7$
(73)

18. 40^2
(78)

19. $\sqrt{81}$
(89)

20. $480 \div 10$
(54)

21. $5n = 240$
(26)

22. $1 \div \frac{1}{3}$
(87)

23. $\frac{3}{4} \times 3$
(86)

24. $\frac{3}{5} \times \square = \frac{60}{100}$
(79)

The table below lists ways Brian can earn extra credit in social studies. Use the table to answer problems 25 and 26.

Extra Credit

Magazine report	35 points
TV report	50 points
Book report	75 points
Museum report	100 points

25. Brian has done a book report, two magazine reports, and a TV report. How many points has he earned?
(11, Inv. 5)

26. Brian needs to earn a total of 400 points. How many more points does he need?
(11, Inv. 5)

27. A bag contains 3 red marbles, 5 white marbles, 2 blue marbles, and 6 orange marbles. A marble is drawn without looking. Find the probability that the marble is orange. Write the answer as a reduced fraction.
(57, 81)

28. The area of this square is 25 square inches.
(53, 72)

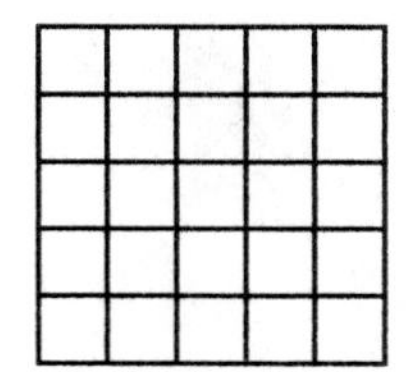

(a) How long is each side?

(b) What is its perimeter?

29. What is the next term of the sequence below? Describe the pattern in words.
(32, Inv. 7)

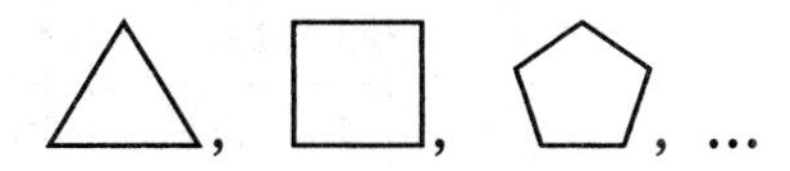

30. (a) List the factors of 16 in order from least to greatest.
(2, 25, 84, 89)

(b) Is the number of factors odd or even?

(c) What is the median of the factors?

(d) What is $\sqrt{16}$?

INVESTIGATION 9

Focus on

Line Graphs

Often we are interested in seeing the changes in data that occur over a period of time. Below we show the normal temperature in the city of Boston for each month of the year.

Normal Boston Temperature

Month	Temp.	Month	Temp.
January	30°F	July	74°F
February	31°F	August	72°F
March	38°F	September	65°F
April	49°F	October	55°F
May	59°F	November	45°F
June	68°F	December	34°F

The temperature is lowest in January and February. Then the weather warms up steadily until summer arrives. It stays warm through August and then cools steadily after that. In December the temperature is almost as low as at the beginning of the year.

To show the change of temperature over time, we can use a **line graph.** We will draw the line graph on a grid. First we label each of the 12 months along the grid's **horizontal axis.** Then we label temperatures from 0°F through 80°F along the grid's **vertical axis.** We label up to 80° on the grid because we need to graph temperatures as high as 74°. We choose our interval to be 10°F on the vertical axis. We could use a smaller interval instead (such as 5°), but then our grid would be bigger. Above each month, we place a dot at a height equal to the normal temperature for that month.

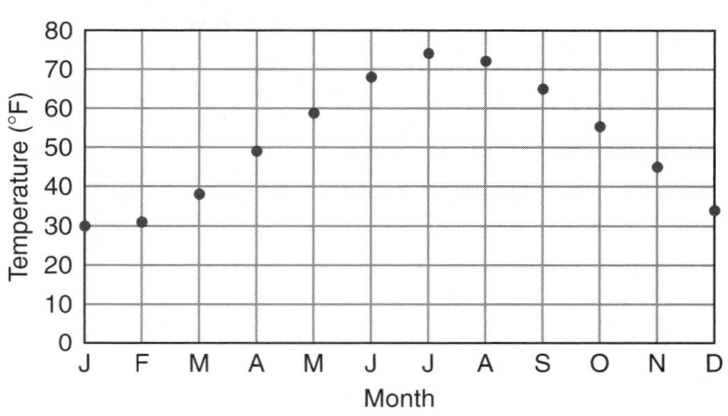

Finally, we connect the dots with line segments to produce our line graph. The **rising line** from January until July shows that the temperature is increasing. The **falling line** from July to December shows that the temperature is decreasing. The line is steepest in spring and in fall. During these times, the normal temperature is changing the most quickly.

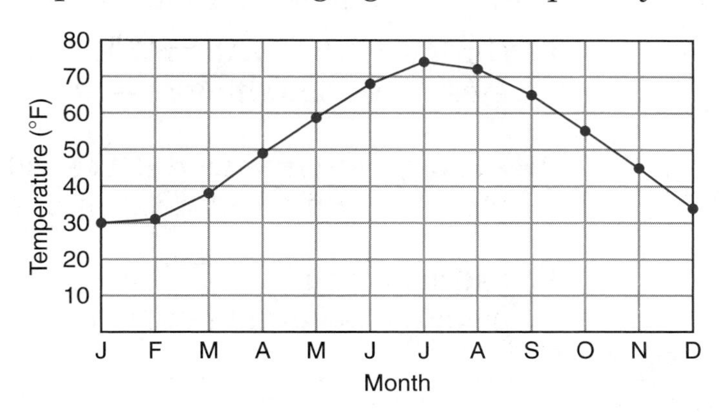

Starting in 1998, Kyle recorded the number of pull-ups he could do at the beginning of each year.

Year	1998	1999	2000	2001	2002	2003
Pull-ups	4	6	10	11	13	12

1. Make a line graph showing how the number of pull-ups Kyle could do changed over this period.

2. In which year did Kyle not record an improvement?

3. In which year did Kyle record the greatest improvement over the previous year? How can you tell?

Mr. Escobar invests in stocks. He has constructed a line graph to show the value of his stocks at the beginning of each year.

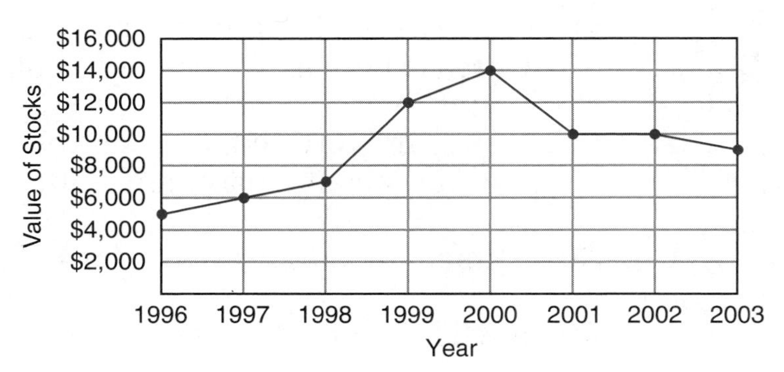

4. How much were Mr. Escobar's stocks worth at the beginning of 2001? About how much were the stocks worth at the beginning of 2003?

5. At the beginning of which year were his stocks worth the most? About how much were they worth then?

6. During which year did his stocks increase in value the most? By about how much?

7. During which year did the value of his stocks decrease the most? By about how much?

8. Estimate the overall change in the value of his stocks from the beginning of 1996 to the beginning of 2003.

We can use a **double-line graph** to show how two or more things change in relation to one another. For example, the double-line graph below shows the change in population of the cities of Austin and Pittsburgh from 1950 to 2000. The legend to the right tells which line belongs to which city.

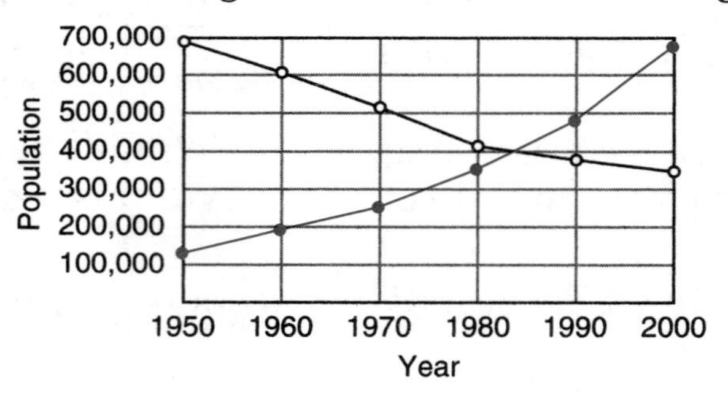

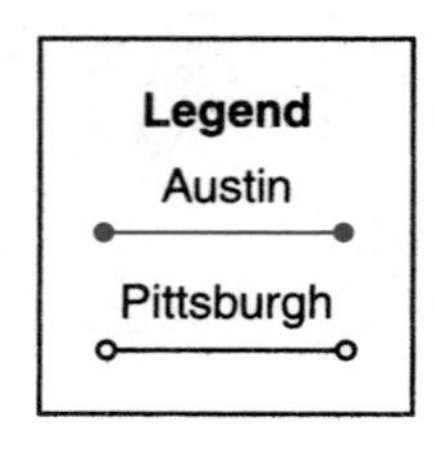

9. Approximately what was Austin's population in 1970?

10. Approximately what was Pittsburgh's highest population between 1950 and 2000?

11. Approximately how much did Pittsburgh's population decrease between 1950 and 2000?

12. Approximately what was the difference between the populations of Austin and Pittsburgh in 2000?

13. Estimate when the populations of the two cities were equal.

14. In what decade did Austin's population increase the most?

Naomi and Takeshi each took 6 quizzes in their science class. Each quiz was worth 10 points. Their scores on the quizzes are summarized in the table below.

Quiz	Naomi	Takeshi
1	6	4
2	7	4
3	8	5
4	9	6
5	8	7
6	6	7

15. Make a double-line graph for the scores. What general patterns do you see?

Extensions

a. Using the Internet or an almanac, find the population of your state in the years 1950, 1960, 1970, 1980, 1990, and 2000. Make a line graph showing the population change by decades.

b. Two of the greatest baseball players of all time played together for the New York Yankees: Babe Ruth and Lou Gehrig. Look up their home-run numbers for each of the years 1925 through 1932, and display the data on a double-line graph.

LESSON

91 Simplifying Improper Fractions

WARM-UP

Facts Practice: 40 Fractions to Reduce (Test I)

Mental Math:

Roman numerals:†

a. Write 13 in Roman numerals.

b. Write VIII in our number system.

Problem Solving:

This table lists the years from 2001 to 2006 and the day of the week on which each year begins. Notice that each year begins one day of the week later than the first day of the previous year until 2005. Since 2004 is a leap year and has an additional day, the year 2005 begins an additional day later. Copy this table and continue it through the year 2015, which begins on a Thursday.

Year	First Day
2001	Monday
2002	Tuesday
2003	Wednesday
2004	Thursday
2005	Saturday
2006	Sunday

NEW CONCEPT

We have learned two ways to simplify fractions. We have converted improper fractions to whole numbers or mixed numbers, and we have reduced fractions. In some cases we need to use **both** ways in order to simplify a fraction. Consider the following story:

After the party some pizza was left over. There was $\frac{3}{4}$ of a pizza in one box and $\frac{3}{4}$ of a pizza in another box. Altogether, how much pizza was in the two boxes?

†In Lessons 91–105, the Mental Math section "Roman numerals" reviews concepts from Appendix Topic A. You may skip these Warm-up problems if you have not covered Appendix Topic A.

In this story about combining, we add $\frac{3}{4}$ to $\frac{3}{4}$.

$$\frac{3}{4} + \frac{3}{4} = \frac{6}{4}$$

We see that the sum is an improper fraction. To convert an improper fraction to a mixed number, we divide the numerator by the denominator and write the remainder as a fraction.

$$\frac{6}{4} \longrightarrow 4\overline{)6} = 1\frac{2}{4} \quad \begin{array}{r} 4 \\ \hline 2 \end{array}$$

The improper fraction $\frac{6}{4}$ is equal to the mixed number $1\frac{2}{4}$. However, $1\frac{2}{4}$ can be reduced.

$$1\frac{2}{4} = 1\frac{1}{2}$$

The simplified answer to $\frac{3}{4} + \frac{3}{4}$ is $1\frac{1}{2}$.

Example 1 Write $\frac{8}{6}$ as a reduced mixed number.

Solution To convert $\frac{8}{6}$ to a mixed number, we divide 8 by 6 and get $1\frac{2}{6}$. Then we reduce $1\frac{2}{6}$ by dividing both terms of the fraction by 2 to get $\mathbf{1\frac{1}{3}}$.

Convert		Reduce
$\frac{8}{6} = 1\frac{2}{6}$	$\longrightarrow$	$1\frac{2}{6} = \mathbf{1\frac{1}{3}}$

Example 2 Add: $1\frac{7}{8} + 1\frac{3}{8}$

Solution We add to get $2\frac{10}{8}$. We convert the improper fraction $\frac{10}{8}$ to $1\frac{2}{8}$ and add it to the 2 to get $3\frac{2}{8}$. Finally, we reduce the fraction to get $\mathbf{3\frac{1}{4}}$.

Add		Convert		Reduce
$1\frac{7}{8} + 1\frac{3}{8} = 2\frac{10}{8}$	$\longrightarrow$	$2\frac{10}{8} = 3\frac{2}{8}$	$\longrightarrow$	$3\frac{2}{8} = \mathbf{3\frac{1}{4}}$

LESSON PRACTICE

Practice set* Simplify each fraction or mixed number:

a. $\frac{6}{4}$ b. $\frac{10}{6}$ c. $2\frac{8}{6}$ d. $3\frac{10}{4}$

e. $\frac{10}{4}$ f. $\frac{12}{8}$ g. $4\frac{14}{8}$ h. $1\frac{10}{8}$

Perform each indicated operation. Simplify your answers.

i. $1\frac{5}{6} + 1\frac{5}{6}$ j. $2\frac{3}{4} + 4\frac{3}{4}$ k. $\frac{5}{3} \times \frac{3}{2}$

l. Each side of this square is $\frac{5}{8}$ inches long. What is the perimeter of the square? Show your work.

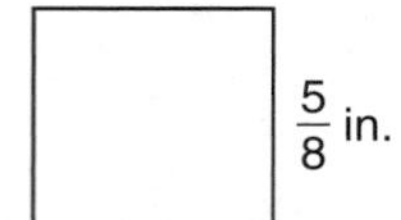

MIXED PRACTICE

Problem set

1. Two fathoms deep is 12 feet deep. How deep is 10 fathoms?
(49)

2. When Jessica baby-sits, she is paid $6.50 per hour. If she baby-sits Saturday from 10:30 a.m. to 3:30 p.m., how much money will she be paid?
(49)

3. Use digits to write the number one hundred fifty-four million, three hundred forty-three thousand, five hundred fifteen.
(52)

4. (a) How many quarter-mile laps does Jim have to run to complete 1 mile?
(87)

   (b) How many quarter-mile laps does Jim have to run to complete 5 miles?

5. Write a fraction equal to $\frac{3}{4}$ that has a denominator of 8. Add that fraction to $\frac{5}{8}$. Remember to convert the answer to a mixed number.
(75, 79)

6. What mixed number names the number of shaded hexagons?
(40, 81)

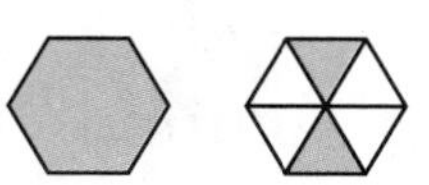

7. Which segment does *not* name a radius of this circle?
(53, 61)

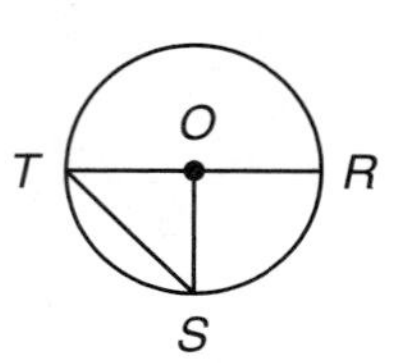

A. $\overline{OR}$ B. $\overline{OS}$

C. $\overline{RT}$ D. $\overline{OT}$

8. (86) Compare: $\frac{1}{2}$ of 2 ◯ $2 \times \frac{1}{2}$

9. (83) What is the shape of a can of beans?

10. (61, 73) *AB* is 3.2 cm. *BC* is 1.8 cm. *CD* equals *BC*. Find *AD*.

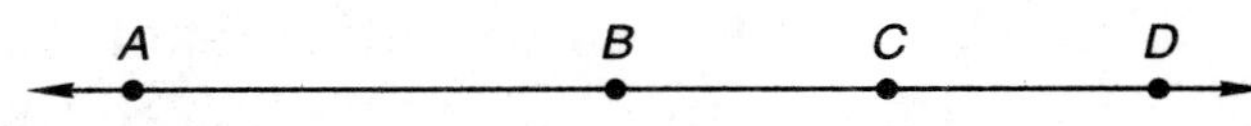

11. (91) $1\frac{3}{4} + 1\frac{3}{4}$

12. (81) $5\frac{7}{8} - 1\frac{3}{8}$

13. (86, 91) $3 \times \frac{3}{8}$

14. (24, 70) $10 − ($1.25 + 35¢)

15. (17) $4.32 × 5

16. (56) 416 × 740

17. (24, 73) 4.51 − (2.3 + 0.65)

18. (34) 960 ÷ 8

19. (54) 80)9600

20. (26, 34) $5m$ = $12.00

21. (76, 91) $\frac{5}{2} \times \frac{2}{3}$

22. (87) $\frac{2}{3} \div \frac{1}{3}$

23. (87) $\frac{2}{3} \div \frac{1}{6}$

Use this information to answer problems 24 and 25:

Tyrone fixed his function machine so that when he puts in a 3, a 9 comes out. When he puts in a 6, an 18 comes out. When he puts in a 9, a 27 comes out.

3 → NUMBER-CHANGING MACHINE → 9
6 → → 18
9 → → 27

24. (Inv. 7) Which of the following does Tyrone's function machine do to the numbers he puts into it?

A. It adds 3.
B. It multiplies by 3.
C. It adds 9.
D. It multiplies by 2 and 3.

25. (Inv. 7) Tyrone put in a number, and a 12 came out. What number did he put in?

26. Assuming that the sequence below repeats with period 3, write the next 5 terms.
(Inv. 7)

4, 4, 1, 4, 4, ...

The days of the week are Sunday, Monday, Tuesday, Wednesday, Thursday, Friday, and Saturday. Make a list of the number of letters in each name. Friday, for instance, has 6 letters and Saturday has 8. Refer to your list of numbers to answer problems 27–30.

27. What number is the median?
(84)

28. What number is the mode?
(84)

29. What is the range?
(84)

30. Find the mean and write it as a mixed number.
(84)

LESSON

92 Dividing by Two-Digit Numbers

WARM-UP

Facts Practice: 40 Fractions to Reduce (Test I)

Mental Math:

a. Reduce: $\frac{6}{8}$, $\frac{6}{9}$, $\frac{6}{12}$ **b.** 10% of \$5000 **c.** $\frac{1}{5}$ of \$5000

d. $\frac{1}{3}$ of 100 **e.** $\sqrt{100}$, × 2, × 50, − 1, ÷ 9

Roman numerals:

f. Write 17 in Roman numerals.

g. Write XII in our number system.

Problem Solving:

Recall that a *permutation* is an ordered arrangement of objects. Adam, Barbara, and Conrad stood side by side to have their picture taken (A, B, C). Then Barbara and Conrad switched places (A, C, B). List the remaining possible side-by-side arrangements.

NEW CONCEPT

In this lesson we will begin dividing by two-digit numbers. Dividing by two-digit numbers is necessary to solve problems like the following:

One hundred forty-four players signed up for soccer. If the players are separated into 12 equal teams, how many players will be on each team?

When we divide by a two-digit number, we continue to follow the four steps of division: divide, multiply, subtract, and bring down. When we divide by two-digit numbers, the "divide" step takes a little more thought, because we have not memorized the two-digit multiplication facts.

Example 1 Divide: 150 ÷ 12

Solution We begin by breaking the division into a smaller division problem. Starting from the first digit in 150, we try to find a number that 12 will divide into at least once. Our first smaller division is $12\overline{)15}$. We see that there is one 12 in 15, so we write "1" above the 5 of 15. Then we multiply, subtract, and bring down.

```
     1
12)150
   12
   --
    30
```

Now we begin a new division. This time we find 12)30. If we are not sure of the answer, we may need to try more than once to find the number of 12's in 30. We find that there are two 12's in 30. We write "2" above the 0 of 150. Then we multiply and subtract. Since there is no digit to bring down, we are finished. The answer is **12 R 6.**

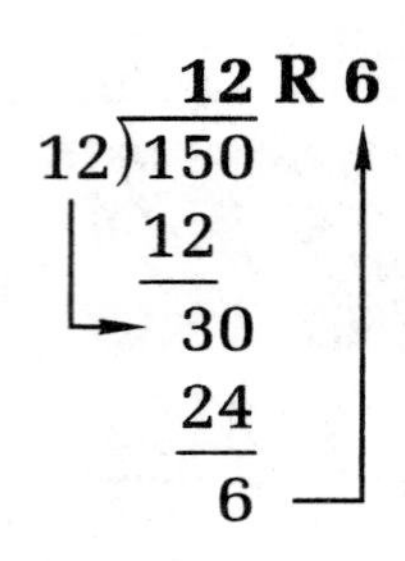

To check our answer, we multiply 12 by 12 and then add the remainder, which is 6.

$$\begin{array}{rl} 12 & \\ \times\ 12 & \\ \hline 144 & \\ +\ \ 6 & \text{remainder} \\ \hline 150 & \text{(check)} \end{array}$$

There are some "tricks" we can use to make dividing by two-digit numbers easier. One trick is to think of dividing by only the first digit.

Example 2 Divide: 32)987

Solution We begin by breaking the division into the smaller division problem 32)98. Instead of thinking, "How many 32's are in 98?" we can use the first-digit trick and think, "How many 3's are in 9?" We see "32)98" but we think "3)9." We try 3 as an answer. Since we are really finding 32)98, we write the 3 above the 8 of 98. Then we multiply 3 by 32, subtract, and bring down.

$$\begin{array}{r} \textbf{30 R 27} \\ 32\overline{)987} \\ 96\downarrow \\ \hline 27 \\ 0 \\ \hline 27 \end{array}$$

Now we begin the new division 32)27. Since there is not even one 32 in 27, we write "0" in the answer; then we multiply and subtract. There are no digits to bring down, so we are finished. The answer is **30 R 27.** We can check our answer by multiplying 30 by 32 and then adding the remainder, 27.

$$\begin{array}{rl} 32 & \\ \times\ 30 & \\ \hline 960 & \\ +\ 27 & \text{remainder} \\ \hline 987 & \text{(check)} \end{array}$$

LESSON PRACTICE

Practice set Divide:

a. $11\overline{)253}$ b. $21\overline{)253}$ c. $31\overline{)403}$

d. $12\overline{)253}$ e. $12\overline{)300}$ f. $23\overline{)510}$

g. One hundred forty-four players signed up for soccer. If the players are separated into 12 equal teams, how many players will be on each team?

Divide. Use the first-digit trick to help with the "divide" step.

h. $30\overline{)682}$ i. $32\overline{)709}$ j. $43\overline{)880}$

k. $22\overline{)924}$ l. $22\overline{)750}$ m. $21\overline{)126}$

n. $21\overline{)654}$ o. $41\overline{)910}$ p. $21\overline{)1290}$

MIXED PRACTICE

Problem set

1. (31, 32) Draw a pair of horizontal line segments. Make them the same length. Then draw two more line segments to make a quadrilateral.

2. (28, 49) Nathan worked on his homework from 3:30 p.m. to 6 p.m. How many minutes did Nathan work on his homework?

3. (67) Write a decimal number equal to the mixed number $3\frac{9}{10}$.

4. (49) If 24 eggs exactly fill 2 cartons, how many eggs will it take to fill 3 cartons?

5. (18) Some 1-inch cubes were used to build this 4-inch cube. How many 1-inch cubes were used?

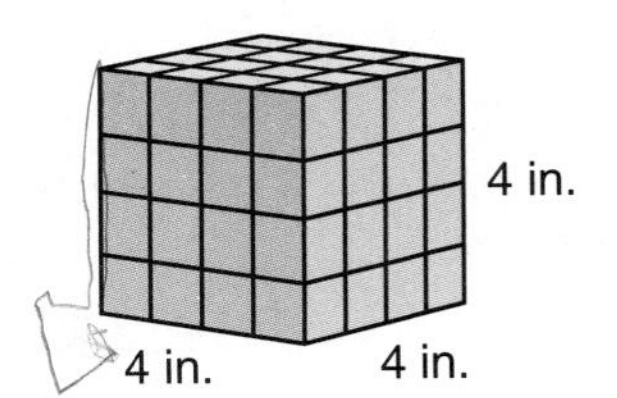

6. (87) (a) How many apples weighing $\frac{1}{3}$ pound each would it take to total 1 pound?

   (b) How many apples weighing $\frac{1}{3}$ pound each would it take to total 4 pounds?

7. (83) Name this shape. How many edges does it have?

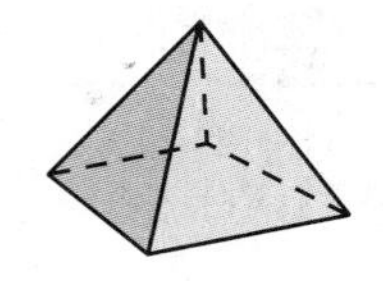

8. (71) Name the shaded portion of this square as a decimal number, as a reduced fraction, and as a percent.

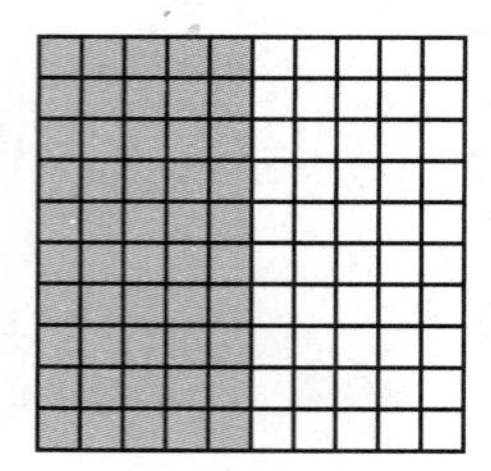

9. (23) Which of these numbers does not equal $\frac{1}{2}$?

A. 0.5 B. 50% C. $\frac{6}{12}$ D. 0.05

10. (61) AB is 40 millimeters. BC is half of AB. CD equals BC. Find AD.

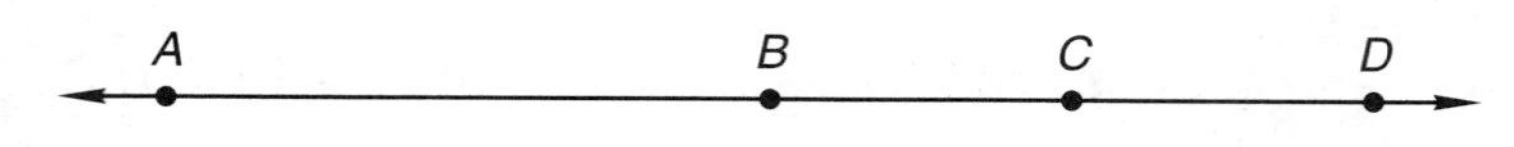

11. (73) $8.7 + 6.25$

12. (73) $12.75 - 4.2$

13. (78) 4^3

14. (17) $8 \times \$125$

15. (89) $\sqrt{100} - \sqrt{64}$

16. (92) $293 \div 13$

17. (92) $24\overline{)510}$

18. (91) $3\frac{5}{8} + 1\frac{7}{8}$

19. (63) $5 - 1\frac{2}{5}$

20. (86) $\frac{1}{3}$ of 5

21. (76) $\frac{3}{4} \times \frac{4}{3}$

22. (87) $\frac{6}{10} \div \frac{1}{5}$

23. (79, 81) Write a fraction equal to $\frac{2}{5}$ that has a denominator of 10. Add that fraction to $\frac{1}{10}$. Remember to reduce your answer.

24. (62, 72) Estimate the area of a window that is 3 ft 10 in. wide and 2 ft 11 in. tall.

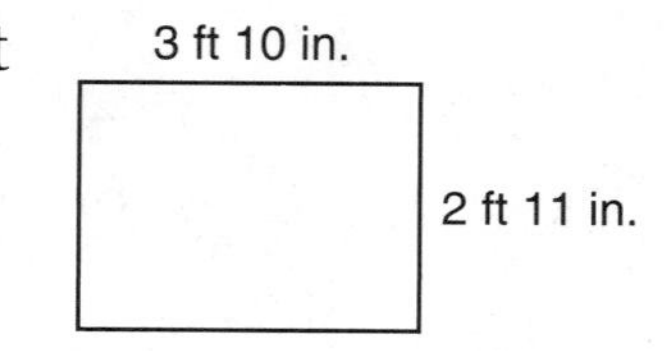

25. (57) A penny, nickel, dime, and quarter are tossed at the same time. Which word best describes the following events: *likely, unlikely, certain,* or *impossible?*

(a) All of the upturned faces are heads.

(b) At least one of the upturned faces is heads.

(c) There is one more heads than there is tails.

Use the information and the table below to answer problems 26–28.

Sumi, Lupe, and Melanie bought treats for the party. Here is a list of the items they purchased.

Groceries

Nuts	$2.19
Mints	$1.19
Cake	$3.87
Ice cream	$1.39

26. *(Inv. 5, 62)* Describe how to estimate the total cost of the items. What is your estimate?

27. *(11, Inv. 5)* What was the total cost of the treats?

28. *(50, Inv. 5)* If the girls share the cost evenly, how much will each girl pay?

29. *(Inv. 7)* Assuming that the sequence below repeats with period 5, write the next 5 terms.

4, 4, 1, 4, 4, ...

30. *(73)* In the 1988 Summer Olympic games in Seoul, South Korea, U.S. athlete Florence Griffith-Joyner won three gold medals in track events. "Flo-Jo," as she was called, finished the 200-meter run in 21.34 seconds, breaking the previous Olympic record of 21.81 seconds. By how much did Florence Griffith-Joyner break the previous Olympic record?

LESSON

93 Comparative Bar Graphs

WARM-UP

Facts Practice: 40 Fractions to Reduce (Test I)

Mental Math:

a. Reduce: $\frac{5}{20}$, $\frac{5}{15}$, $\frac{5}{10}$ **b.** 7(34) **c.** 436 ÷ 4

d. 50% of \$100 **e.** 50% of \$10 **f.** 50% of \$1

g. $\frac{1}{3}$ of 6, × 2, + 1, × 5, − 1, ÷ 6

Roman numerals:

h. Write 11 in Roman numerals.

i. Write IX in our number system.

Problem Solving:

Alba rolled two number cubes. The total was 5. Copy this table and write all the ways Alba could have rolled 5.

First Cube	Second Cube

NEW CONCEPT

Comparative bar graphs can be used to display two or more sets of related data.

Example The average daily high temperature in January and July for five cities is displayed in the comparative vertical bar graph below.

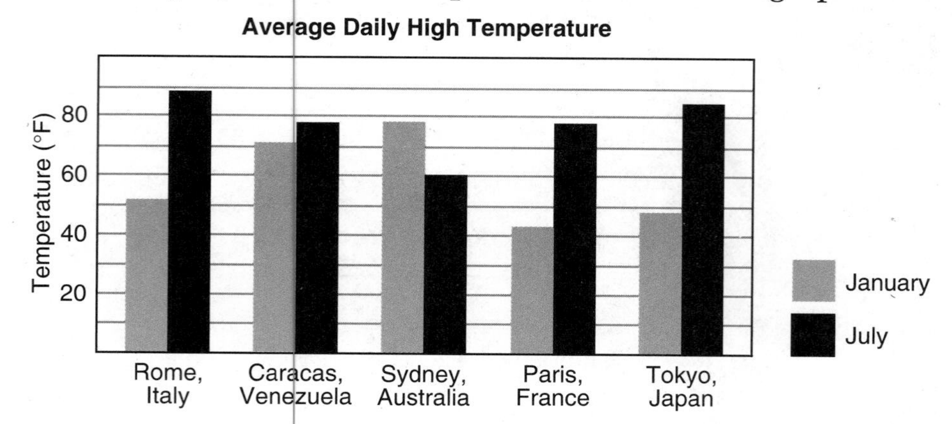

(a) In which city was the average July high temperature highest?

(b) In which city was the average January high temperature lowest?

(c) Which city had the least difference between these temperatures? Do you know why?

(d) For which city is the average January high temperature greater than the average July high temperature? Do you know why?

Solution (a) The tallest black bar appears above **Rome, Italy.** The average July high temperature is about 89°F in Rome.

(b) The shortest blue bar appears above **Paris, France.** The average January high temperature is about 42°F in Paris.

(c) The smallest difference in heights of the bars occurs above **Caracas, Venezuela.** Caracas is near the equator, and temperatures in locations near the equator do not vary much throughout the year.

(d) We look for the city that has a blue bar that is taller than its black bar. We find **Sydney, Australia.** Australia is warmer in January than in July because it is south of the equator. South of the equator, January is in the summer and July is in the winter.

LESSON PRACTICE

Practice set Lana, Alice, Ted, and José each take two quizzes. Each quiz has ten questions. The scores on the quizzes are shown in the table below.

Student	Quiz 1	Quiz 2
Lana	8	8
Alice	3	6
Ted	6	7
José	7	10

Make a comparative **horizontal** bar graph to show the scores. There should be two bars for each student.

MIXED PRACTICE

Problem set

1. (77, 85) The saying "A pint's a pound the world around" means that a pint of water weighs about a pound. About how much does 2 quarts of water weigh?

2. (49) If 3 of them cost \$2.55, how much would 4 of them cost?

3. (46) If 300 marbles will fill a carton, how many marbles will make the carton $\frac{1}{2}$ full?

4. (71) Name the shaded portion of this group as a decimal number, as a reduced fraction, and as a percent.

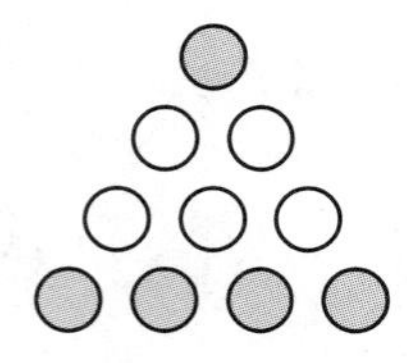

5. (87) (a) How many plums weighing $\frac{1}{5}$ pound each would it take to total 1 pound?

(b) How many plums weighing $\frac{1}{5}$ pound each would it take to total 3 pounds?

6. (8) Write the following sentence using digits and symbols:

When nine is subtracted from twelve, the difference is three.

7. (86) Compare: $\frac{2}{3}$ of 3 ◯ $3 \times \frac{2}{3}$

8. (18) If $3n = 18$, then $2n + 5$ equals which of the following?

A. 23 B. 17 C. 31 D. 14

9. (83) This shape is a special type of rectangular solid. Every face of the shape is a square. What word names this shape?

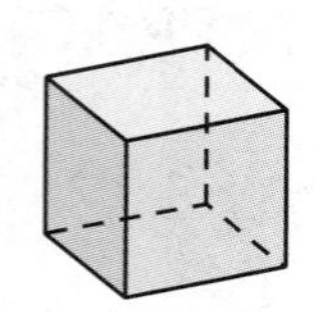

10. (32, 61) Which of these angles appears to be a right angle?

A. $\angle AOB$ B. $\angle BOC$

C. $\angle COD$ D. $\angle AOC$

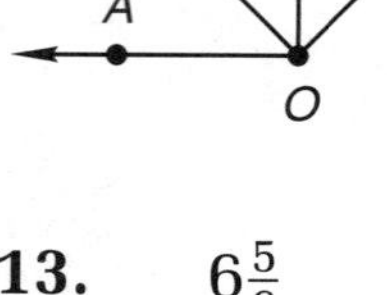

11. (91) $1\frac{3}{5} + 2\frac{4}{5}$

12. (90) $4\frac{5}{8} - \frac{1}{8}$

13. (41) $6\frac{5}{6} - 1\frac{5}{6}$

14. (87) $1 \div \frac{1}{8}$

15. (90) $\frac{8}{10} \times \frac{5}{10}$

16. (87) $\frac{1}{5} \div \frac{1}{10}$

17. (24, 73) $12.34 - (5.67 - 0.8)$

18. (13, 26) $(\$20 - \$6.55) \div 5$

19. (70) $10 \times 56¢$

20. (18, 29) $6 \times 78 \times 900$

21. (92) $31\overline{)970}$

22. (78, 89) $9^2 - \sqrt{9}$

23. Write fractions equal to $\frac{3}{4}$ and $\frac{1}{6}$ that have denominators of 12. Then add the fractions.
(79)

Look at the picture below. Then answer problems 24–26.

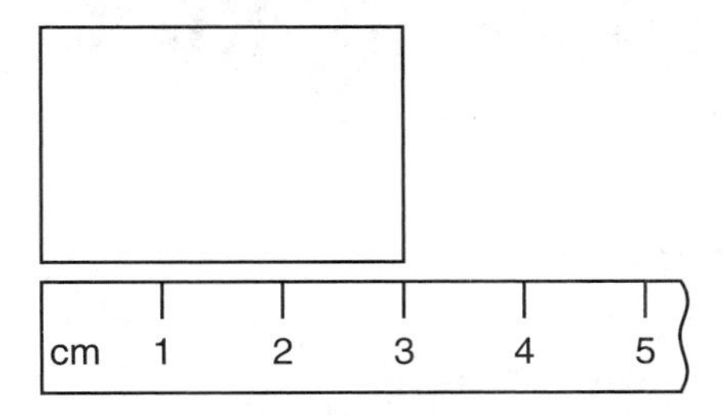

24. How long is the rectangle?
(44)

25. The rectangle is 1 centimeter longer than it is wide. What is the perimeter of the rectangle?
(53)

26. What is the area of the rectangle?
(72)

27. (a) Write the next three terms of the repeating sequence below.
(Inv. 7, 88)

(b) What is the period of the sequence?

(c) What transformation is shown in the sequence?

E, Ш, Ǝ, m, E, ____, ____, ____, . . .

Refer to the spinner to answer problems 28–30.

28. If you spin this spinner 60 times, about how many times would you expect it to stop on 2?
(Inv. 6)

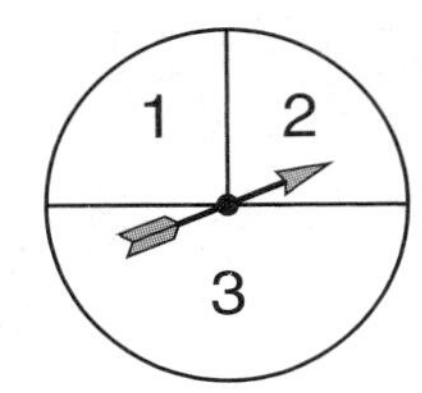

29. What percent of the spinner's face is region 2?
(71)

30. What decimal part of the spinner's face is region 3?
(71)

LESSON

94 Using Estimation When Dividing by Two-Digit Numbers

WARM-UP

Facts Practice: 40 Fractions to Reduce (Test I)

Mental Math:

a. Reduce: $\frac{3}{6}$, $\frac{3}{9}$, $\frac{3}{12}$ **b.** $\frac{1}{3}$ of 15 **c.** $\frac{2}{3}$ of 15

d. 50% of 15 **e.** $\sqrt{81}$, × 5, − 1, ÷ 4, + 1, ÷ 4, − 3

Roman numerals:

f. Write 20 in Roman numerals.

g. Write XIV in our number system.

Problem Solving:

Two cups equal a pint, and two pints equal a quart. Two quarts equal a half gallon. Two half gallons equal one gallon. A quart of milk was poured out of a full gallon container. How many pints of milk were still in the container?

NEW CONCEPT

In Lesson 92 we learned a trick to help us divide by two-digit numbers. The problems in that lesson were chosen so that using the first digit to guess the division answer would work. However, this method does not always work. In this lesson we will learn another strategy for two-digit division.

Using the first-digit trick for 19)59, we would follow this process:

We see:	We think:	We try the guess, but the guess is too large:
$19\overline{)59}$ with ? above	$1\overline{)5}$ with (5) circled above	$19\overline{)59}$ with 5 above; (95) circled below

Our guess, 5, is incorrect because there are not five 19's in 59. Our guess is too large. So we will **estimate.** To estimate, we mentally round both numbers to the nearest 10. Then we use the first-digit trick with the rounded numbers.

We see:	We round:	We think:	We try:
$19\overline{)59}$ →	$20\overline{)60}$ →	$2\overline{)6}$ with ③ →	$\begin{array}{r} 3 \text{ R } 2 \\ 19\overline{)59} \\ \underline{57} \\ 2 \end{array}$

Example Divide: $19\overline{)595}$

Solution We begin by breaking the division into the smaller division problem $19\overline{)59}$. We round to $20\overline{)60}$ and focus on the first digits, $2\overline{)6}$. We guess 3, so we write the "3" above the 9 of 59. Then we multiply 3 by 19, subtract, and bring down. The next division is $19\overline{)25}$. We may estimate to help us divide. We write "1" in the answer; then we multiply and subtract.

$$\begin{array}{r} 31 \text{ R } 6 \\ 19\overline{)595} \\ \underline{57} \\ 25 \\ \underline{19} \\ 6 \end{array}$$

The answer is **31 R 6.** To check our answer, we multiply 31 by 19 and add the remainder, which is 6.

LESSON PRACTICE

Practice set* Divide:

a. $19\overline{)792}$ **b.** $30\overline{)600}$ **c.** $29\overline{)121}$

d. $29\overline{)900}$ **e.** $48\overline{)829}$ **f.** $29\overline{)1210}$

g. $28\overline{)896}$ **h.** $18\overline{)782}$ **i.** $39\overline{)1200}$

MIXED PRACTICE

Problem set

1. (80) List all of the prime numbers less than 50 that end with the digit 1.

2. (20) What number is missing in this division problem?

$$\square \div 8 = 24$$

3. (49) Sofia ran 660 yards in 3 minutes. At this rate, how many yards would she run in 6 minutes?

4. Write a decimal number equal to the mixed number $4\frac{9}{10}$.
(71)

5. Seventy-six trombone players led the parade. If they marched in 4 equal rows, how many were in each row?
(21)

6. (a) A dime is what fraction of a dollar?
(87)

(b) How many dimes are in $1?

(c) How many dimes are in $4?

7. Which of the following means, "How many 19's are in 786?"
(92)

A. $19 \div 786$ B. $786 \div 19$ C. 19×786

8. (a) How many $\frac{1}{4}$'s are in 1?
(87)

(b) How many $\frac{1}{3}$'s are in 1?

9. What word names this shape?
(83)

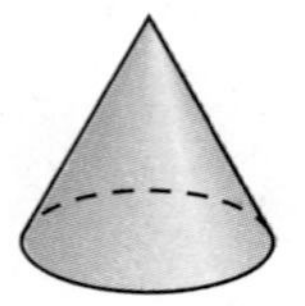

10. If $\overline{LN}$ is perpendicular to $\overline{JM}$, then $\angle JNL$ is what type of angle?
(31, 32, 61)

A. acute B. right C. obtuse

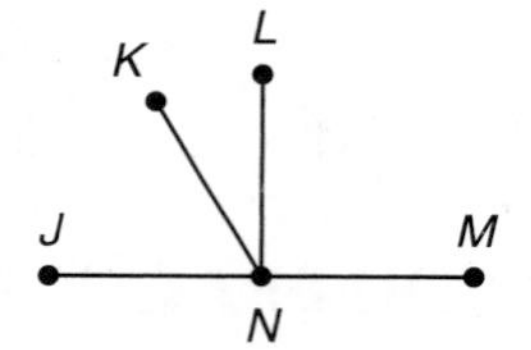

11. $63.75 + $1.48 + 59¢ + $5
(70)

12. $1010 - (101 - 10)$
(24)

13. $3.48 × 7
(17)

14. 25^2
(78)

15. $19\overline{)786}$
(94)

16. $\sqrt{36} + \sqrt{64}$
(89)

17. $38\overline{)1200}$
(94)

18. $\frac{5}{6} + \frac{5}{6} + \frac{5}{6}$
(91)

19. $\frac{5}{6} \times 3$
(86, 91)

20. Reduce: $\frac{8}{12}$
(90)

21. $3 - \left(2 - \frac{1}{4}\right)$
(24, 63)

22. $\frac{1}{3}$ of $\frac{3}{4}$
(76)

23. Write a fraction equal to $\frac{2}{3}$ that has a denominator of 12. Subtract that fraction from $\frac{11}{12}$. Remember to reduce the answer.
(79, 81)

The graph below shows Jeff's height on his birthday from ages 9 to 14. Use this graph to answer problems 24 and 25.

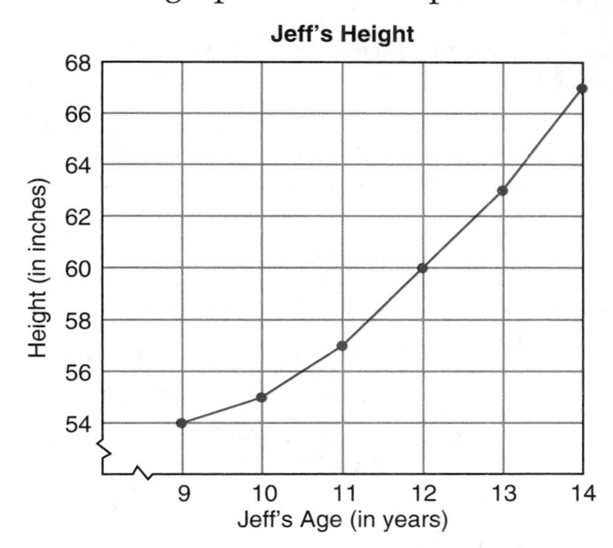

24. How many inches did Jeff grow between his twelfth and fourteenth birthdays?
(Inv. 9)

25. On which birthday was Jeff 5 feet tall?
(Inv. 9)

The sides of this square are one yard long. Since 1 yard equals 3 feet, the sides are also 3 feet long. Refer to this figure to answer problems 26–30.

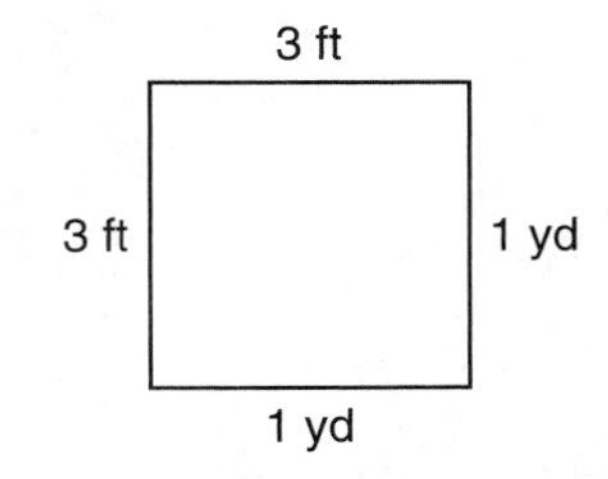

26. Which of these terms does not describe the figure?
(32, 45)

A. rectangle B. parallelogram

C. pentagon D. regular quadrilateral

27. (a) What is the perimeter of the square in feet?
(53)

(b) What is the perimeter in yards?

28. (a) What is the area of the square in square feet?
(72)

(b) What is the area in square yards?

29. Compare: 1 yd ◯ 3 ft
(74)

30. Compare: 1 sq. yd ◯ 9 sq. ft
(72, 74)

LESSON

95 Reciprocals

WARM-UP

Facts Practice: 40 Fractions to Reduce (Test I)

Mental Math:

a. What is the reduced mixed number for $\frac{10}{4}$?

b. What is the reduced mixed number for $\frac{10}{6}$?

c. What is the reduced mixed number for $\frac{10}{8}$?

d. $\frac{1}{5}$ of 15 **e.** $\frac{2}{5}$ of 15 **f.** $\frac{3}{5}$ of 15

Roman numerals:

g. Write 15 in Roman numerals.

h. Write XXVI in our number system.

Problem Solving:

A line of symmetry divides a figure into mirror images. A square has four lines of symmetry. At right we show two of these lines of symmetry. Draw a square and show all four lines of symmetry.

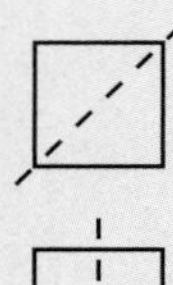

NEW CONCEPT

If we switch the numerator and denominator of a fraction, the new fraction is the **reciprocal** of the first fraction. The reciprocal has the same terms, but their positions are reversed. When we switch the positions of the numerator and denominator, we **invert** the fraction.

$$\text{The reciprocal of } \frac{2}{3} \text{ is } \frac{3}{2}.$$

$$\text{The reciprocal of } \frac{3}{2} \text{ is } \frac{2}{3}.$$

Whole numbers also have reciprocals. Recall that a whole number may be written as a fraction by writing a 1 under the whole number. So the whole number 2 may be written as $\frac{2}{1}$. To find the reciprocal of $\frac{2}{1}$, we invert the fraction and get $\frac{1}{2}$.

$$\text{Since } 2 = \frac{2}{1}, \text{ the reciprocal of 2 is } \frac{1}{2}.$$

Notice that the product of $\frac{1}{2}$ and 2 is 1.

$$\frac{1}{2} \times 2 = 1$$

The product of any number and its reciprocal is 1.

$$\frac{2}{3} \times \frac{3}{2} = \frac{6}{6} = 1 \qquad \frac{1}{2} \times \frac{2}{1} = \frac{2}{2} = 1$$

Notice that reciprocals appear when we ask these division questions:

How many $\frac{1}{2}$'s are in 1? Answer: 2 $\left(\text{or } \frac{2}{1}\right)$

How many $\frac{1}{3}$'s are in 1? Answer: 3 $\left(\text{or } \frac{3}{1}\right)$

How many $\frac{1}{4}$'s are in 1? Answer: 4 $\left(\text{or } \frac{4}{1}\right)$

How much of 4 is in 1? Answer: $\frac{1}{4}$

The reciprocal also appears as the answer to this question:

How many $\frac{2}{3}$'s are in 1? Answer: $1\frac{1}{2}$ $\left(\text{or } \frac{3}{2}\right)$

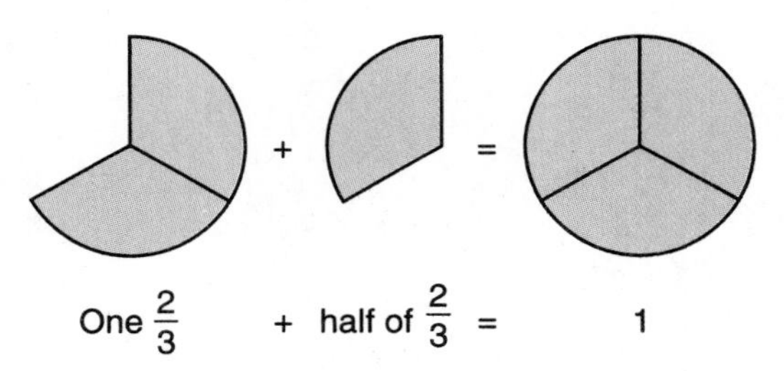

When we divide 1 by any number (except 0), the answer is the reciprocal of the number.

Example 1 What is the reciprocal of $\frac{5}{6}$?

Solution The reciprocal of $\frac{5}{6}$ is $\mathbf{\frac{6}{5}}$. We leave the answer as an improper fraction.

Example 2 What is the product of $\frac{1}{3}$ and its reciprocal?

Solution The reciprocal of $\frac{1}{3}$ is $\frac{3}{1}$. To find the product, we multiply.

$$\frac{1}{3} \times \frac{3}{1} = \mathbf{1}$$

The product of any fraction and its reciprocal is 1.

Example 3 What is the reciprocal of 4?

Solution To find the reciprocal of a whole number, we may first write the whole number as a fraction by writing a 1 under it. To write 4 as a fraction, we write $\frac{4}{1}$. The reciprocal of $\frac{4}{1}$ is $\mathbf{\frac{1}{4}}$.

Example 4 Divide: $1 \div \frac{3}{4}$

Solution This problem means, "How many $\frac{3}{4}$'s are in 1?" When we divide 1 by any number other than zero, the quotient is the reciprocal. So the answer to this division is the reciprocal of $\frac{3}{4}$, which is $\frac{4}{3}$, or $1\frac{1}{3}$. We check the answer by multiplying the quotient $\frac{4}{3}$ by the divisor $\frac{3}{4}$.

$$\frac{4}{3} \times \frac{3}{4} = \frac{12}{12} = 1$$

The result is the original dividend, 1, so the answer is correct.

LESSON PRACTICE

Practice set Write the reciprocal of each number in problems **a–l.** Leave improper fractions as improper fractions.

a. $\frac{4}{5}$ **b.** $\frac{6}{5}$ **c.** 3 **d.** $\frac{7}{8}$

e. $\frac{3}{8}$ **f.** 5 **g.** $\frac{3}{10}$ **h.** $\frac{5}{12}$

i. 2 **j.** $\frac{1}{5}$ **k.** 10 **l.** 1

m. How many $\frac{3}{5}$'s are in 1? **n.** Divide: $1 \div \frac{4}{5}$

o. Think of a fraction and write it down. Then write its reciprocal. Multiply the two fractions. What is the product? (Be sure to show your work.)

MIXED PRACTICE

Problem set

1. (50) These three boxes of nails weigh 35 lb, 42 lb, and 34 lb. If some nails are moved from the heaviest box to the other two boxes so that all three boxes weigh the same, how much will each box weigh?

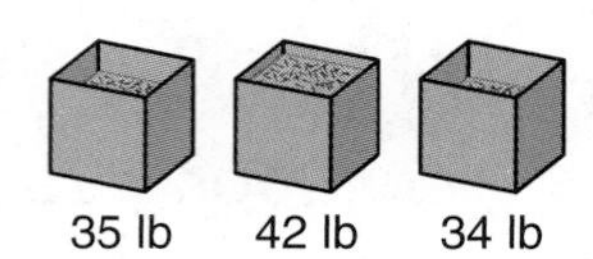

2. (11) Each finger of the human hand is formed by three bones except for the thumb, which is formed by two bones. The palm contains five bones, one leading to each finger. Not counting the bones in the wrist, the hand contains how many bones?

3. (71) Name the shaded portion of this square as a decimal number, as a reduced fraction, and as a percent.

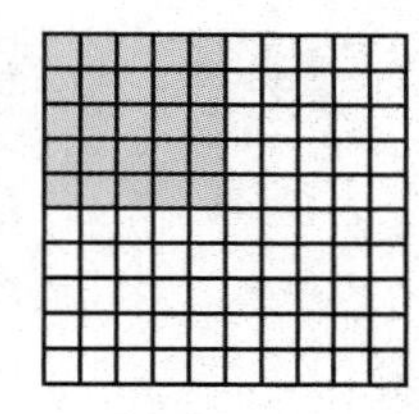

4. (95) What is the product of $\frac{2}{3}$ and its reciprocal?

5. (87) (a) A quarter is what fraction of a dollar?

(b) How many quarters equal \$1?

(c) How many quarters equal \$5?

6. (95) What is the reciprocal of $\frac{3}{4}$? What is the product of $\frac{3}{4}$ and its reciprocal?

7. (92) Which of the following means, "How many 25's are there in 500?"

A. 25 ÷ 500 B. 500 ÷ 25 C. 25 × 500

8. (95) (a) What is the reciprocal of 6?

(b) What is the reciprocal of $\frac{1}{4}$?

9. (31, 32, 61) If $\overline{LN}$ is perpendicular to $\overline{JM}$, then which of these angles is an acute angle?

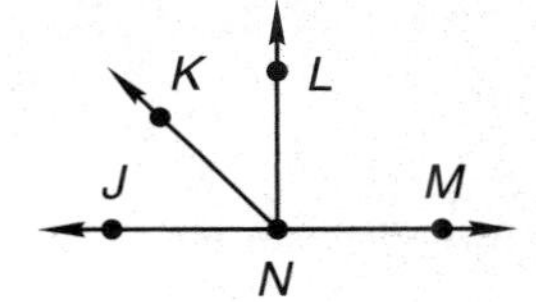

A. $\angle LNM$ B. $\angle JNL$

C. $\angle KNL$ D. $\angle KNM$

10. (13, 26) (\$20 − \$4.72) ÷ 8

11. (56) 160 × \$1.25

12. (24, 73) 25.45 − (1.4 + 0.28)

13. (78) 100^2

14. (94) $31\overline{)140}$

15. (26, 94) $27x = 567$

16. (90) Reduce: $\frac{15}{25}$

17. (91) $1\frac{5}{6} + 1\frac{5}{6}$

18. (81) $4\frac{5}{6} - 1\frac{1}{6}$

19. (86) $\frac{3}{8}$ of 24

20. (86, 91) $3 \times \frac{4}{5}$

21. (87) $\frac{9}{10} \div \frac{1}{10}$

22. (79) Write fractions equal to $\frac{3}{4}$ and $\frac{1}{6}$ that have denominators of 12. Subtract the smaller fraction from the larger fraction.

23. Divide 123 by 10 and write the answer as a mixed number.
(58)

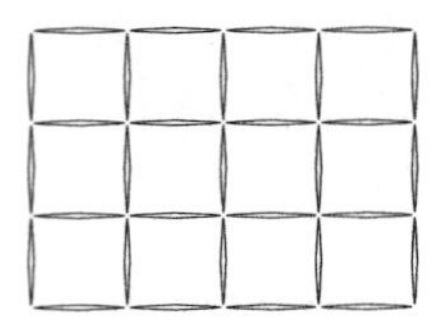

Isabella used toothpicks to make this rectangle. Refer to this rectangle to answer problems 24 and 25.

24. How many toothpicks form the perimeter of this rectangle?
(53)

25. The rectangle closes in an area covered with small squares. How many small squares cover the area of the rectangle?
(72)

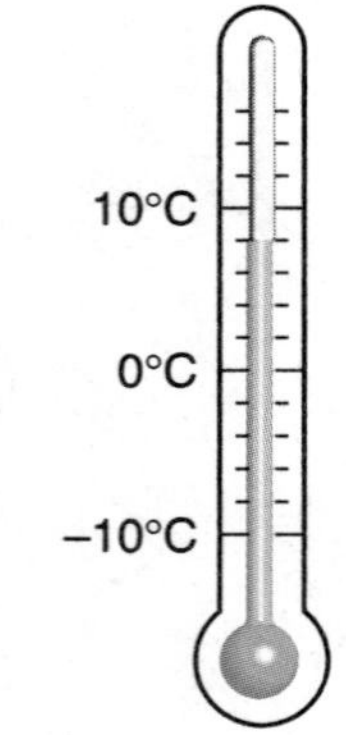

26. Robert awoke on a cool fall morning and looked at the thermometer outside his window. What temperature is indicated on the thermometer?
(27)

27. Copy the thermometer on your paper. Along the right-hand side of the thermometer, write the Fahrenheit temperatures for 10°C, 0°C, and –10°C. (Recall that a difference of 10°C is equal to a difference of 18°F.)
(27)

28. $\sqrt{100} - \sqrt{36}$
(89)

29. Maggie tossed two standard number cubes. She needs to roll a 12 to win the game. What word best describes her chances of rolling 12 in one try?
(Inv. 6)

A. certain
B. likely
C. unlikely
D. impossible

30. Which of these Venn diagrams illustrates the relationship between squares (S) and rhombuses (R)? Explain your answer.
(45, Inv. 8)

A. S R
B. S R
C. S R
D. R S

LESSON

96 Using Reciprocals to Divide Fractions

WARM-UP

Facts Practice: 40 Fractions to Reduce (Test I)

Mental Math:

a. What is the reduced mixed number for $\frac{14}{4}$?

b. What is the reciprocal of $\frac{5}{6}$?

c. $\frac{1}{4}$ of 12 **d.** $\frac{2}{4}$ of 12 **e.** $\frac{3}{4}$ of 12

Roman numerals:

f. Write 34 in Roman numerals.

g. Write XXIII in our number system.

Problem Solving:

Kerry is wearing a necklace with 30 beads strung in a red-white-blue-red-white-blue pattern. If she counts beads in the direction shown, starting with red, what will be the color of the one hundredth bead?

NEW CONCEPT

Reciprocals can help us solve division problems such as the following:

$$\frac{1}{2} \div \frac{2}{3}$$

This problem means "How many $\frac{2}{3}$'s are in $\frac{1}{2}$?" However, since $\frac{2}{3}$ is more than $\frac{1}{2}$, the answer is less than 1. So we change the question to

"How much of $\frac{2}{3}$ is in $\frac{1}{2}$?"

"How much of [circle with 2 of 3 parts shaded] is in [circle with 1 of 2 parts shaded] ?"

This problem is different from the problems we have been solving. To solve this problem, we will use another method.

This method uses reciprocals to help us find the answer. We begin by asking a different question: "How many $\frac{2}{3}$'s are in 1?" Once we know how many $\frac{2}{3}$'s are in 1, then we can find how much of $\frac{2}{3}$ is in $\frac{1}{2}$.

Step 1: How many $\frac{2}{3}$'s are in 1? The answer is $\frac{3}{2}$, which is the reciprocal of $\frac{2}{3}$.

Step 2: The number of $\frac{2}{3}$'s in $\frac{1}{2}$ is *half* the number of $\frac{2}{3}$'s in 1. So we multiply $\frac{3}{2}$ by $\frac{1}{2}$.

$$\frac{1}{2} \times \frac{3}{2} = \frac{3}{4}$$

This method changes the division problem into a multiplication problem. Instead of dividing $\frac{1}{2}$ by $\frac{2}{3}$, we end up multiplying $\frac{1}{2}$ by the reciprocal of $\frac{2}{3}$.

$$\frac{1}{2} \div \frac{2}{3} = ?$$

$$\frac{1}{2} \times \frac{3}{2} = \frac{3}{4}$$

Example 1 Divide: $\frac{2}{3} \div \frac{1}{2}$

Solution We are finding the number of $\frac{1}{2}$'s in $\frac{2}{3}$. The number of $\frac{1}{2}$'s in 1 is $\frac{2}{1}$. So the number of $\frac{1}{2}$'s in $\frac{2}{3}$ is $\frac{2}{3}$ of $\frac{2}{1}$. We multiply $\frac{2}{3}$ by the reciprocal of the second fraction, $\frac{1}{2}$. We simplify the answer $\frac{4}{3}$ to get $\mathbf{1\frac{1}{3}}$.

$$\frac{2}{3} \div \frac{1}{2}$$

$$\frac{2}{3} \times \frac{2}{1} = \frac{4}{3}$$

$$= \mathbf{1\frac{1}{3}}$$

Example 2 Divide: $2 \div \frac{2}{3}$

Solution We are finding the number of $\frac{2}{3}$'s in 2. The number of $\frac{2}{3}$'s in 1 is $\frac{3}{2}$. So the number of $\frac{2}{3}$'s in 2 is twice that many. We write the whole number 2 as the fraction $\frac{2}{1}$. Then we multiply $\frac{2}{1}$ by the reciprocal of $\frac{2}{3}$. Finally, we simplify the answer and find that the number of $\frac{2}{3}$'s in 2 is **3**.

$$\frac{2}{1} \div \frac{2}{3}$$

$$\frac{2}{1} \times \frac{3}{2} = \frac{6}{2}$$

$$= \mathbf{3}$$

LESSON PRACTICE

Practice set* Divide:

a. $\frac{1}{3} \div \frac{1}{2}$ **b.** $\frac{2}{3} \div \frac{3}{4}$ **c.** $\frac{2}{3} \div \frac{1}{4}$

d. $\frac{1}{2} \div \frac{1}{3}$ **e.** $\frac{3}{4} \div \frac{2}{3}$ **f.** $3 \div \frac{3}{4}$

g. $2 \div \frac{1}{3}$ **h.** $3 \div \frac{2}{3}$ **i.** $10 \div \frac{5}{6}$

j. How many $\frac{1}{3}$'s are in $\frac{3}{4}$?

k. How much of $\frac{3}{4}$ is in $\frac{1}{3}$?

MIXED PRACTICE

Problem set

1. (37) Draw two circles. Shade $\frac{1}{2}$ of one circle and $\frac{2}{3}$ of the other circle.

2. (71, 76) James gave Robert half of a candy bar. Robert gave his sister half of what he had. What fraction of the whole candy bar did Robert's sister get? What percent of the whole candy bar did she get?

3. (86) How much is $\frac{2}{3}$ of one dozen?

4. (62) Estimate the product of 712 and 490 by rounding both numbers to the nearest hundred before multiplying.

5. (52) Use digits to write the number ninety-three million, eight hundred fourteen thousand, two hundred.

6. (87) Which of these means, "How many one tenths are there in three?"

A. $\frac{1}{10} \div 3$ B. $3 \div \frac{1}{10}$ C. $\frac{1}{10} \div \frac{3}{10}$

7. (79) Write fractions equal to $\frac{1}{4}$ and $\frac{1}{5}$ that have denominators of 20. Then add the fractions.

8. (96) (a) $1 \div \frac{1}{10}$ (b) $3 \div \frac{1}{10}$

9. (15) Recall that the multiples of a number are the numbers we say when counting by that number. The first four multiples of 2 are 2, 4, 6, and 8. What are the first four multiples of 3?

10. (46, 86) The blossom of the saguaro cactus is the state flower of Arizona. A saguaro cactus can weigh as much as 10 tons. About $\frac{3}{4}$ of a saguaro's weight comes from the water it stores inside of it. If a saguaro cactus weighs 10 tons, about how much of its weight is water?

11. (61) *AB* is 3 cm. *BC* is 4 cm. *AD* is 10 cm. Find *CD.*

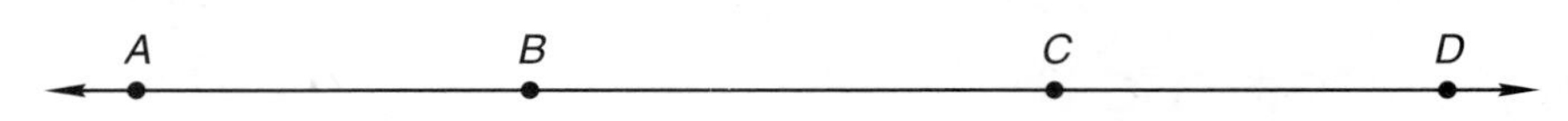

12. (71) Name the shaded portion of this square as a decimal number, as a reduced fraction, and as a percent.

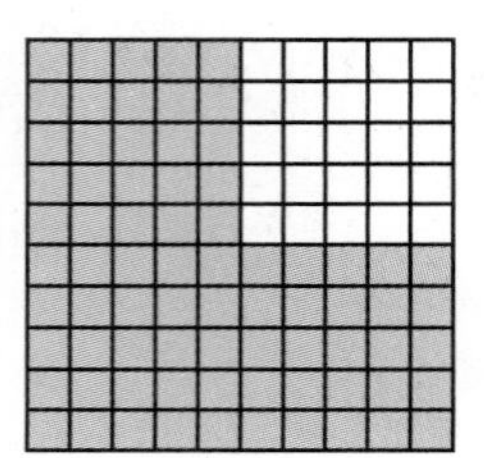

13. (96) $\frac{1}{3} \div \frac{1}{4}$

14. (96) $\frac{1}{4} \div \frac{1}{3}$

15. (96) $3 \div \frac{1}{2}$

16. (73) $m + 1.4 = 3.75$

17. (73) $m - 1.4 = 3.75$

18. (79) $\frac{1}{10} \times \square = \frac{10}{100}$

19. (70) $20 \times 47¢ = \$$_____

20. (94) $568 \div 15$

21. (54) $30\overline{)427}$

22. (26, 34) $6m = \$30.24$

23. (76, 86, 91) $5 \times \left(\frac{2}{3} \times \frac{1}{2}\right)$

24. (43, 63) $5 - \left(1\frac{1}{4} + 2\right)$

25. (78, 89) Compare: $\sqrt{100} \bigcirc 5^2$

At Walton School there are 15 classrooms. The numbers of students in each classroom are listed below. Use this information to answer problems 26–28.

20, 18, 30, 20, 22, 28, 31, 20, 27, 30, 26, 31, 20, 24, 28

26. (84) What is the mode of the number of students in the classrooms?

27. (84) What is the range?

28. (84) What is the median number of students in the classrooms?

In this figure triangles *ABC* and *ADC* are congruent. Refer to the figure to answer problems 29 and 30.

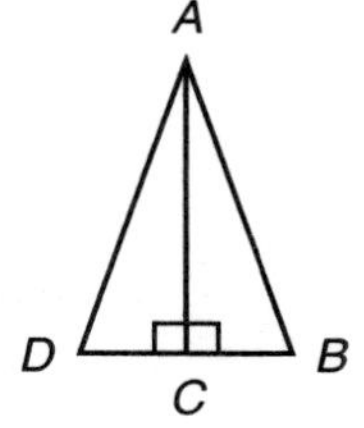

29. (36, 61) Classified by sides, what type of triangle is triangle *ABD*?

30. (61, 88) What single transformation would move triangle *ABC* to the position of triangle *ADC*?

LESSON 97 Ratios

WARM-UP

Facts Practice: 40 Fractions to Reduce (Test I)

Mental Math:

a. Simplify: $\frac{7}{6}$, $\frac{8}{6}$, $\frac{9}{6}$ **b.** What is the reciprocal of $\frac{1}{3}$?

c. $\frac{1}{3}$ of 100 **d.** $33\frac{1}{3} + 33\frac{1}{3}$ **e.** $\frac{2}{3}$ of 100

f. 50 × 10, ÷ 2, − 10, ÷ 4, + 3, ÷ 9

Roman numerals:

g. Write 35 in Roman numerals.

h. Write XXXIV in our number system.

Problem Solving:

George was down to three clean socks, one red, one white, and one blue. How many combinations of two socks can George make from these three socks?

For each combination of two socks, George could choose between two permutations of the socks. For example, George could wear a red sock on his left foot and a white sock on his right foot (R, W), or he could switch the socks (W, R). List all the permutations of two socks George could make.

NEW CONCEPT

A **ratio** is a way of describing a relationship between two numbers.

If there are 12 boys and 18 girls in a class, then the ratio of boys to girls in the class is 12 to 18.

We often write ratios as fractions. We write the terms of the ratio in order from top to bottom.

The ratio "12 to 18" is written $\frac{12}{18}$.

We read the ratio $\frac{12}{18}$ by saying "twelve to eighteen."

We reduce ratios just as we reduce fractions. Since 12 and 18 are both divisible by 6, we divide each term of $\frac{12}{18}$ by 6.

$$\frac{12 \div 6}{18 \div 6} = \frac{2}{3}$$

So the ratio of boys to girls in the class is $\frac{2}{3}$ ("two to three"). This means that for every two boys in the class, there are three girls.

Example There were 12 girls and 16 boys in the class. What was the ratio of boys to girls?

Solution First we place the numbers in the correct order. We are asked for the ratio of boys to girls. Since we follow the order from top to bottom, we write the number of boys as the numerator and the number of girls as the denominator.

$$\frac{\text{boys}}{\text{girls}} \quad \frac{16}{12}$$

Unlike fractions, we do not write ratios as mixed numbers. The top number of a ratio may be greater than the bottom number. However, we do reduce ratios. Since the terms of the ratio, 16 and 12, are both divisible by 4, we reduce the ratio as follows:

$$\frac{16 \div 4}{12 \div 4} = \frac{\mathbf{4}}{\mathbf{3}}$$

The ratio of boys to girls in the class was $\frac{4}{3}$.

LESSON PRACTICE

Practice set There were 20 prairie dogs and 30 jackrabbits in Henry's backyard.

a. What was the ratio of jackrabbits to prairie dogs?

b. What was the ratio of prairie dogs to jackrabbits?

There were 8 red socks and 10 blue socks in George's drawer.

c. What was the ratio of red socks to blue socks?

d. What was the ratio of blue socks to red socks?

MIXED PRACTICE

Problem set

1. (97) There were 15 pennies and 10 nickels in Kordell's drawer. What was the ratio of pennies to nickels in his drawer?

2. (6, 41) Write this sentence using digits and symbols:

The sum of one fourth and one fourth is one half.

3. (49) Cynthia had 4 one-dollar bills, 3 quarters, 2 dimes, and 1 nickel. If she spent half of her money, how much money does she have left?

4. How many $\frac{1}{8}$'s are in $\frac{1}{2}$?
(87)

5. Name the number of shaded circles as a decimal number and as a reduced mixed number.
(71, 81)

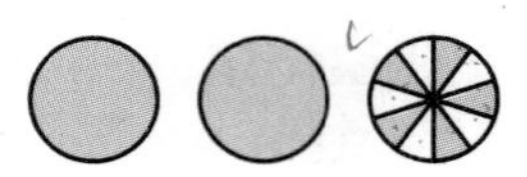

6. When the decimal number eleven and twelve hundredths is subtracted from twelve and eleven hundredths, what is the difference?
(68, 73)

7. (a) A quart is what fraction of a gallon?
(85, 86)

(b) How many quarts are in 1 gallon?

(c) How many quarts are in 4 gallons?

8. Write fractions equal to $\frac{2}{3}$ and $\frac{2}{5}$ that have denominators of 15. Then subtract the smaller fraction from the larger fraction.
(79)

9. Name the point marked by the arrow as a decimal number and as a fraction.
(66)

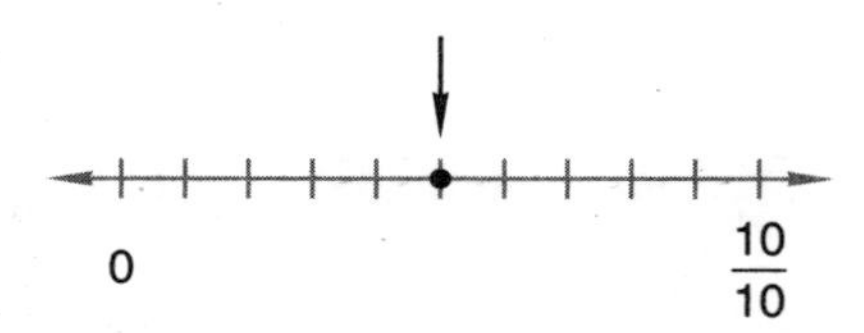

10. Compare: $\frac{1}{2} \div 2 \bigcirc 2 \div \frac{1}{2}$
(96)

11. AB is 30 millimeters. CD is 40 millimeters. AD is 90 millimeters. Find BC.
(61)

A B C D

12. $3 \div \frac{2}{3}$
(96)

13. $\frac{2}{3} \div 3$
(96)

14. $\frac{7}{10} + \frac{7}{10}$
(91)

15. $43.15 + 8.69 + 7.2 + 5.0$
(73)

16. (\$10 − 19¢) ÷ 9
(24, 70)

17. 6 × 72¢ = \$_____
(70)

18. 35^2
(78)

19. $24\overline{)500}$
(92)

20. Reduce: $\frac{50}{100}$
(90)

21. $12y = 1224$
(26, 92)

22. $5\frac{3}{4} - \left(3 - 1\frac{3}{4}\right)$
(63, 81)

23. $1\frac{1}{4} + 1\frac{1}{4} + 1\frac{1}{4} + 1\frac{1}{4}$
(59)

24. $\frac{3}{10} = \frac{\square}{100}$
(79)

25. (a) What is the length of each side of this square?
(44, 53)

(b) What is the perimeter of this square?

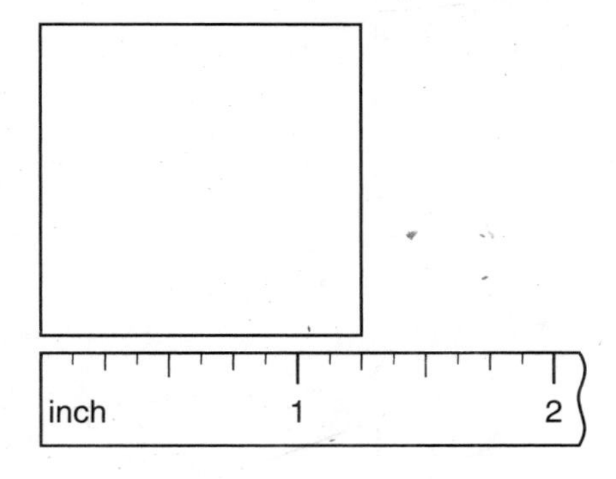

26. If the area of a square is 64 square inches, then what is the length of each side?
(72, 89)

27. What number is $\frac{1}{64}$ of 640?
(86, 92)

28. What are the next three terms in this Fibonacci sequence?
(Inv. 7)

1, 1, 2, 3, 5, 8, 13, 21, _____, _____, _____, ...

29. (a) List the factors of 64 from least to greatest.
(2, 25, 84, 89)

(b) Is the number of factors an odd or even number?

(c) What is the median of the factors?

(d) What is $\sqrt{64}$?

30. There are 50 stars and 13 stripes on the United States flag. What is the ratio of stripes to stars on the flag?
(97)

LESSON

98 Negative Numbers

WARM-UP

Facts Practice: 40 Fractions to Reduce (Test I)

Mental Math:

a. Simplify: $\frac{6}{4}$, $\frac{7}{4}$, $\frac{8}{4}$ b. What is the reciprocal of $\frac{3}{4}$? ... of $\frac{1}{4}$?

c. 50% of $20 d. 25% of $20 e. 10% of 100

f. $\frac{1}{3}$ of 21, × 2, + 1, ÷ 3, × 6, + 2, ÷ 4

Roman numerals:

g. Write 18 in Roman numerals.

h. Write XVII in our number system.

Problem Solving:

Some 1-inch cubes were used to build this rectangular solid. How many 1-inch cubes were used?

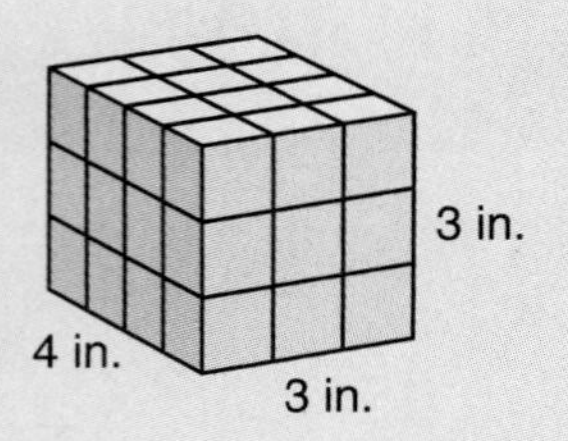

NEW CONCEPT

Numbers that are greater than zero are **positive numbers.** Numbers that are less than zero are **negative numbers.** Zero is neither positive nor negative. On the number line below, we show both positive and negative numbers. We write negative numbers with a minus sign in front. Point A is at –3, which we read as "negative three."

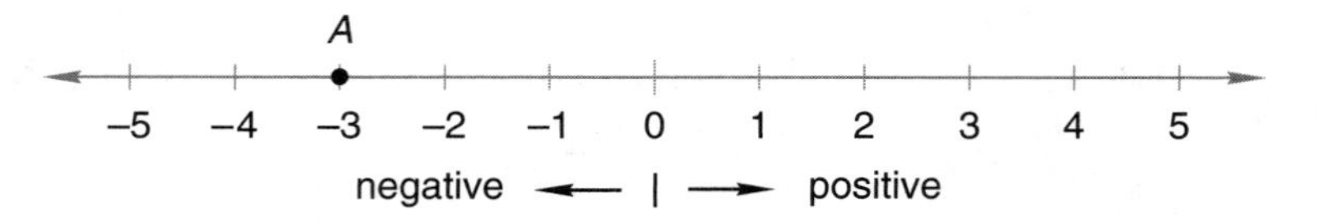

One place we might see negative numbers is on a thermometer. On a very cold day the temperature might drop below zero. If the temperature is four degrees below zero, we might say the temperature is "minus four."

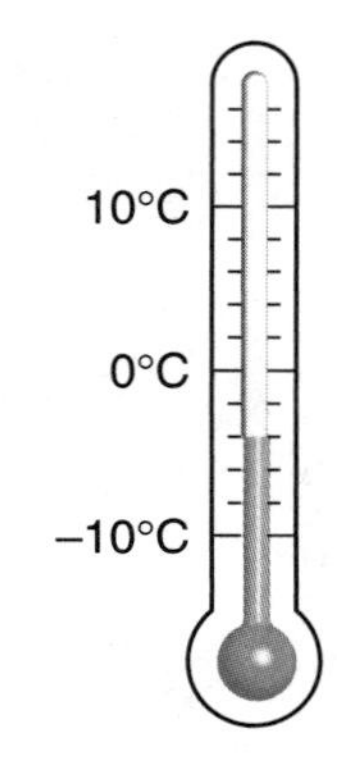

We also use negative numbers in situations where money is borrowed. Suppose Brad has no money and borrows five dollars from his brother. Now suppose Brad spends the money he borrowed. Brad would then have less than no money, because he would still owe his brother five dollars. We can use the number –5 to describe Brad's debt in dollars.

Example 1 Point *B* represents what number on this number line?

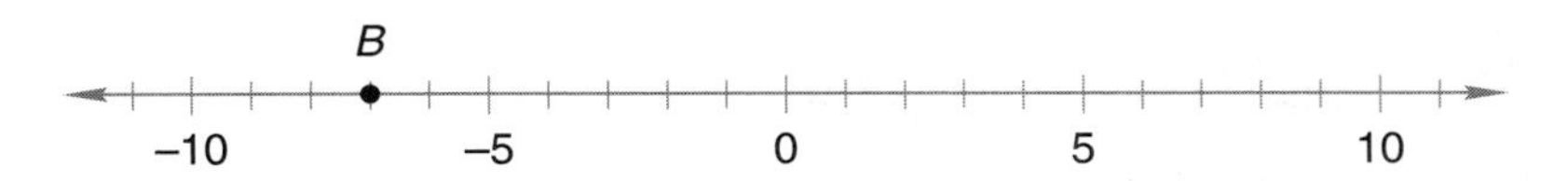

Solution We see that the distance between tick marks is one unit. Counting back from zero, we find that point *B* represents **–7.**

Example 2 What temperature is indicated on this thermometer?

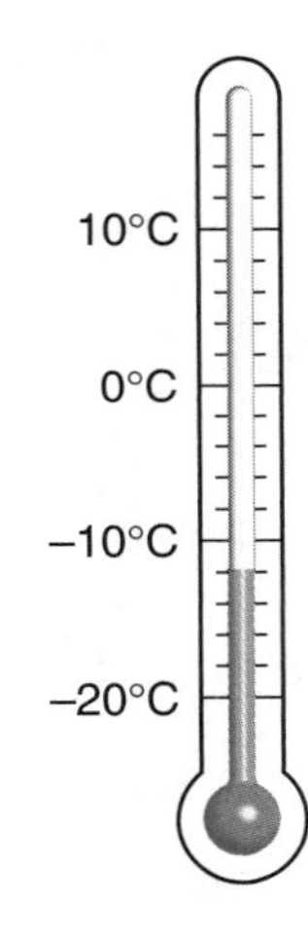

Solution The distance between tick marks is two degrees. Counting down from 0°, we find that the thermometer indicates a temperature of **–12°C.**

Example 3 Use a negative number to indicate a debt of 200 dollars.

Solution **–200 dollars**

LESSON PRACTICE

Practice set

a. Draw a number line and put a dot at negative four. Label the point with the letter *D.*

b. Point *F* represents what number on this number line?

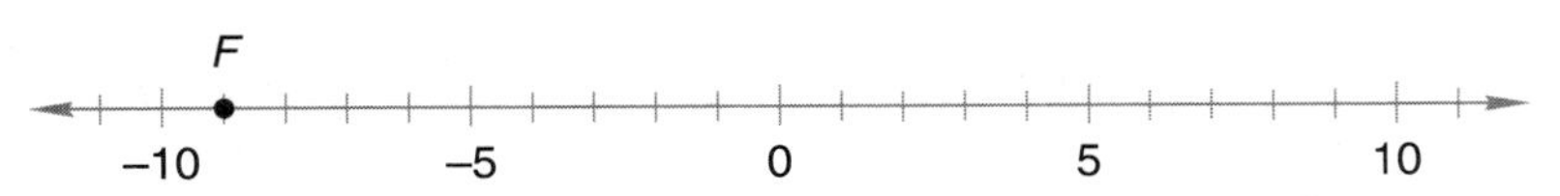

c. Use digits and symbols to write the temperature that is twelve degrees below zero on the Fahrenheit scale.

d. What temperature is shown on this thermometer?

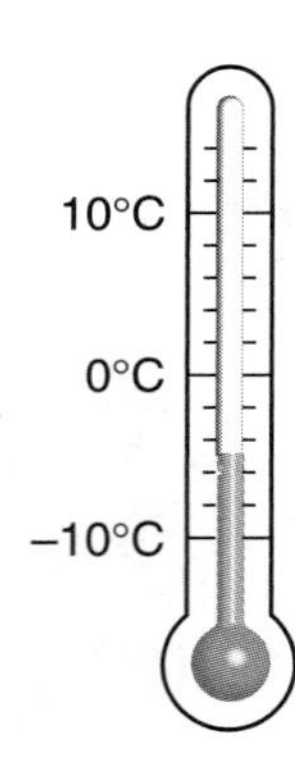

e. Use a negative number to indicate a debt of 20 dollars.

f. Which number is neither negative nor positive?

MIXED PRACTICE

Problem set

1. There were 12 dogs and 8 cats at the class pet show. What was the ratio of cats to dogs at the show?
(97)

2. (a) What is the name of this solid?
(83)
(b) How many faces does it have?

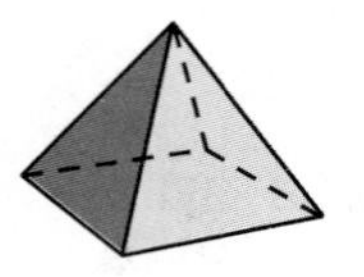

3. Juan lives 1.2 miles from school. How far does he travel going from home to school and back home?
(73)

4. One day in January the temperature at 9:00 p.m. in Juneau, Alaska, was 2°F. By 11:00 p.m. the temperature had dropped 5°. What was the temperature at 11:00 p.m.?
(98)

5. Which arrow could be pointing to $2\frac{1}{3}$ on the number line below?
(38)

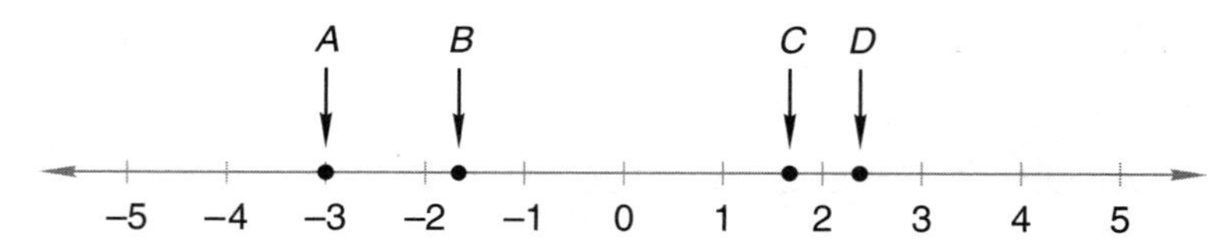

6. Which arrow in problem 5 is pointing to negative three?
(98)

7. If $\overline{LN}$ is perpendicular to $\overline{JM}$, then which of these angles is obtuse?
(31, 32, 61)

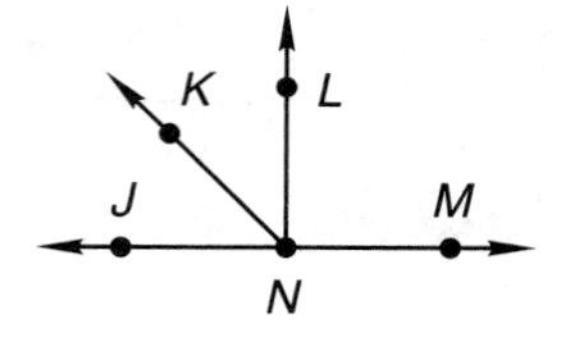

A. ∠*JNK*  B. ∠*KNL*

C. ∠*KNM* D. ∠*LNM*

8. 6.5 + 2.47 + 0.875
(73)

9. 4.26 + 8.0 + 15.9
(73)

10. 23.45 − 1.2
(73)

11. 0.367 − 0.1
(73)

12. $1.25 × 7
(17)

13. 750 × 608
(56)

14. 364 ÷ 16
(94)

15. $7.20 ÷ 20
(54)

16. (59) $3\frac{1}{2} + 1\frac{1}{2}$

17. (41) $5\frac{8}{15} - 4\frac{7}{15}$

18. (63) $6 - 1\frac{1}{3}$

19. (86) $1 \times \frac{5}{7}$

20. (86) $\frac{4}{5}$ of 25

21. (96) $\frac{3}{4} \div \frac{2}{3}$

22. (79) $\frac{7}{10} = \frac{\square}{100}$

23. (90) Reduce: $\frac{30}{100}$

24. (27) This thermometer shows the temperature on a warm summer morning. What is the temperature shown?

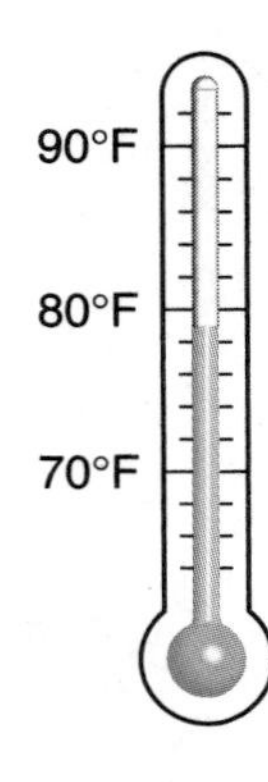

25. (78, 89) $9^2 + \sqrt{81}$

26. (82, 90) (a) Find the greatest common factor (GCF) of 70 and 100.

(b) Use the GCF of 70 and 100 to reduce $\frac{70}{100}$.

27. (76) Compare:

(a) $\frac{1}{2} \times \frac{1}{3} \bigcirc \frac{1}{2}$

(b) $\frac{1}{2} \times \frac{1}{3} \bigcirc \frac{1}{3}$

28. (98) This thermometer shows the temperature on a cold winter evening. What is the temperature shown?

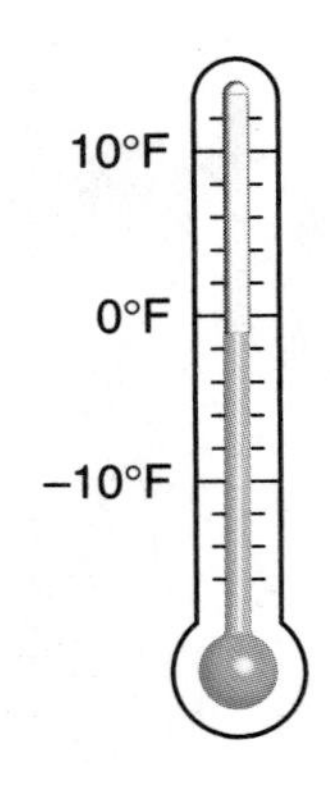

29. (80) Which of these numbers is a composite number?

A. 3 B. 5 C. 7 D. 9

30. (69) Arrange these decimal numbers in order from least to greatest:

0.376 0.037 0.38 0.367

LESSON

99 Adding and Subtracting Whole Numbers and Decimal Numbers

WARM-UP

Facts Practice: 40 Fractions to Reduce (Test I)

Mental Math:

a. Simplify: $\frac{10}{8}$, $\frac{11}{8}$, $\frac{12}{8}$

b. What is the reciprocal of $\frac{1}{2}$? ... of 2?

c. $\frac{1}{8}$ of 100

d. $12\frac{1}{2} + 12\frac{1}{2} + 12\frac{1}{2}$

e. $\frac{3}{8}$ of 100

f. $\sqrt{64}$, × 6, ÷ 8, × 4, ÷ 3

Roman numerals:

g. Write 37 in Roman numerals.

h. Write XXXII in our number system.

Problem Solving:

Two cups equal a pint. Two pints equal a quart. Two quarts equal a half gallon. Two half gallons equal a gallon. If a gallon container of water is used to fill a half-gallon container, a quart container, a pint container, and a cup container, how much water will be left in the gallon container?

NEW CONCEPT

Sometimes we need to add whole numbers and decimal numbers in the same problem. Here is an example:

The Kubats hired a carpenter to cut an opening in their wall and to install a new door. The carpenter needed to order a frame for the door to cover the thickness of the wall. The carpenter knew that the siding was 1 inch thick, the wall stud was 3.5 inches thick, and the drywall was 0.5 inches thick.

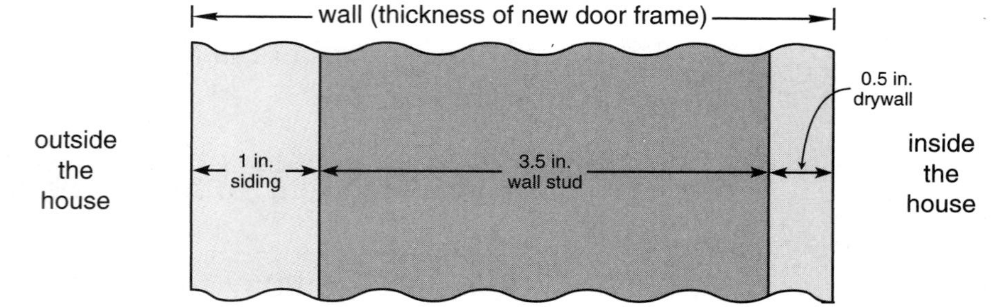

To find the thickness of the wall, the carpenter writes 1 inch as 1.0 inches, aligns the decimal points of all three measurements, and adds. He finds that a 5.0-inch-thick door frame is needed.

$$\begin{array}{r} 1.0 \text{ in.} \\ 3.5 \text{ in.} \\ + \ 0.5 \text{ in.} \\ \hline 5.0 \text{ in.} \end{array}$$

The carpenter wrote the whole number 1 as the decimal number 1.0 so that he could align the decimal points before adding. Since a decimal point marks the end of a whole number, we may add a decimal point to the back (right-hand side) of a whole number. After placing the decimal point, we may also attach zeros to make arithmetic with the whole number easier.

When adding whole numbers to decimal numbers, it might help to remember the game "Pin the Tail on the Donkey." The tail belongs on the back of the donkey, and the decimal point belongs on the back of the whole number. Remember this rule:

"Pin the decimal point on the back of the whole number."

Example 1 Add: 6.2 + 3 + 4.25

Solution To add digits with the same place value, we align decimal points. In this problem the whole number 3 has the same place value as the 6 and the 4. We place a decimal point to the right of the 3 and align decimal points. We may fill empty decimal places with zeros if we wish.

$$\begin{array}{r} 6.20 \\ 3.00 \\ + \ 4.25 \\ \hline \mathbf{13.45} \end{array}$$

Example 2 Subtract: 24.6 − 8

Solution We place a decimal point to the right of the whole number 8 and then align decimal points before subtracting. We may fill the empty decimal place with a zero if we wish.

$$\begin{array}{r} 24.6 \\ - \ \ 8.0 \\ \hline \mathbf{16.6} \end{array}$$

Example 3 Which digit in 4.65 is in the same place as the 2 in 12?

Solution The 2 in 12 is in the ones place. In 4.65 a decimal point separates the ones place and the tenths place, marking the end of the whole number and the beginning of the fraction. So the **4** in 4.65 is in the same place as the 2 in 12.

LESSON PRACTICE

Practice set* Find each sum or difference:

a. 4.3 + 2

b. 12 + 1.2

c. 6.4 + 24

d. 4 + 1.3 + 0.6

e. 5.2 + 0.75 + 2

f. 56 + 75.4

g. 8 + 4.7 + 12.1

h. 9 + 4.8 + 12

i. 4.75 − 2

j. 12.4 − 5

k. Which digit in 24.7 is in the same place as the 6 in 16?

l. Compare: 12 ◯ 12.0

MIXED PRACTICE

Problem set

1. (97) There were 50 boys and 60 girls on the playground. What was the ratio of girls to boys on the playground?

2. (37, 71, 81) The pizza was sliced into 6 equal pieces. Martin ate 2 pieces. What fraction of the pizza did he eat? What percent of the pizza did he eat?

3. (49) Artichokes were on sale. Five of them cost $1. At this rate, what would be the price for a dozen artichokes?

4. (35, 99) Maria ran 100 yards in 13.8 seconds. Mike ran 1 second slower than Maria. How long did it take Mike to run 100 yards?

5. (38, 66, 81) Name point x on the number line below as both a decimal number and a reduced mixed number.

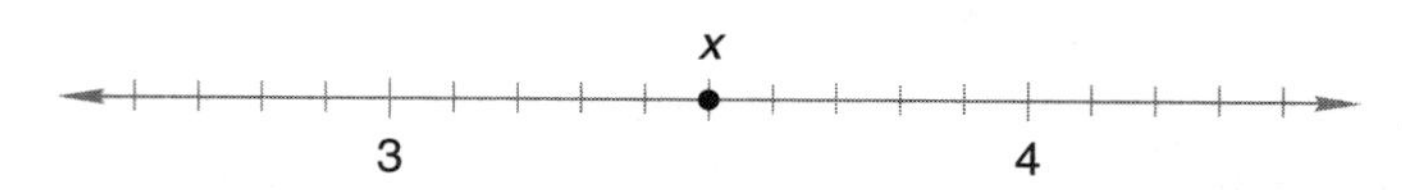

6. (78) If $10n = 100$, then n^2 equals what number?

7. (68) Write the decimal number one thousand, six hundred twenty and three tenths.

8. The doorway was 6 ft 10 in. tall and 2 ft 11 in. wide. Estimate the area of the doorway.
(62, 72)

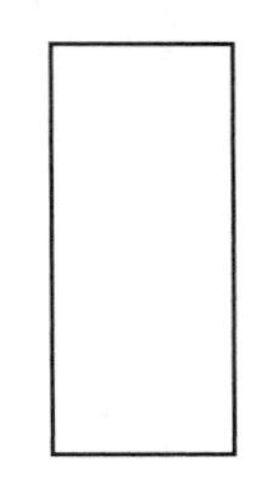

9. Write a fraction equal to $\frac{3}{4}$ that has a denominator of 8. Then subtract that fraction from $\frac{7}{8}$.
(79)

10. Is \$7.13 closer to \$7 or \$8?
(69)

11. *QT* is 100 mm. *QR* is 23 mm. *RS* equals *QR*. Find *ST*.
(61)

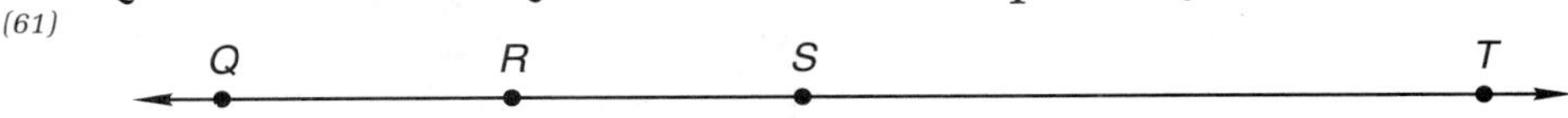

12. $3.4 + 5$
(99)

13. $7.25 - 7$
(99)

14. $\sqrt{25} - \sqrt{16}$
(89)

15. 60^2
(78)

16. $28\overline{)952}$
(94)

17. $\$18.27 \div 9$
(34)

18. $4\frac{5}{8} + 1\frac{7}{8}$
(91)

19. $5 - \left(2\frac{3}{5} - 1\right)$
(43, 63)

20. $\frac{3}{4} \times \frac{1}{3}$
(90)

21. $\frac{3}{4} \div 3$
(96)

22. $\frac{9}{10} = \frac{\square}{100}$
(79)

23. Reduce: $\frac{20}{100}$
(90)

Use this information to answer problems 24 and 25:

Tyrone fixed his function machine so that when he puts in a 12, a 6 comes out. When he puts in a 10, a 4 comes out. When he puts in an 8, a 2 comes out.

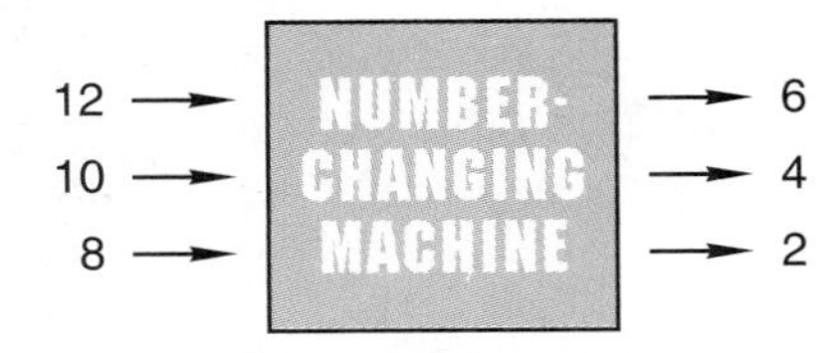

24. What rule does the machine use?
(Inv. 7)

A. It subtracts 2.
B. It divides by 2.
C. It subtracts 6.
D. It adds 6.

25. If Tyrone puts in a 6, what number will come out?
(Inv. 7)

26. Use digits and symbols to write the temperature that is ten degrees below zero Fahrenheit.
(98)

27. There is a pattern to the differences between successive terms of this sequence:
(Inv. 7)

3, 4, 7, 12, 19, ...

Assuming the pattern of differences continues, find the next three terms of the sequence.

28. A standard number cube is rolled once.
(Inv. 6)

(a) What is the probability that the upturned face is 3 or less?

(b) Describe a different event that has the same probability.

29. Write two billion, six hundred million in expanded notation using powers of 10.
(78)

30. Compare: 2 liters ◯ 1 gallon
(85)

LESSON

100 Simplifying Decimal Numbers

WARM-UP

Facts Practice: 40 Fractions to Reduce (Test I)

Mental Math:

a. Simplify: $\frac{15}{10}$, $\frac{20}{10}$, $\frac{25}{10}$ **b.** What is the reciprocal of 3? ... of $\frac{3}{5}$?

c. 25% of \$100 **d.** 25% of \$10 **e.** 25% of \$1

Roman numerals:

f. Write 27 in Roman numerals.

g. Write XXVIII in our number system.

Problem Solving:

The uppercase letter A has one line of symmetry. Write the first five letters of the alphabet in uppercase and show the line of symmetry for each letter.

NEW CONCEPT

When we write numbers, we usually write them in simplest form. To simplify a number, we change the form of the number, but we do not change the value of the number. For example, we simplify fractions by reducing. Often, we can simplify decimal numbers by removing unnecessary zeros. We will explain this by simplifying 0.20.

The decimal number 0.20 has a 2 in the tenths place and a 0 in the hundredths place. The zero in the hundredths place means "no hundredths." If we remove this zero from 0.20, we get 0.2. The number 0.2 also has a 2 in the tenths place and "no hundredths." Thus, 0.20 equals 0.2. We say that 0.20 simplifies to 0.2.

We can remove zeros from the front of whole numbers and from the back of decimal numbers. We may remove zeros until we come to a digit that is not a zero or until we come to a decimal point. Below we have simplified 02.0100, 20.0, and 0.200 by removing unnecessary zeros.

02.0100	20.0	0.200
2.01	20	0.2 or .2

In the center example, we took two steps to simplify 20.0. After removing the unnecessary zero, we also removed the decimal point. A decimal point can be removed when there is no fraction part to a number.

To simplify 0.200, we removed the two trailing zeros, leaving 0.2 as the simplified form. We can also remove the zero in front, leaving .2 as the simplified form. The numbers 0.2 and .2 are equal, and both forms are correct. However, if the whole-number part of a decimal number is zero, it is customary to write a zero in the ones place, which is what we will do in this book. Note that calculators also display a zero in the ones place of such numbers.

In some situations we might want to attach zeros to a decimal number. The decimal point of a decimal number determines place value, not the number of digits. So attaching zeros at the end of a decimal number does not change place values.

Example 1 Otis added 3.75 to 2.75 and found that the sum was 6.50. Simplify the sum.

Solution We may remove the ending zero(s) of a decimal number.

$$6.50 = \mathbf{6.5}$$

Example 2 Attach a zero to the end of 5 without changing the value of the number.

Solution If we attach a zero to 5 without using a decimal point, we get 50, which does not equal 5. So we write the whole number 5 with a decimal point and then attach a zero.

$$5 = \mathbf{5.0}$$

LESSON PRACTICE

Practice set Simplify each decimal number:

a. 03.20 **b.** 0.320 **c.** 32.00 **d.** 3.020

Simplify each answer:

e. $\begin{array}{r} 3.65 \\ +\ 6.35 \\ \hline \end{array}$ **f.** $\begin{array}{r} 23.16 \\ -\ 19.46 \\ \hline \end{array}$ **g.** $\begin{array}{r} 4.23 \\ -\ 3.18 \\ \hline \end{array}$

h. Attach a zero to the end of 2.5 without changing its value.

i. Attach a zero to the end of 6 without changing its value.

MIXED PRACTICE

Problem set

1. James counted 60 peas and 20 carrot slices on his plate. What was the ratio of carrot slices to peas on his plate?
(97)

2. A package of 10 hot dogs costs $1.25. At that price, what would be the cost of 100 hot dogs?
(49)

3. Three fourths of the 28 students finished the test early. How many students finished the test early? What percent of the students finished the test early?
(46, 71)

4. This rectangle was formed with pins 1 inch long.
(53, 72)

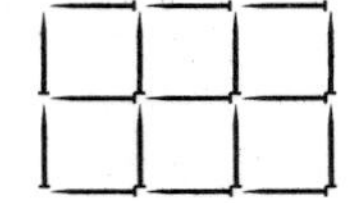

(a) How many pins form the perimeter?

(b) How many small squares cover this rectangle?

5. Attach a zero to the end of 8 without changing the value of the number.
(100)

6. (a) Which arrow could be pointing to $7\frac{3}{4}$ on the number line below?
(38, 98)

(b) Which arrow could be pointing to negative 2?

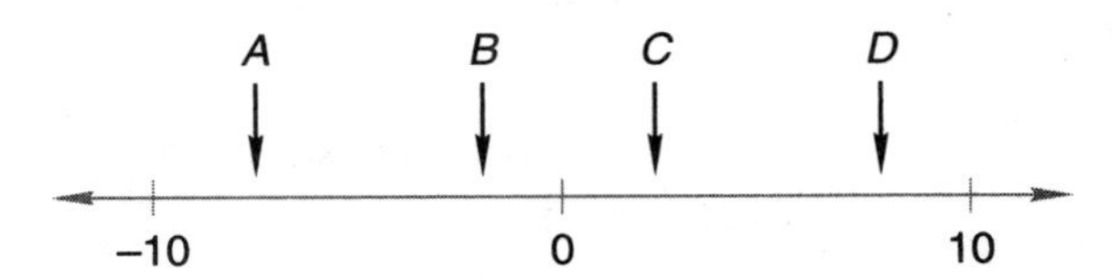

7. Write fractions equal to $\frac{5}{6}$ and $\frac{3}{4}$ that have denominators of 12. Then subtract the smaller fraction from the larger fraction.
(79)

8. The giraffe stood 5 meters tall. Five meters is how many centimeters?
(74)

9. *AB* is 40 mm. *BC* is half of *AB*. *CD* equals *BC*. Find *AD*.
(61)

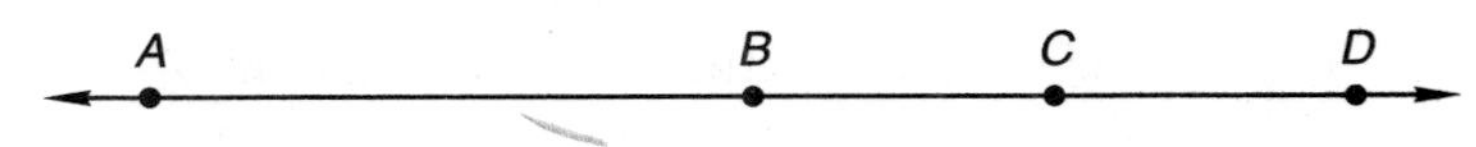

10. $6.2 + 3 + 4.25$
(99)

11. $10^3 - 10^2$
(78)

12. (99) $6.37 - 6$

13. (56) 234×506

14. (29) $10 \times \$1.75$

15. (54) $\$17.50 \div 10$

16. (79) $\frac{1}{50} = \frac{\square}{100}$

17. (90) Reduce: $\frac{40}{100}$

18. (89) $\sqrt{64}$

19. (26, 94) $16w = 832$

20. (91) $\frac{5}{9} + \frac{5}{9} + \frac{5}{9}$

21. (76) $\frac{9}{10} \times \frac{9}{10}$

22. (96) $\frac{2}{3} \div \frac{3}{4}$

23. (96) $3 \div \frac{3}{4}$

24. (74) The flagpole is 10 yards tall. The flagpole is how many feet tall?

25. (28) How many months is it from May 1 of one year to January 1 of the next year?

26. (83) A rectangular prism has

(a) how many faces?

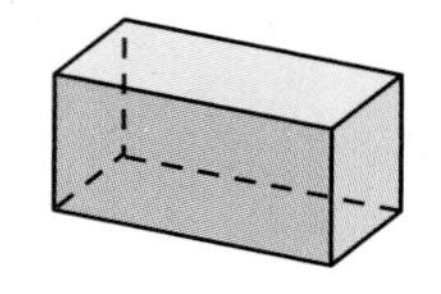

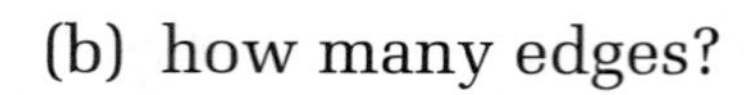

(b) how many edges?

27. (98) Water freezes at 0°C. What temperature on the Celsius scale is five degrees colder than the freezing temperature of water?

Thirty children were asked whether they have a sister and whether they have a brother. The Venn diagram below records their responses. Use this information to answer problems 28–30.

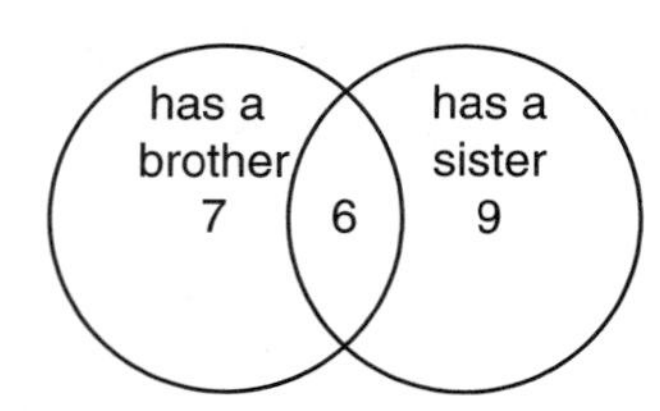

28. (Inv. 8) How many children have a brother?

29. (Inv. 8) How many have a sister but not a brother?

30. (Inv. 8) The numbers in the circles do not add to 30. What does that mean?

INVESTIGATION 10

Focus on

Graphing Points on a Coordinate Plane

If we draw two perpendicular number lines so that they intersect at their zero points, we create an area that is divided into four sections. Any point within this area can be named with two numbers, one from each number line. Here we show some examples:

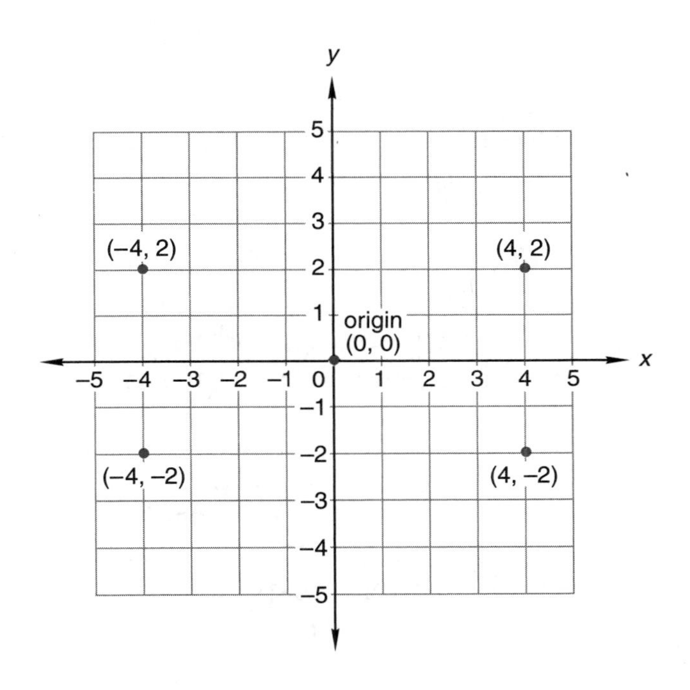

The horizontal number line is called the **x-axis**, and the vertical number line is called the **y-axis.** The numbers in parentheses are called **coordinates,** which give a point's "address." Coordinates are taken from the scales on the x-axis and y-axis. The first number in parentheses gives a point's horizontal position. The second number gives the point's vertical position. The point where the x-axis and y-axis intersect is called the **origin.** Its coordinates are (0, 0). The flat area that contains the two axes and all the points between them is a **coordinate plane.**

Refer to this coordinate plane to answer problems 1–5:

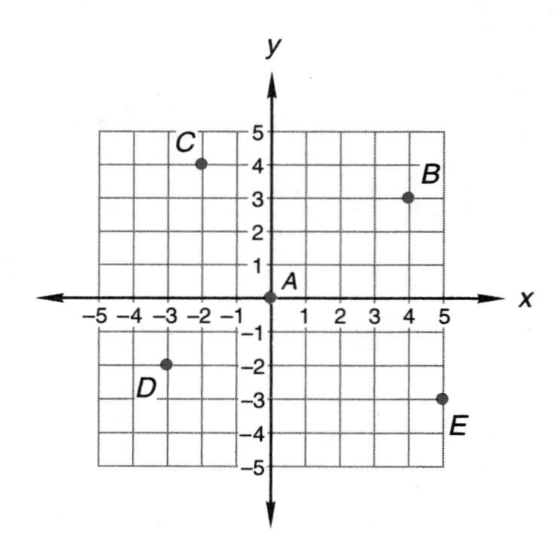

1. The coordinates of point A are (0, 0). What is the name for this point?

Write the coordinates of each of these points:

2. point B 3. point C 4. point D 5. point E

Below is a design drawn on a coordinate plane. To draw the design, we could start at (0, 4), draw a segment to (−3, −4), and then continue through the pattern back to (0, 4). We would connect the points in this order:

(0, 4) → (−3, −4) → point F →
point G → point H → (0, 4)

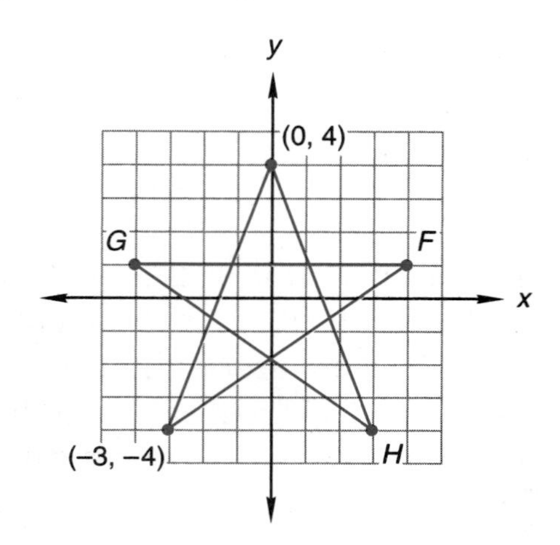

Write the coordinates of each of these points from the star design above:

6. point F 7. point G 8. point H

Activity: *Graphing Designs*

Material needed:

- graph paper or copies of Activity Master 22 (available in *Saxon Math 6/5 Assessments and Classroom Masters*)

Note: If you use graph paper, you will need to draw your own x-axis and y-axis. Be sure to draw each axis on a grid line and not between grid lines.

a. Graph each of the following points. Then make a design by connecting the points in alphabetical order. Complete the design by drawing a segment from the last point back to the first point.

$I(1, 3)$	$M(-2, -4)$
$J(-2, 3)$	$N(1, -4)$
$K(-4, 1)$	$O(3, -2)$
$L(-4, -2)$	$P(3, 1)$

b. Draw a straight-line design on graph paper. Be sure each corner of the design is at a point where grid lines meet. Then create directions for another student to re-create your design. Your directions should consist of the coordinates of each corner, listed in an order that will complete the design.

LESSON

101 Rounding Mixed Numbers to the Nearest Whole Number

WARM-UP

Facts Practice: 50 Fractions to Simplify (Test J)

Mental Math:

a. Describe how to estimate the sum of 29 and 19.

b. Simplify: $\frac{10}{3}$, $\frac{10}{4}$, $\frac{10}{5}$

c. 50% of 50, + 50, + 2, ÷ 7, + 3, ÷ 7

Roman numerals:

d. Write 33 in Roman numerals.

e. Compare: 19 ◯ XXI

Problem Solving:

The multiples of 7 are 7, 14, 21, 28, 35, ... We can use multiples of 7 to help us count days of the week. Seven days after Monday is Monday. Fourteen days after Monday is Monday again. So 15 days after Monday is just 1 day after Monday. What day is 30 days after Monday? ... 50 days after Saturday? ... 78 days after Tuesday?

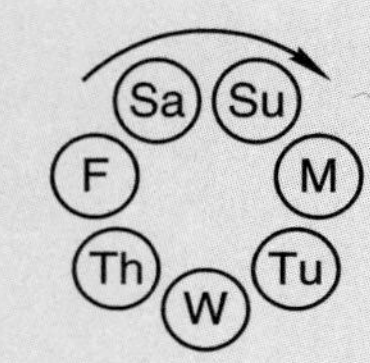

NEW CONCEPT

The mixed number $7\frac{3}{4}$ is between 7 and 8. To round $7\frac{3}{4}$ to the nearest whole number, we decide whether $7\frac{3}{4}$ is nearer 7 or nearer 8. To help us understand this question, we can use this number line:

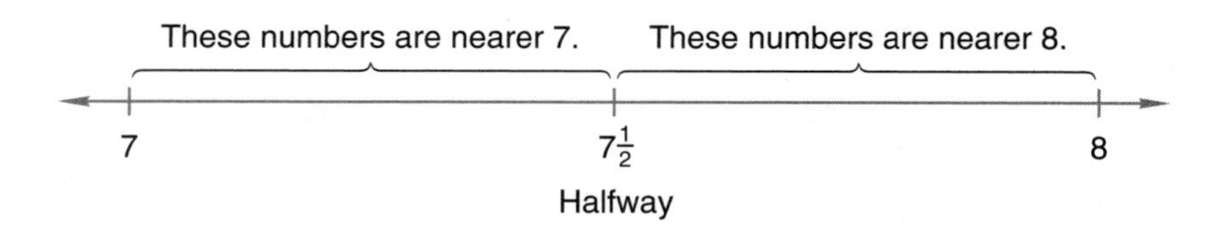

We see that $7\frac{1}{2}$ is halfway between 7 and 8. Since $7\frac{3}{4}$ is between $7\frac{1}{2}$ and 8, we know that $7\frac{3}{4}$ is nearer 8 than 7. So $7\frac{3}{4}$ rounds up to 8.

Example 1 Round $6\frac{2}{5}$ to the nearest whole number.

Solution The mixed number $6\frac{2}{5}$ is between 6 and 7. We need to decide whether it is nearer 6 or nearer 7. The number $6\frac{1}{2}$ is halfway between 6 and 7. The number $6\frac{2}{5}$ is less than $6\frac{1}{2}$ because the numerator of $\frac{2}{5}$ is less than half the denominator. So we round $6\frac{2}{5}$ down to **6.**

Example 2 Estimate the area of this rectangle by rounding the length and the width to the nearest inch before multiplying.

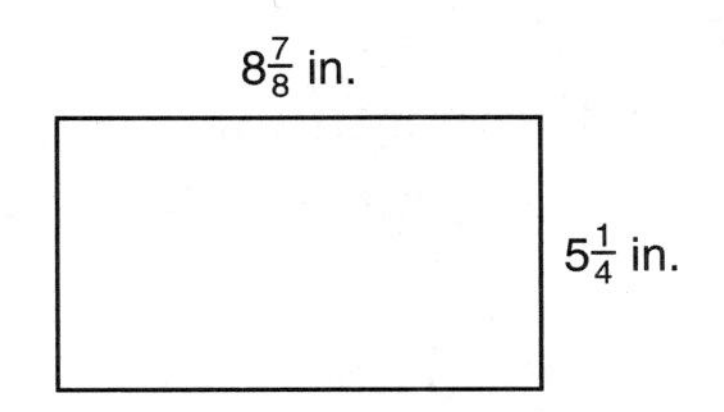

Solution We round $8\frac{7}{8}$ inches to 9 inches, and we round $5\frac{1}{4}$ inches to 5 inches. Then we multiply.

$$9 \text{ in.} \times 5 \text{ in.} = \mathbf{45\ sq.\ in.}$$

LESSON PRACTICE

Practice set Round each mixed number to the nearest whole number:

a. $3\frac{2}{3}$ **b.** $7\frac{1}{8}$ **c.** $6\frac{3}{5}$

d. $6\frac{1}{4}$ **e.** $12\frac{5}{6}$ **f.** $25\frac{3}{10}$

g. Estimate the product of $9\frac{4}{5}$ and $5\frac{1}{3}$.

h. Estimate the sum of $36\frac{5}{8}$ and $10\frac{9}{10}$.

i. Estimate the perimeter of the rectangle in example 2.

MIXED PRACTICE

Problem set

1. *(97)* There were 60 deer and 40 antelope playing on the range. What was the ratio of deer to antelope playing on the range?

2. *(53, 74)* If a side of a regular octagon is 25 centimeters long, then the perimeter of the octagon is how many meters?

3. *(28)* What year was five decades before 1826?

4. (86) What number is $\frac{3}{4}$ of 100?

5. (44) Write the length of this line segment as a number of millimeters and as a number of centimeters.

mm 10 20 30 40

cm 1 2 3 4

6. (66) If the segment in problem 5 were cut in half, then each small segment would be how many centimeters long?

7. (69) Is \$8.80 closer to \$8 or to \$9?

8. (101) Estimate the difference when $7\frac{3}{4}$ is subtracted from $18\frac{7}{8}$.

9. (74) The kite was at the end of 240 feet of string. How many yards of string had been let out?

10. (61) *AB* is 60 mm. *BC* is half of *AB*. *CD* is one third of *AB*. Find *AD*.

11. (99) $4 + 8.57 + 12.3$

12. (99) $16.37 - 12$

13. (29) $\$3.58 \times 10$

14. (78) 24^2

15. (94) $\frac{4300}{25}$

16. (26, 94) $14w = \$20.16$

17. (89) $\sqrt{9} + \sqrt{16}$

18. (79) Write fractions equal to $\frac{5}{6}$ and $\frac{1}{4}$ that have denominators of 12. Then subtract the smaller fraction from the larger fraction.

19. (91) $6\frac{3}{5} + 1\frac{3}{5}$

20. (81) $8\frac{5}{6} - 1\frac{1}{6}$

21. (90) $\frac{2}{10} \times \frac{5}{10}$

22. (96) $2 \div \frac{4}{5}$

23. (79) $\frac{9}{50} = \frac{\square}{100}$

Use this information to answer problems 24 and 25:

Becky ran two races at the track meet. She won the 100-meter race with a time of 13.8 seconds. In the 200-meter race she came in second with a time of 29.2 seconds.

24. *(Inv. 5, 99)* In the 200-meter race the winner finished 1 second faster than Becky. What was the winning time?

25. *(11, Inv. 5)* Becky earned points for her team. At the track meet first place earns 5 points, second place earns 3 points, and third place earns 1 point. How many points did Becky earn?

26. *(90)* Reduce: $\frac{50}{100}$

27. *(Inv. 10)* Write the coordinates of each point labeled on the coordinate plane at right.

(a) A (b) B

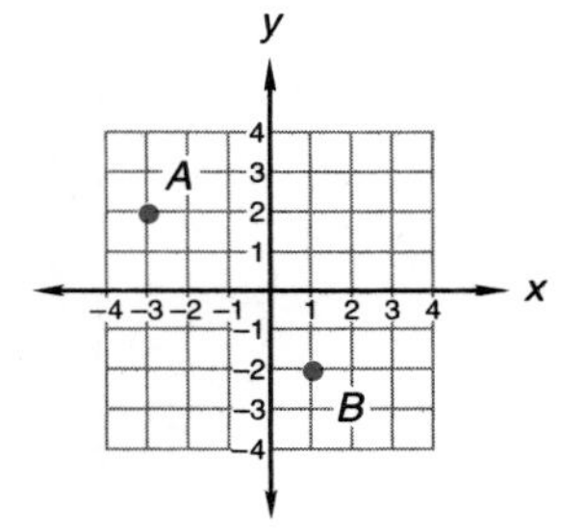

28. *(84)* Becky has run the 100-meter race five times. Her times in seconds are listed below. What is the median of Becky's 100-meter race times?

14.0, 13.8, 13.7, 13.9, 14.1

29. *(72)* A square-foot floor tile is 12 inches on each side. One square foot is how many square inches?

30. *(44, 53)* Use an inch ruler to find the length and width of this rectangle. Then calculate the perimeter of the rectangle.

LESSON

102 Subtracting Decimal Numbers Using Zeros

WARM-UP

Facts Practice: 50 Fractions to Simplify (Test J)

Mental Math:

a. Describe how to estimate the sum of $6\frac{7}{8}$ and $4\frac{5}{6}$.
b. How many ounces are in one pound? ... two pounds?
c. Simplify: $\frac{4}{6}$, $\frac{8}{6}$, $\frac{9}{6}$
d. $\frac{1}{3}$ of 15, × 2, + 2, × 2, ÷ 3, + 1, ÷ 3, ÷ 3

Roman numerals:

e. Write 38 in Roman numerals.
f. Compare: XXIX ◯ 30

Problem Solving:

If two standard number cubes are rolled, many pair combinations are possible. Here are some of the possible combinations:

(1, 1), (1, 2), (1, 3), (1, 4), (1, 5), (1, 6),
(2, 2), (2, 3), (2, 4), (2, 5), (2, 6)

List the rest of the possible combinations. In all, how many combinations are possible?

NEW CONCEPT

For some subtraction problems we need to add decimal places to perform the subtraction. If we subtract 0.23 from 0.4, we find there is an "empty" place in the problem.

$$\begin{array}{r} 0.4\_ \\ -\ 0.23 \\ \hline \end{array}$$ ← empty place

We fill the empty place with a zero. Then we subtract.

$$\begin{array}{r} 0.\overset{3}{\cancel{4}}\,{}^{1}0 \\ -\ 0.2\,3 \\ \hline 0.1\,7 \end{array}$$

Example 1 Subtract: 0.4 − 0.231

Solution We set up the problem by lining up the decimal points, remembering to write the first number on top. We fill empty places with zeros. Then we subtract.

$$\begin{array}{r} 0.\overset{3}{\cancel{4}}\overset{9}{\cancel{\overset{1}{0}}}\overset{1}{0} \\ -\ 0.231 \\ \hline \mathbf{0.169} \end{array}$$

Example 2 Subtract: 3 − 1.23

Solution This problem is similar to subtracting $1.23 from $3. We place the decimal point to the right of the 3, fill the decimal places with zeros, and subtract.

$$\begin{array}{r} \overset{2}{\cancel{3}}.\overset{9}{\cancel{\overset{1}{0}}}\overset{1}{0} \\ -\ 1.23 \\ \hline \mathbf{1.77} \end{array}$$

LESSON PRACTICE

Practice set* Subtract:

a. 0.3 − 0.15 **b.** 0.3 − 0.25

c. 4.2 − 0.42 **d.** 3.5 − 0.35

e. 10 − 6.5 **f.** 6.5 − 4

g. 1 − 0.9 **h.** 1 − 0.1

i. 1 − 0.25 **j.** 2.5 − 1

k. Angela poured 1.2 liters of soda pop from a full 2-liter container. How much soda pop was left in the container? Show your work.

MIXED PRACTICE

Problem set

1. (31, 45) Draw two parallel segments that are horizontal. Make the upper segment longer than the lower segment. Connect the endpoints of the segments to form a quadrilateral. What kind of quadrilateral did you draw?

2. (77, 85) "A pint's a pound the world around" means that a pint of water weighs about a pound. About how much does a gallon of water weigh?

3. (101) Estimate the sum of $7\frac{1}{5}$ and $3\frac{7}{8}$ by rounding both numbers to the nearest whole number before adding.

4. (50) There are 43 people waiting in the first line and 27 people waiting in the second line. If some of the people in the first line move to the second line so that the same number of people are in each line, then how many people will be in each line?

5. (78) If $25m = 100$, then m^2 equals what number?

6. (71) Name the shaded portion of this square as a decimal number, as a reduced fraction, and as a percent.

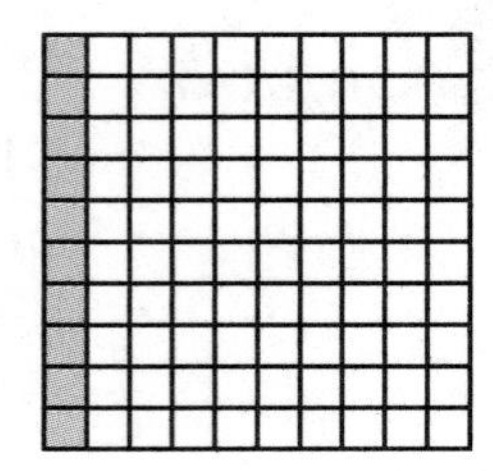

7. (75, 79) Write fractions equal to $\frac{1}{5}$ and $\frac{7}{8}$ that have denominators of 40. Then add the fractions. Remember to convert the answer to a mixed number.

8. (68, 69) Compare: one tenth ◯ ten hundredths

9. (15) The first four multiples of 2 are 2, 4, 6, and 8. What are the first four multiples of 6?

10. (53) This rectangle was made with pins 1 inch long.

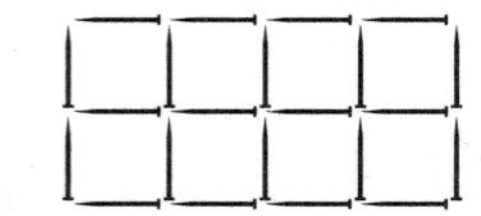

(a) The length of the rectangle is about how many inches?

(b) The perimeter of the rectangle is about how many inches?

11. (61) *AB* is 60 mm. *BC* is half of *AB*. *CD* is half of *BC*. Find *AD*.

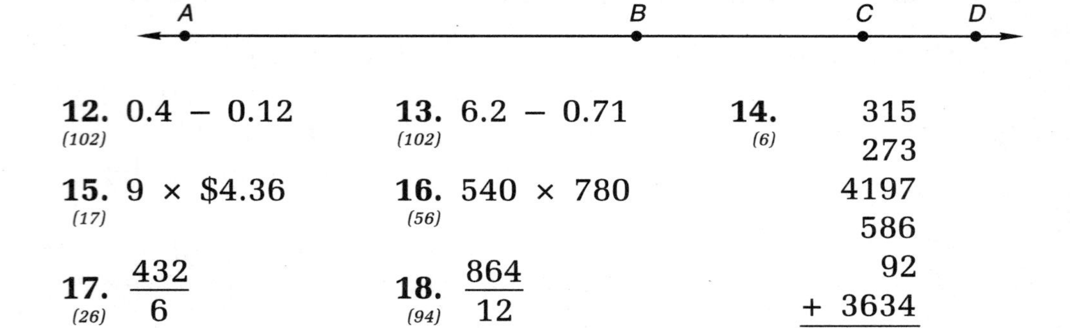

12. (102) $0.4 - 0.12$

13. (102) $6.2 - 0.71$

14. (6)
$$\begin{array}{r} 315 \\ 273 \\ 4197 \\ 586 \\ 92 \\ +\ 3634 \\ \hline \end{array}$$

15. (17) $9 \times \$4.36$

16. (56) 540×780

17. (26) $\dfrac{432}{6}$

18. (94) $\dfrac{864}{12}$

19. (63, 75) $5 - \left(1\frac{2}{3} + 1\frac{2}{3}\right)$

20. (59, 76) $\frac{5}{6} \times \left(3 \times \frac{2}{5}\right)$

21. $2 \div \frac{1}{3}$ *(96)*

22. $\frac{1}{3} \div 2$ *(96)*

23. $\frac{12}{50} = \frac{\square}{100}$ *(79)*

This graph shows how Darren spends his time each school day. Use the information in this graph to answer problems 24 and 25.

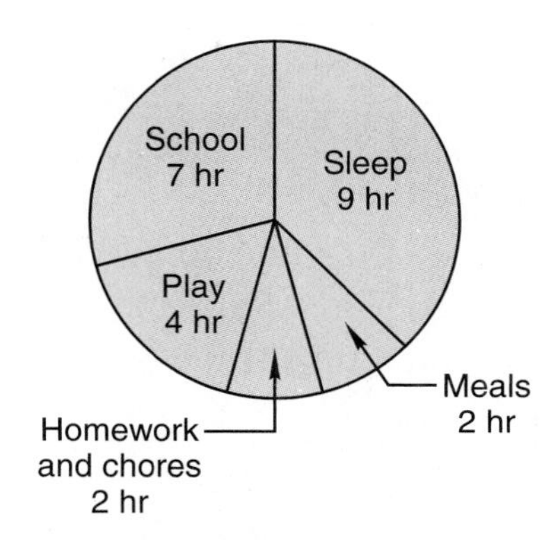

24. What is the total number of hours shown in the graph? *(Inv. 8)*

25. What fraction of the day does Darren spend sleeping? *(Inv. 8, 90)*

26. Freddy poured 1.4 liters of soda pop from a full 2-liter container. How much soda pop was left in the container? *(102)*

27. Write the next four terms of this counting sequence: *(Inv. 7)*

..., 2.5, 2.8, 3.1, 3.4, _____, _____, _____, _____, ...

28. How many blocks were used to build this rectangular solid? *(18)*

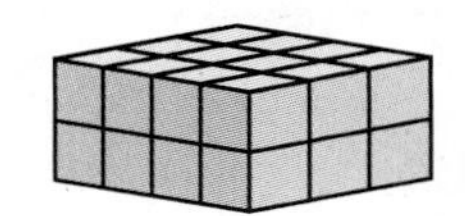

29. A hamburger stand sells chocolate, strawberry, and vanilla milkshakes. Requests for milkshakes over a two-day period are recorded below. *(Inv. 6)*

Flavor	Orders
chocolate	23
strawberry	16
vanilla	41

Estimate the probability that someone who orders a milkshake will ask for chocolate.

30. Arrange these numbers in order from least to greatest: *(69, 98)*

1, 0.1, −1, 1.1

LESSON 103 Volume

WARM-UP

Facts Practice: 50 Fractions to Simplify (Test J)

Mental Math:

a. Describe how to estimate the product of $7\frac{3}{4}$ and $6\frac{1}{3}$.
b. How many feet are in one yard? ... 100 yards?
c. Simplify: $\frac{6}{8}, \frac{9}{8}, \frac{12}{8}$
d. $\sqrt{100}$, × 5, + 4, ÷ 9, × 7, + 2, ÷ 4

Roman numerals:

e. Write 36 in Roman numerals.
f. Compare: XXIII ◯ 23

Problem Solving:

How many 1-inch cubes would be needed to build a cube with edges 2 inches long?

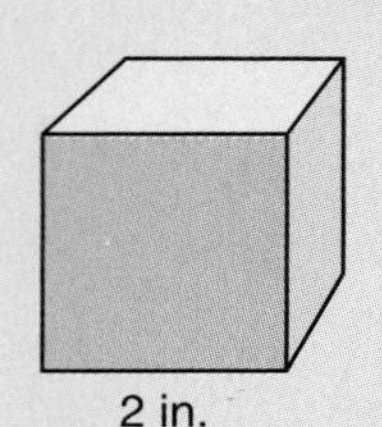

NEW CONCEPT

The **volume** of an object is the amount of space the object occupies. Geometric figures that occupy space include cubes, spheres, cones, cylinders, pyramids, and combinations of these shapes. In this lesson we will concentrate on finding the volume of rectangular solids.

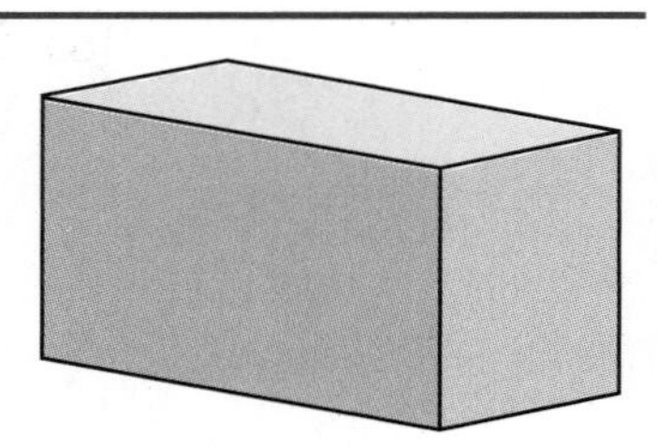
Rectangular solid

The units we use to measure volume are **cubic units.** Here we illustrate a cubic centimeter and a cubic inch:

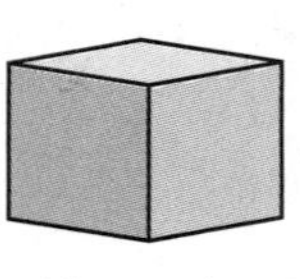
1 cubic centimeter

1 cubic inch

The Problem Solving in this lesson's Warm-up is a way of asking for a volume of a cube. The volume of the cube equals the number of 1-inch cubes needed to build the larger cube.

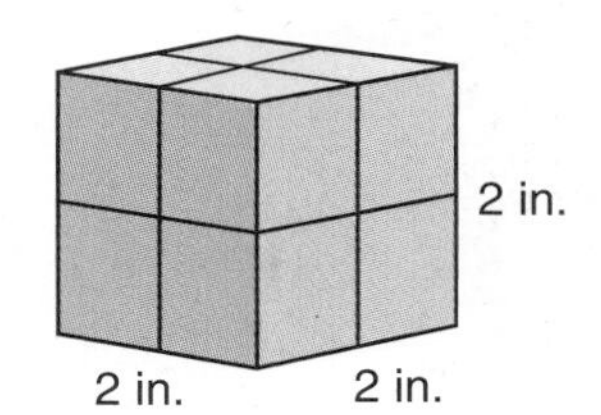

The larger cube is 2 inches long, 2 inches wide, and 2 inches high. We see that the cube is built from 8 1-inch cubes. Each 1-inch cube occupies 1 cubic inch of space. So the volume of the cube is 8 cubic inches.

Example 1 Find the volume of this rectangular solid.

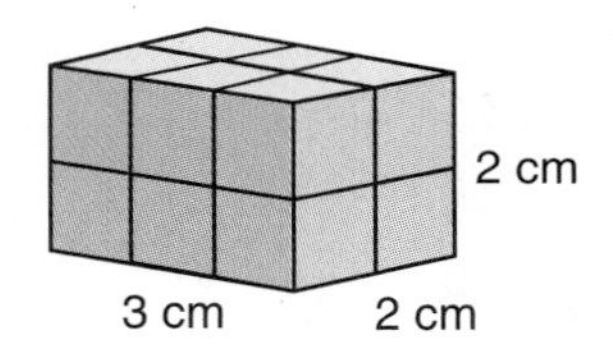

Solution The solid is 3 cm long, 2 cm wide, and 2 cm high. There are 6 cubes in each layer of the solid. The solid has 2 layers, so there are 12 cubes in all. Since the cubes are 1-cm cubes, the volume is **12 cubic centimeters.**

Example 2 What is the volume of this solid?

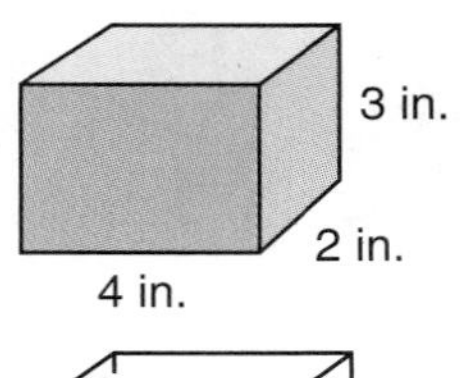

Solution The solid is 4 inches long, 2 inches wide, and 3 inches high. For the bottom layer, we imagine a 4-by-2 rectangle of 1-inch cubes, which is 8 cubes. Three layers are needed for the whole solid. Since $3 \times 8 = 24$, the volume is **24 cubic inches.**

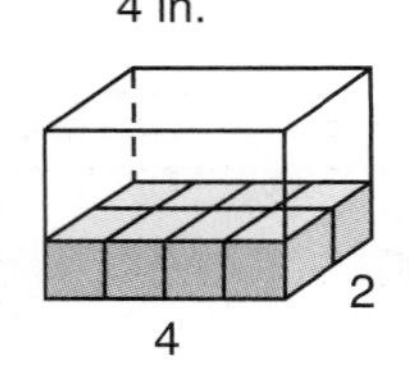

Example 3 As Dion ate breakfast he estimated the volume of the cereal box. What is the approximate volume of the box?

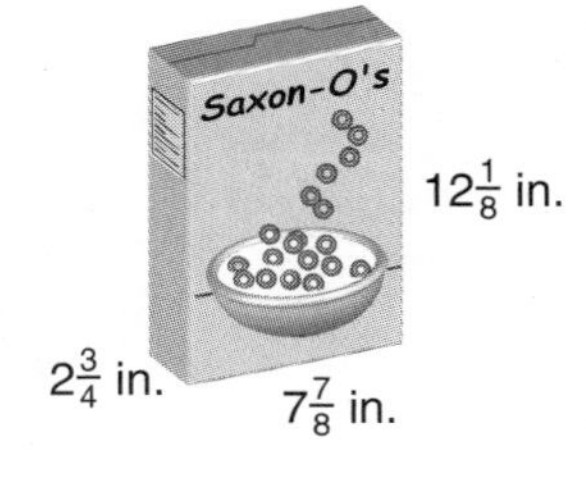

Solution We round the length, width, and height to the nearest inch. The base is about 3 inches by 8 inches, so almost 24 1-inch cubes can fit on the base. The box is about 12 inches tall, so we multiply 24 by 12. We find that the volume of the cereal box is about **288 cubic inches.**

LESSON PRACTICE

Practice set Find the volume of each rectangular solid:

a.

4 in.
4 in. 2 in.

b.

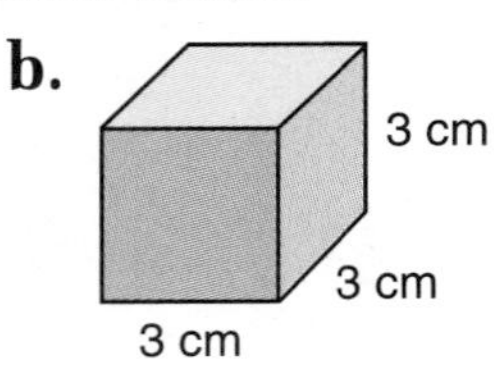

c.

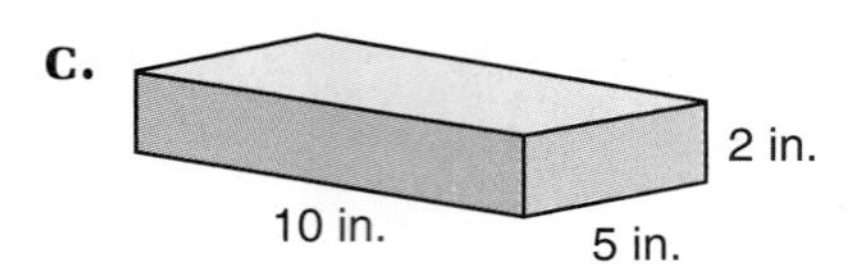

d.

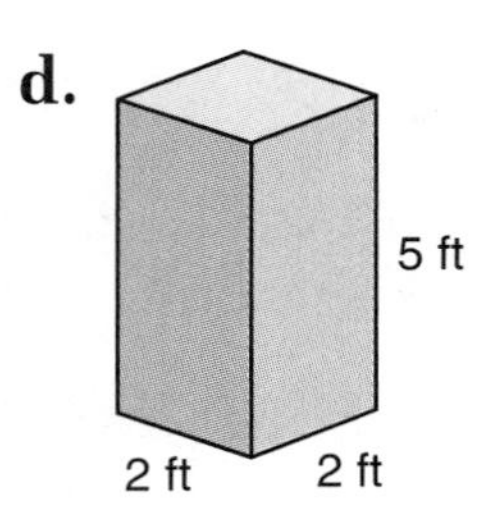

e. Ella's closet is 3 feet wide, 2 feet deep, and 8 feet high. How many boxes that are 1-foot cubes could Ella fit into her closet?

f. The cracker box was $2\frac{3}{4}$ inches wide, $5\frac{1}{4}$ inches long, and $7\frac{7}{8}$ inches high. What is the approximate volume of the box?

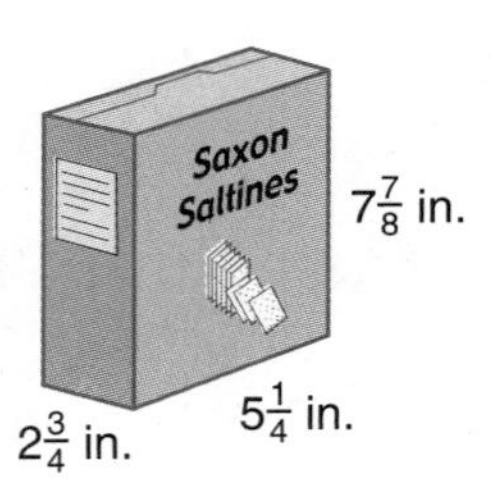

MIXED PRACTICE

Problem set

1. (97) The room was cluttered with 15 magazines and 25 newspapers. What was the ratio of magazines to newspapers cluttering the room?

2. (46, 71) About $\frac{1}{3}$ of the weight of a banana is the weight of the peel. If a banana weighs 12 ounces, then the weight of the peel would be about how many ounces? About what percent of the weight of a banana is the weight of the peel?

3. (57, 80) What is the probability that a standard number cube, when rolled, will stop with a prime number on top?

4. (71, 81) Name the total number of shaded circles as a decimal number and as a reduced mixed number.

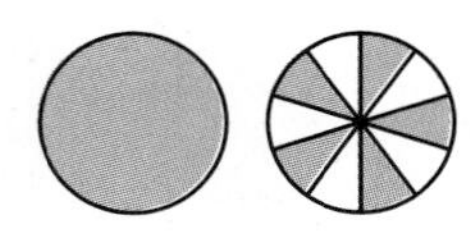

5. Which digit in 1.234 is in the same place as the 6 in 56.78?
(64)

6. If the radius of a wheel is 30 centimeters, then how many centimeters is its diameter?
(53)

7. Estimate the quotient when $9\frac{2}{3}$ is divided by $4\frac{5}{6}$.
(101)

8. Is \$12.65 closer to \$12 or to \$13?
(69)

9. Which arrow could be pointing to 5.8 on this number line?
(66)

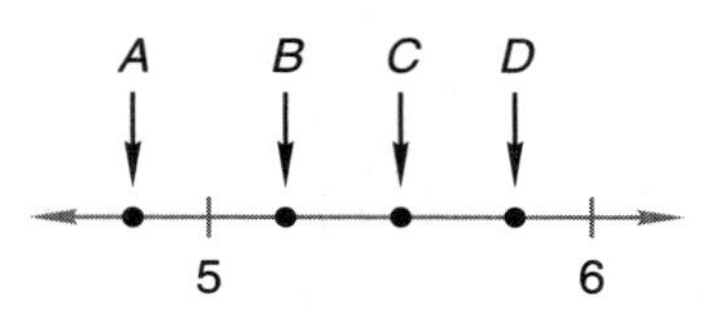

10. Estimate the area of this rectangle by first rounding the length and width to the nearest inch.
(72, 101)

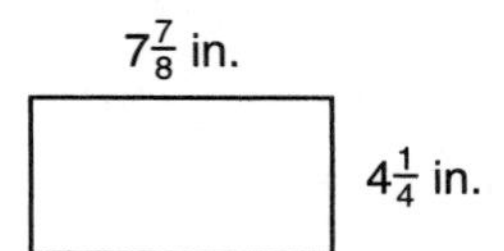

11. *QT* is 10 cm. *QR* is 4 cm. *RS* is half of *QR*. Find *ST*.
(61)

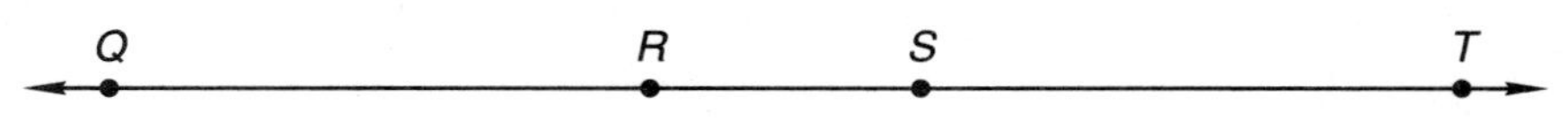

12. $3 - 2.35$
(102)

13. $10 - 4.06$
(102)

14. $4.35 + 12.6 + 15$
(99)

15. $7 \times 47 \times 360$
(18, 56)

16. 2^5
(78)

17. \$47.00 ÷ 20
(54)

18. $\sqrt{25} - \sqrt{9}$
(89)

19. $16x = 2112$
(26, 94)

20. $3\frac{2}{3} + \left(2 - \frac{2}{3}\right)$
(59, 63)

21. $\frac{1}{2} \times \left(4 \times \frac{1}{4}\right)$
(24, 86)

22. $1 \div \frac{7}{5}$
(96)

23. $\frac{3}{2} \div \frac{2}{3}$
(96)

24. $\frac{4}{10} \times \frac{5}{10}$
(90)

25. $\frac{1}{25} = \frac{\square}{100}$
(79)

26. Reduce: $\frac{500}{1000}$
(90)

27. (a) What is the volume of this rectangular solid?
(83, 103)

(b) How many faces does it have?

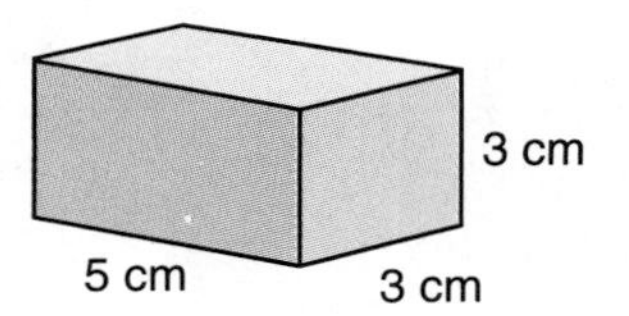

28. Write the coordinates of each point:
(Inv. 10)

(a) *A* (b) *B*

(c) *C* (d) *D*

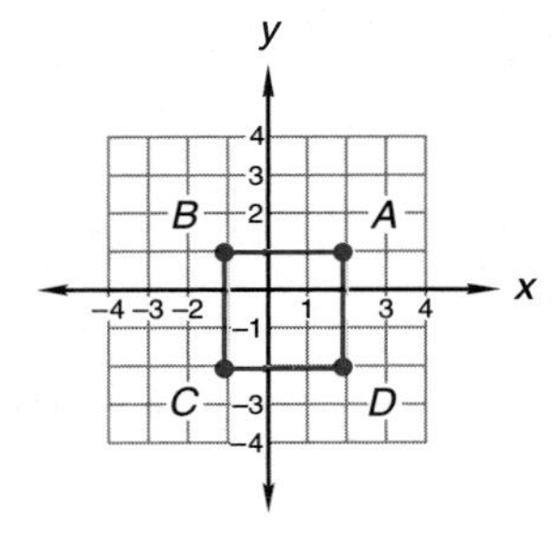

Use an inch ruler to measure the sides of this triangle. Refer to the illustration and measurements to answer problems 29 and 30.

29. (a) How many inches long is each side of the triangle?
(44, 53)

(b) What is the perimeter of the triangle?

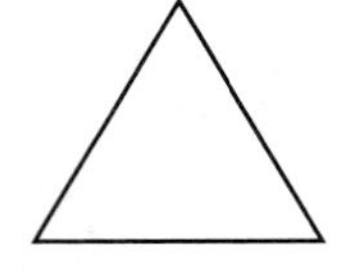

30. Classify the triangle by sides and by angles.
(36)

LESSON

104 Rounding Decimal Numbers to the Nearest Whole Number

WARM-UP

Facts Practice: 50 Fractions to Simplify (Test J)

Mental Math:

a. Describe how to estimate the quotient when $9\frac{5}{8}$ is divided by $1\frac{9}{10}$.

b. Simplify: $\frac{12}{10}$, $\frac{15}{10}$, $\frac{25}{10}$

Roman numerals:

c. Write 29 in Roman numerals.

d. Compare: 26 ◯ XXIV

Problem Solving:

If we multiply $2 \times 2 \times 2$, the product is the cube number 8. If we multiply $3 \times 3 \times 3$, the product is the cube number 27. What are the next two cube numbers?

NEW CONCEPT

In previous problem sets we have answered questions such as the following:

Is \$7.56 closer to \$7 or \$8?

When we answer this question, we are rounding \$7.56 to the nearest dollar. This is an example of rounding a decimal number to the nearest whole number. Using rounded numbers helps us to estimate.

A number written with digits after the decimal point is not a whole number. It is between two whole numbers. We will learn how to find which of the two whole numbers it is nearer. A number line can help us understand this idea.

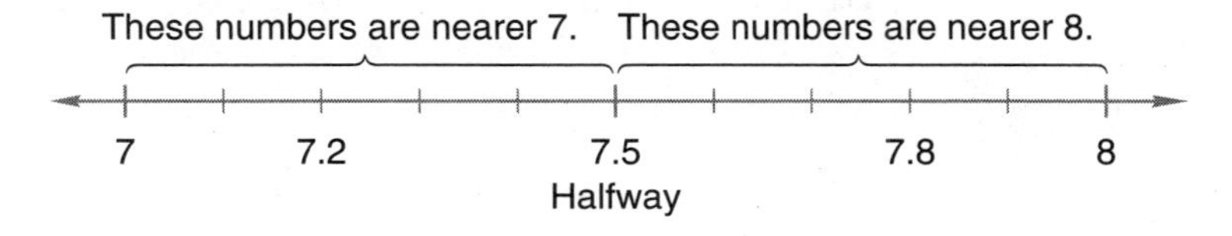

The decimal number 7.5 is halfway from 7 to 8. It is the same distance from 7.5 to 7 as it is from 7.5 to 8. The number 7.2 is less than halfway, so it is nearer 7. The number 7.8 is more than halfway, so it is nearer 8.

Example 1 Round 7.6 to the nearest whole number.

Solution The decimal number 7.6 is greater than 7 but is less than 8. Halfway from 7 to 8 is 7.5. Since 7.6 is more than halfway, we round up to the whole number **8.** We can see on this number line that 7.6 is closer to 8 than it is to 7.

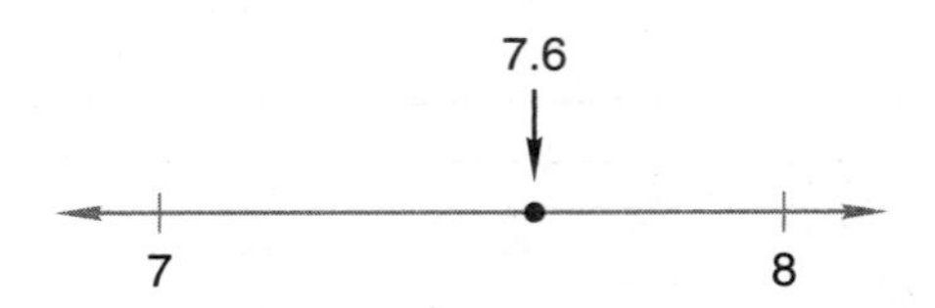

Example 2 Estimate the product of 8.78 and 6.12.

Solution Rounding decimal numbers with two decimal places is similar to rounding money. The decimal number 8.78 rounds to the whole number 9 just as $8.78 rounds to $9. Likewise, 6.12 rounds to the whole number 6. We multiply 9 by 6 and find that the product of 8.78 and 6.12 is about **54.**

Example 3 Estimate the area of this rectangle:

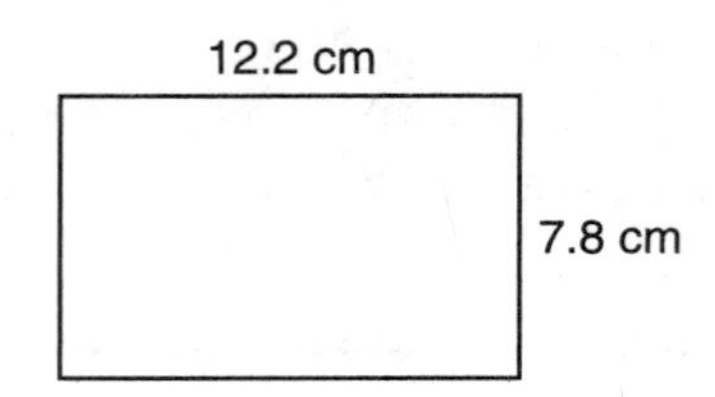

Solution We round the length to 12 cm and the width to 8 cm. Then we multiply.

$$12 \text{ cm} \times 8 \text{ cm} = \mathbf{96 \text{ sq. cm}}$$

LESSON PRACTICE

Practice set* Round each money amount to the nearest dollar:

a. $6.24 **b.** $15.06 **c.** $118.59

d. Estimate the sum of $12.89 and $6.95.

Round each decimal number to the nearest whole number:

e. 4.75 **f.** 12.3 **g.** 96.41

h. 7.4 **i.** 45.7 **j.** 89.89

k. Estimate the product of 9.8 and 6.97.

l. Stephanie ran one lap in 68.27 seconds. Round her time to the nearest second.

MIXED PRACTICE

Problem set

1. (45) Draw a quadrilateral with two pairs of parallel sides and no right angles.

2. (49, 97) In Sovann's class there are twice as many boys as there are girls. There are 18 boys in the class.

(a) How many girls are in the class?

(b) How many students are in the class?

(c) What is the ratio of boys to girls in the class?

3. (84) Marcia's last seven test scores were 85, 90, 90, 80, 80, 80, and 75.

(a) Arrange the seven scores in order from lowest to highest.

(b) What is the median of the scores?

(c) What is the mode of the scores?

4. (13, 76) Write this sentence using digits and symbols:

The product of one half and one third is one sixth.

5. (68) Which digit is in the tenths place in 142.75?

6. (96) Compare: $\frac{1}{2} \div \frac{1}{3} \bigcirc \frac{1}{3} \div \frac{1}{2}$

7. (71) Draw four circles the same size. Shade 25% of the first circle, 50% of the second circle, 75% of the third circle, and 100% of the fourth circle.

8. (101) Round $4\frac{3}{10}$ to the nearest whole number.

9. (104) (a) Round \$10.49 to the nearest dollar.

(b) Round \$9.51 to the nearest dollar.

10. (15) The first five multiples of 2 are 2, 4, 6, 8, and 10. What are the first five multiples of 7?

11. Which arrow could be pointing to 7.2 on this number line?
(66)

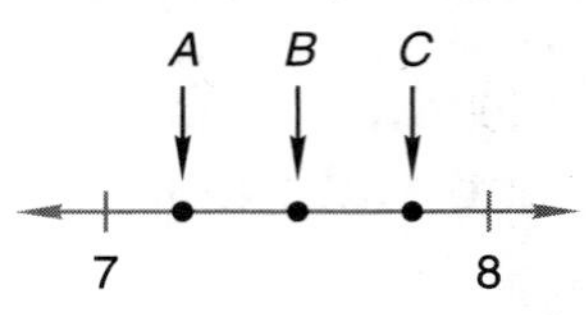

12. Estimate the area of this rectangle by first rounding the length and width to the nearest centimeter.
(72, 104)

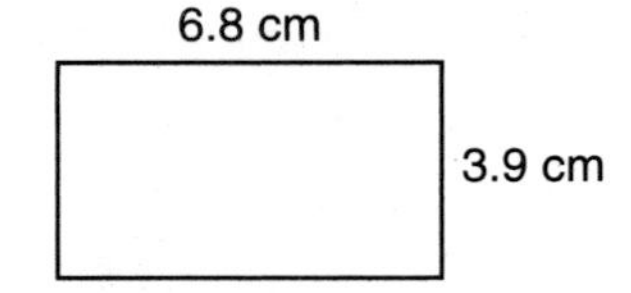

13. 6.4 + 2.87 + 4
(99)

14. ($16 − $5.74) ÷ 6
(13, 24)

15. $5.64 × 10
(29)

16. 976 × 267
(55)

17. All these ratios are equal. What is the quotient of each division?
(97)

$$\frac{640}{32}, \frac{320}{16}, \frac{160}{8}, \frac{80}{4}$$

18. Write a fraction equal to $\frac{2}{3}$ with a denominator of 9. Then add $\frac{7}{9}$ to the fraction you wrote. Remember to convert the sum to a mixed number.
(75, 79)

19. $5\frac{2}{3} + \left(3 - \frac{1}{3}\right)$
(63, 75)

20. $2 \times \left(\frac{1}{2} \times \frac{1}{3}\right)$
(76, 86)

21. $\frac{3}{10}$ of 30
(86)

22. $\frac{4}{25} = \frac{\square}{100}$
(79)

This map has been divided into a grid to make towns easier to find. Use the map to answer problems 23–25.

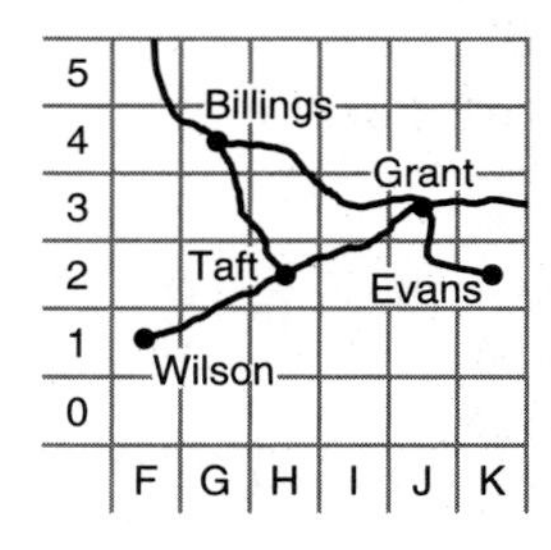

23. We find Taft in region H2. In which region do we find Billings?
(Inv. 10)

A. G4   B. F4   C. H2   D. F5

24. What town do we find in region J3?
(Inv. 10)

25. What letter and number show where to find Evans?
(Inv. 10)

26. $10^2 - \sqrt{100}$
(78, 89)

27. Write the coordinates of each point:
(Inv. 10)

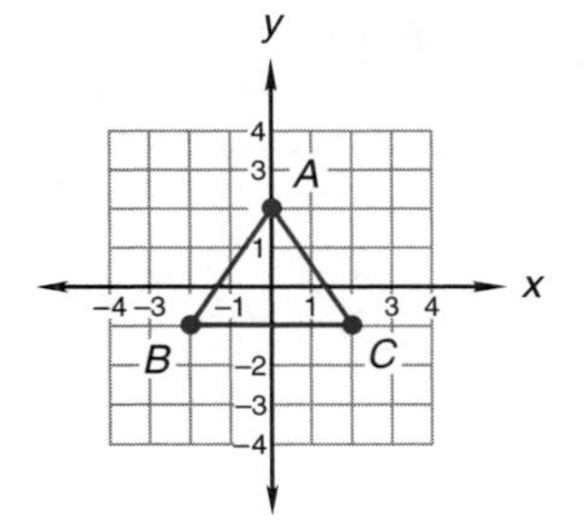

(a) A

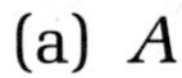

(b) B

(c) C

28. Determine a possible pattern for this sequence, and draw the next figure.
(Inv. 7)

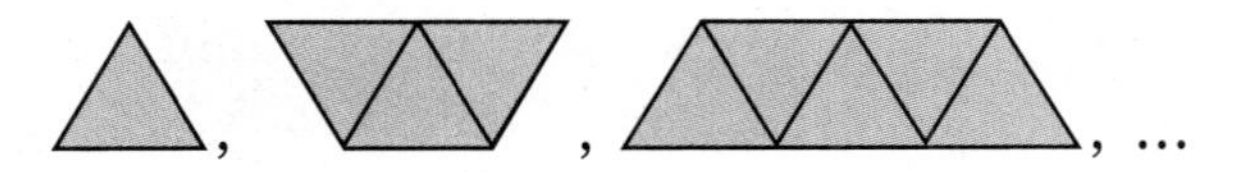

29. (a) What is the volume of a box of tissues with the dimensions shown?
(83, 103)

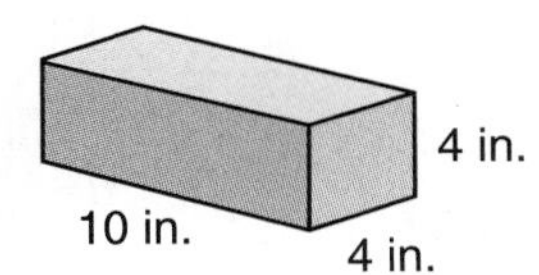

(b) How many vertices does the box have?

30. About 26% of the people living in Oklahoma are under age 18. Write 26% as a reduced fraction.
(71, 91)

LESSON

105 Symmetry

WARM-UP

Facts Practice: 50 Fractions to Simplify (Test J)

Mental Math:

a. Describe how to estimate the cost of 8 yards of fabric if the price of the fabric is $6.95 per yard.

b. $\frac{1}{5}$ of $20 **c.** $\frac{2}{5}$ of $20 **d.** $\frac{4}{5}$ of $20

e. $\sqrt{49}$, × 8, − 1, ÷ 5, − 1, × 4, + 2, ÷ 6

Roman numerals:

f. Write 32 in Roman numerals.

g. Compare: XXXVI ◯ 34

Problem Solving:

Print the last three letters of the alphabet in uppercase. Which of the three letters has no lines of symmetry? Which of the letters has just one line of symmetry? Which of the letters has two lines of symmetry?

NEW CONCEPT

The Problem Solving in today's Warm-up asked you to identify **lines of symmetry** in letters of the alphabet. In this lesson we will look for lines of symmetry in other figures as well. Figures that can be divided into *mirror images* by at least one line of symmetry are said to have **reflective symmetry**. If a mirror is placed upright along a line of symmetry, the reflection in the mirror appears to complete the figure.

Example 1 Here we show a regular triangle and a regular pentagon. Find the number of lines of symmetry in each figure.

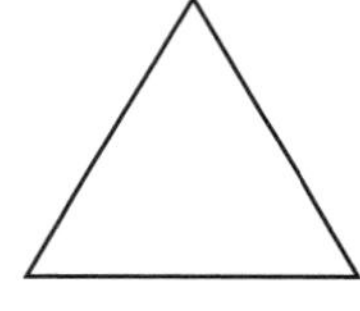

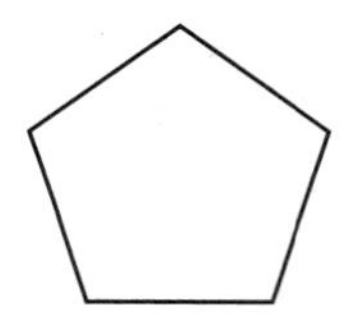

Solution A line of symmetry divides a figure into mirror images. In each of these figures, a line of symmetry passes through a vertex and splits the opposite side into two segments of equal length.

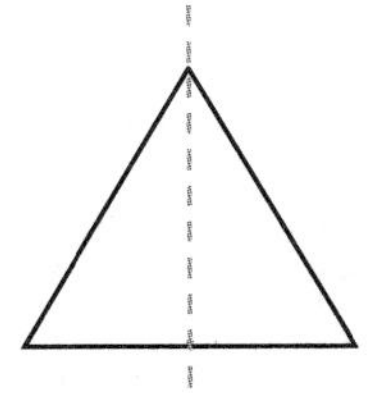 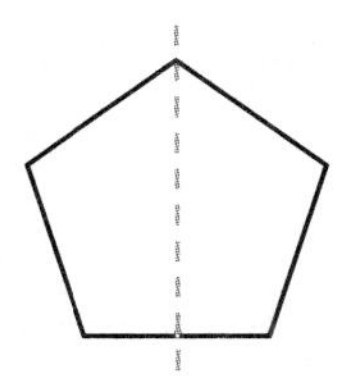

Since these polygons are regular, we find a line of symmetry through each vertex of the polygon.

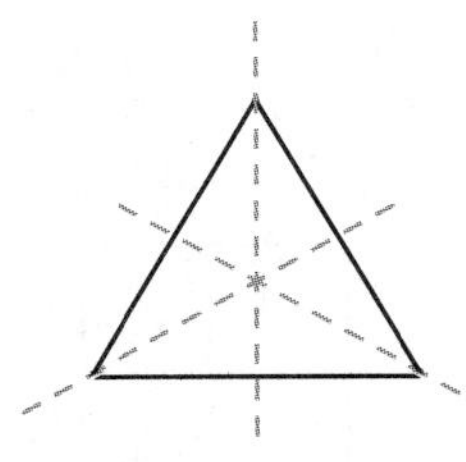 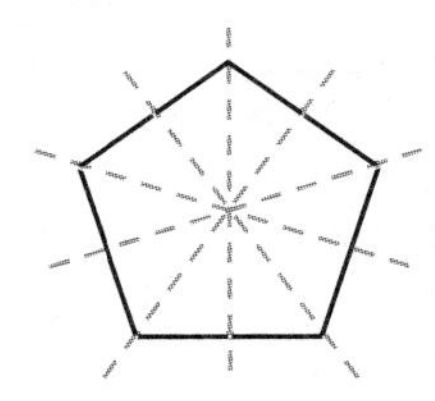

So the regular triangle has **three** lines of symmetry, and the regular pentagon has **five** lines of symmetry.

Example 2 Here we show a regular quadrilateral and a regular hexagon. Find the number of lines of symmetry in each figure.

 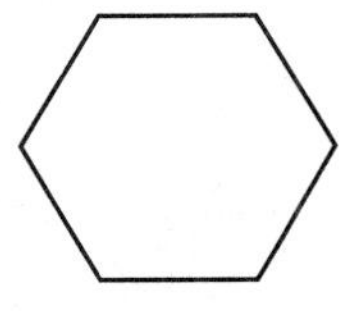

Solution There is a line of symmetry that passes through a vertex and its opposite vertex.

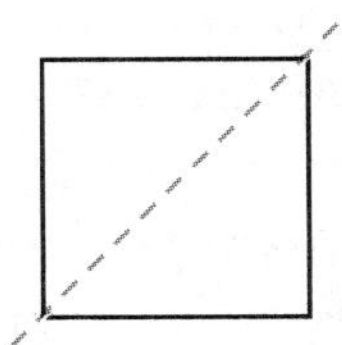 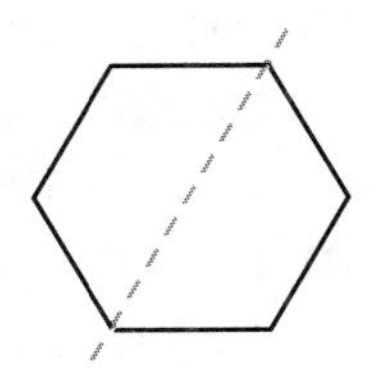

We find two of these lines of symmetry for the square and three for the hexagon.

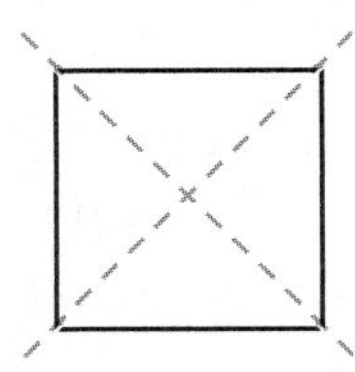 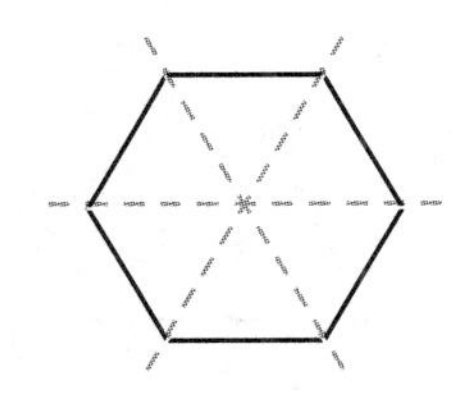

In addition to the lines of symmetry through the vertices, there are lines of symmetry through the sides of these figures.

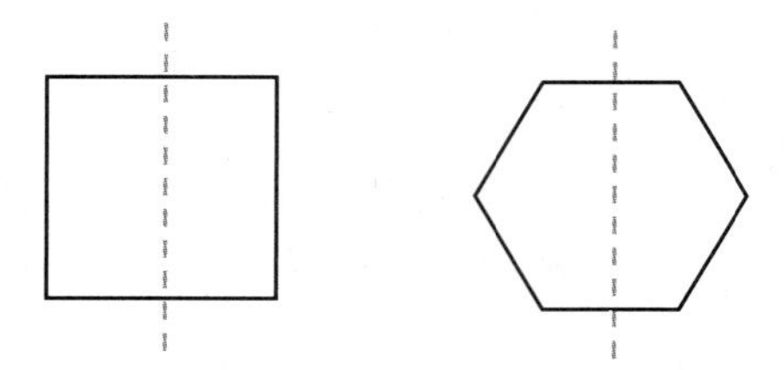

Again we find two such lines of symmetry for the square and three for the hexagon.

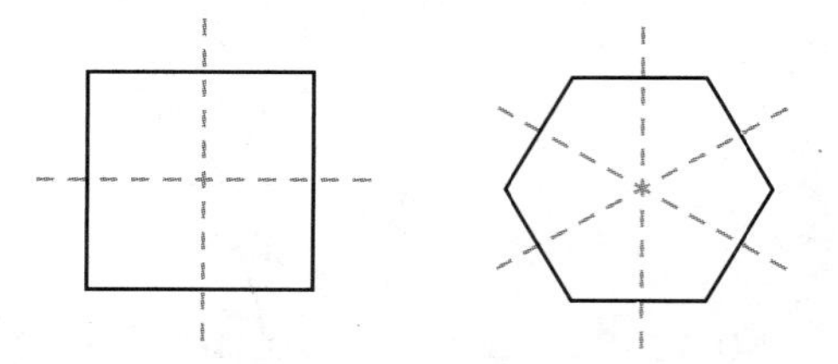

In all we find **four** lines of symmetry for the square and **six** lines of symmetry for the hexagon.

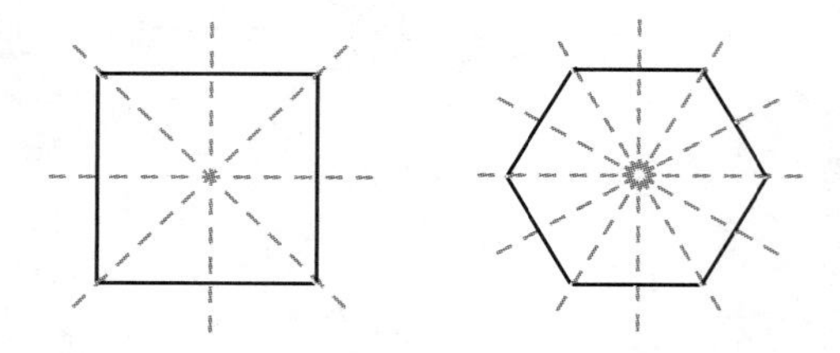

Besides having reflective symmetry, regular polygons also have **rotational symmetry.** A figure has rotational symmetry if it regains its original orientation more than once during a full turn. For example, if we rotate an equilateral triangle one third of a turn, the triangle reappears in its original orientation.

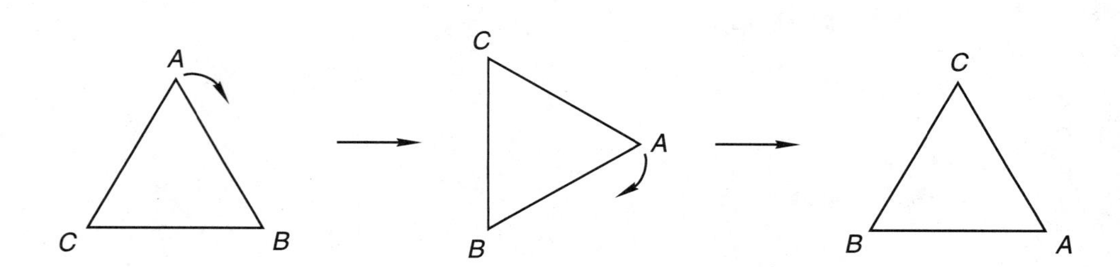

Example 3 Which of these letters has rotational symmetry?

M A T H

Solution If you turn your book as you look at the letters, you will see **H** reappear in its proper orientation after half a turn. It is the only letter of these four with rotational symmetry.

LESSON PRACTICE

Practice set Draw each figure and all its lines of symmetry:

a. 

b. 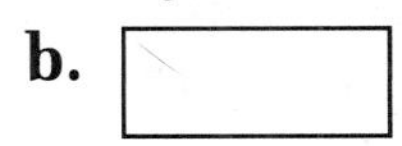

c. 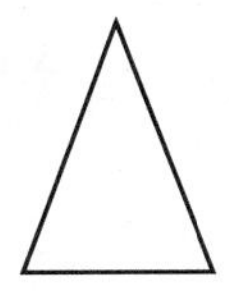

d. 

e. A regular octagon has how many lines of symmetry?

f. Which of these letters has rotational symmetry?

P L U S

MIXED PRACTICE

Problem set

1. (97) The ratio of boys to girls in the auditorium was 4 to 5. If there were 40 boys in the auditorium, how many girls were there? (*Hint:* In this problem the ratio 4 to 5 means that for every 4 boys there were 5 girls.)

2. (30) This circle is divided into tenths. How many tenths does it take to equal one whole?

3. (70) Tony had six coins in his pockets totaling 43¢. How many of the coins were nickels?

4. (68) Emma finished the race in ten and twenty-three hundredths seconds. Use digits to write that number of seconds.

5. (49) If 20 comic books cost \$50, how many comic books could you buy with \$100?

6. (79, 90) Write a fraction equal to $\frac{1}{2}$ that has a denominator of 10. Then subtract that fraction from $\frac{9}{10}$. Remember to reduce the answer.

7. (50) Inez and Felicia had three days to read a book. Inez read 40 pages the first day, 60 pages the second day, and 125 pages the third day. Felicia read the same book, but she read an equal number of pages each of the three days. How many pages did Felicia read each day?

8. (62) Estimate the cost of 12 comic books priced at \$1.95 each.

9. (104) Estimate the quotient when 20.8 is divided by 6.87 by rounding both decimal numbers to the nearest whole number before dividing.

10. (53, 72, Inv. 10) (a) The area of rectangle $ABCD$ is how many square units?

(b) The perimeter of the rectangle is how many units?

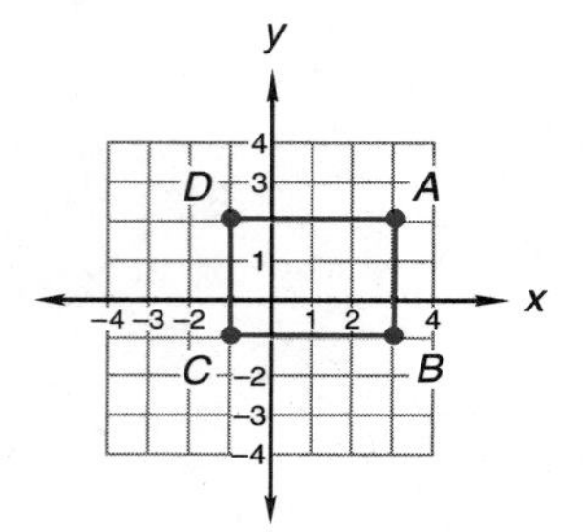

Refer to quadrilateral $ABCD$ to answer problems 11 and 12.

11. (32, 61) Recall that a right angle is sometimes marked with a square in the corner. Both $\angle CDA$ and $\angle DCB$ are right angles. Which angle appears to be acute?

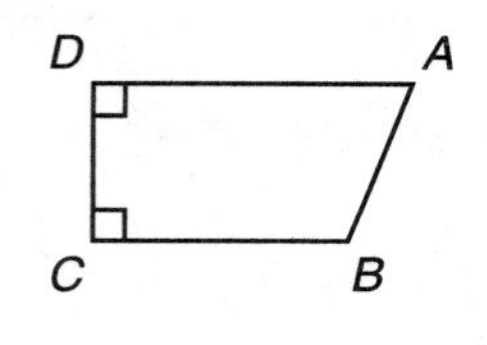

12. (45) What type of quadrilateral is quadrilateral $ABCD$?

13. (90) $\frac{1}{100} + \frac{9}{100}$

14. (90) $\frac{63}{100} - \frac{13}{100}$

15. (90) $\frac{5}{10} \times \frac{5}{10}$

16. (96) $\frac{3}{5} \div \frac{3}{4}$

17. (99) $3.76 + 12 + 6.8$

18. (102) $12 - 1.25$

19. $\sqrt{64} + \sqrt{36}$
(89)

20. 31^2
(78)

21. $28\overline{)5964}$
(94)

22. $14m = 5964$
(26, 94)

23. $\frac{3}{20} \times \square = \frac{15}{100}$
(79)

24. $\frac{7}{25} = \frac{\square}{100}$
(79)

Use this information to answer problems 25 and 26:

> *Tyrone fixed his function machine so that when he puts in a 24, an 8 comes out. When he puts in a 12, a 4 comes out. When he puts in a 6, a 2 comes out.*

24 → NUMBER-CHANGING MACHINE → 8
12 → → 4
6 → → 2

25. What rule does the function machine use?
(Inv. 7)

A. It divides by 3.
B. It multiplies by 3.
C. It divides by 2.
D. It subtracts 8.

26. If Tyrone puts in a 30, what number will come out?
(Inv. 7)

27. Draw a regular quadrilateral and show its lines of symmetry.
(53, 105)

28. (a) What is the volume of a box of cereal with these dimensions?
(83, 103)

(b) How many edges does the box have?

10 in.
7 in. 2 in.

29. From (2, 1) to (5, 1) is three units. How many units is it from (5, 1) to (5, –3)?
(Inv. 10)

30. Which of these letters has rotational symmetry?
(105)

T E N

LESSON

106 Reading and Ordering Decimal Numbers Through Ten-Thousandths

WARM-UP

Facts Practice: 50 Fractions to Simplify (Test J)

Mental Math:

a. Describe how you would estimate the cost of each pound of nails if a 50-pound box of nails cost $29.85.

b. How many pounds are in one ton? ... two tons? ... half a ton?

c. $\frac{1}{3}$ of 100

d. $\frac{2}{3}$ of 100

e. 20 × 30, + 40, ÷ 10

Roman numerals:†

f. Write CCX in our number system.

g. Write LXXX in our number system.

Problem Solving:

Fifty pennies fill a penny roll. Forty nickels fill a nickel roll. How many penny rolls are equal in value to two nickel rolls?

NEW CONCEPT

We have used bills and coins to help us understand place value. As we move to the right on the chart below, we see that each place is one tenth of the value of the place to its left.

tens place	ones place		tenths place	hundredths place	thousandths place	ten-thousandths place
___	___	.	___	___	___	___
$10 bills	$1 bills		dimes	pennies	mills	

The third place to the right of the decimal point is the thousandths place. Its value is $\frac{1}{1000}$. We do not have a coin that is $\frac{1}{1000}$ of a dollar, but we do have a name for $\frac{1}{1000}$ of a dollar. A thousandth of a dollar is a **mill.** Ten mills equal one penny.

†In Lessons 106–120 the Mental Math section "Roman numerals" reviews concepts from Appendix Topic B. You may skip these Warm-up problems if you have not covered Appendix Topic B.

To name decimal numbers with three decimal places, we use the word *thousandths*. To name numbers with four decimal places, we use *ten-thousandths.*

Example 1 Use words to name 12.625.

Solution **twelve and six hundred twenty-five thousandths**

Example 2 Round 7.345 to the nearest whole number.

Solution The number 7.345 is a number that is 7 plus a fraction. So it is more than 7 but less than 8. We need to decide whether it is nearer 7 or nearer 8.

Remember that zeros at the end of a decimal number do not change the value of the number. So the halfway point between 7 and 8 may be named using any number of decimal places.

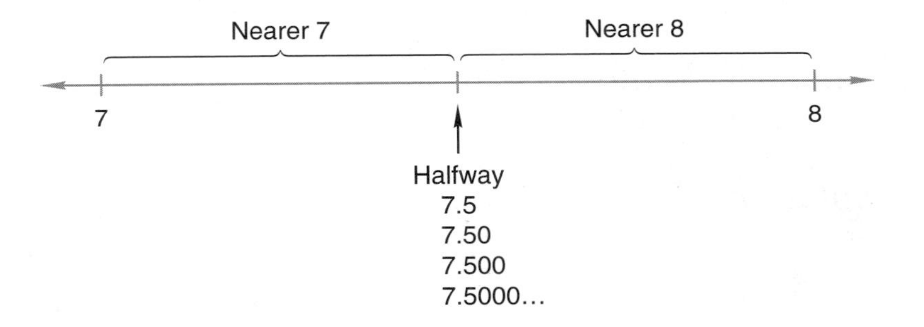

Since 7.500 is halfway between 7 and 8, the number we are rounding, 7.345, is less than halfway.

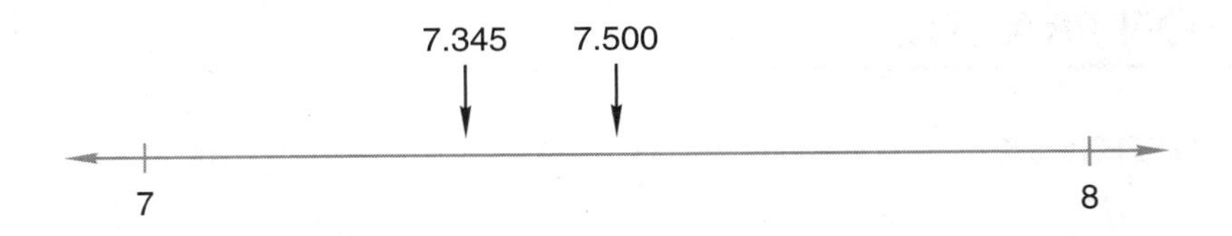

So 7.345 rounds down to the whole number **7.**

Example 3 Compare: 4.5 ◯ 4.456

Solution The comparison is easier to see if the numbers have the same number of decimal places. We will attach zeros to 4.5 so that it has the same number of decimal places as 4.456. We see that 4.5 is greater.

$$\mathbf{4.500 > 4.456}$$

Example 4 Arrange these decimal numbers in order from least to greatest:

0.45, 0.457, 0.5, 0.0475

Solution The number 0.0475 has the most decimal places (four), but that does not mean it is the greatest number of the set. The size of a decimal number is determined by place value, not by the number of digits. We attach zeros to the other numbers so that all the numbers have four decimal places. That makes ordering the numbers easier.

0.4500, 0.4570, 0.5000, 0.0475

We now arrange the numbers from least to greatest.

0.0475, 0.4500, 0.4570, 0.5000

As a final step we remove the unnecessary zeros.

0.0475, 0.45, 0.457, 0.5

Example 5 Write 0.0475 as an unreduced fraction. Then name both numbers.

Solution A decimal number with four decimal places can be written as a fraction with a denominator of 10,000.

$$0.0475 = \frac{475}{10{,}000}$$

Both numbers are named **four hundred seventy-five ten-thousandths.**

LESSON PRACTICE

Practice set Use words to name each number:

a. 6.875 **b.** 0.025 **c.** 0.1625

Round each decimal number to the nearest whole number:

d. 4.375 **e.** 2.625 **f.** 1.3333

g. Compare: 0.375 ◯ 0.0375

h. Arrange these numbers in order from least to greatest:

0.15, 0.1025, 0.125, 0.1

i. Use digits to write one hundred twenty-five thousandths.

MIXED PRACTICE

Problem set

1. (49) Milton was given a $100 gift certificate. If he could buy 6 games with $25, how many games could he buy with his $100 gift certificate?

2. (102) A meter is 100 centimeters, so a centimeter is one hundredth of a meter (0.01 meter). A meterstick broke into two parts. One part was 0.37 meter long. How long was the other part?

3. (71, 90) Name the total shaded portion of these two squares as a decimal number and as a reduced mixed number.

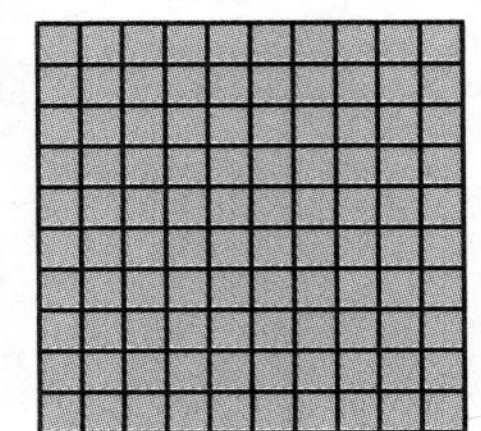 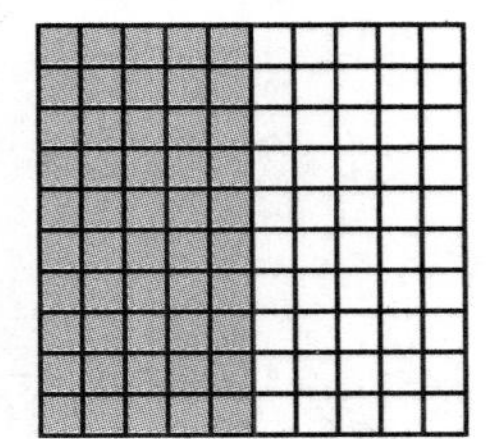

4. (106) Estimate the product of 8.33 and 7.667 by rounding both decimal numbers to the nearest whole number before multiplying.

5. (15) What are the first five multiples of 8?

6. (46, 97) Three fifths of the 30 students in the class were girls.

(a) How many girls were in the class?

(b) How many boys were in the class?

(c) What was the ratio of boys to girls in the class?

7. (62) Estimate the sum of $8.96, $12.14, and $4.88 by rounding each amount to the nearest dollar before adding.

8. (106) Write 5.375 with words.

9. (53, 72, Inv. 10) (a) The perimeter of the square is how many units?

(b) The area of the square is how many square units?

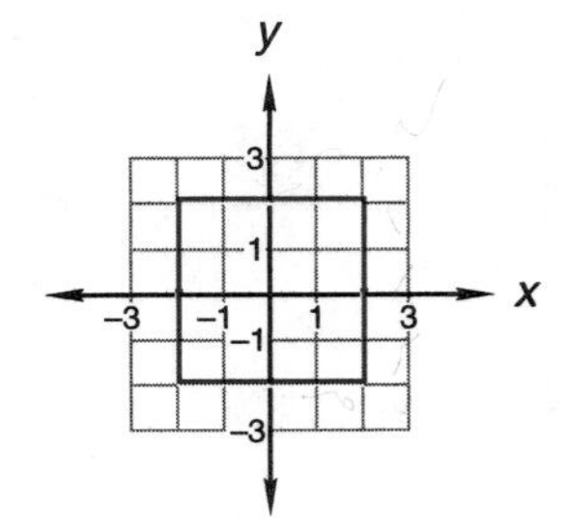

10. Arrange these numbers in order from least to greatest:
(106)

0.96, 0.875, 0.9, 1

11. $4\frac{3}{8} + 1\frac{3}{8}$
(81)

12. $3\frac{7}{10} + \frac{3}{10}$
(59)

13. $4 - 1\frac{3}{10}$
(63)

14. 1.23 + 0.4567 + 0.5
(73)

15. 4 − 1.3
(102)

16. 8 × 57 × 250
(18, 55)

17. 5 × $7.25
(17)

18. 8)$26.00
(26)

19. 436 ÷ 21
(94)

20. 16)5040
(94)

21. $5 \times \frac{3}{10}$
(86, 91)

22. $5 \div \frac{2}{3}$
(96)

23. Write fractions equal to $\frac{1}{6}$ and $\frac{1}{8}$ that have denominators of 24. Then add the fractions.
(79)

This graph shows the fraction of students in a class who have hair of a certain color. Use this graph to answer problems 24 and 25.

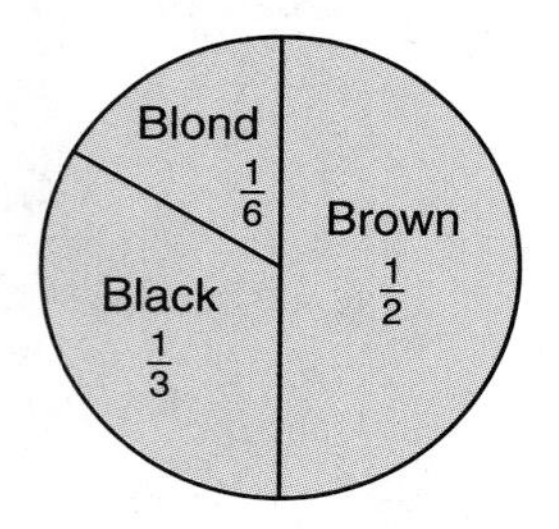

24. There are 30 students in the class. How many students have black hair? What percent of the students have black hair?
(71, Inv. 8)

25. Which two groups, taken together, total one half of the class?
(Inv. 8)

A. black and brown

B. brown and blond

C. blond and black

26. Compare: $\sqrt{81} \bigcirc 3^2$
(78, 89)

27. (a) What is the volume of a cube with the measurements shown?
(83, 103)

(b) What is the shape of each surface of the cube?

4 in.

28. Print in uppercase the eighth letter of the alphabet, and show its lines of symmetry.
(105)

29. For exercise Diana walked around the park. She walked around the park 4 times Monday, 6 times Tuesday, and 7 times Wednesday. Diana walked around the park an average of how many times each day? Write your answer as a mixed number.
(50, 58)

30. Mary spent most of one day hiking up Giant Mountain in Adirondack Park in New York. During the hike she drank about three pints of water. About how many ounces of water did Mary drink?
(85)

A. 32 oz B. 48 oz C. 64 oz D. 100 oz

LESSON

107 Using Percent to Name Part of a Group

WARM-UP

Facts Practice: 50 Fractions to Simplify (Test J)

Mental Math:

a. Describe how you would estimate the cost of 9.8 gallons of gas at $\$1.49\frac{9}{10}$ per gallon.

b. How many quarts are in one gallon? ... four gallons? ... half a gallon?

c. 50% of $40 **d.** 25% of $40 **e.** 10% of $40

Roman numerals:

f. Write CXC in our number system.

g. Write LXV in our number system.

Problem Solving:

If a coin is flipped, there are two possible outcomes: heads (H) or tails (T). If a coin is flipped twice, there are four possible outcomes: heads then heads (H, H), heads then tails (H, T), tails then heads (T, H), or tails then tails (T, T). How many outcomes are possible for a coin that is flipped three times? List all the possible outcomes, starting with heads then heads then heads (H, H, H).

NEW CONCEPT

Percent is a word that means "out of 100." If we read that 50 percent of all Americans drive cars, we understand that 50 out of every 100 Americans drive cars. Likewise, the statement "Ten percent of the population is left-handed" means that 10 out of every 100 people are left-handed. When we say "percent," we speak as though there were 100 in the group. However, we may say "percent" even when there are more than or less than 100 in the group.

Like fractions, percents name parts of a whole. We have used fraction manipulatives to learn the percents that are equivalent to some fractions. In this lesson we will learn how to find percents for other fractions by renaming the fraction with a denominator of 100.

Example 1 If 8 of the 20 students are boys, what percent of the students are boys?

Solution If we write the number of boys over the total number of students in the group, we get 8 boys over 20 total. If we multiply this fraction by a name for 1 so that the denominator becomes 100, the numerator will be the percent. So we multiply by $\frac{5}{5}$.

$$\frac{8 \text{ boys}}{20 \text{ total}} \times \frac{5}{5} = \frac{40 \text{ boys}}{100 \text{ total}}$$

This means that if there were 100 students, there would be 40 boys. Thus, **40 percent** of the students are boys.

Example 2 There were 400 pieces of candy in all. If 60 pieces were chocolate, what percent of the candy was chocolate?

Solution We have the fraction 60 chocolates over 400 total. We can partially reduce this fraction ratio to make the denominator equal 100. We do this by dividing each term by 4.

$$\frac{60 \text{ chocolates} \div 4}{400 \text{ total} \quad \div 4} = \frac{15 \text{ chocolates}}{100 \text{ total}}$$

When the denominator is 100, the top number is the percent. Thus, **15 percent** of the candy was chocolate.

Instead of using the word *percent,* we may use the percent sign (%). Using the percent sign, we write 15 percent as **15%.**

Some fractions are not easily renamed as parts of 100. Let's suppose that $\frac{1}{6}$ of the students earned an A on the test. What percent of the students earned an A?

$$\frac{1}{6} = \frac{?}{100}$$

Since 100 is not a multiple of 6, there is no whole number by which we can multiply the numerator and denominator of $\frac{1}{6}$ to rename it with a denominator of 100. However, we can find $\frac{1}{6}$ of 100% by multiplying and then dividing.

$$\frac{1}{6} \times 100\% = \frac{100\%}{6}$$

$$\begin{array}{r} 16\frac{4}{6}\% \\ 6\overline{)100\%} \\ \underline{6} \\ 40 \\ \underline{36} \\ 4 \end{array} \quad 16\tfrac{4}{6}\% = 16\tfrac{2}{3}\%$$

We find that $\frac{1}{6}$ equals $16\frac{2}{3}\%$.

Example 3 The team won $\frac{2}{3}$ of its games. Find the percent of its games the team won.

Solution We first multiply $\frac{2}{3}$ by 100%.

$$\frac{2}{3} \times 100\% = \frac{200\%}{3}$$

Then we divide 200% by 3 and write the quotient as a mixed number.

$$\begin{array}{r} 66\frac{2}{3}\% \\ 3\overline{)200\%} \\ \underline{18} \\ 20 \\ \underline{18} \\ 2 \end{array}$$

The team won $\mathbf{66\frac{2}{3}\%}$ of its games.

LESSON PRACTICE

Practice set

a. If 120 of the 200 students are girls, then what percent of the students are girls?

b. If 10 of the 50 pieces of candy are green, then what percent of the pieces of candy are green?

c. Sixty out of 300 is like how many out of 100?

d. Forty-eight out of 200 is what percent?

e. Thirty out of 50 is what percent?

f. If half of the people ate lunch, then what percent of the people ate lunch?

g. Five minutes is $\frac{1}{12}$ of an hour. Five minutes is what percent of an hour?

MIXED PRACTICE

Problem set

1. (99) Loretha swam 100 meters in 63.8 seconds. Julie swam 100 meters 1 second faster than Loretha. How long did it take Julie to swim 100 meters?

2. (101) Estimate the area of this rectangle.

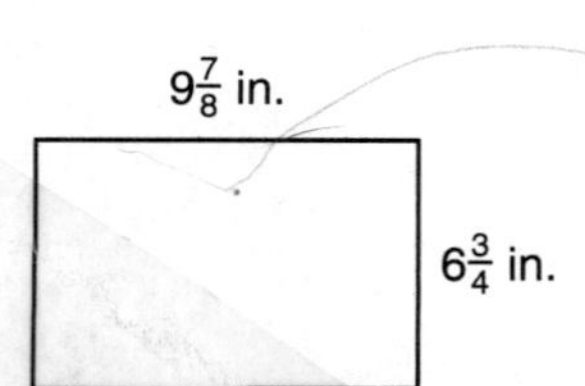

3. The camel could carry 245 kilograms. If each bundle of straw weighed 15 kilograms, how many full bundles of straw could the camel carry?
(21, 22)

4. Estimate the total cost of 8 books priced at \$6.98 each by rounding the cost per book to the nearest dollar before multiplying.
(104)

5. If 60 of the 200 students are girls, then what percent of the students are girls?
(107)

6. Compare: $\frac{1}{10} + \frac{1}{10} \bigcirc 0.1 + 0.1$
(71)

7. Estimate the quotient when 19.8 is divided by 3.875.
(101)

8. If a bag contains 50 pieces of candy and 10 of the pieces are green, then what percent of the candy pieces are green?
(107)

9. Write a fraction equal to $\frac{1}{3}$ that has the same denominator as the fraction $\frac{1}{6}$. Then add the fraction to $\frac{1}{6}$. Remember to reduce your answer.
(79, 81)

10. (a) The perimeter of the blue rectangle is how many units?
(53, 72)

(b) The area of the blue rectangle is how many square units?

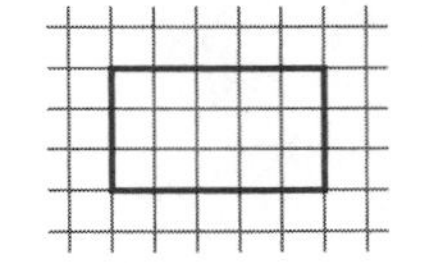

11. *QT* equals 9 centimeters. *QR* equals *RS* equals *ST*. Find *QR*.
(61)

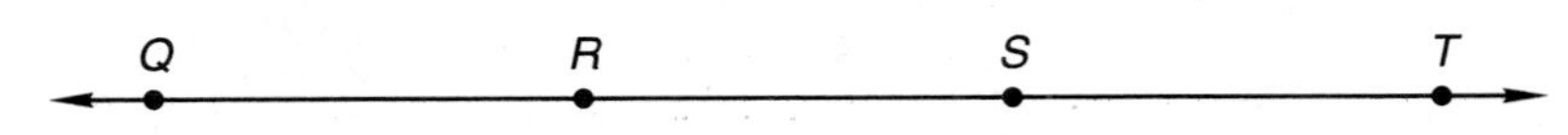

12. Which arrow could be pointing to 1.3275?
(106)

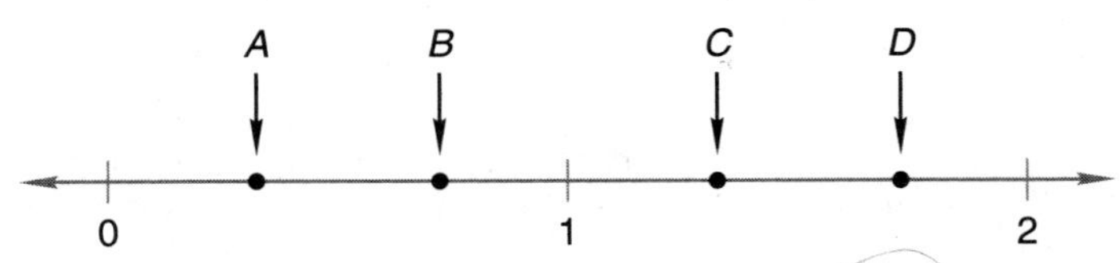

13. $\frac{31}{100} + \frac{29}{100}$
(90)

14. $5 - 3\frac{7}{10}$
(63)

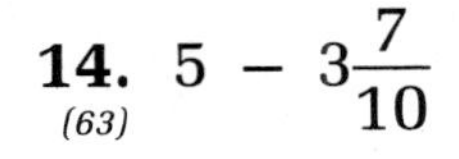

15. $5 - 3.7$
(102)

16. $10 \times \$3.65$
(29)

17. 468×579
(55)

18. $\$36.50 \div 10$
(54)

19. $5\overline{)8765}$
(26)

20. $640 \div 32$
(94)

21. $\frac{3}{10} \times \frac{7}{10}$
(76)

22. $4 \div \frac{3}{5}$
(96)

The table below shows how many votes each student candidate received in the class election. Use the table to answer problems 23–25.

Election Results

Miguel	𝍸 𝍸 ||
Debbie	𝍸 ||
Patrick	𝍸
Golda	𝍸 |||

23. How many votes did Miguel receive?
(Inv. 8)

24. What fraction of the votes did Golda receive?
(Inv. 8, 90)

25. A student in the class noticed that there could have been a four-way tie in the election. If there had been a four-way tie, how many votes would each of the four students have received?
(50, Inv. 8)

26. Reduce: $\frac{25}{100}$
(90)

27. $10^3 - \sqrt{100}$
(78, 89)

28. Triangle *ABC* is which type of triangle?
(36)

A. acute
B. right
C. obtuse
D. regular

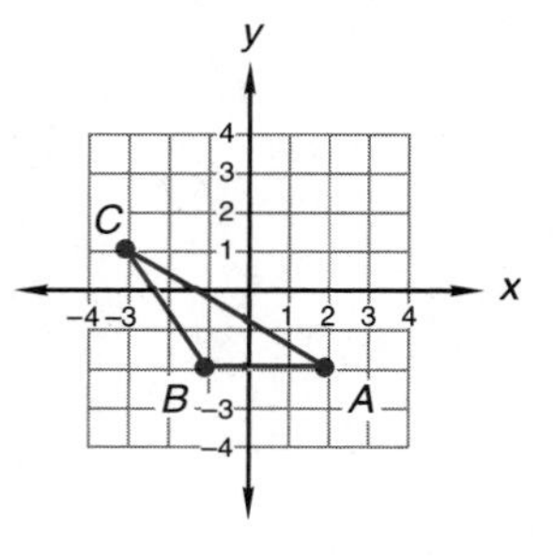

29. Write the coordinates of each vertex of ΔABC.
(Inv. 10)

30. Estimate the volume of a box of crackers with the dimensions shown.
(101, 103)

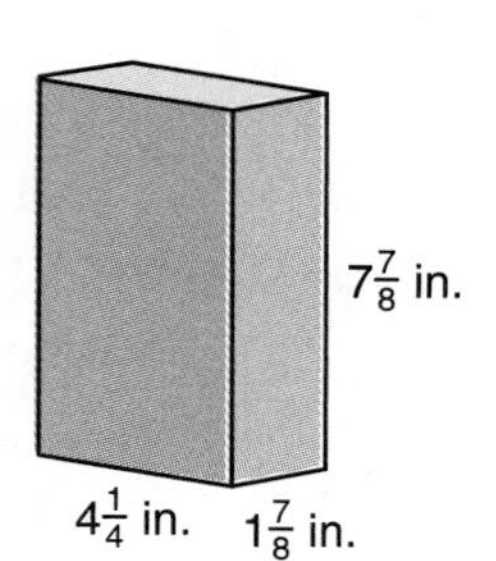

LESSON 108 Schedules

WARM-UP

Facts Practice: 50 Fractions to Simplify (Test J)

Mental Math:

a. Describe how you would estimate the cost of 11.17 gallons of gas at $\$1.39\frac{9}{10}$ per gallon.

b. How many years are in one century? ... ten centuries?

c. $\frac{1}{4}$ of $80 **d.** $\frac{3}{4}$ of $80 **e.** 50% of $\frac{1}{2}$

Roman numerals:

f. Write CL in our number system.

g. Write LXVI in our number system.

Problem Solving:

How many 1-inch cubes would be needed to build a cube with edges 3 inches long?

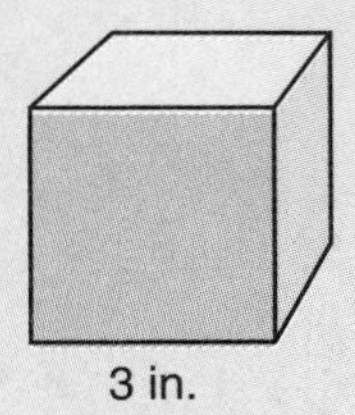

NEW CONCEPT

In this lesson we will practice reading **schedules.** A schedule is a list of times and events that shows when the events are planned to happen.

Example 1 The schedule of events for the state track meet was shown in the program. Deardra qualified to run both the 300-meter low hurdles and the 200-meter dash. Her second race starts how many minutes after the start of her first race?

10:45 a.m.	400-meter relay
12:00 p.m.	100-meter high hurdles
12:15 p.m.	110-meter high hurdles
12:30 p.m.	100-meter dash
12:55 p.m.	400-meter dash
Intermission	
2:00 p.m.	1600-meter run
3:10 p.m.	300-meter low hurdles
3:25 p.m.	300-meter intermediate hurdles
3:40 p.m.	200-meter dash
4:10 p.m.	1600-meter relay

Solution The 300-meter low hurdles race is scheduled for 3:10 p.m., and the 200-meter dash is scheduled for 3:40 p.m. If the events are held as scheduled, Deardra's second race will start **30 minutes** after the start of her first race.

One type of schedule is a travel itinerary. An itinerary lists starting locations and destinations together with planned departure and arrival times.

Example 2 Frank planned a round-trip flight from Oklahoma City to Indianapolis. Here is Frank's flight itinerary:

	Depart		Arrive	
Aug 22	6:11 a	Okla. City	8:09 a	Chicago
Aug 22	9:43 a	Chicago	10:38 a	Indianapolis
Aug 29	9:58 a	Indianapolis	11:03 a	St. Louis
Aug 29	12:04 p	St. Louis	1:33 p	Okla. City

Frank needs to change planes on his way to Indianapolis and on his way back to Oklahoma City. In which cities does he change planes? How much time does Frank have in the schedule to make those plane changes?

Solution Frank's trip to Indianapolis has two legs: one from Oklahoma City to Chicago, with a scheduled arrival at 8:09 a.m., and one from Chicago to Indianapolis, with a scheduled departure at 9:43 a.m. So on Frank's trip to Indianapolis, he stops in **Chicago** and has **1 hour 34 minutes** in the schedule to change planes.

On Frank's return trip, the first leg has a scheduled arrival in **St. Louis** at 11:03 a.m. The second leg has a scheduled departure at 12:04 p.m. So Frank has **1 hour 1 minute** in the schedule to change planes in St. Louis.

LESSON PRACTICE

Practice set Refer to the track-meet schedule in example 1 to answer problems **a** and **b.**

a. Luis qualified for the 1600-meter run. He usually starts warming up 45 minutes before the start of the race. At what time should Luis start his warm-up?

b. Lashlie is the leading qualifier in both the 100-meter and 200-meter dashes. How much time is scheduled between the start of those two events?

Use the flight itinerary in example 2 to answer problems **c** and **d**.

c. Frank's departure from Indianapolis is how many days after his arrival?

d. For his flight to Indianapolis, Frank wants to get to the Oklahoma City airport one hour before the scheduled take-off. The drive from Frank's home to the airport usually takes half an hour. About what time should Frank leave home to drive to the airport?

A. 4:00 a.m. B. 4:30 a.m. C. 5:00 a.m. D. 5:30 a.m.

e. James rode the train from Chicago to Springfield. Here is the schedule for the train he boarded:

Station	Arrive	Depart
Chicago, IL		10:45 a
Joliet, IL	11:55 a	11:55 a
Bloomington, IL	02:05 p	02:35 p
Springfield, IL	03:50 p	03:55 p
St. Louis, MO	05:40 p	

From the time the train departs Chicago until the time it arrives in Springfield is how many hours and minutes?

MIXED PRACTICE

Problem set

1. (52) Bobby weighs forty-five million, four hundred fifty-four thousand, five hundred milligrams. Use digits to write that number of milligrams.

2. (49) What is the total cost of 2 items at \$1.26 each, 3 items at 49¢ each, plus a total tax of 24¢?

3. (73) Flora rode her bike 2.5 miles from her house to the library. How far did she ride going to the library and back home?

4. (18) If $4y = 20$, then $2y - 1$ equals what number?

5. (27) The arrow is pointing to what number on this scale?

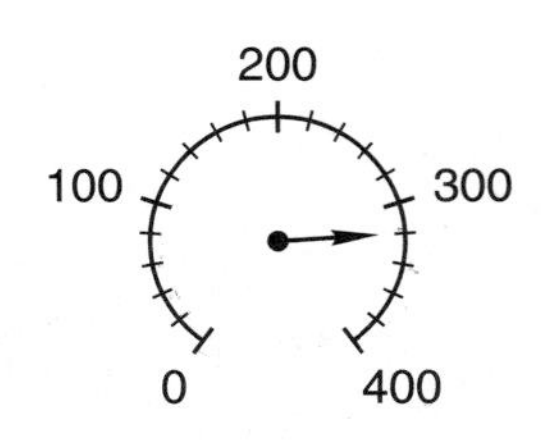

6. Fifteen of the 25 students in the class are boys.
(97, 107)

(a) What percent of the students are boys?

(b) What is the ratio of boys to girls in the class?

7. Estimate the sum of 12.7 and 8.167 by rounding both numbers to the nearest whole number before adding.
(101)

8. Write the reduced fraction that equals 80%.
(71, 90)

9. Compare: 50% ◯ $\frac{1}{2}$
(71)

10. 45^2
(78)

11. Use words to name the number 76.345. Which digit is in the tenths place?
(106)

12. A blue rectangle is drawn on this grid.
(53, 72)

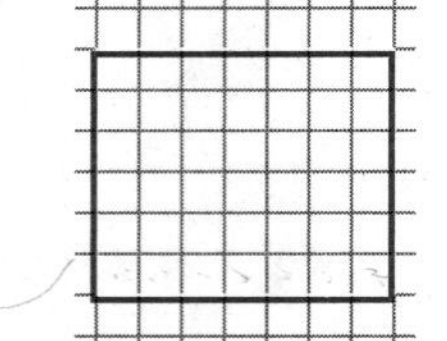

(a) The perimeter of the rectangle is how many units?

(b) The area of the rectangle is how many square units?

13. *WX* is 48 mm. *XY* is half of *WX*. *YZ* equals *XY*. Find *WZ*.
(61)

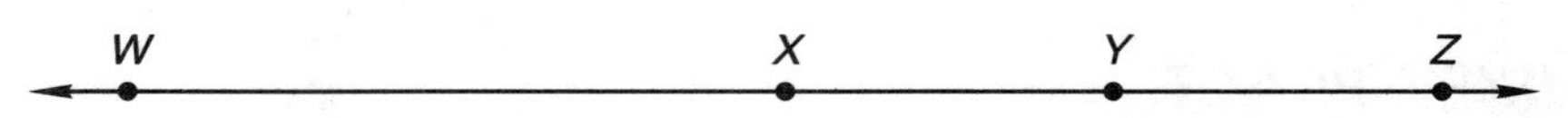

14. 2.386 + 1.2 + 16.25 + 10
(99)

15. 4.2 − (3 − 0.45)
(24, 102)

16. \$37.05 ÷ 15
(94)

17. Write a fraction equal to $\frac{1}{2}$ that has the same denominator as $\frac{1}{6}$. Then add the fraction to $\frac{1}{6}$. Remember to reduce your answer.
(79, 81)

18. $\frac{1}{2} \div \frac{2}{3}$
(96)

19. $\frac{3}{10} \times \frac{3}{10}$
(76)

20. $\frac{4}{11} + \frac{5}{11}$
(41)

21. $4\frac{5}{7} - \frac{1}{7}$
(43)

22. Five sixths of the two dozen juice bars were strawberry. How many of the juice bars were strawberry?
(46)

This table shows how many students received certain scores out of a possible 20 on the test. Use the table to answer problems 23–26.

Test Results

Score	Number of Students
20	4
19	4
18	5
17	6
16	3
15	2

23. *(Inv. 8, 84)* Which score was made by the greatest number of students?

24. *(Inv. 8, 84)* If 25 students took the test, how many students got fewer than 15 correct?

25. *(Inv. 8, 84)* If the lowest score was 13, what was the range of the scores?

26. *(Inv. 8, 84)* If all 25 scores were listed in order like this:

20, 20, 20, 20, 19, 19, ...

which score would be in the middle of the list?

27. *(103)* What is the volume of a closet that is 5 feet wide, 2 feet deep, and 8 feet high?

28. *(74, 107)* Two feet is what percent of a yard?

29. *(105)* This star has how many lines of symmetry?

30. *(32)* The star has how many sides? What kind of polygon is the star?

LESSON

109 Multiplying Decimal Numbers

WARM-UP

Facts Practice: 50 Fractions to Simplify (Test J)

Mental Math:

a. Jorge's car traveled 298 miles on 9.78 gallons of gas. Describe how to estimate the number of miles Jorge's car traveled on each gallon of gas.

b. $\frac{1}{8}$ of 80 **c.** $\frac{3}{8}$ of 80 **d.** 25% of 80

e. $\sqrt{81}$, × 10, − 2, ÷ 2, + 1, ÷ 5

Roman numerals:

f. Write CV in our number system.

g. Write XL in our number system.

Problem Solving:

How many seconds are in one day?

NEW CONCEPT

What is one tenth of one tenth? We will use pictures to answer this question.

The first picture at right is a square. The square represents one whole, and each column is one tenth of the whole. We have shaded one tenth of the whole.

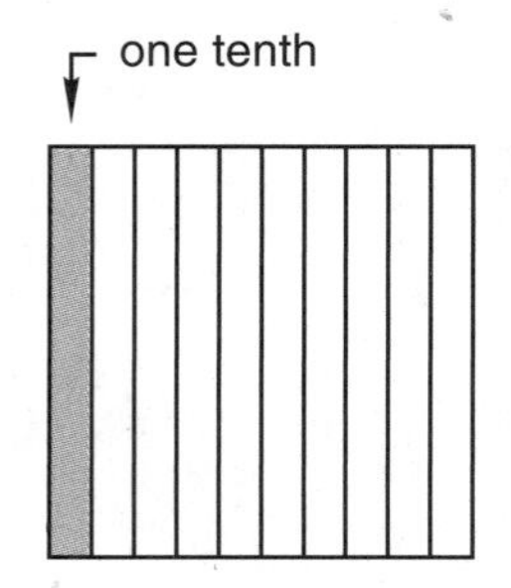

To find one tenth of one tenth, we divide each tenth into ten parts. In the second picture at right, we show each column divided into ten parts. One small square is shaded. We have shaded one tenth of one tenth of the whole. The shaded part is **one hundredth** of the whole.

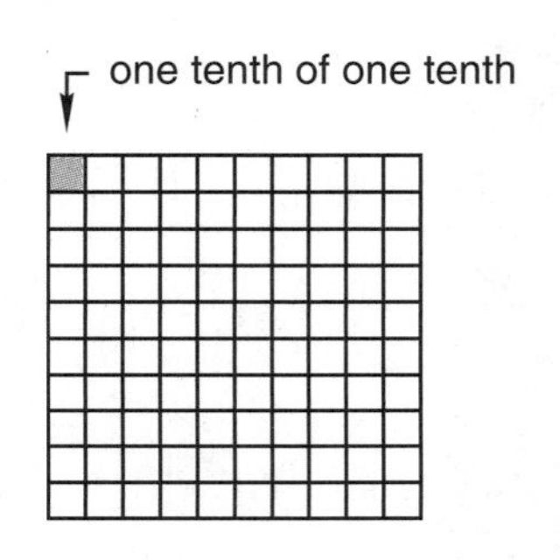

When we find one tenth of one tenth, we are multiplying. Here we show the problem written as a multiplication equation:

$$\frac{1}{10} \times \frac{1}{10} = \frac{1}{100}$$

We can also write the same problem using decimal numbers, like this:

$$\begin{array}{r} 0.1 \\ \times\ 0.1 \\ \hline 0.01 \end{array}$$

When we set up a decimal multiplication problem, we do not line up the decimal points as we do in addition and subtraction. We just set up the problem as though it were a whole-number problem and then multiply. To place the decimal point in the answer, we first count the total number of decimal places in both factors. Then we insert a decimal point in the answer so that it has the same total number of decimal places as the factors.

Copy and **study** the following examples and solutions:

Examples and *Solutions*

$$\begin{array}{rl} \overset{1}{} & \\ 0.12 & \text{2 digits to right of decimal point} \\ \times \quad 6 & \text{0 digits to right of decimal point} \\ \hline 0.72 & \text{2 digits to right of decimal point} \end{array}$$

$$\begin{array}{rl} \overset{1}{} & \\ 25 & \text{0 digits to right of decimal point} \\ \times\ 0.3 & \text{1 digit to right of decimal point} \\ \hline 7.5 & \text{1 digit to right of decimal point} \end{array}$$

$$\begin{array}{rl} \overset{4}{} & \\ 0.15 & \text{2 digits to right of decimal point} \\ \times \quad 0.9 & \text{1 digit to right of decimal point} \\ \hline 0.135 & \text{3 digits to right of decimal point} \end{array}$$

The rule for multiplying decimal numbers is, **"Multiply, then count."** We **multiply** the digits; then we **count** the total number of decimal places in the factors. Then, starting from the right side of the answer, we count over that many digits and mark the decimal point.

In the chart below we have summarized the rules of decimal arithmetic for adding, subtracting, and multiplying.

Decimals Chart

Operation	+ or −	×
Memory cue	line up . ± . .	×; then count ._ × ._ .__
You may need to ... • Place a decimal point on the end of whole numbers. • Fill empty places with zero.		

LESSON PRACTICE

Practice set* Multiply:

a. 0.3×4 **b.** 3×0.6 **c.** 0.12×12 **d.** 1.4×0.7

e. 0.3×0.5 **f.** 1.2×3

g. 1.5×0.5 **h.** 0.25×1.1

i. Compare: $\frac{3}{10} \times \frac{3}{10} \bigcirc 0.3 \times 0.3$

j. What is the area of this square?

0.8 cm

MIXED PRACTICE

Problem set

1. (109) Copy the decimals chart in this lesson.

2. (107) Forty of Maggie's 50 answers were correct. What percent of Maggie's answers were correct?

3. (109) Compare: $\frac{1}{10} \times \frac{1}{10} \bigcirc 0.1 \times 0.1$

4. (28) What time is 35 minutes before midnight?

5. (68) Use digits to write the decimal number one hundred one and one hundred one thousandths.

6. (50) Three small blocks of wood are balanced on one side of a scale with a 100-gram weight and a 500-gram weight on the other side. If each block weighs the same, what is the weight of each block?

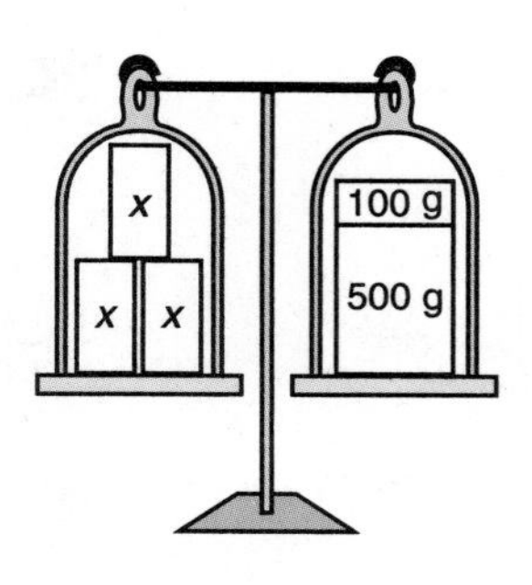

7. (15) What are the first five multiples of 10?

8. (104) Estimate the difference of \$23.07 and \$6.94 by rounding both amounts to the nearest dollar and then subtracting.

9. (53, 72) A rectangle is drawn on this grid.

(a) How many units is the perimeter of the rectangle?

(b) How many square units is the area of the rectangle?

10. (71, 90) (a) Write the reduced fraction equal to 10%.

(b) Write the reduced fraction equal to 20%.

11. (99) $32.3 + 4.96 + 7.5 + 11$

12. (24, 102) $1 - (1.36 - 0.8)$

13. (109) 12×1.2

14. (109) 0.15×0.9

15. (109) 0.16×10

16. (26, 94) $13m = 3705$

17. (26) $6\overline{)\$8.76}$

18. (94) $980 \div 28$

19. (41) $1\frac{3}{5} + 1\frac{1}{5}$

20. (91) $4\frac{3}{10} + 1\frac{2}{10}$

21. (41) $4\frac{3}{10} - 1\frac{2}{10}$

22. (79) Write fractions equal to $\frac{2}{3}$ and $\frac{1}{2}$ that have denominators of 6. Then subtract the smaller fraction from the larger fraction.

23. (90) $\frac{3}{10} \times \frac{1}{3}$

24. (96) $\frac{3}{4} \div \frac{3}{5}$

25. (96) $\frac{3}{10} \div 3$

26. (72) The floor of a room that is 12 feet wide and 15 feet long will be covered with tiles that are 1 foot square. How many tiles are needed?

27. (53) Baseboard will be nailed around the edge of the floor described in problem 26. Ignoring the door opening, how many feet of baseboard are needed?

28. (103) What is the volume occupied by a refrigerator with the dimensions shown?

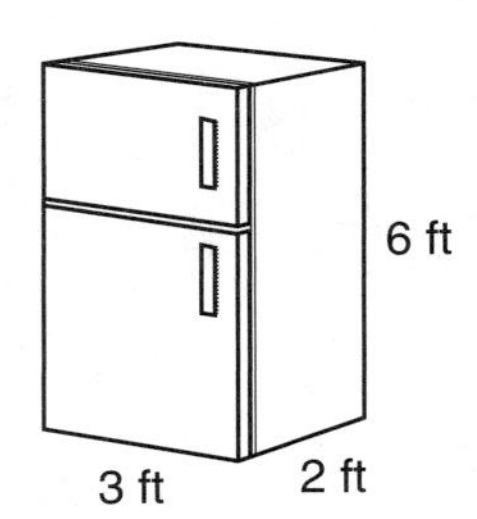

Below is a schedule of one day's soccer matches during the 2000 Summer Olympics in Australia. Refer to this schedule to answer problems 29 and 30.

Sydney Time	Event	Venue
5:00 p.m.–7:00 p.m.	Women: Australia vs. Germany	Bruce Stadium, Canberra
5:00 p.m.–7:00 p.m.	Women: Sweden vs. Brazil	Melbourne Cricket Ground
6:30 p.m.–8:30 p.m.	Men: Nigeria vs. Honduras	Hindmarsh Stadium
7:00 p.m.–9:00 p.m.	Men: Cameroon vs. Kuwait	Brisbane Cricket Ground
8:00 p.m.–10:00 p.m.	Men: USA vs. Czech Republic	Bruce Stadium, Canberra
8:00 p.m.–10:00 p.m.	Men: Australia vs. Italy	Melbourne Cricket Ground

29. (108) How much time is allowed in the schedule for each soccer game?

30. (108) How much time is allowed between games when more than one game is played at a venue?

LESSON

110 Multiplying Decimal Numbers: Using Zeros as Placeholders

WARM-UP

Facts Practice: 50 Fractions to Simplify (Test J)

Mental Math:

a. Describe how to estimate the product of $8\frac{3}{4}$ and $5\frac{1}{4}$.

b. How many centimeters are in one meter? ... ten meters?

c. Simplify: $\frac{6}{9}, \frac{12}{9}, \frac{24}{9}$

d. $\frac{1}{6}$ of 30, $\times$ 5, + 2, $\div$ 3, $\times$ 4, $\div$ 6

Roman numerals:

e. Write MCXX in our number system.

f. Write XLV in our number system.

Problem Solving:

Which uppercase letters of the alphabet have a vertical line of symmetry?

NEW CONCEPT

When we multiply decimal numbers, we follow the rule "Multiply, then count." We count the total number of decimal places in the factors. Then, starting from the right-hand end of the product, we count over the same number of places and mark the decimal point. Sometimes there are more decimal places in the factors than there are digits in the product. Look at this problem, for example:

$$\begin{array}{r} 0.3 \\ \times\ 0.3 \\ \hline .\_9 \end{array}$$

There are two digits to the right of the decimal points in the factors. So we count over two places in the product, but there is only one digit.

To complete the multiplication, we use a rule from the bottom of the decimals chart in Lesson 109. We "fill empty places with zero." Then we add a zero to the left of the decimal point.

$$\begin{array}{r} 0.3 \\ \times\ 0.3 \\ \hline 0.09 \end{array}$$

Add a zero to the left of the decimal point.

Fill the empty place with zero.

Changing the problem 0.3 × 0.3 to a fraction problem may help us understand why we use zeros as placeholders. Since 0.3 equals $\frac{3}{10}$, we may write the multiplication problem like this:

$$\frac{3}{10} \times \frac{3}{10} = \frac{9}{100}$$

The product $\frac{9}{100}$ may be written as the decimal number 0.09.

Example Multiply: 0.12 × 0.3

Solution We set up the problem as though it were a whole-number problem. We follow the rule "Multiply, then count." We "fill empty places with zero" and get the product **0.036.**

$$\begin{array}{r} 0.12 \\ \times \quad 0.3 \\ \hline 36 \end{array}$$ 3 digits to the right of the decimal points

.036 Count over 3 places; fill the empty place with zero.

LESSON PRACTICE

Practice set* Multiply:

a. $\begin{array}{r} 0.25 \\ \times \quad 0.3 \\ \hline \end{array}$ **b.** $\begin{array}{r} 0.12 \\ \times \ 0.12 \\ \hline \end{array}$ **c.** $\begin{array}{r} 0.125 \\ \times \quad 0.3 \\ \hline \end{array}$ **d.** $\begin{array}{r} 0.05 \\ \times \ 0.03 \\ \hline \end{array}$

e. 0.03 × 0.3 **f.** 3.2 × 0.03 **g.** 0.6 × 0.16

h. 0.12 × 0.2 **i.** 0.01 × 0.1 **j.** 0.07 × 0.12

k. What is the area of this rectangle?

0.4 m

0.2 m

MIXED PRACTICE

Problem set

1. (104) Estimate the product of 5.375 and 3.8 by rounding both numbers to the nearest whole number before multiplying.

2. (107) The football team played 10 games and won 5. What percent of the games did the team win?

3. (71, 90) (a) Write the reduced fraction that equals 30%.

(b) Write the reduced fraction that equals 40%.

4. Two fifths of the 100 passengers stayed in the subway cars until the last stop. How many of the 100 passengers got off the subway cars before the last stop?
(46, 60)

5. Name the length of this segment as a number of centimeters and as a number of millimeters.
(66)

mm 10 20 30 40

cm 1 2 3 4

6. If the segment in problem 5 were cut into thirds, each third would be how many centimeters long?
(46, 74)

7. Write fractions equal to $\frac{5}{6}$ and $\frac{3}{4}$ that have denominators of 12. Then add the fractions. Remember to convert the sum to a mixed number.
(75, 79)

8. A hexagon is drawn on this grid.
(53, 72)

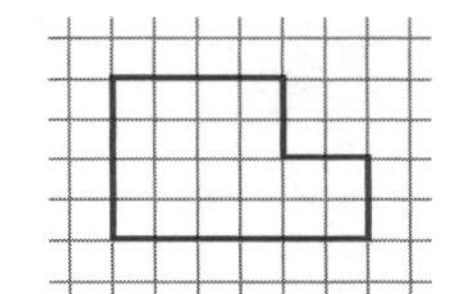

(a) How many units is the perimeter of this hexagon?

(b) How many square units is the area of the hexagon?

9. In rectangle *ABCD,* which segment is parallel to $\overline{AB}$?
(31, 61)

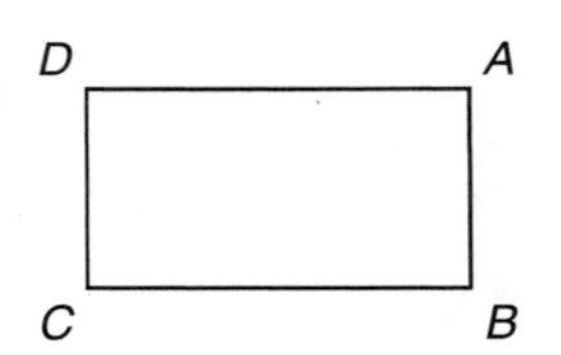

10. In rectangle *ABCD,* which two segments are perpendicular to $\overline{AB}$?
(31, 61)

11. Write 0.375 as an unreduced fraction. Then use words to name the number.
(68)

12. $6 - 4.32$
(102)

13. 0.12×0.11
(110)

14. 0.04×0.28
(110)

15. 10×0.25
(109)

16. $19x = 3705$
(26, 94)

17. $\sqrt{400}$
(89)

18. 30^2
(78)

19. $\frac{5}{13} + \frac{10}{13}$
(91)

20. $\frac{11}{12} - \frac{7}{12}$
(90)

21. $1 \times \frac{5}{6}$
(86)

22. $2 \div \frac{5}{6}$
(96)

23. $\frac{5}{6} \div 2$
(96)

This pie chart shows the percent of students in the class who made certain grades in math. Use this graph to answer problems 24–26.

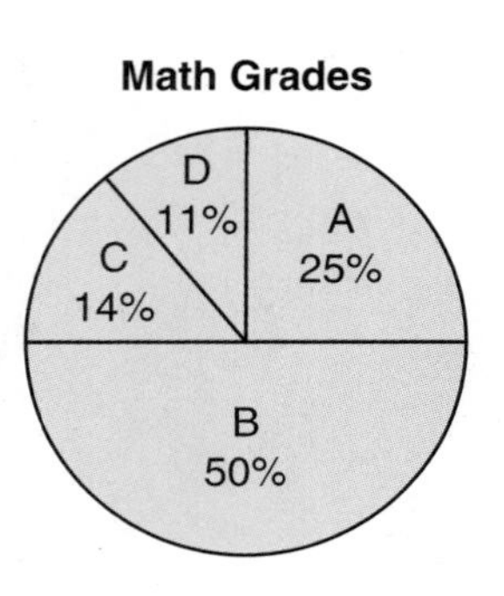

24. *(Inv. 8)* Add the percents shown on the graph. What is the total?

25. *(Inv. 8)* What grade was made by $\frac{1}{4}$ of the students?

26. *(Inv. 6)* If the teacher draws a test from a stack of tests without looking, what is the probability that the test will have the grade B?

27. *(Inv. 7)* Draw the next term of this sequence:

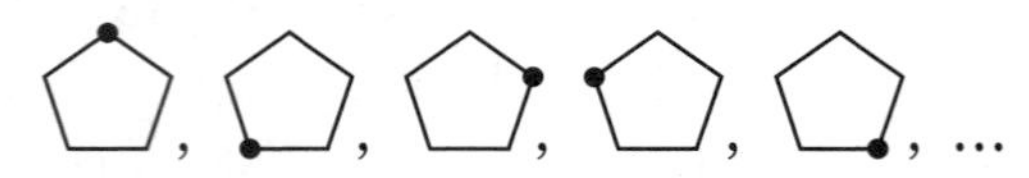

28. *(88)* What transformation changes the terms of the sequence in problem 27?

The three runners below received medals in the men's 100-meter run at the 2000 Summer Olympic Games in Sydney, Australia. Refer to this information to answer problems 29 and 30.

Runner	Country	Time
Ato Bolden	Trinidad and Tobago	9.99 seconds
Maurice Greene	United States	9.87 seconds
Obadele Thompson	Barbados	10.04 seconds

29. *(69)* Write the last names of the runners in the order of their finish, starting with the first-place runner.

30. *(69, 73)* The first-place runner ran how many seconds faster than the third-place runner?

INVESTIGATION 11

Focus on

Scale Drawings

A **scale drawing** is a picture or diagram of a figure that has the same shape as the figure but is a different size. Below is a scale drawing of the bedroom shared by Jane and Alicia. Notice the legend to the right of the picture. It shows that 1 centimeter in the picture represents 4 feet in the actual bedroom. The equivalence 1 cm = 4 ft is called the **scale.**

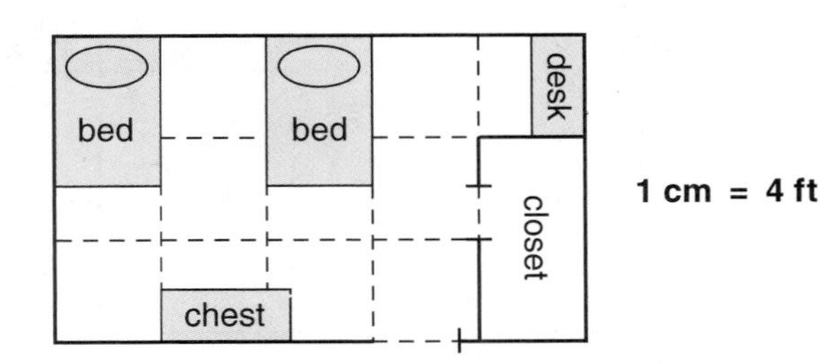

Since 1 cm in the picture represents 4 ft in the actual bedroom, we also know the following relationships:

2 cm represents 8 ft (since $2 \times 4 = 8$)

3 cm represents 12 ft (since $3 \times 4 = 12$)

4 cm represents 16 ft (since $4 \times 4 = 16$)

5 cm represents 20 ft (since $5 \times 4 = 20$)

If we measure the picture, we find that it is 5 cm long and 3 cm wide. This means that the actual bedroom is 20 ft long and 12 ft wide.

1. What is the actual distance between the beds?

2. What is the actual length and width of the closet?

3. What is the actual area of the entire room? What is the area if you subtract the area of the closet?

Measurements in the picture may be fractions of centimeters. For example, a measurement of 0.5 cm $\left(\frac{1}{2}\text{ cm}\right)$ represents $\frac{1}{2}$ of 4 ft. (Remember that the word *of,* when used with fractions, tells us to multiply.)

$$\frac{1}{2}\text{ of } 4\text{ ft} = \frac{1}{2} \times 4\text{ ft} = 2\text{ ft}$$

4. What is the actual length and width of the beds?

5. What is the actual length and width of the desk?

6. What actual length does a measurement of $1\frac{1}{4}$ cm represent? Can you identify an object in the picture that is about that long?

Andrew is on the corner of Wilson and 3rd Avenue. His position is marked by the "X" on the scale drawing below. Andrew's house is halfway between Taft and Lincoln on 5th Avenue; it is marked by the symbol ⌂.

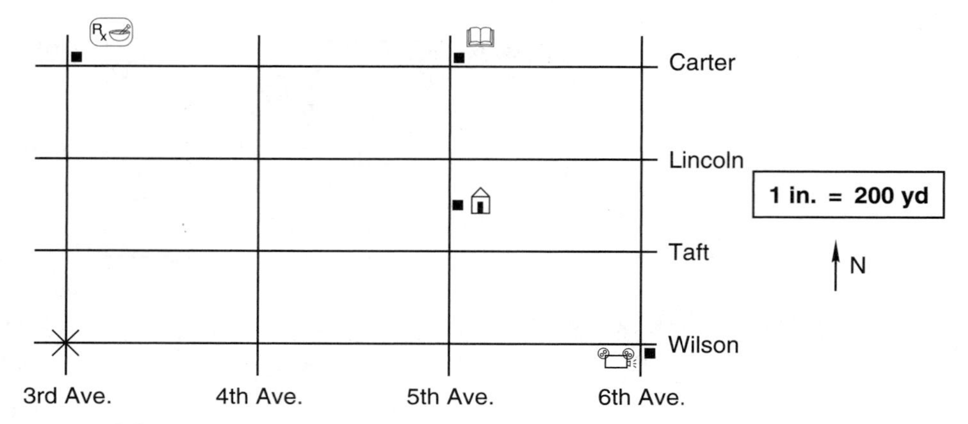

For problems 7–10 below, assume Andrew travels only along the streets shown.

7. How far is Andrew from the movie theatre ( ) at the corner of Wilson and 6th Avenue?

8. How far is he from the drugstore ( ) on the corner of Carter and 3rd Avenue?

9. How far is he from the library ( ) on the corner of Carter and 5th Avenue? Describe three different routes he could take that all give the least distance.

10. How far is Andrew from his house?

11. Measure the straight-line distance in inches between Andrew's starting point and the corner of Carter and 5th Avenue. From this measurement, estimate the actual straight-line distance in yards.

A familiar type of scale drawing is a map. On a certain map of New York City, the scale is 6 cm = 1 mi. This means that 6 centimeters on the map represents 1 mile of actual distance.

12. What length on the map corresponds to an actual distance of 3 miles? What length on the map corresponds to an actual distance of $\frac{1}{2}$ mile?

13. What fraction of a mile corresponds to 1 cm on the map? What fraction of a mile is represented by 5 cm?

14. What length on the map represents an actual distance of $2\frac{1}{6}$ miles?

15. Part of the New York City map is shown below. Estimate the actual distance between Frederick Douglass Boulevard and Lenox Avenue as a fraction of a mile. (Use the shortest distance between the two roads.)

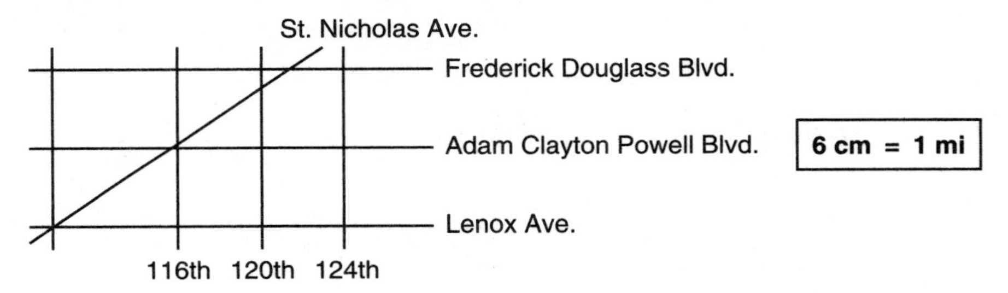

Extensions

a. Draw a scale picture of the kitchen in your house. Include the stove, refrigerator, and other important items. Make your scale 1 in. = 2 ft.

b. Obtain a street map of your city or a nearby city. Using the legend on the map, estimate the shortest distance between your school and a park of your choice, using the road system rather than a straight-line distance. Describe the route you chose.

c. We can make **scale models** of 3-dimensional figures. Model trains and action figures are examples of scale models. Using cardboard and glue or tape, make a scale model of the barn below. Use the scale 1 cm = 4 ft. Note that the front and the back of the barn are pentagons.

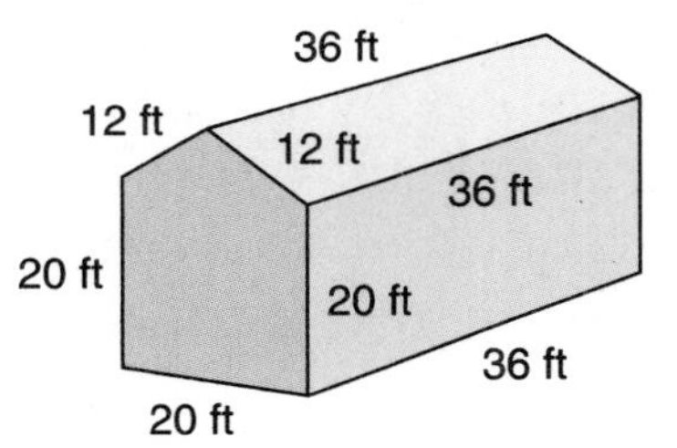

LESSON

111 Multiplying Decimal Numbers by 10, by 100, and by 1000

WARM-UP

Facts Practice: 30 Percents to Write as Fractions (Test K)

Mental Math:

a. Describe how to estimate the quotient when $7\frac{4}{5}$ is divided by $3\frac{3}{4}$.
b. How many meters are in one kilometer? ... one tenth of a kilometer?
c. 50% of \$10 **d.** 25% of \$10 **e.** 10% of \$10
f. $\frac{1}{3}$ of 60, × 2, + 2, ÷ 6, × 4, + 2, ÷ 2

Roman numerals:

g. Write MMXC in our number system.
h. Compare: XLIV ◯ 45

Problem Solving:

A roll of dimes contains 50 dimes. A roll of quarters contains 40 quarters. How many rolls of dimes are equal in value to two rolls of quarters?

NEW CONCEPT

Each place in our decimal number system is assigned a particular value. The value of each place is 10 times greater each time we move one place to the left. So when we multiply a number by 10, the digits all shift one place to the left. For example, when we multiply 34 by 10, the 3 shifts from the tens place to the hundreds place, and the 4 shifts from the ones place to the tens place. We fill the ones place with a zero.

```
   3   4 .
  ↙   ↙
3   4   0 .      (10 × 34 = 340)
```

Shifting digits to the left can help us quickly multiply decimal numbers by 10, 100, or 1000. Here we show a decimal number multiplied by 10.

```
0 . 3   4
  ↙   ↙
3 . 4            (10 × 0.34 = 3.4)
```

We see that the digit 3 moved to the other side of the decimal point when it shifted one place to the left. The decimal point holds steady while the digits move. Although it is the digits that change places when the number is multiplied by 10, we can produce the same result by moving the decimal point in the opposite direction.

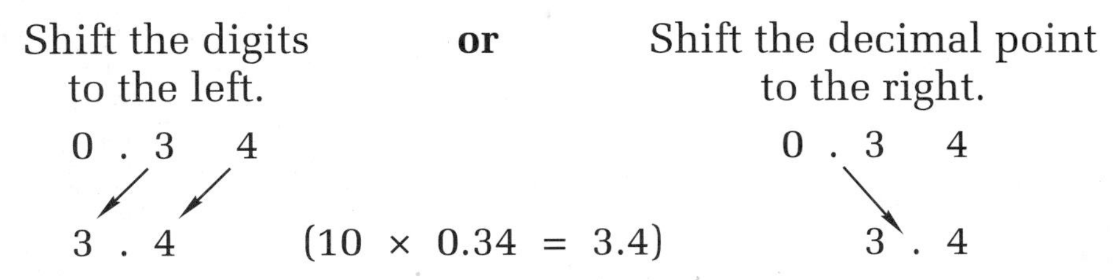

When we multiply by 10, we may simply shift the decimal point one place to the right.

Since 100 is 10×10, multiplying by 100 is like multiplying by 10 *twice.* When we multiply by 100, we may shift the decimal point *two* places to the right.

Since 1000 is $10 \times 10 \times 10$, we may shift the decimal point *three* places to the right when we multiply by 1000.

The number of places we shift the decimal point is the same as the number of zeros we see in 10 or 100 or 1000.

Example Multiply: 1.234×100

Solution To multiply mentally by 100, we may shift the decimal point two places to the right. The product is **123.4.**

$$1.234 \times 100 = 123.4$$

LESSON PRACTICE

Practice set Multiply:

a. 1.234×10 **b.** 1.234×1000 **c.** 0.1234×100

d. 0.345×10 **e.** 0.345×100 **f.** 0.345×1000

g. 5.67×10 **h.** 5.67×1000 **i.** 5.67×100

MIXED PRACTICE

Problem set

1. (50) In three classrooms there were 23 students, 25 students, and 30 students. If the students in the three classrooms were rearranged so that there were an equal number of students in each room, how many students would there be in each classroom?

2. (35) Genghis Khan was born in 1167. In 1211 he invaded China. How old was he then?

3. (a) Write the reduced fraction equal to 25%.
(71, 90)
(b) Write the reduced fraction equal to 50%.

4. (a) List the first six multiples of 6.
(15)
(b) List the first four multiples of 9.
(c) Which two numbers appear in both lists?

5. Name the shaded portion of this square as a percent, as a decimal number, and as a reduced fraction.
(71)

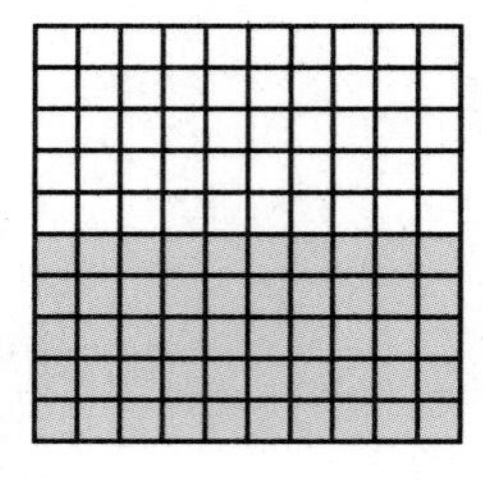

6. Name the shape of a basketball.
(83)

7. How many months are in $1\frac{1}{2}$ years?
(28)

8. (a) How many units long is the perimeter of this shape?
(53, 72)
(b) How many square units is the area of this shape?

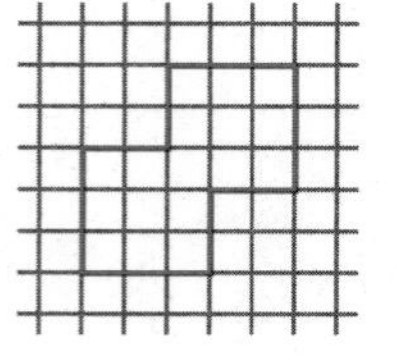

9. *QR* is 45 mm. *RS* is one third of *QR*. *QT* is 90 mm. Find *ST*.
(61)

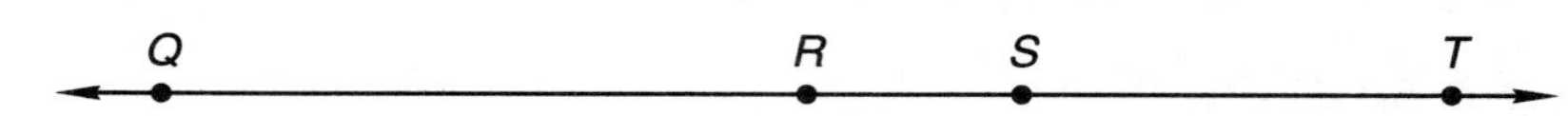

For problems 10 and 11, multiply mentally by shifting the decimal point.

10. 1.23 × 10
(111)

11. 3.42 × 1000
(111)

12. Use words to name this sum:
(68, 106)

$$15 + 9.67 + 3.292 + 5.5$$

13. 4.3 − 1.21
(102)

14. 0.14 × 0.6
(110)

15. 48 × 0.7
(109)

16. 0.735×10^2
(78, 111)

17. Write a fraction equal to $\frac{3}{4}$ that has the same denominator as $\frac{3}{8}$. Then add the fraction to $\frac{3}{8}$. Remember to convert your answer to a mixed number.
(75, 79)

18. $16\overline{)4000}$
(94)

19. $18.00 ÷ 10
(54)

20. (91) $\frac{7}{11} + \frac{8}{11}$

21. (90) $3\frac{7}{12} + \frac{1}{12}$

22. (81) $5\frac{9}{10} - 5\frac{3}{10}$

23. (91) $\frac{7}{2} \times \frac{1}{2}$

24. (96) $\frac{2}{3} \div \frac{1}{4}$

25. (96) $3 \div \frac{3}{4}$

26. (89) Compare: $\sqrt{9} + \sqrt{16} \bigcirc \sqrt{9 + 16}$

27. (107) The names of two of the 12 months begin with the letter A. What percent of the names of the months begin with the letter A?

Elizabeth studied this list of flights between Los Angeles and Philadelphia. Refer to this list to answer problems 28–30.

Los Angeles to Philadelphia	
Depart	Arrive
6:15 A	2:34 P
10:10 A	6:33 P
12:56 P	9:15 P
3:10 P	11:19 P

Philadelphia to Los Angeles	
Depart	Arrive
7:55 A	10:41 A
10:00 A	12:53 P
1:30 P	4:17 P
5:40 P	8:31 P

28. (108) Elizabeth wants to arrive in Philadelphia before 8 p.m. However, she does not want to wake up very early to catch a flight. Which departure time is Elizabeth likely to choose?

29. (108) For her return flight Elizabeth would like to leave as late as possible and still arrive in Los Angeles by 9:00 p.m. Which departure time is Elizabeth likely to choose?

30. (108) According to the times on the schedule, the 10:10 a.m. flight from Los Angeles to Philadelphia arrives at 6:33 p.m., yet the 10:00 a.m. flight from Philadelphia arrives in Los Angeles at 12:53 p.m. What do you suppose accounts for the apparent difference of 5 hours 30 minutes in the durations of the two flights?

LESSON

112 Finding the Least Common Multiple of Two Numbers

WARM-UP

Facts Practice: 30 Percents to Write as Fractions (Test K)

Mental Math:

a. Describe how to estimate the cost of 98 tickets at $2.50 each.

b. How many ounces are in one pound? ... two pounds? ... half a pound?

c. $\frac{1}{10}$ of 30 **d.** $\frac{3}{10}$ of 30 **e.** $\frac{9}{10}$ of 30

f. $\sqrt{100}$, ÷ 2, × 7, + 1, ÷ 6, × 4, ÷ 2

Roman numerals:

g. Write MCL in our number system.

h. Compare: 96 ◯ XCIV

Problem Solving:

Victor dropped a rubber ball and found that each bounce was half as high as the previous bounce. He dropped the ball from 8 feet, measured the height of each bounce, and recorded the results in a table. Copy this table and complete it through the fifth bounce.

Heights of Bounces

First	4 ft
Second	
Third	
Fourth	
Fifth	

NEW CONCEPT

Here we list the first few multiples of 4 and 6:

Multiples of 4: 4, 8, (12), 16, 20, (24), 28, 32, (36), ...

Multiples of 6: 6, (12), 18, (24), 30, (36), ...

We have circled the multiples that 4 and 6 have in common. The smallest number that is a multiple of both 4 and 6 is 12.

The smallest number that is a multiple of two or more numbers is called the **least common multiple** of the numbers. The letters **LCM** are sometimes used to stand for "least common multiple."

Example Find the least common multiple (LCM) of 6 and 8.

Solution We begin by listing the first few multiples of 6 and 8. Then we circle the multiples they have in common.

Multiples of 6: 6, 12, 18, (24), 30, 36, 42, (48), ...

Multiples of 8: 8, 16, (24), 32, 40, (48), ...

As we see above, the *least* of the common multiples of 6 and 8 is **24.**

LESSON PRACTICE

Practice set* Find the least common multiple (LCM) of each pair of numbers:

a. 2 and 3 **b.** 3 and 5 **c.** 5 and 10

d. 2 and 4 **e.** 3 and 6 **f.** 6 and 10

g. The denominators of $\frac{5}{8}$ and $\frac{3}{10}$ are 8 and 10. What is the least common multiple of 8 and 10?

MIXED PRACTICE

Problem set

1. (77) A small car weighs about one ton. Most large elephants weigh four times that much. About how many pounds would a large elephant weigh?

2. (74) The Arctic Ocean is almost completely covered with the polar ice cap, which averages about 10 feet thick. About how many inches thick is the polar ice cap?

3. (21, 111) What is the total cost of 10 movie tickets priced at $5.25 each?

4. (68) Which digit in 375.246 is in the hundredths place?

5. (32) Draw a pentagon.

6. (71) Write 12.5 as a mixed number.

7. (71) Name the shaded portion of this square as a percent, as a decimal number, and as a reduced fraction.

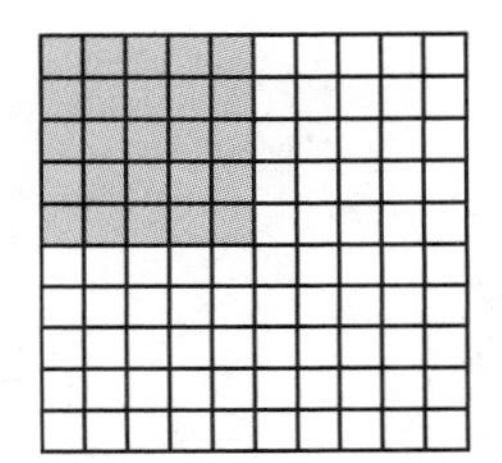

8. (83) Name the shape of an aluminum can.

9. (36, 53) What is the perimeter of this equilateral triangle?

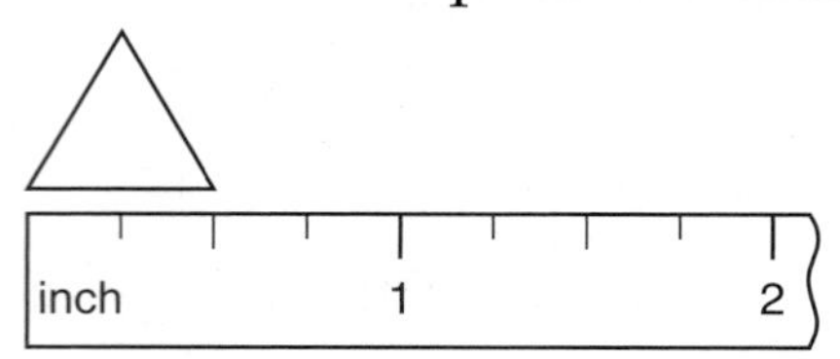

10. (112) Find the least common multiple (LCM) of 6 and 9.

11. (53, 61) If $\overline{OM}$ measures 15 mm, then what is the measure of $\overline{LN}$?

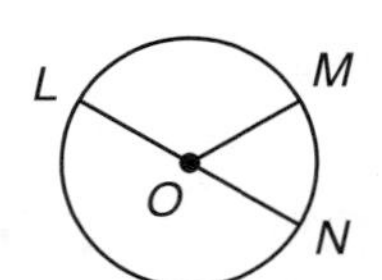

12. (61, 102) WX is 4.2 cm. XY is 3 cm. WZ is 9.2 cm. Find YZ.

W X Y Z

13. (99) $4.38 + 7.525 + 23.7 + 9$

14. (24, 102) $5 - (4.3 - 0.21)$

15. (109) 3.6×40

16. (110) 0.15×0.5

17. (111) 10×0.125

18. (26) $4w = 300$

19. (54) $40\overline{)3000}$

20. (94) $25\overline{)3300}$

21. (63, 75) $3\frac{3}{7} + \left(5 - 1\frac{2}{7}\right)$

22. (41, 86) $1\frac{1}{2} - \left(3 \times \frac{1}{2}\right)$

23. (79) Write fractions equal to $\frac{1}{4}$ and $\frac{2}{3}$ that have denominators of 12. Then subtract the smaller fraction from the larger fraction.

On this grid different shapes are centered at certain points where the grid lines intersect. For example, there is a circle at point B1 where the grid lines labeled "B" and "1" intersect. Use this grid to answer problems 24 and 25.

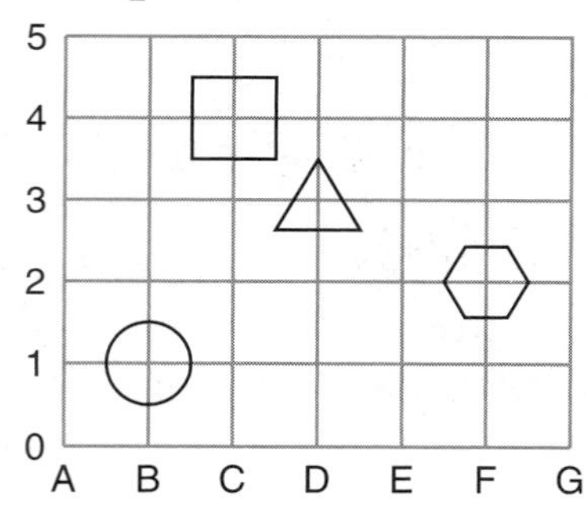

24. (32, Inv. 10) Name the shape at point D3.

25. (32, Inv. 10) What letter and number names the point where there is a hexagon?

26. Compare: $3^2 + 4^2 \bigcirc 5^2$
(78)

27. Find the percent equivalent to $\frac{1}{8}$ by multiplying 100% by $\frac{1}{8}$. Write the result as a mixed number with the fraction reduced.
(107)

Karen's flight schedule between Oklahoma City and Indianapolis is shown below. Refer to this schedule to answer problems 28–30.

FLIGHT 41 Thu, Aug 22	6:11a–8:09a plane change	Oklahoma City (OKC)	Chicago (ORD)
FLIGHT 11 operated by partner airline Thu, Aug 22	9:43a–10:38a	Chicago (ORD)	Indianapolis (IND) total duration: 4h 27min
FLIGHT 327 Thu, Aug 29	9:58a–11:03a plane change	Indianapolis (IND)	St Louis (STL)
FLIGHT 337 Thu, Aug 29	12:04p–1:33p	St Louis (STL)	Oklahoma City (OKC) total duration: 3h 35min

28. The first leg of Karen's flight to Indianapolis takes her to Chicago. How much time is in the schedule for changing planes in Chicago?
(108)

29. The times listed in the schedule are gate-to-gate times, from the time the plane pushes away from the gate at departure to the time the plane pulls into the gate at arrival. Find the total of the gate-to-gate times for the two legs from Oklahoma City to Indianapolis.
(108)

30. The total of the gate-to-gate times for the two return legs to Oklahoma City is how many minutes less than for the outbound flights? What might account for the difference in travel time?
(108)

LESSON

113 Writing Mixed Numbers as Improper Fractions

WARM-UP

Facts Practice: 30 Percents to Write as Fractions (Test K)

Mental Math:

a. Describe how to estimate the cost of 8.9 gallons of gasoline at $\$1.79\frac{9}{10}$ per gallon.

b. Simplify: $\frac{8}{12}, \frac{9}{12}, \frac{15}{12}$

c. 25% of 12

d. 50% of 19

e. 75% of 12

f. $\frac{1}{6}$ of 24, × 5, + 1, ÷ 3, × 8, − 2, ÷ 9

Roman numerals:

g. Write CXLV in our number system.

h. Compare: MD ◯ 2000

Problem Solving:

How many 1-inch cubes would be needed to build a rectangular solid 5 inches long, 4 inches wide, and 3 inches high?

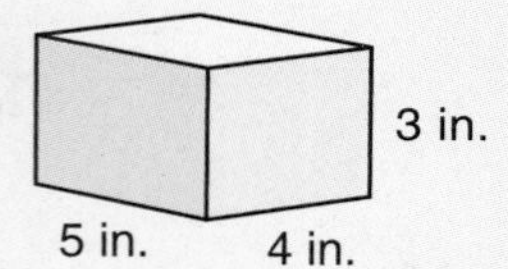

NEW CONCEPT

The picture below shows $1\frac{1}{2}$ shaded circles. How many half circles are shaded?

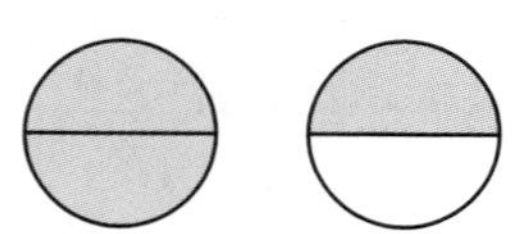

Three halves are shaded. We may name the number of shaded circles as the mixed number $1\frac{1}{2}$ or as the improper fraction $\frac{3}{2}$.

$$1\frac{1}{2} = \frac{3}{2}$$

We have converted improper fractions to mixed numbers by dividing. In this lesson we will practice writing mixed numbers as improper fractions. We will use this skill later when we learn to multiply and divide mixed numbers.

To help us understand changing mixed numbers into fractions, we can draw pictures. Here we show the number $2\frac{1}{4}$ using shaded circles:

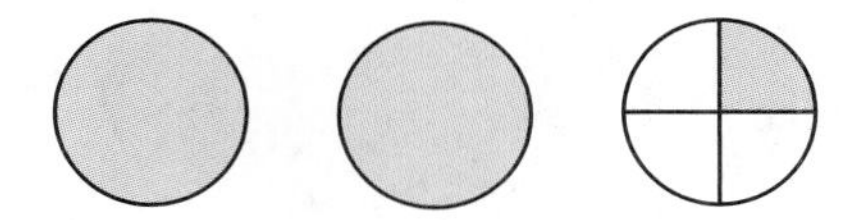

To show $2\frac{1}{4}$ as an improper fraction, we divide the whole circles into the same-size pieces as the divided circle. In this example we divide each whole circle into fourths.

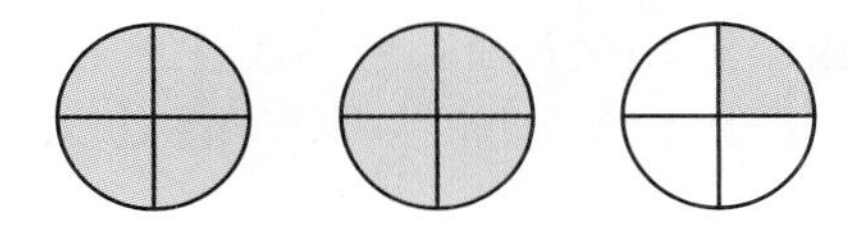

Now we count the total number of fourths that are shaded. We see that $2\frac{1}{4}$ equals the improper fraction $\frac{9}{4}$.

Example 1 Name the number of shaded circles as an improper fraction and as a mixed number.

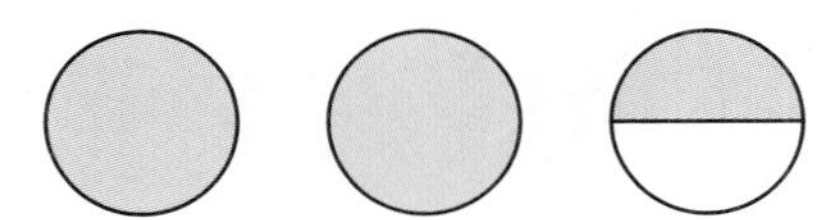

Solution To show the improper fraction, we divide the whole circles into the same-size pieces as the divided circle (in this case, halves). The improper fraction is $\mathbf{\frac{5}{2}}$. The mixed number is $\mathbf{2\frac{1}{2}}$.

$$\frac{2}{2} + \frac{2}{2} + \frac{1}{2} = \frac{5}{2} = 2\frac{1}{2}$$

Example 2 Change $2\frac{1}{3}$ to an improper fraction.

Solution One way to find an improper fraction equal to $2\frac{1}{3}$ is to draw a picture that illustrates $2\frac{1}{3}$.

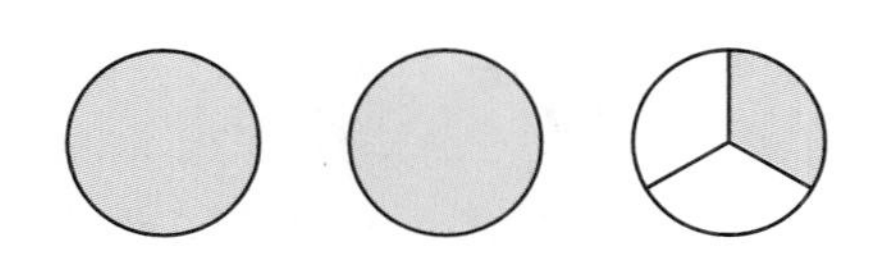

We have shaded 2 whole circles and $\frac{1}{3}$ of a circle. Now we divide each whole circle into thirds and count the total number of thirds.

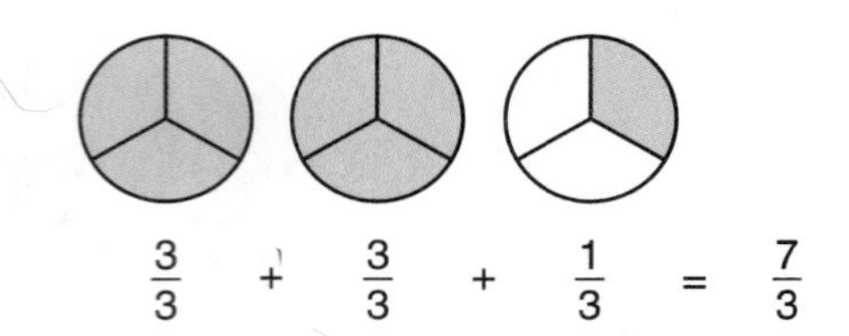

$$\frac{3}{3} + \frac{3}{3} + \frac{1}{3} = \frac{7}{3}$$

We see that seven thirds are shaded. So an improper fraction equal to $2\frac{1}{3}$ is $\frac{7}{3}$.

It is not necessary to draw a picture. We could remember that each whole is $\frac{3}{3}$. So the 2 of $2\frac{1}{3}$ is equal to $\frac{3}{3} + \frac{3}{3}$, which is $\frac{6}{3}$. Then we add $\frac{6}{3}$ to $\frac{1}{3}$ and get $\frac{7}{3}$.

LESSON PRACTICE

Practice set* For problems **a–c,** name the number of shaded circles as an improper fraction and as a mixed number.

a. 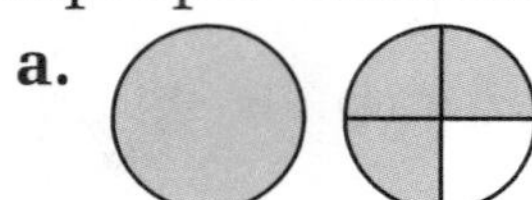

b. 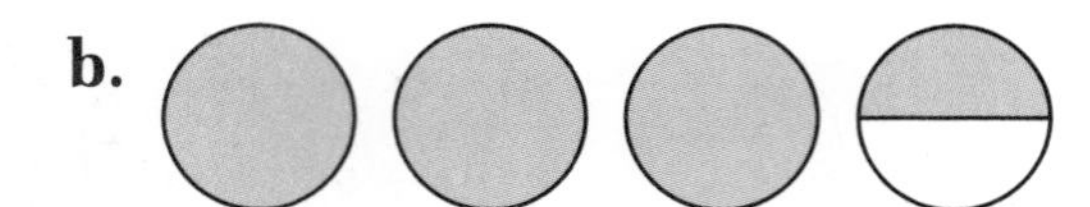

c. 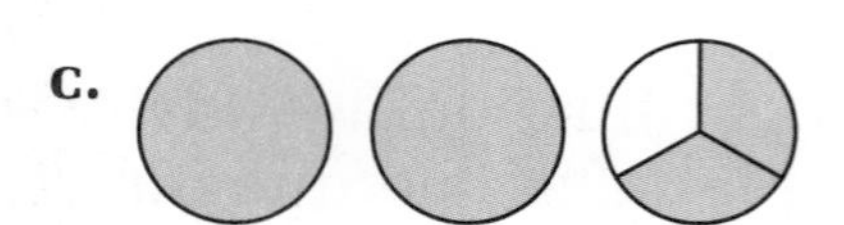

Change each mixed number to an improper fraction:

d. $4\frac{1}{2}$ **e.** $1\frac{2}{3}$ **f.** $2\frac{3}{4}$ **g.** $3\frac{1}{8}$

MIXED PRACTICE

Problem set

1. (50) On a five-day trip the Jansens drove 1400 miles. What was the average number of miles the Jansens drove on each of the five days?

2. (62) Estimate the product of 634 and 186 by rounding both numbers to the nearest hundred before multiplying.

3. (71, 79) (a) $\frac{1}{10} = \frac{\square}{100}$

(b) What percent equals the fraction $\frac{1}{10}$?

4. The weight of an object on the Moon is about $\frac{1}{6}$ of the weight of the same object on Earth. A person on Earth who weighs 108 pounds would weigh about how many pounds on the Moon?
(46)

5. Name the total number of shaded circles as an improper fraction and as a mixed number.
(113)

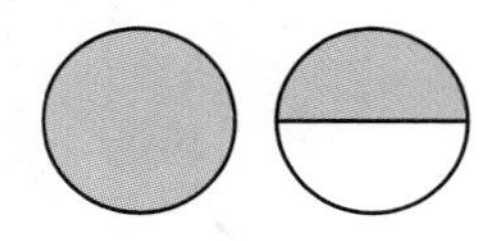

6. (a) What is the perimeter of this 1-inch square?
(53, 72)

(b) What is the area of this square?

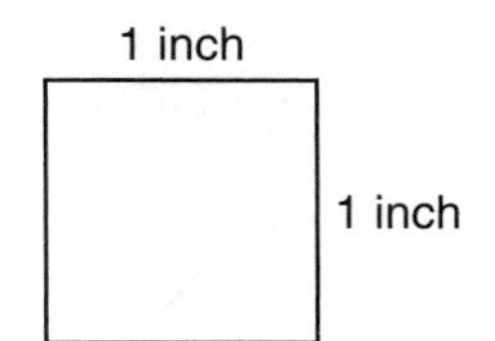

7. What fraction of a year is 3 months? What percent of a year is 3 months?
(28, 81, 107)

8. (a) Name this shape.
(83)

(b) How many faces does the shape have?

9. The denominators of $\frac{1}{6}$ and $\frac{1}{4}$ are 6 and 4. What is the least common multiple (LCM) of the denominators?
(112)

10. To what mixed number is the arrow pointing?
(38, 81)

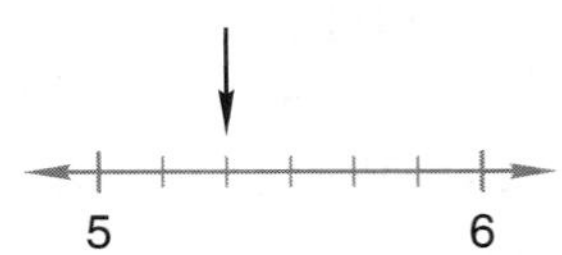

11. $4.239 + 25 + 6.79 + 12.5$
(99)

12. $6.875 - (4 - 3.75)$
(24, 102)

13. $\begin{array}{r} 3.7 \\ \times\ 0.8 \\ \hline \end{array}$
(109)

14. $\begin{array}{r} 0.125 \\ \times\ 100 \\ \hline \end{array}$
(111)

15. $\begin{array}{r} 0.32 \\ \times\ 0.04 \\ \hline \end{array}$
(110)

16. $\frac{408}{17}$
(94)

17. $27\overline{)705}$
(94)

18. $5\overline{)\$17.70}$
(26)

19. (43) $3\frac{7}{10} + 4$

20. (81) $5\frac{5}{8} + \frac{1}{8}$

21. (63) $7 - 4\frac{3}{10}$

22. (86) $\frac{5}{6}$ of 4

23. (76) $\frac{3}{8} \times \frac{1}{2}$

24. (96) $\frac{3}{8} \div \frac{1}{2}$

25. (79) Write fractions equal to $\frac{1}{6}$ and $\frac{1}{4}$ that have denominators of 12. Then add the fractions.

26. (103) What is the volume of a chest of drawers with the dimensions shown?

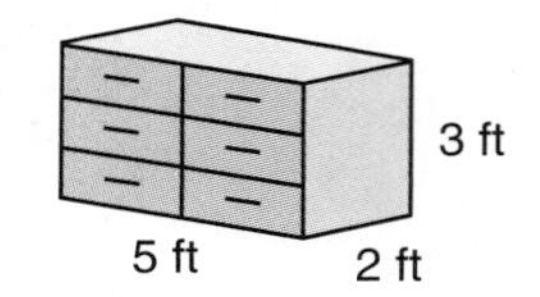

27. (72) What is the area of the top of the chest described in problem 26?

Lillian is planning a trip from San Diego to San Luis Obispo. The schedules for the trains she plans to take are printed below. Use this information to answer problems 28–30.

STATION	#29			#48	
San Diego	Dp	9:30 A		Ar	7:50 P
Anaheim	↓	11:26 A		↑	5:51 P
Los Angeles		12:30 P			4:55 P
Ventura		2:21 P			2:39 P
Santa Barbara		3:10 P			1:40 P
Solvang		4:05 P			12:25 P
San Luis Obispo		5:30 P			11:10 A
Paso Robles	Ar	6:20 P		Dp	10:00 A

28. (108) The trip from San Diego to San Luis Obispo takes how long?

29. (108) Train #48 stops in Santa Barbara for 15 minutes before continuing. At what time does the train depart from Santa Barbara?

30. (21, 108) The distance between San Diego and San Luis Obispo is about 320 miles. From departure to arrival, the train travels about how many miles each hour?

A. 30 miles B. 40 miles C. 50 miles D. 60 miles

LESSON

114 Problems with No Solutions or Many Solutions

WARM-UP

Facts Practice: 30 Percents to Write as Fractions (Test K)

Mental Math:

a. Describe how to estimate the product of $5\frac{3}{4}$ and $6\frac{7}{8}$.

b. How many inches are in one foot? ... two and a half feet?

c. $\frac{1}{8}$ of 24   d. $\frac{3}{8}$ of 24   e. $\frac{5}{8}$ of 24

f. 25% of 40, + 2, × 2, + 1, ÷ 5, × 3, + 1, ÷ 8, − 2

Roman numerals:

g. Write MDC in our number system.

h. Compare: MDXX ◯ 1520

Problem Solving:

Tamara had 24 square color tiles on her desk. She arranged them into a rectangle made up of one row of 24 tiles. Then she arranged them into a new rectangle made up of two rows of 12 tiles.

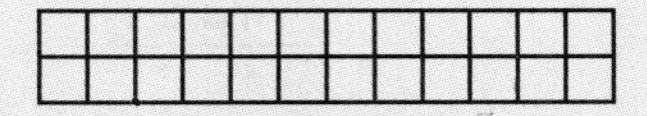

Draw two more rectangles Tamara could make using all 24 tiles.

NEW CONCEPT

When we are trying to solve problems, we are usually looking for one specific answer. Look at this problem:

> *Mary has $5 more than Chandra. Together they have $25. How much money does each girl have?*

Mary having $15 and Chandra having $10 is a solution to the problem. And this solution is the only answer: no other splitting of $25 results in two amounts that differ by five dollars. But not all problems have a single answer. Some problems have no answer, while others have more than one answer.

Example 1 Find two consecutive whole numbers whose sum is 8.

Solution Consecutive whole numbers are whole numbers next to each other, such as 5 and 6. We check the pairs of whole numbers whose sum is 8. These pairs are 1 and 7; 2 and 6; 3 and 5; and 4 and 4. None of these pairs has consecutive whole numbers. So there is **no solution** to this problem. We can also see that there is no solution in a different way. Each pair of consecutive whole numbers must consist of one even number and one odd number. An even number plus an odd number is always odd. Thus, their sum cannot be 8.

Example 2 The length and width of a certain rectangle are whole numbers. If the area of the rectangle is 24 square units, what are the dimensions of the rectangle?

Solution We know that the area of a rectangle equals its length times its width, so we find whole numbers that have a product of 24. One answer is **8 by 3.** A second answer is **12 by 2.** Thus, there is more than one solution to this problem. (You will be asked to find more answers to this problem in the Lesson Practice section.)

Example 3 A class has two types of games. Game A requires 3 players, and game B requires 2 players. How many of each type of game can 12 students play so that each student plays in only one game?

Solution First we will consider the two situations in which only one type of game is played. This happens if *only* game A is played or *only* game B is played.

Game A requires 3 players. Twelve students can be divided into 4 groups of 3. So one solution is **4 of game A and 0 of game B.**

Game B requires 2 players. Twelve students can be divided into 6 groups of 2. So another solution is **0 of game A and 6 of game B.**

We start a table to record the results. You will be asked to find other combinations in the Lesson Practice section.

Combinations of Game A and Game B for 12 Students

Number of Game A	Number of Game B
4	0
0	6

LESSON PRACTICE

Practice set

a. Find two consecutive whole numbers whose sum is 9.

b. Find two consecutive whole numbers whose sum is 10.

c. In example 2, two combinations of length and width were found for a rectangle with an area of 24 square units: 8 by 3 and 12 by 2. Find two more combinations of whole-number lengths and widths.

d. In example 3, two combinations of games were found for 12 students to play. Copy the table in the example 3 solution, and try some different combinations of games for the 12 students. Find all possible combinations.

MIXED PRACTICE

Problem set

1. *(Inv. 3, 37)* Draw a circle and shade all but $\frac{1}{3}$ of it. What percent of the circle is shaded?

2. *(74)* Which of these units of length would probably be used to measure the length of a room?

A. inches B. feet

C. miles D. light-years

3. *(105)* Which of these does *not* show a line of symmetry?

A. 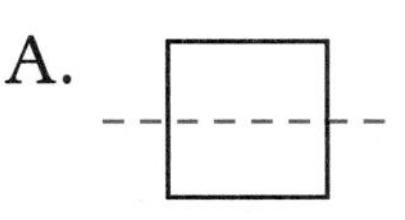 B. 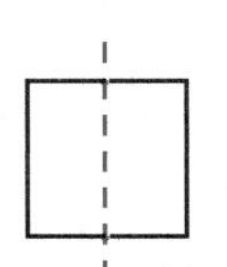 C. 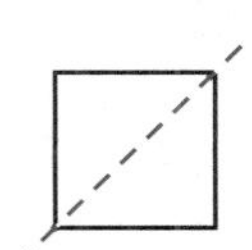 D. 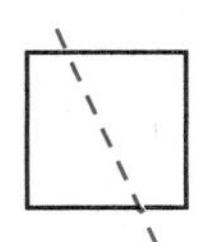

4. *(21)* Michael's car can travel 28 miles on one gallon of gas. How far can his car travel on 16 gallons of gas?

5. *(113)* Write $1\frac{3}{4}$ as an improper fraction.

6. *(79, 81)* Write a fraction equal to $\frac{1}{2}$ that has the same denominator as $\frac{5}{6}$. Then subtract the fraction from $\frac{5}{6}$. Remember to reduce your answer.

7. *(112)* The denominators of $\frac{3}{8}$ and $\frac{5}{6}$ are 8 and 6. What is the least common multiple (LCM) of the denominators?

8. (57) What fraction names the probability that with one spin the spinner will stop on sector A?

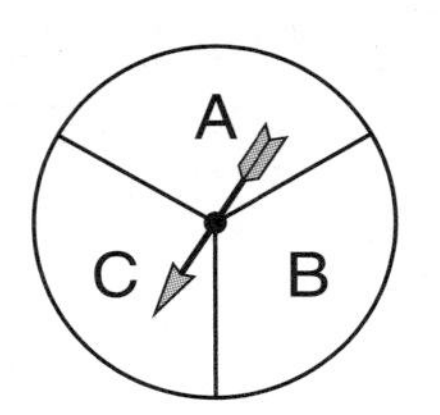

9. (57) What is the probability that with one spin the spinner will stop on sector B?

10. (61) QS is 6 cm. RS is 2 cm. RT is 6 cm. Find QT.

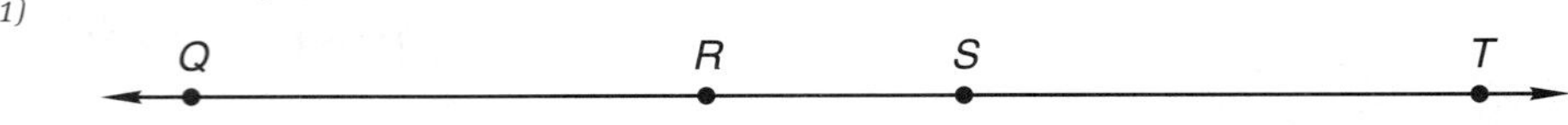

11. (99) $45 + 16.7 + 8.29 + 4.325$

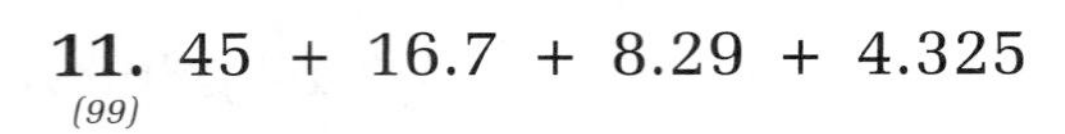

12. (24, 99) $4.2 - (3.2 - 1)$

13. (110) 0.75×0.05

14. (109) 0.6×38

15. (111) 100×7.5

16. (92) $\$24.36 \div 12$

17. (78, 94) $4600 \div 5^2$

18. (81) $6\frac{9}{10} - \frac{1}{10}$

19. (75) $5\frac{4}{9} + 3\frac{5}{9}$

20. (96) $4 \div \frac{1}{8}$

21. (90) $4 \times \frac{1}{8}$

22. (97) At the Little League baseball game there were 18 players and 30 spectators. What was the ratio of players to spectators at the game?

23. (107) (a) What percent of the rectangle is shaded?

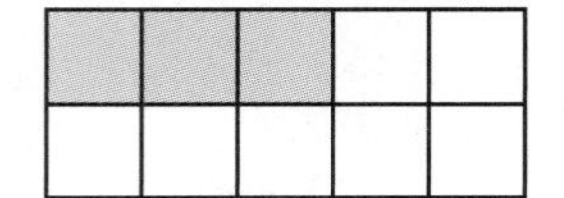

(b) What percent of the rectangle is not shaded?

24. (71, 90) (a) Write the reduced fraction equal to 60%.

(b) Write the reduced fraction equal to 70%.

25. (53) A loop of string can be arranged to form a rectangle that is 12 inches long and 6 inches wide. If the same loop of string is arranged to form a square, what would be the length of each side of the square?

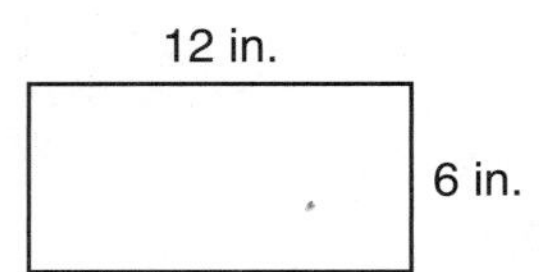

26. (a) What is the area of the rectangle pictured in problem 25?
(72)

(b) What is the area of the square described in problem 25?

27. Find the percent equivalent to $\frac{1}{6}$ by multiplying 100% by $\frac{1}{6}$ and writing the answer as a mixed number with the fraction reduced.
(107)

Two squares form this hexagon. Refer to this figure for problems 28–30.

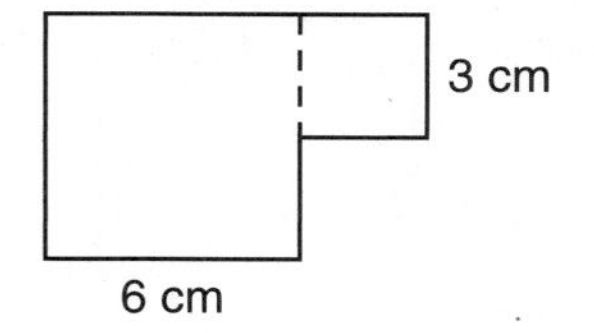

28. What is the area of each square?
(72)

29. Combine the areas of the two squares to find the area of the hexagon.
(72)

30. If the squares were separated, their perimeters would be 12 cm and 24 cm respectively. However, the perimeter of the hexagon is not the sum of the perimeters of the squares, because one side of the small square and part of one side of the large square are not part of the perimeter of the hexagon. Copy the hexagon on your paper, and show the length of each of the six sides. What is the perimeter of the hexagon?
(53)

LESSON

115 Area, Part 2

WARM-UP

Facts Practice: 30 Percents to Write as Fractions (Test K)

Mental Math:

a. The sides of a square are 5 inches long. What is the perimeter of the square? What is the area of the square?

b. How many ounces are in a pound? ... two and a half pounds?

c. 25% of 80 **d.** 50% of 80 **e.** 75% of 80

Roman numerals:

f. Write MMIV in our number system.

g. Compare: 92 ◯ LXXXII

Problem Solving:

This uppercase letter B has a horizontal line of symmetry. List the other uppercase letters of the alphabet that have a horizontal line of symmetry.

NEW CONCEPT

Recall that we calculate the area of a rectangle by multiplying its length and width. In this lesson we will calculate the area of figures that can be divided into rectangles.

Example Two rectangles are joined to form a hexagon. What is the area of the hexagon?

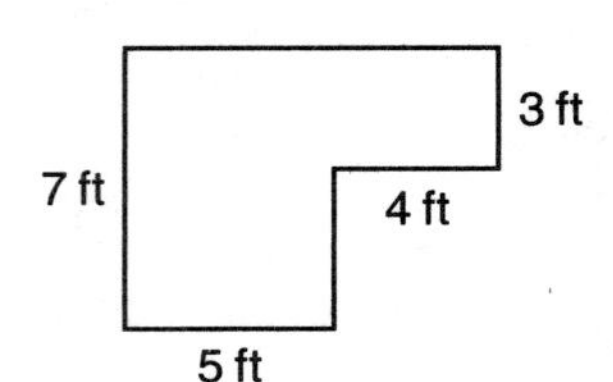

Solution The hexagon can be divided into two rectangles. We find the area of each rectangle, then add the areas to find the area of the hexagon.

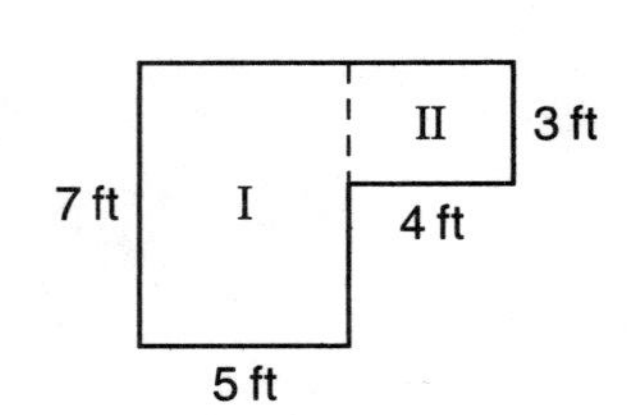

Area I	7 ft × 5 ft =	35 sq. ft
+ Area II	4 ft × 3 ft =	12 sq. ft
Combined area		**47 sq. ft**

LESSON PRACTICE

Practice set Copy each figure on your paper. Then find the area of each figure by dividing it into two rectangles and adding the areas of the parts.

a.

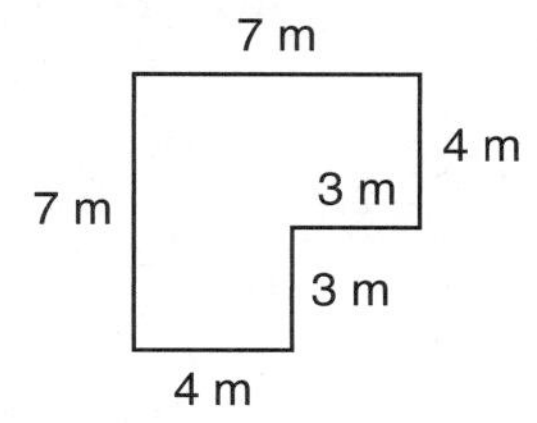

b.

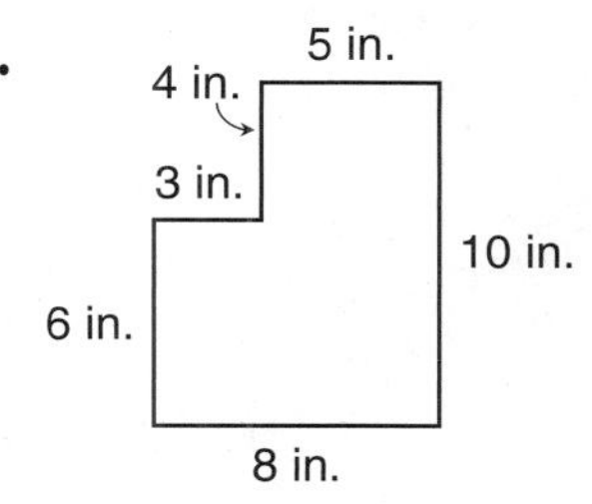

c.

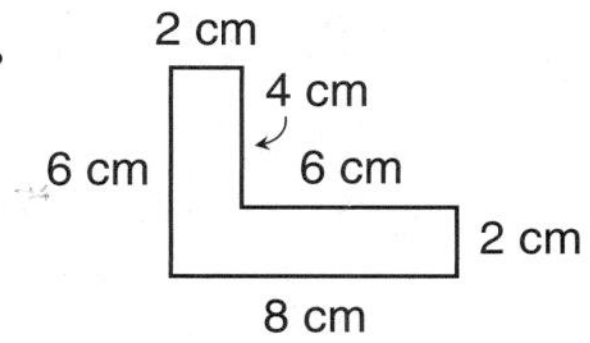

d.

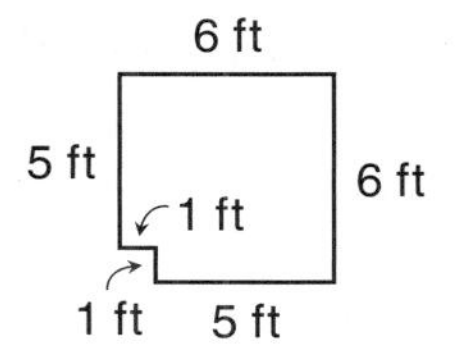

MIXED PRACTICE

Problem set

1. Jack was 48 inches tall. The giant was 24 feet tall. How many feet taller than Jack was the giant?
(35, 74)

2. Name the total number of shaded circles below as an improper fraction and as a mixed number.
(113)

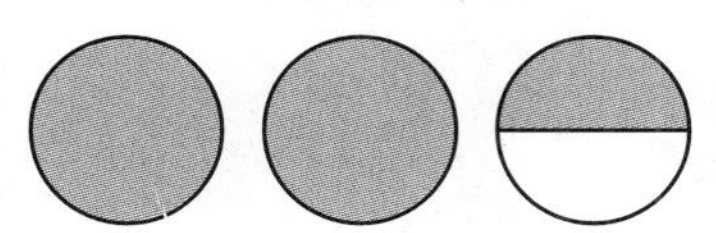

3. What fraction names the probability that with one spin the spinner will stop on sector A?
(57)

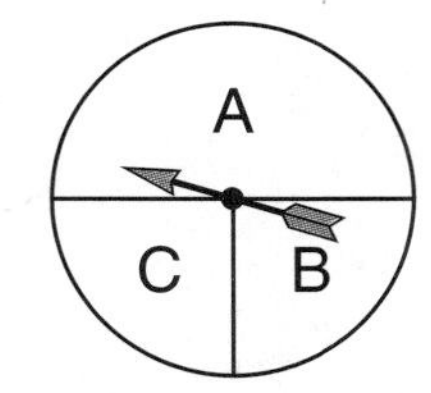

4. What is the probability that with one spin the spinner will stop on sector B?
(57)

5. To what mixed number is the arrow pointing?
(38)

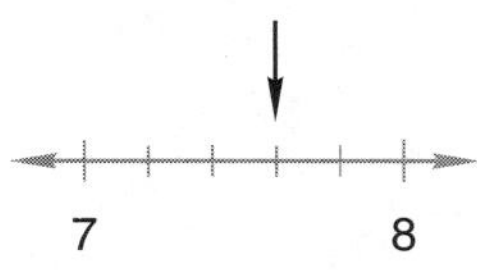

6. What time is $1\frac{1}{2}$ hours after 11:40 a.m.?
(28)

7. Which pair of fractions has the same denominator?
(Inv. 2)

A. $\frac{1}{3}, \frac{1}{4}$ B. $\frac{4}{3}, \frac{4}{2}$ C. $\frac{1}{4}, \frac{3}{4}$

8. The denominators of $\frac{2}{5}$ and $\frac{2}{3}$ are 5 and 3. Find the least common multiple (LCM) of the denominators.
(112)

9. Estimate the perimeter of this rectangle.
(53, 104)

3.98 m

2.96 m

10. Estimate the area of this rectangle.
(104)

11. $42.98 + 50 + 23.5 + 0.025$
(99)

12. How much greater than 5.18 is 6? Use words to write your answer.
(68, 102)

13. 0.375×10
(111)

14. 0.14×0.06
(110)

15. 7.8×19
(109)

16. $2340 \div 30$
(54)

17. $18\overline{)2340}$
(94)

18. $7\overline{)8765}$
(26)

19. $\frac{5}{6} + 1\frac{5}{6}$
(91)

20. $7\frac{5}{8} - 7\frac{1}{8}$
(90)

21. $\frac{4}{5} \times \frac{2}{3}$
(76)

22. $\frac{4}{5} \div \frac{2}{3}$
(96)

23. $\frac{2}{5} = \frac{\square}{15}$
(79)

24. $\frac{2}{3} = \frac{\square}{15}$
(79)

25. In problems 23 and 24 you made fractions equal to $\frac{2}{5}$ and $\frac{2}{3}$ with denominators of 15. Add the fractions you made. Remember to convert the answer to a mixed number.
(91)

26. What is the perimeter of this regular pentagon?
(53)

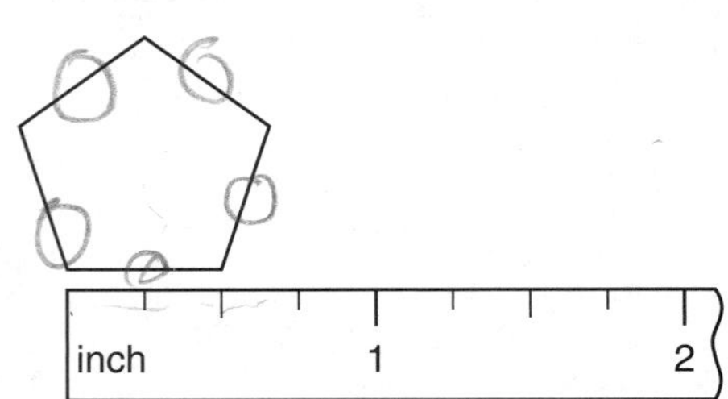

27. A regular pentagon has how many lines of symmetry?
(105)

28. What is the area of this hexagon?
(115)

29. What is the perimeter of the hexagon?
(53)

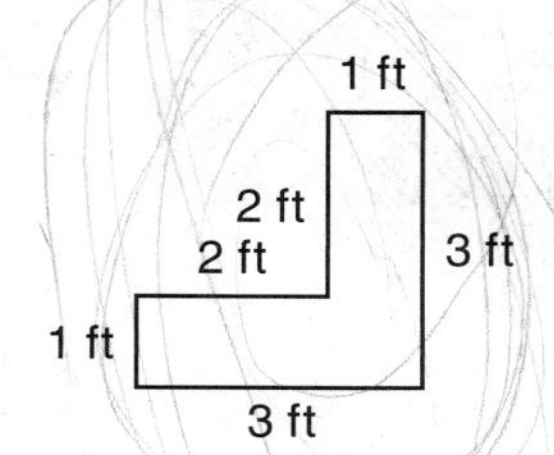

30. What fraction of a square mile is a field that is $\frac{1}{2}$ mile long and $\frac{1}{4}$ mile wide?
(76)

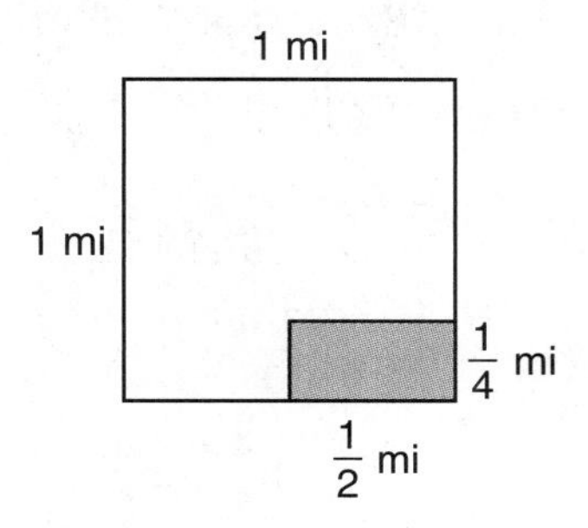

LESSON

116 Finding Common Denominators to Add, Subtract, and Compare Fractions

WARM-UP

Facts Practice: 30 Percents to Write as Fractions (Test K)

Mental Math:

a. A rectangle is 6 inches long and 4 inches wide. What is its perimeter? What is its area?

b. How many seconds are in one minute? ... two and a half minutes?

c. 10% of $300 **d.** 10% of $30 **e.** 10% of $3

f. $\sqrt{16}$, × 5, − 6, ÷ 7, + 8, × 9, ÷ 10

Roman numerals:

g. Write MDCCLXXVI in our number system.

h. Compare: CCCIV ◯ 340

Problem Solving:

The local newspaper sells advertising for $20 per column inch per day. An ad that is 2 columns wide and 4 inches long is 8 column inches (2 × 4 = 8) and costs $160 each day (8 × $20). What would be the total cost of running a 3-column by 8-inch ad for two days?

NEW CONCEPT

The fractions $\frac{1}{4}$ and $\frac{3}{4}$ have common denominators. The fractions $\frac{1}{2}$ and $\frac{1}{4}$ do not have common denominators. Fractions have common denominators if their denominators are equal.

Common denominators	Different denominators
$\frac{1}{4} \longleftrightarrow \frac{3}{4}$	$\frac{1}{2} \longleftrightarrow \frac{1}{4}$

To add or subtract fractions that have different denominators, we first change the name of one or more of the fractions so that they have common denominators. The least common multiple (LCM) of the denominators is the least common denominator of the fractions. The denominators of $\frac{1}{2}$ and $\frac{1}{4}$ are 2 and 4. The LCM of 2 and 4 is 4. So the least common denominator for halves and fourths is 4.

Example 1 Add: $\frac{1}{2} + \frac{1}{4}$

Solution Since $\frac{1}{2}$ and $\frac{1}{4}$ have different denominators, we change the name of $\frac{1}{2}$ so that both fractions have a denominator of 4. We change $\frac{1}{2}$ to fourths by multiplying by $\frac{2}{2}$, which gives us $\frac{2}{4}$.

$$\frac{1}{2} \times \frac{2}{2} = \frac{2}{4}$$

Then we add $\frac{2}{4}$ and $\frac{1}{4}$ to get $\frac{3}{4}$.

$$\frac{2}{4} + \frac{1}{4} = \mathbf{\frac{3}{4}}$$

Example 2 Subtract:

$$\begin{array}{r} 3\frac{1}{2} \\ -\ 1\frac{1}{6} \\ \hline \end{array}$$

Solution We work with the fraction part of each mixed number first. The denominators are 2 and 6. We can change halves to sixths. We multiply $\frac{1}{2}$ by $\frac{3}{3}$ and get $\frac{3}{6}$.

$$\frac{1}{2} \times \frac{3}{3} = \frac{3}{6}$$

Then we subtract and reduce the answer.

$$\begin{array}{r} 3\frac{3}{6} \\ -\ 1\frac{1}{6} \\ \hline 2\frac{2}{6} \end{array} = \mathbf{2\frac{1}{3}}$$

Example 3 Compare: $\frac{3}{4} \bigcirc \frac{7}{8}$

Solution Rewriting fractions with common denominators can help us compare fractions. The denominators are 4 and 8. We change fourths to eighths by multiplying by $\frac{2}{2}$.

$$\frac{3}{4} \times \frac{2}{2} = \frac{6}{8}$$

We see that $\frac{6}{8}$ is less than $\frac{7}{8}$. So we have the following:

$$\mathbf{\frac{3}{4} < \frac{7}{8}}$$

Example 4 Add: $\frac{1}{3} + \frac{1}{2}$

Solution For this problem we need to rename *both* fractions. The denominators are 3 and 2. The LCM of 3 and 2 is 6. So the least common denominator for thirds and halves is sixths. We rename the fractions and then add.

$$\begin{array}{r} \frac{1}{3} \times \frac{2}{2} = \frac{2}{6} \\ + \frac{1}{2} \times \frac{3}{3} = \frac{3}{6} \\ \hline \mathbf{\frac{5}{6}} \end{array}$$

LESSON PRACTICE

Practice set* Find each sum or difference. As you work the problems, follow these steps:

1. Figure out what the new denominator should be.
2. Change the name of one or both fractions.
3. Add or subtract the fractions.
4. Reduce the answer when possible.

a. $\frac{1}{2} + \frac{1}{8}$ **b.** $\frac{1}{2} - \frac{1}{4}$ **c.** $\frac{3}{4} + \frac{1}{8}$

d. $\frac{2}{3} - \frac{1}{9}$ **e.** $\frac{1}{3} + \frac{1}{4}$ **f.** $\frac{1}{2} - \frac{1}{3}$

g. $3\frac{1}{4} + 2\frac{1}{2}$ **h.** $2\frac{1}{8} + 5\frac{1}{2}$ **i.** $3\frac{1}{2} - 1\frac{1}{6}$ **j.** $2\frac{3}{4} - 2\frac{1}{2}$

k. $5\frac{5}{8} + 1\frac{1}{4}$ **l.** $3\frac{1}{2} + 1\frac{1}{3}$ **m.** $4\frac{3}{4} - 1\frac{2}{3}$ **n.** $4\frac{1}{2} - 1\frac{1}{5}$

MIXED PRACTICE

Problem set

1. (37, 107) Draw a circle. Shade all but $\frac{1}{6}$ of it. What percent of the circle is shaded?

2. (35) In 1875 Bret Harte wrote a story about the California Gold Rush of 1849. How many years after the Gold Rush did he write the story?

3. (57, 107) (a) What is the chance of the spinner stopping on 4 with one spin?

(b) What is the probability that with one spin the spinner will stop on a number less than 4?

4. (105) Which of these does not show a line of symmetry?

A. 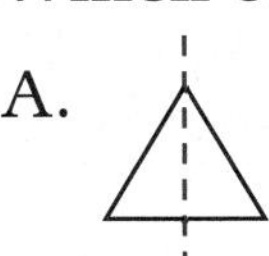 B. 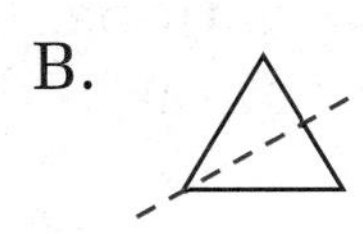 C. 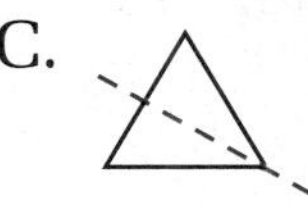 D. 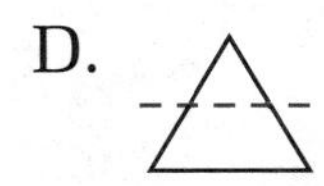

5. (116) Compare these fractions. First write the fractions with common denominators.

$$\frac{2}{3} \bigcirc \frac{5}{6}$$

6. (113) Name the total number of shaded circles as an improper fraction and as a mixed number.

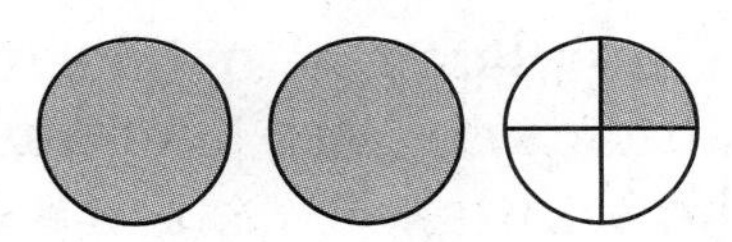

7. (97) Alberto counted 100 cars and 60 trucks driving by the school. What was the ratio of trucks to cars that Alberto counted driving by the school?

8. (53, 73) What is the perimeter of this square?

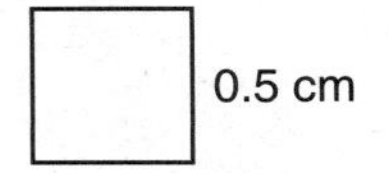

9. (109) What is the area of the square in problem 8?

10. (61) *AC* is 70 mm. *BC* is 40 mm. *BD* is 60 mm. Find *AD*.

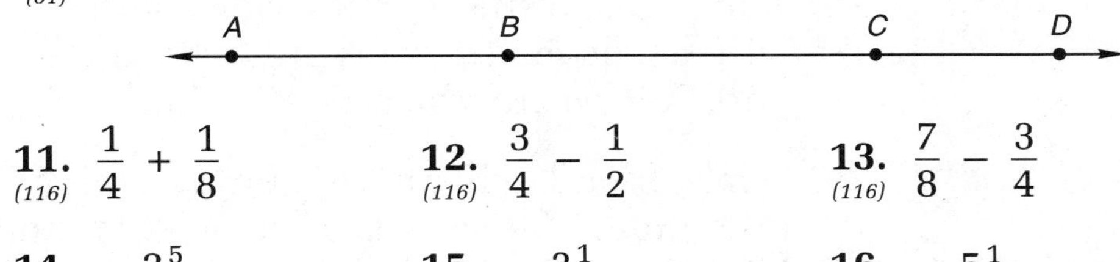

11. (116) $\frac{1}{4} + \frac{1}{8}$

12. (116) $\frac{3}{4} - \frac{1}{2}$

13. (116) $\frac{7}{8} - \frac{3}{4}$

14. (116) $2\frac{5}{8} - 1\frac{1}{2}$

15. (116) $3\frac{1}{2} - 2\frac{1}{8}$

16. (116) $5\frac{1}{6} + 1\frac{1}{3}$

17. (86, 91) $\frac{3}{5} \times 3$

18. (96) $3 \div \frac{3}{5}$

19. (111) 6.5×100

20. (109) 4.6×80

21. (110) 0.18×0.4

22. (54) $10\overline{)\$13.20}$

23. (92) $12\overline{)\$13.20}$

24. (94) $1470 \div 42$

25. (32, 61) Which angle in quadrilateral *ABCD* is an obtuse angle?

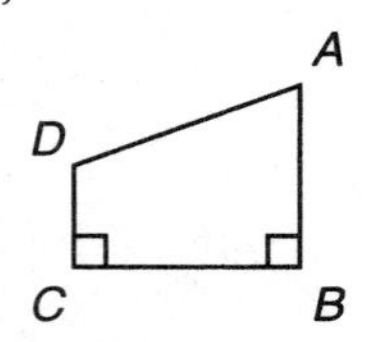

26. (116) Add these fractions. First rename the fractions so that they have a common denominator of 12.

$$\frac{1}{4} + \frac{2}{3}$$

27. (115) What is the area of this figure?

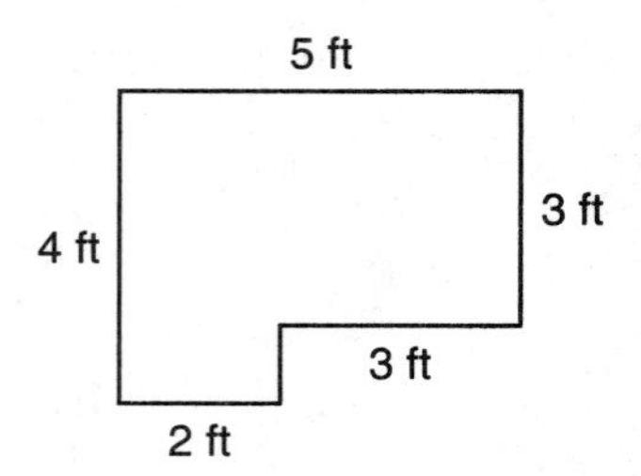

Tim planned to take the train from Fort Collins, where he attends college, to Union Station in Denver. From Union Station he will take a taxi to a job interview, meet a friend for dinner, and then return to Fort Collins at night. Use this information and the train schedule below to answer problems 28–30.

Cheyenne • Fort Collins • Denver

6		Connecting Train Number		6
11:30A	Dp	Cheyenne, WY	Ar	12:30A
12:20P	↓	Fort Collins, CO	↑	11:40P
12:40P		Loveland, CO		11:30P
1:00P		Longmont, CO		11:10P
1:35P		Boulder, CO	Ar	10:35P
2:05P	Ar	Denver, CO –Union Station	Dp	9:00P

28. (108) Tim plans to study on the ride south to Denver. How long will he have for studying between the departure from Fort Collins and the arrival in Denver?

29. (108) Tim's friend will drive Tim to Union Station from the restaurant. The drive takes about 20 minutes. Tim wants to be at Union Station an hour before departure. By what time should Tim and his friend leave the restaurant?

30. (108) If the college campus is a 5-minute walk from the Fort Collins train station, can Tim be back on campus by midnight? Explain your answer.

LESSON

117 Dividing a Decimal Number by a Whole Number

WARM-UP

Facts Practice: 30 Percents to Write as Fractions (Test K)

Mental Math:

a. Describe how to estimate the total cost of 32.6 liters of gasoline at $\$0.49\frac{9}{10}$ per liter.

b. How many minutes are in one hour? ... three and a half hours?

c. 25% of $16 d. 50% of $16 e. 75% of $16

f. $\frac{1}{5}$ of 20, × 4, − 4, ÷ 4, + 4, × 4

Roman numerals:

g. Write MCM in our number system.

h. Compare: 110 ◯ XC

Problem Solving:

Shirts are on sale for 10% off. The regular price is $30. Ten percent of $30 is $3. So $3 is taken off the regular price, making the sale price $27. Pants are on sale for 25% off. The regular price is $40. What is the sale price?

NEW CONCEPT

Dividing a decimal number by a whole number is like dividing money by a whole number. The decimal point in the quotient is directly above the decimal point inside the division box. In the chart below, "÷ by whole (W)" means "division by a whole number." The memory cue "up" reminds us where to place the decimal in the quotient. (We will later learn a different rule for dividing by a decimal number.)

Decimals Chart

Operation	+ or −	×	÷ by whole (W)
Memory cue	line up . ± . .	×; then count ._ × ._ .__	up $W\overline{)\ .}$ with . above
You may need to ... • Place a decimal point on the end of whole numbers. • Fill empty places with zero.			

We sometimes need to use one or more zeros as placeholders when dividing decimal numbers. Here we show this using money.

Suppose $0.12 is shared equally by 3 people. The division will look like this. Notice that the decimal point in the quotient is directly above the decimal point in the dividend. We fill empty places with zero and see that each person will receive $0.04.

$$\begin{array}{r} \$\ .\ 4 \\ 3\overline{)\$0.12} \\ \underline{12} \\ 0 \end{array} \longrightarrow \begin{array}{r} \$0.04 \\ 3\overline{)\$0.12} \\ \underline{12} \\ 0 \end{array}$$

decimal point "up"

Example 1 Divide: $2\overline{)4.8}$

Solution We are dividing by 2, which is a whole number. We remember the memory cue "up" and place the decimal point in the answer directly above the decimal point inside the division box. Then we divide.

$$\begin{array}{r} \mathbf{2.4} \\ 2\overline{)4.8} \\ \underline{4} \\ 0\,8 \\ \underline{8} \\ 0 \end{array}$$

Example 2 Divide: $3\overline{)0.42}$

Solution We place the decimal point in the answer "straight up." Then we divide.

$$\begin{array}{r} \mathbf{0.14} \\ 3\overline{)0.42} \\ \underline{3} \\ 12 \\ \underline{12} \\ 0 \end{array}$$

Example 3 Divide: $0.15 \div 3$

Solution We rewrite the problem using a division box. The decimal point in the answer is "straight up." We divide and remember to fill empty places with zeros.

$$\begin{array}{r} \mathbf{0.05} \\ 3\overline{)0.15} \\ \underline{15} \\ 0 \end{array}$$

Example 4 Divide: $0.0024 \div 3$

Solution We rewrite the problem using a division box. The decimal point in the answer is "straight up." We divide and remember to fill empty places with zeros.

$$\begin{array}{r} \mathbf{0.0008} \\ 3\overline{)0.0024} \end{array}$$

LESSON PRACTICE

Practice set* Divide:

a. $4\overline{)0.52}$ b. $6\overline{)3.6}$ c. $0.85 \div 5$

d. $5\overline{)7.5}$ e. $5\overline{)0.65}$ f. $2.1 \div 3$

g. $4\overline{)0.16}$ h. $0.35 \div 7$ i. $5\overline{)0.0025}$

j. $0.08 \div 4$ k. $6\overline{)0.24}$ l. $0.0144 \div 3$

m. A gallon is about 3.78 liters. About how many liters is half a gallon? (Find half by dividing by 2.)

MIXED PRACTICE

Problem set

1. Write the following sentence using digits and symbols:
(6, 116)

   The sum of one sixth and one third is one half.

2. Gilbert scored half of his team's points. Juan scored 8 fewer points than Gilbert. The team scored 36 points. How many points did Juan score?
(49)

3. In the northern hemisphere the first day of winter is on or very near December 21. The first day of summer is 6 months later. The first day of summer is on or very near what date?
(28)

4. (a) What is the probability that with one spin the spinner will stop on a number greater than one?
(57, 107)

   (b) What is the chance of spinning a three with one spin?

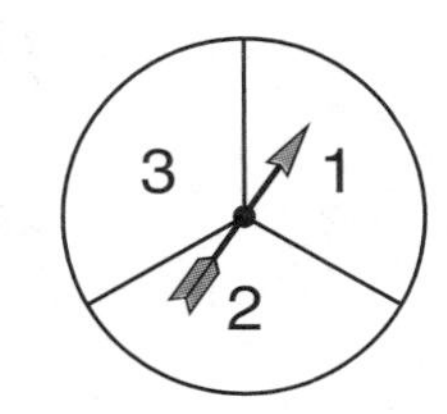

5. Name the shaded portion of this rectangle as a fraction, as a decimal, and as a percent.
(71)

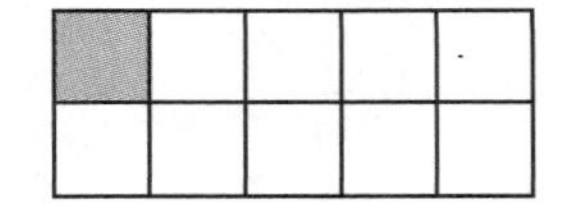

6. If each side of an octagon is 6 inches long, then the perimeter of the octagon is how many feet?
(32, 53)

7. Name the total number of shaded circles as an improper fraction and as a mixed number.
(113)

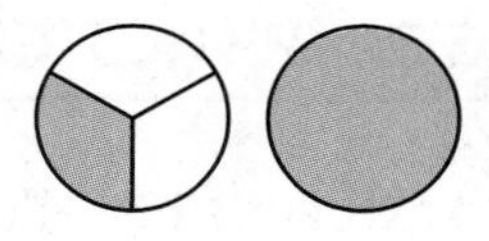

8. What is the largest four-digit odd number that uses the digits 7, 8, 9, and 0 only once each?
(2)

Refer to rectangle *ABCD* at right to answer problems 9 and 10.

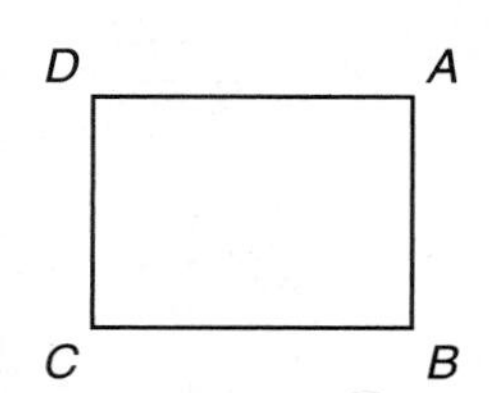

9. In rectangle *ABCD,* which segment is parallel to $\overline{AB}$?
(31, 61)

10. In the rectangle, *AB* is 3 cm and *BC* is 4 cm.
(53, 72)

(a) What is the perimeter of the rectangle?

(b) What is the area of the rectangle?

11. *KL* is 56 mm. *LM* is half of *KL*. *MN* is half of *LM*. Find *KN*.
(61)

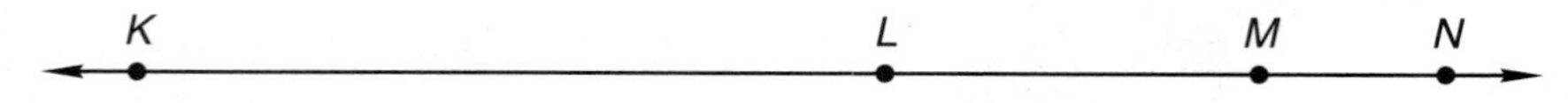

12. 16 + 3.17 + 49 + 1.125
(99)

13. How much greater is 3.42 than 1.242?
(102)

14. 4.3 × 100
(111)

15. 6.4 × 3.7
(109)

16. 0.36 × 0.04
(110)

17. $2\overline{)3.6}$
(117)

18. $7\overline{)0.0049}$
(117)

19. 1.35 × 90
(109)

20. $2\frac{1}{8} + 1\frac{3}{4}$
(116)

21. $\frac{1}{3} + \frac{1}{6}$
(116)

22. $\frac{7}{10} - \frac{1}{2}$
(116)

23. $3\frac{9}{10} - \frac{1}{5}$
(116)

24. $4 \times \frac{3}{2}$
(86, 91)

25. $\frac{3}{4} \div \frac{1}{4}$
(96)

26. Reduce: $\frac{18}{144}$
(90)

27. Find the sum of $3\frac{1}{5}$ and $2\frac{1}{2}$ by first rewriting the fractions with 10 as the common denominator.
(116)

28. To finish covering the floor of a room, Abby needed a rectangular piece of floor tile 6 inches long and 3 inches wide.
(72, 76)

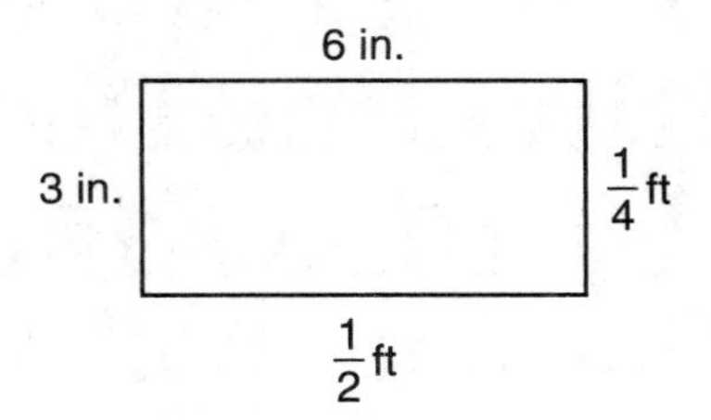

(a) What is the area of this rectangle in square inches?

(b) What is the area of the rectangle in square feet?

A 2-inch by 2-inch square is joined with a 5-inch by 5-inch square to form a hexagon. Refer to the figure to answer problems 29 and 30.

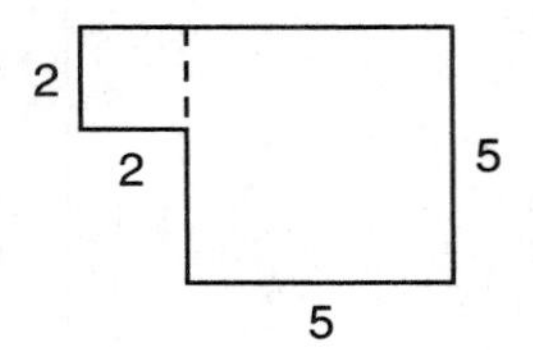

29. What is the area of the hexagon?
(115)

30. Copy the hexagon and show the lengths of all six sides. Then find the perimeter of the hexagon.
(53)

LESSON

118 Using Zero as a Placeholder • Dividing Decimal Numbers by 10, by 100, and by 1000

WARM-UP

Symmetry Activity:

Materials needed: pencil and paper, mirror or other reflective surface, scissors (optional)

The word BOB has a horizontal line of symmetry because each of its letters has a horizontal line of symmetry. -BOB-

If we fold under or cut away the upper half of the word along the line of symmetry, the lower half of the word looks like this:

By placing the paper against a mirror, we can make the upper half of the word "reappear."

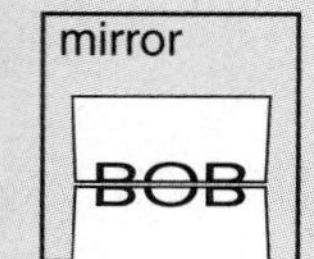

Make up a word that has a horizontal line of symmetry, and try the activity for yourself. How does this "trick" work?

NEW CONCEPTS

Using zero as a placeholder

We usually do not write remainders with decimal division problems. The procedure we will follow for now is to continue dividing until the "remainder" is zero. In order to continue the division, we may need to attach extra zeros to the decimal number that is being divided. **Remember, attaching extra zeros to the back of a decimal number does not change the value of the number.**

Example 1 Divide: $0.6 \div 5$

Solution The first number goes inside the division box. The decimal point is straight up. As we divide, we attach a zero and keep dividing.

$$\begin{array}{r} \mathbf{0.12} \\ 5\overline{)0.60} \\ \underline{5} \\ 10 \\ \underline{10} \\ 0 \end{array}$$

Example 2 Divide: $0.3 \div 4$

Solution As we divide, we attach zeros and keep dividing. We fill empty places in the quotient with zero.

$$\begin{array}{r} \mathbf{0.075} \\ 4\overline{)0.300} \\ \underline{28} \\ 20 \\ \underline{20} \\ 0 \end{array}$$

Example 3 Divide: $3.4 \div 10$

Solution As we divide, we attach a zero to 3.4 and continue dividing. Notice that the same digits appear in the quotient and dividend, but in different places.

$$\begin{array}{r} \mathbf{0.34} \\ 10\overline{)3.40} \\ \underline{3\,0} \\ 40 \\ \underline{40} \\ 0 \end{array}$$

Dividing decimal numbers by 10, by 100, and by 1000

When we divide a number by 10, we find that the answer has the same digits, but the digits have shifted one place to the right.

$$\begin{array}{r} 34. \\ 10\overline{)340.} \end{array} \qquad \begin{array}{r} .34 \\ 10\overline{)3.40} \end{array}$$

We can use this pattern to find the answer to a decimal division problem when the divisor is 10. The shortcut is very similar to the method we use when multiplying a decimal number by 10. In both cases it is the digits that are shifting places. However, we can make the digits appear to shift places by shifting the decimal point instead. To divide by 10, we shift the decimal point one place to the left.

$$3.4 \div 10 = .34$$

Dividing by 100 is like dividing by 10 twice. When we divide by 100, we shift the decimal point two places to the left. When we divide by 1000, we shift the decimal point three places to the left. We shift the decimal point the same number of places as there are zeros in the number we are dividing by (10 or 100 or 1000). We can remember which way to shift the decimal point if we keep in mind that dividing a number into 10 or 100 or 1000 parts produces **smaller** numbers. As a decimal point moves to the left, the value of the number becomes smaller and smaller.

Example 4 Mentally divide 3.5 by 100.

Solution When we divide by 10 or 100 or 1000, we can find the answer mentally without performing the division algorithm. To divide by 100, we shift the decimal point two places. We know that the answer will be less than 3.5, so we remember to shift the decimal point to the left. We fill the empty place with a zero.

$$3.5 \div 100 = \mathbf{0.035}$$

LESSON PRACTICE

Practice set* Divide:

a. $0.6 \div 4$ b. $0.12 \div 5$ c. $0.1 \div 4$

d. $0.1 \div 2$ e. $0.4 \div 5$ f. $1.4 \div 8$

g. $0.5 \div 4$ h. $0.6 \div 8$ i. $0.3 \div 4$

Mentally perform the following divisions:

j. $2.5 \div 10$ k. $32.4 \div 10$ l. $2.5 \div 100$

m. $32.4 \div 100$ n. $2.5 \div 1000$ o. $32.4 \div 1000$

p. $12 \div 10$ q. $12 \div 100$ r. $12 \div 1000$

MIXED PRACTICE

Problem set

1. Which of these shows two parallel line segments that are not horizontal?
(31)

A. B.  C.  D.

2. Estimate the product of $6\frac{1}{10}$ and $4\frac{7}{8}$ by rounding both numbers to the nearest whole number before multiplying.
(101)

3. How many 12¢ pencils can Julie buy with one dollar?
(21, 22)

4. Which of these figures does not show a line of symmetry?
(105)

A. B. 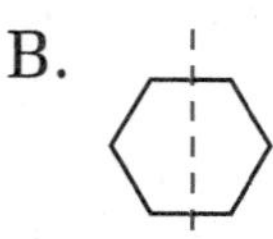 C. 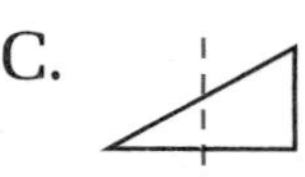 D. 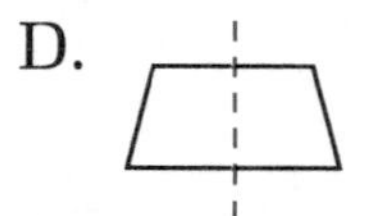

5. (107) The first roll knocked down 3 of the 10 bowling pins. What percent of the pins were still standing?

6. (71, 90) (a) Write the fraction equal to 4%.

(b) Write the fraction equal to 5%.

7. (113) Name the total number of shaded circles as an improper fraction and as a mixed number.

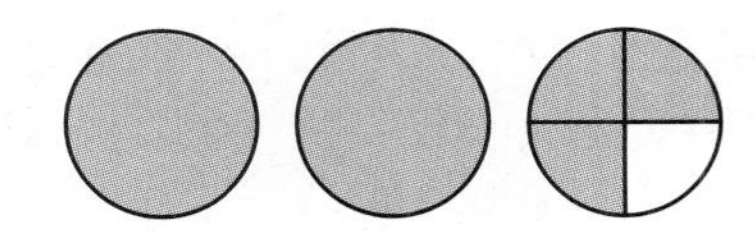

8. (76, 113) Write the mixed number $1\frac{3}{8}$ as an improper fraction. Then multiply the improper fraction by $\frac{1}{2}$. What is the product?

9. (32) A stop sign has the shape of an 8-sided polygon. What is the name for a polygon that has 8 sides?

10. (116) Arrange these numbers in order from least to greatest:

$$\frac{5}{3}, \frac{5}{6}, \frac{5}{5}$$

11. (53, 110, 117) The perimeter of this square is 1.2 meters.

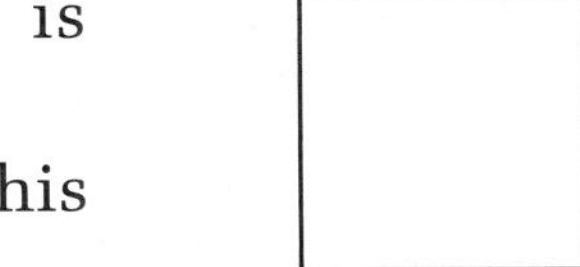

(a) How long is each side of this square?

(b) What is the area of this square?

12. (99) $49.35 + 25 + 3.7$

13. (68, 102) Subtract 1.234 from 2. Use words to write the answer.

14. (117) $0.0125 \div 5$

15. (111) 4.2×100

16. (110) 0.5×0.17

17. (118) $0.6 \div 4$

18. (118) $0.6 \div 10$

19. (118) $4\overline{)1.8}$

20. (116) $3\frac{1}{9} + \frac{1}{3}$

21. (116) $\frac{1}{3} + \frac{5}{6}$

22. (116) $\frac{7}{8} - \frac{1}{4}$

23. (116) $4\frac{1}{2} - 1\frac{3}{10}$

24. (86, 91) $6 \times \frac{2}{3}$

25. (96) $6 \div \frac{2}{3}$

26. Subtract $1\frac{1}{3}$ from $2\frac{3}{4}$. First write the fractions with 12 as
(116) the common denominator.

27. Divide mentally:
(118)
(a) 3.5 ÷ 100 (b) 87.5 ÷ 10

28. Compare: $\sqrt{81} + \sqrt{100} \bigcirc 9^2 + 10^2$
(78, 89)

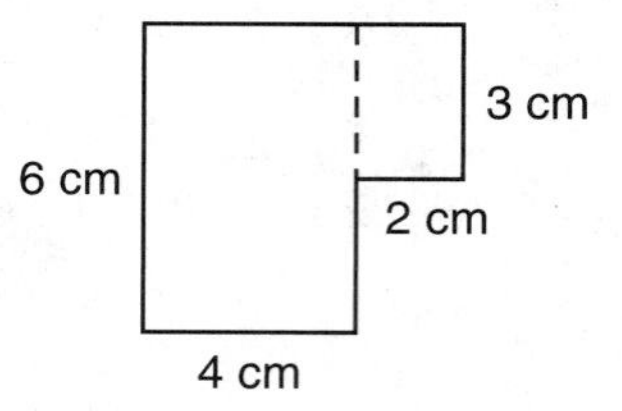

A 2-cm by 3-cm rectangle is joined to a 4-cm by 6-cm rectangle to form this hexagon. Refer to the figure to answer problems 29 and 30.

29. What is the area of the hexagon?
(115)

30. Copy the hexagon and show the lengths of all six sides.
(53) Then find the perimeter of the hexagon.

LESSON

119 Dividing by a Decimal Number

WARM-UP

Facts Practice: 30 Percents to Write as Fractions (Test K)

Mental Math:

a. Describe how to find the perimeter and area of a sheet of paper that is 11 inches long and $8\frac{1}{2}$ inches wide.

b. How much is $\frac{1}{3}$ of \$60? **c.** How much is $\frac{1}{3}$ less than \$60?

d. How much is $\frac{1}{3}$ more than \$60?

Roman numerals:

e. Write MCMLXIX in our number system.

f. Compare: MCMXCIX ◯ MM

Problem Solving:

Find the four uppercase letters that have both a horizontal and a vertical line of symmetry.

NEW CONCEPT

We have practiced dividing decimal numbers by whole numbers. In this lesson we will practice dividing decimal numbers by decimal numbers.

Look at these two problems. They are different in an important way.

$$3\overline{)0.12} \qquad 0.3\overline{)0.12}$$

The problem on the left is division **by a whole number.** The problem on the right is division **by a decimal number.**

When dividing by a decimal number with pencil and paper, we take an extra step. Before dividing, we shift the decimal points so that we are dividing by a whole number instead of by a decimal number.

$$0.3\overline{)0.12}$$

We move the decimal point of the divisor so that it becomes a whole number. Then we move the decimal point of the dividend the same number of places. The decimal point in the quotient will be straight up from the new location of the dividend's decimal point. To remember how to divide by a decimal number, we may think, "Over, over, and up."

up

$0.3\overline{)0.12}$

over over

To help us understand why this procedure works, we will write "0.12 divided by 0.3" with a division bar.

$$\frac{0.12}{0.3}$$

Notice that we can change the divisor, 0.3, into a whole number by multiplying by 10. So we multiply by $\frac{10}{10}$ to make an equivalent division problem.

$$\frac{0.12}{0.3} \times \frac{10}{10} = \frac{1.2}{3}$$

Multiplying by $\frac{10}{10}$ moves both decimal points "over." Now the divisor is a whole number and we can divide.

$$3\overline{)1.2} = 0.4$$

We will add this memory cue to the decimals chart. In the last column, "÷ by decimal (*D*)" means "division by a decimal number."

Decimals Chart

Operation	+ or −	×	÷ by whole (*W*)	÷ by decimal (*D*)
Memory cue	line up	×; then count	up	over, over, up
You may need to ... • Place a decimal point on the end of whole numbers. • Fill empty places with zero.				

Example Divide: $0.6\overline{)2.34}$

Solution We are dividing by the decimal number 0.6. We change 0.6 into a whole number by moving its decimal point "over." We also move the decimal point in the dividend "over." The decimal point in the quotient will be "straight up" from the new location of the decimal point in the division box.

$$
\begin{array}{r}
\mathbf{3.9} \\
0.6\overline{)2.34} \\
\underline{18} \\
54 \\
\underline{54} \\
0
\end{array}
$$

LESSON PRACTICE

Practice set* Divide:

a. $0.3\overline{)1.2}$ b. $0.3\overline{)0.42}$ c. $1.2\overline{)0.24}$

d. $0.4\overline{)0.24}$ e. $0.4\overline{)5.6}$ f. $1.2\overline{)3.6}$

g. $0.6\overline{)2.4}$ h. $0.5\overline{)0.125}$ i. $1.2\overline{)2.28}$

MIXED PRACTICE

Problem set

1. Copy the decimals chart in this lesson.
(119)

2. What is the average of 5, 6, 7, 8, and 9?
(50)

3. In the forest there were lions and tigers and bears. There were 24 bears. If there were twice as many lions as tigers and twice as many tigers as bears, how many lions were there?
(49)

4. Joey has $18.35. John has $22.65. They want to put their money together to buy a car that costs $16,040. How much more money do they need?
(49)

5. Which of these figures does not show a line of symmetry?
(105)

A. B. C. D.

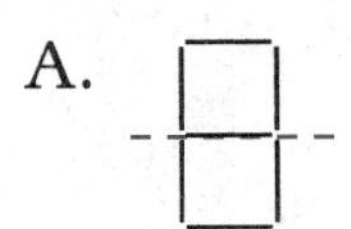

6. Write the mixed number $3\frac{1}{3}$ as an improper fraction. Then multiply the improper fraction by $\frac{3}{4}$. Remember to simplify your answer.
(76, 91, 113)

Refer to quadrilateral *ABCD* to answer problems 7 and 8.

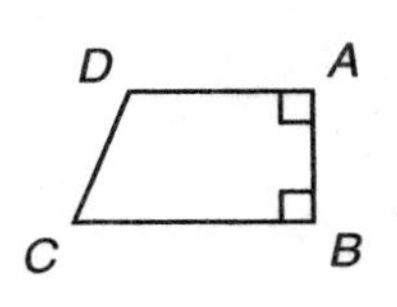

7. In quadrilateral *ABCD,* which angle appears to be an obtuse angle?
(32, 61)

8. What type of quadrilateral is quadrilateral *ABCD?*
(45)

9. (116) $3\frac{1}{2} + 1\frac{1}{3}$

10. (116) $2\frac{1}{6} + 1\frac{1}{2}$

11. (116) $5\frac{5}{6} - 1\frac{1}{2}$

12. (116) $4\frac{2}{3} - 1\frac{1}{4}$

13. (117) $6\overline{)0.0144}$

14. (118) $5\overline{)1.2}$

15. (118) $12\overline{)0.18}$

16. (119) $0.3\overline{)0.24}$

17. (119) $0.5\overline{)1.0}$

18. (119) $1.2\overline{)0.180}$

19. (118) Divide mentally:

(a) $0.5 \div 10$

(b) $0.5 \div 100$

20. (24, 102) $(3 - 1.6) - 0.16$

21. (110) 0.12×0.30

22. (111) 0.12×10

23. (109) 7.6×3.9

24. (86, 91) $4 \times \frac{3}{8}$

25. (96) $4 \div \frac{3}{8}$

26. (53, 116) What is the perimeter of this rectangle?

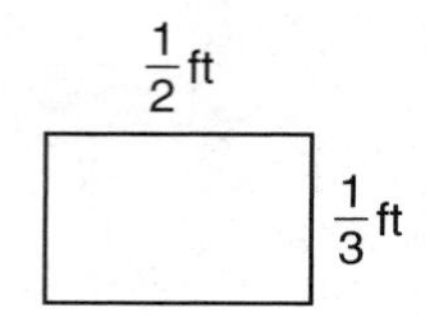

27. (76) What is the area of this rectangle?

28. (103) What is the volume of a room that is 10 feet wide, 12 feet long, and 8 feet high?

Two squares are joined to form this hexagon. Refer to the figure to answer problems 29 and 30.

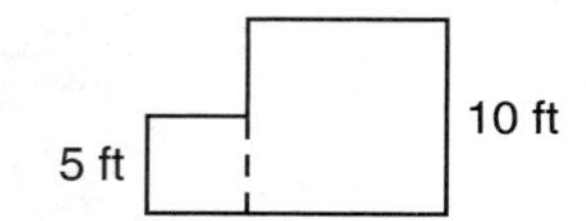

29. (115) What is the area of the hexagon?

30. (53) Copy the hexagon and show the lengths of the six sides. Then find the perimeter of the hexagon.

LESSON

120 Multiplying Mixed Numbers

WARM-UP

Facts Practice: 30 Percents to Write as Fractions (Test K)

Mental Math:

a. Describe how to find the average length of three used pencils.
b. What number is 10% of 20?
c. What number is 10% more than 20?
d. What number is 10% less than 20?
e. $\frac{1}{4}$ of 80, × 3, + 3, ÷ 9, × 4, − 1, ÷ 3

Roman numerals:

f. Write the current year in Roman numerals.
g. Write DCIV in our number system.

Problem Solving:

Sandra used small cubes to build larger cubes. She built a 2-by-2-by-2 cube and a 3-by-3-by-3 cube. On your paper, draw a 4-by-4-by-4 cube. How many small cubes are needed to build it?

NEW CONCEPT

To multiply mixed numbers, we change the mixed numbers to improper fractions before we multiply.

$$2\frac{1}{2} \times 1\frac{2}{3}$$

Change mixed numbers to improper fractions first.

$$\frac{5}{2} \times \frac{5}{3} = \frac{25}{6} \qquad \frac{25}{6} = 4\frac{1}{6}$$

Then multiply. Then simplify.

Example 1 Multiply: $\frac{1}{5} \times 4\frac{1}{2}$

Solution First we write the mixed number as an improper fraction. When both numbers are written as fractions, we multiply. We find that $\frac{1}{5}$ of $4\frac{1}{2}$ is $\mathbf{\frac{9}{10}}$.

$$\frac{1}{5} \times 4\frac{1}{2}$$

$$\frac{1}{5} \times \frac{9}{2} = \mathbf{\frac{9}{10}}$$

Example 2 Multiply: $3 \times 2\frac{1}{3}$

Solution We write both numbers as improper fractions; then we multiply.

$$3 \times 2\frac{1}{3}$$

$$\frac{3}{1} \times \frac{7}{3} = \frac{21}{3} = \mathbf{7}$$

We simplified the result to find that the product is **7.** We found our answer by multiplying. We find the same answer if we add:

$$2\frac{1}{3} + 2\frac{1}{3} + 2\frac{1}{3} = 6\frac{3}{3} = 7$$

LESSON PRACTICE

Practice set* Multiply:

a. $1\frac{1}{2} \times 1\frac{3}{4}$ **b.** $3\frac{1}{2} \times 1\frac{2}{3}$ **c.** $3 \times 2\frac{1}{2}$

d. $4 \times 3\frac{2}{3}$ **e.** $\frac{1}{3} \times 2\frac{1}{3}$ **f.** $\frac{1}{6} \times 2\frac{5}{6}$

MIXED PRACTICE

Problem set

1. Copy the decimals chart from Lesson 119.
(119)

2. Name this shape:
(83)

3. Write the following sentence using digits and symbols:
(4, 15)

The sum of two and two equals the product of two and two.

4. Which of these is not equal to $\frac{1}{2}$?
(71, 100)

A. 0.5 B. 50% C. 0.50 D. 0.05

5. Estimate the sum of $3\frac{1}{3}$ and $7\frac{3}{4}$ by rounding both numbers to the nearest whole number before adding.
(101)

6. Lillian can type 2 pages in 1 hour. At that rate, how long will it take her to type 100 pages?
(49)

7. In rectangle $ABCD$, $\overline{BC}$ is twice the length of $\overline{AB}$. Segment AB is 3 inches long.
(53, 72)

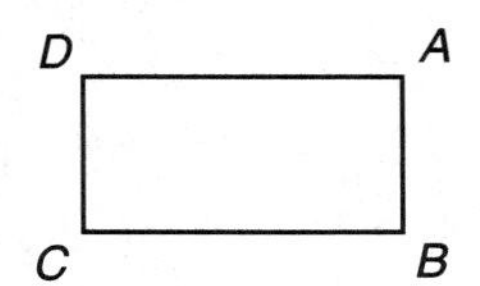

(a) What is the perimeter of the rectangle?

(b) What is the area of the rectangle?

8. Emilio is about to roll a standard number cube.
(57, 80, 107)

(a) What is the probability that he will get a prime number in one roll?

(b) What is the chance that he will not get a prime number in one roll?

9. An octagon has how many more sides than a pentagon?
(32)

10. What is the average of 2, 4, 6, and 8?
(50)

11. QR equals RS. ST is 5 cm. RT is 7 cm. Find QT.
(61)

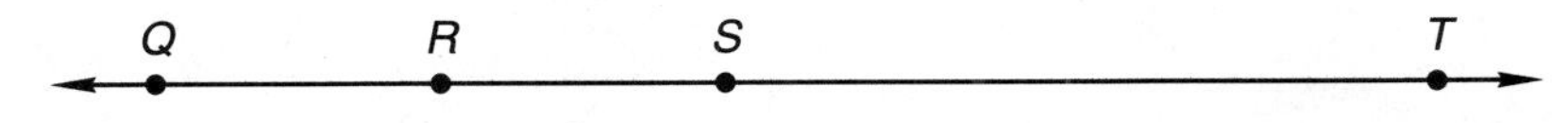

12. $38.248 + 7.5 + 37.23 + 15$
(99)

13. $\$6 - (\$1.49 - 75¢)$
(24, 70)

14. 2.4×100
(111)

15. 0.24×0.12
(110)

16. 2.4×5.7
(109)

17. $8\overline{)0.1000}$
(117)

18. $0.5\overline{)4.35}$
(119)

19. $1.2\overline{)1.44}$
(119)

20. $3\frac{1}{3} + 7\frac{3}{4}$
(116)

21. $\frac{3}{7} + \frac{1}{2}$
(116)

22. $6\frac{14}{15} - 1\frac{1}{5}$
(116)

23. $\frac{4}{5} - \frac{1}{3}$
(116)

24. $\frac{1}{2} \times 3\frac{1}{3}$
(120)

25. $4 \times 2\frac{1}{2}$
(120)

26. (109) What is the area of a bedroom that is 3 meters wide and 4.5 meters long?

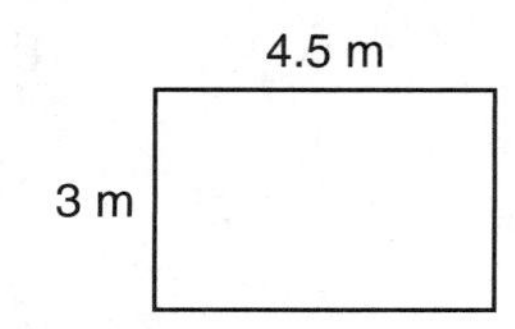

27. (103, 109) What is the volume of a drawer that is 2 ft by 1.5 ft by 0.5 ft?

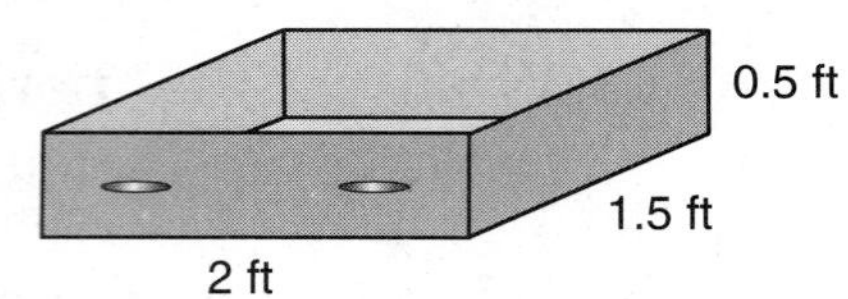

Refer to the figure to answer problems 28–30.

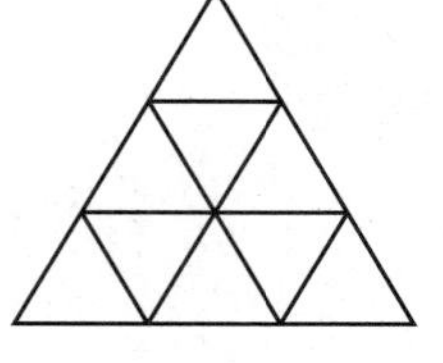

28. (53) The perimeter of each small equilateral triangle is 6 inches. What is the perimeter of the large equilateral triangle?

29. (107) The area of one small triangle is what percent of the area of the large triangle?

30. (1, Inv. 7) Shown below is a sequence of triangle patterns. Draw the next triangle in the pattern on your paper. How many small triangles form the large triangle in your drawing?

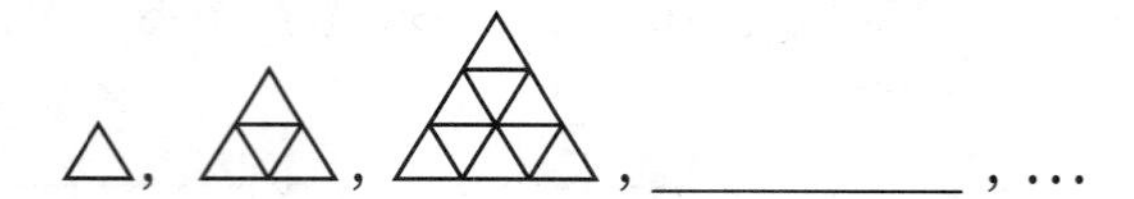

INVESTIGATION 12

Focus on

Tessellations

Archaeologists know that people have been using tiles to make mosaics and to decorate homes, temples, and other buildings since about 4000 B.C. The Romans called these tiles *tesselae,* from which we get the word **tessellation** (tiling). A tessellation is the repeated use of shapes to fill a flat surface without gaps or overlaps. Below are some examples of tessellations. We say that the polygons in these figures *tessellate;* in other words, they tile a plane.

Figure 1

Figure 2

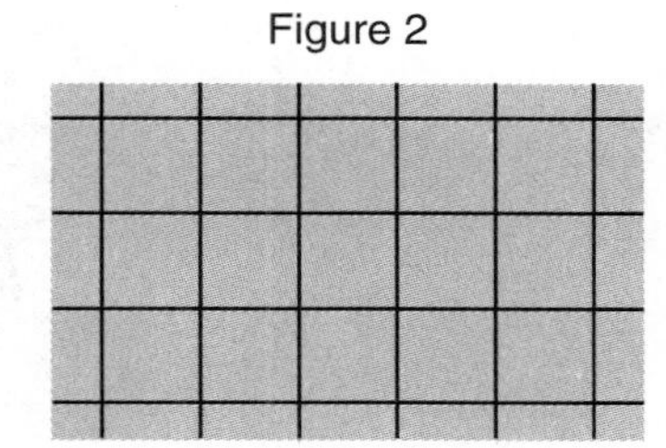

These tessellations are called **regular tessellations** because one type of regular polygon is used. Although the same shape is used repeatedly in regular tessellations, the orientation of the shape may vary from tile to tile. In figure 1, for example, we see that the triangle rotates 180° with each repetition.

Now look at a vertex in each figure, and count the number of polygons that meet at the vertex. Notice that a certain number of polygons meet at each vertex in each tessellation.

1. How many triangles meet at each vertex in figure 1?
2. How many squares meet at each vertex in figure 2?

Only a few regular polygons tessellate. Here is an example of a regular polygon that does not tessellate:

Regular pentagon

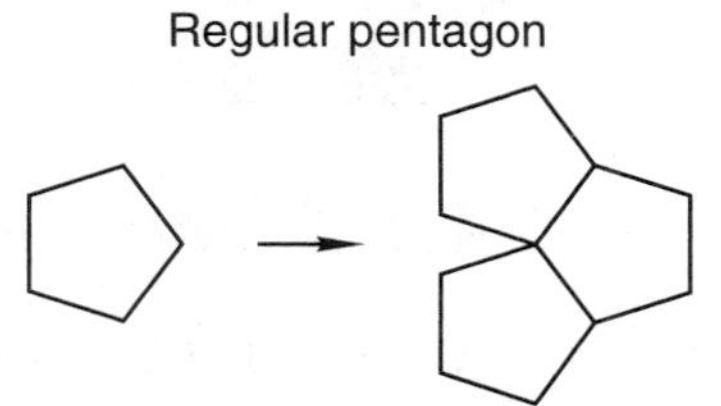

We see that the regular pentagon on the left will not fit into the gap formed by the other pentagons. Therefore, a regular pentagon does not tessellate.

3. Which of these regular polygons tessellates? Draw a tessellation that uses that polygon.

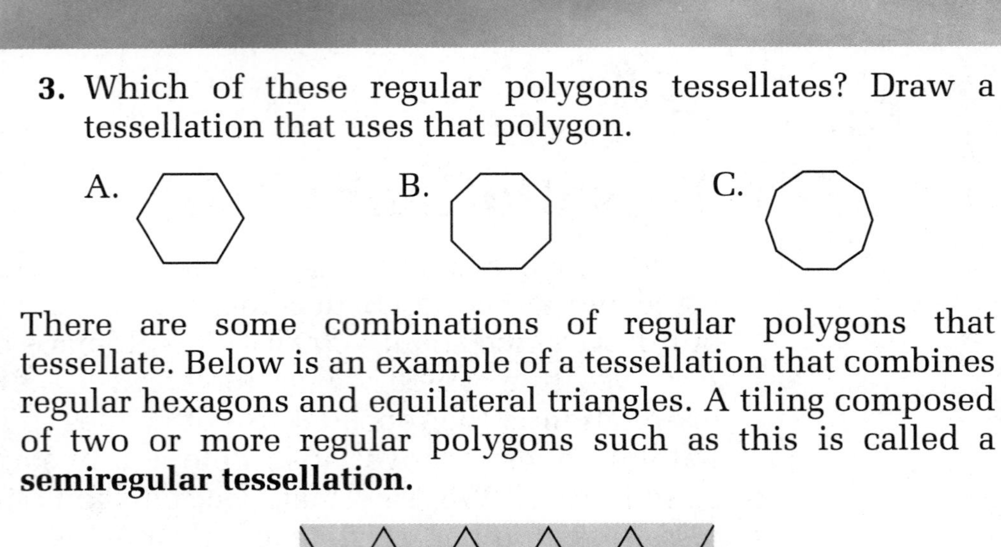

There are some combinations of regular polygons that tessellate. Below is an example of a tessellation that combines regular hexagons and equilateral triangles. A tiling composed of two or more regular polygons such as this is called a **semiregular tessellation.**

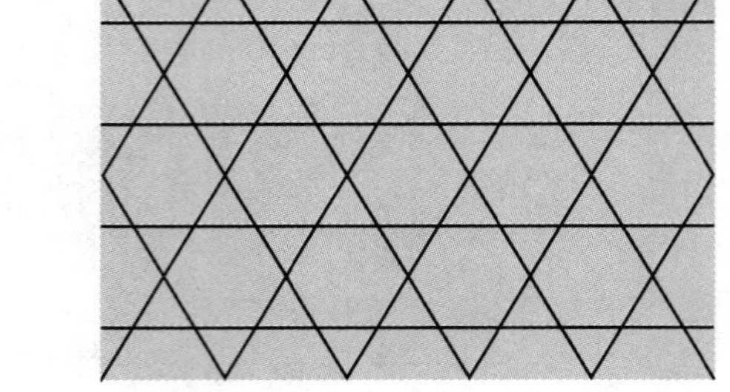

4. Which two of these regular polygons could combine to tile a plane? Draw a picture that shows the tessellation.

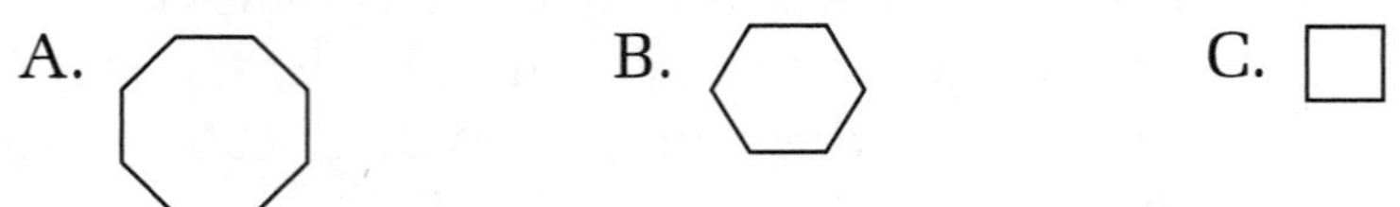

Many polygons that are not regular polygons can tile a plane. In fact, every triangle can tile a plane, and every quadrilateral can tile a plane. Here is an example using each type of polygon:

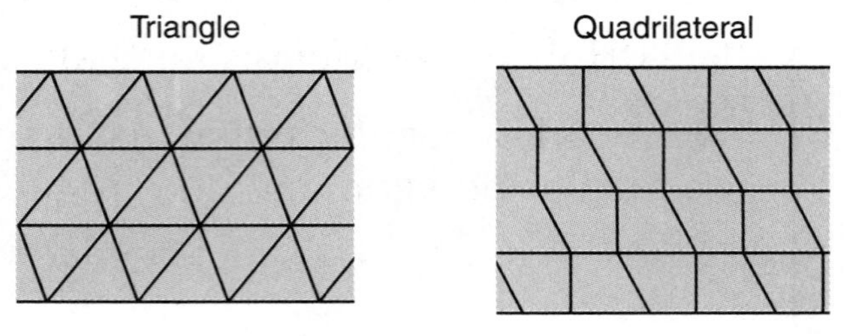

Activity: *Triangle and Quadrilateral Tessellations*

Materials needed for each student:

- scissors
- copy of Activity Master 23 (masters available in *Saxon Math 6/5 Assessments and Classroom Masters*)

5. Carefully cut out the triangles on Activity Master 23. On your desk, arrange the triangles like tiles so that the vertices of six triangles meet at a point and the sides align without gaps or overlapping. Do not flip (reflect) the triangles to make them fit.

6. Carefully cut out the quadrilaterals on Activity Master 23. When tiling with quadrilaterals, arrange the quadrilaterals so that the vertices of four quadrilaterals meet at a point.

Some polygons that tessellate can be carefully altered and fitted together to form intricate tessellations. In the example below we start with an equilateral triangle and alter one side by "cutting out a piece of the triangle." Then we attach the cut-out piece to another side of the triangle. The resulting figure tiles a plane.

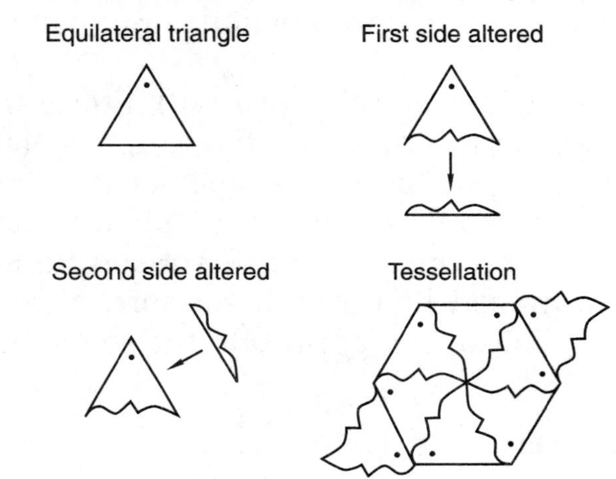

In the next example, we start with a square. We alter one side of the square and then make the corresponding alteration to the opposite side. Then we alter a third side of the square and make the corresponding alteration to the remaining side. The resulting figure tessellates.

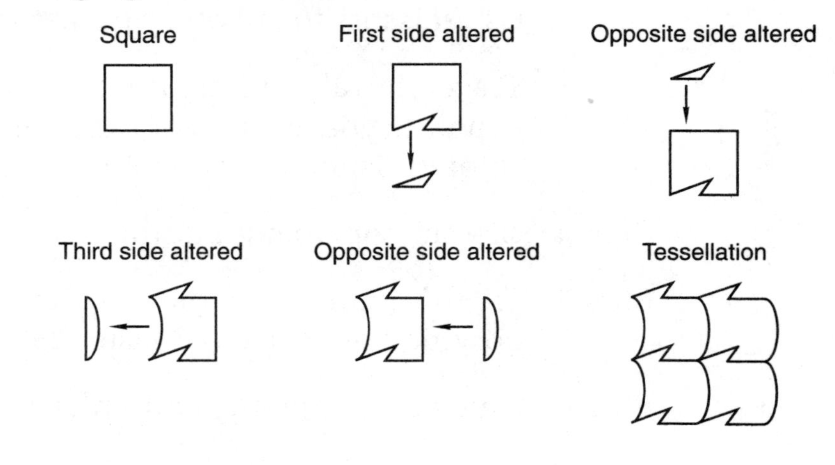

Activity: *Creating Tessellations with Altered Figures*

Materials needed by each student:

- copy of Activity Master 24 (masters available in *Saxon Math 6/5 Assessments and Classroom Masters*)
- ruler
- several sheets of unlined paper
- scissors
- glue or tape
- colored pencils or crayons (optional)

In this activity you will alter a triangle or a square and then use the resulting figure to create a tessellation. First choose one of the two shapes at the bottom of Activity Master 24. Trace that figure onto a blank sheet of paper, using a ruler to keep the sides of the traced figure straight. Then cut out the traced figure with scissors. Now follow the set of directions below that applies to the shape you chose.

Triangle

Step 1: Alter one side of the triangle by cutting a section from the shape. Be sure to cut out only one section. (Do not cut several pieces from the shape.)

Step 2: Tape the cut-out section to another side of the figure. Use scissors to cut away excess tape.

Step 3: Trace the altered figure 8 to 12 times onto blank paper. If you wish, color the figures you traced with colored pencils or crayons.

Step 4: Use scissors to cut out the traced figures.

Step 5: Fit the figures together to tile a portion of the box provided on Activity Master 24.

Step 6: Glue or tape the tiles into place.

Square

Step 1: Alter one side of the square by cutting a section from the shape. Be sure to cut out only one section. (Do not cut several pieces from the shape.)

Step 2: Tape the cut-out section to the opposite side of the figure. Use scissors to cut away excess tape.

Step 3: (optional) Repeat steps 1 and 2 to alter the remaining two sides of the figure.

Step 4: Trace the altered figure 8 to 12 times onto blank paper. If you wish, color the figures you traced with colored pencils or crayons.

Step 5: Use scissors to cut out the traced figures.

Step 6: Fit the figures together to tile a portion of the box provided on Activity Master 24.

Step 7: Glue or tape the tiles into place.

Extensions

a. Find examples of tessellations in floor tiles at school or at home. Trace or copy the patterns, and bring them to class to display.

b. Search the Internet for information about tessellations. Share pictures and/or information you found with the rest of the class.

APPENDIX

Additional Topics and Supplemental Practice

TOPIC A Roman Numerals Through 39

NEW CONCEPT

Roman numerals were used by the ancient Romans to write numbers. Today Roman numerals are still used to number such things as book chapters, movie sequels, and Super Bowl games. We might also find Roman numerals on clocks and buildings.

Some Roman numerals are

I which stands for 1

V which stands for 5

X which stands for 10

The Roman numeral system does not use place value. Instead, the values of the numerals are added or subtracted, depending on their position. For example,

II means 1 plus 1, which is 2 (II does not mean "11")

Below we list the Roman numerals for the numbers 1 through 20. Study the patterns.

1 = I	11 = XI
2 = II	12 = XII
3 = III	13 = XIII
4 = IV	14 = XIV
5 = V	15 = XV
6 = VI	16 = XVI
7 = VII	17 = XVII
8 = VIII	18 = XVIII
9 = IX	19 = XIX
10 = X	20 = XX

The multiples of 5 are 5, 10, 15, 20, ... The numbers that are one less than these (4, 9, 14, 19, ...) have Roman numerals that involve subtraction.

4 = IV ("one less than five")

9 = IX ("one less than ten")

14 = XIV (ten plus "one less than five")

19 = XIX (ten plus "one less than ten")

In each case where a smaller Roman numeral (I) precedes a larger Roman numeral (V or X), we subtract the smaller number from the larger number.

Example (a) Write XXVII in our number system.[†]

(b) Write 34 in Roman numerals.

Solution (a) We can break up the Roman numeral and see that it equals 2 tens plus 1 five plus 2 ones.

XX V II

20 + 5 + 2 = **27**

(b) We think of 34 as "30 plus 4."

30 + 4

XXX IV

So the Roman numeral for 34 is **XXXIV.**

LESSON PRACTICE

Practice set Write the Roman numerals for 1 to 39 in order.

[†]The modern world has adopted the Hindu-Arabic number system with the digits 0, 1, 2, 3, 4, 5, 6, 7, 8, 9, and base 10 place value. For simplicity we refer to the Hindu-Arabic system as "our number system."

TOPIC

B Roman Numerals Through Thousands

NEW CONCEPT

We have practiced using these Roman numerals:

I V X

With these numerals we can write counting numbers up to XXXIX (39). To write larger numbers, we must use the Roman numerals L (50), C (100), D (500), and M (1000). The table below shows the different Roman numeral "digits" we have learned, as well as their respective values.

NUMERAL	VALUE
I	1
V	5
X	10
L	50
C	100
D	500
M	1000

Example Write each Roman numeral in our number system:

(a) LXX (b) DCCL (c) XLIV (d) MMI

Solution (a) LXX is $50 + 10 + 10$, which is **70.**

(b) DCCL is $500 + 100 + 100 + 50$, which is **750.**

(c) XLIV is "10 less than 50" plus "1 less than 5"; that is, $40 + 4 =$ **44.**

(d) MMI is $1000 + 1000 + 1$, which is **2001.**

LESSON PRACTICE

Practice set Write each Roman numeral in our number system:

a. CCCLXII **b.** CCLXXXV **c.** CD

d. XLVII **e.** MMMCCLVI **f.** MCMXCIX

TOPIC

C Base 5

NEW CONCEPT

Our **base 10 number system** uses place value and the digits 0, 1, 2, 3, 4, 5, 6, 7, 8, and 9 to write numbers. The value of each place is ten times the value of the next-smaller place. Some U.S. coins and bills match our base 10 system. Ten pennies equals a dime; ten dimes equals a dollar, and so on. Using no more than nine of each of the coins and bills shown below, we can make any money amount from 1¢ to 99,999¢ ($999.99).

A different set of U.S. money matches the **base 5 number system.** Five pennies equals a nickel, and five nickels equals a quarter.

A base 5 system uses only the digits 0, 1, 2, 3, and 4 to write numbers, and the value of each place is only five times the value of the next-smaller place. Using no more than four pennies, four nickels, and four quarters, we can make any money amount from 1¢ to 124¢. However, when we write 124¢ in base 5 we do not use the number 124. Here is why. The first three places in base 5 are the ones place, the fives place, and the twenty-fives place. (It may be easier to think of these places as the pennies place, the nickels place, and the quarters place.)

Base 5 Place Values

25's place	5's place	1's place
____	____	____
(quarters)	(nickels)	(pennies)

To make 124¢ requires 4 quarters, 4 nickels, and 4 pennies. So the number 124 changed to the base 5 system looks like this:

444 (base 5)

To change a number to base 5, think of how many pennies, nickels, and quarters it would take to make the same number of cents. Remember to use no more than four of any coin. Also remember that you may need to use one or more zeros when you write a number in base 5, just as in base 10.

Example Change the number 15 from base 10 to base 5.

Solution We think of 15 as the money amount 15¢. We can make 15¢ by using 3 nickels and 0 pennies. So 15 in base 5 is written as **30 (base 5).**

LESSON PRACTICE

Practice set Change each of these base 10 numbers to base 5.

a. 31 **b.** 51

c. 10 **d.** 100

e. 38 **f.** 86

Supplemental Practice Problems for Selected Lessons

This appendix contains additional practice problems for concepts presented in selected lessons. It is very important that no problems in the regular problem sets be omitted to make room for these problems. Saxon math is designed to produce long-term retention through repeated exposure to concepts in the problem sets. The problem sets provide enough initial exposure to concepts for most students. However, if a student continues to have difficulty with certain concepts, some of the problems in this appendix can be assigned as remedial exercises.

Lesson 5 Use words to name each number:

1. 44

2. 55

3. 110

4. 312

5. 426

6. $5.37

7. $211.25

8. $608

9. $76.27

10. $9.01

Use digits to write each number:

11. one hundred fourteen

12. two hundred forty

13. seven hundred thirty-two

14. six hundred seven

15. eight hundred sixteen

16. three hundred eighty-four dollars

17. four hundred eighteen dollars

18. one hundred eighty dollars and fifty cents

19. five hundred eight dollars and fifteen cents

20. six hundred fifty dollars

Lesson 6 Add:

1. 3 + 6 + 7 + 8 + 4 + 1
2. 5 + 4 + 3 + 7 + 8 + 6
3. 12 + 4 + 23 + 17 + 8
4. 16 + 24 + 58 + 7 + 9
5. 56 + 9 + 31 + 18 + 7
6. 324 + 472
7. 589 + 723
8. 487 + 706
9. 312 + 58
10. 936 + 87
11. 43 + 246 + 97
12. 517 + 49 + 327
13. 625 + 506 + 84
14. 315 + 287 + 589
15. 643 + 420 + 708
16. 36 + 24 + 275 + 9
17. 513 + 68 + 8 + 45
18. 178 + 215 + 24 + 9
19. 47 + 6 + 428 + 14
20. 351 + 157 + 68 + 5

Lesson 7 Use digits to write each number:

1. seven thousand, two hundred fifty-four
2. twelve thousand, six hundred twenty-five
3. eleven thousand, five hundred eighty
4. twenty-one thousand, three hundred
5. fifty-six thousand, two hundred eight
6. eighteen thousand, seven hundred
7. one hundred seventy-five thousand
8. two hundred ten thousand, five hundred

9. three hundred fifty-six thousand, two hundred

10. nine hundred eighty thousand

Use words to name each number:

11. 6500

12. 4210

13. 1760

14. 8112

15. 21,000

16. 12,500

17. 40,800

18. 118,000

19. 210,600

20. 125,200

Lesson 9 Subtract:

1. $67 - 48$

2. $50 - 36$

3. \$71 − \$63

4. $413 - 242$

5. $531 - 50$

6. \$736 − \$643

7. $345 - 137$

8. $512 - 34$

9. \$650 − \$552

10. $300 - 256$

11. $580 - 74$

12. \$400 − \$ 23

13. $504 - 132$

14. $710 - 68$

15. \$800 − \$743

Lesson 13 Find each sum or difference:

1. \$3.00 + \$2.45

2. \$6.58 + \$4.00

3. \$5.29 + \$4.71

4. \$9.15 + \$10.00

5. \$15.75 + \$8.28

6. \$27.80 + \$6.00

7. \$0.48 + \$0.76
8. \$7.00 + \$12.99
9. \$12.00 + \$8.20
10. \$0.45 + \$0.55
11. \$6.54 − \$1.49
12. \$8.29 − \$1.29
13. \$3.18 − \$2.57
14. \$5.06 − \$0.27
15. \$3.00 − \$1.25
16. \$5.00 − \$4.36
17. \$12.57 − \$5.00
18. \$10.00 − \$8.54
19. \$1.00 − \$0.92
20. \$5.00 − \$4.95

Lesson 17 Multiply:

1. 23 × 7
2. 6 × 43
3. 57 × 4
4. 8 × 36
5. 70 × 6
6. 4 × 78
7. 96 × 8
8. 7 × 905
9. 89 × 6
10. 8 × 709

11. $\begin{array}{r} \$57 \\ \times \quad 4 \\ \hline \end{array}$
12. $\begin{array}{r} \$34 \\ \times \quad 5 \\ \hline \end{array}$
13. $\begin{array}{r} \$2.78 \\ \times \quad 6 \\ \hline \end{array}$
14. $\begin{array}{r} \$8.70 \\ \times \quad 3 \\ \hline \end{array}$
15. $\begin{array}{r} \$3.45 \\ \times \quad 9 \\ \hline \end{array}$
16. $\begin{array}{r} 708 \\ \times \quad 8 \\ \hline \end{array}$

Lesson 18 Multiply:

1. 5 × 9 × 4
2. 5 × 8 × 3
3. 7 × 6 × 5
4. 9 × 4 × 6
5. 7 × 5 × 8
6. 5 × 4 × 3 × 2
7. 3 × 3 × 3 × 3
8. 2 × 5 × 2 × 5
9. 6 × 4 × 2 × 0
10. 3 × 5 × 4 × 6
11. 20 × 7 × 5
12. 4 × 9 × 25

13. $6 \times 30 \times 5$

14. $50 \times 5 \times 8$

15. $5 \times 7 \times 12$

16. $8 \times 10 \times 6$

17. $54 \times 9 \times 0$

18. $5 \times 5 \times 24$

19. $2 \times 5 \times 10$

20. $7 \times 75 \times 4$

Lesson 22 Divide. Write each answer with a remainder.

1. $3\overline{)10}$

2. $4\overline{)33}$

3. $7\overline{)30}$

4. $8\overline{)51}$

5. $6\overline{)53}$

6. $5\overline{)32}$

7. $\frac{17}{2}$

8. $\frac{26}{3}$

9. $\frac{35}{10}$

10. $\frac{28}{5}$

11. $\frac{55}{8}$

12. $\frac{70}{9}$

13. $35 \div 6$

14. $32 \div 7$

15. $32 \div 9$

16. $23 \div 4$

17. $17 \div 6$

18. $35 \div 10$

Lesson 24 Simplify:

1. $8 - (6 - 2)$

2. $8 - (6 + 2)$

3. $8 - (6 \div 2)$

4. $8 \div (6 - 2)$

5. $8 \div (6 + 2)$

6. $(24 \div 6) \div 2$

7. $24 \div (6 \div 2)$

8. $(24 \div 6) - 2$

9. $24 \div (6 - 2)$

10. $(24 \div 6) + 2$

11. $24 \div (6 + 2)$

12. $(24 \div 6) \times 2$

13. $24 \div (6 \times 2)$

14. $(36 \div 6) \div 3$

15. $36 \div (6 \div 3)$

16. $(36 - 6) \times 3$

17. $36 - (6 \times 3)$

18. $(36 \div 6) - 3$

19. $36 + (12 - 6) + 3$

20. $(36 + 12) - (6 + 3)$

Lesson 26 Divide:

1. $2\overline{)136}$	**2.** $2\overline{)356}$	**3.** $3\overline{)234}$
4. $3\overline{)\$4.56}$	**5.** $3\overline{)\$5.67}$	**6.** $4\overline{)\$1.24}$
7. $4\overline{)248}$	**8.** $4\overline{)356}$	**9.** $5\overline{)120}$
10. $5\overline{)\$2.30}$	**11.** $6\overline{)\$4.32}$	**12.** $6\overline{)\$8.76}$
13. $7\overline{)511}$	**14.** $7\overline{)847}$	**15.** $7\overline{)903}$
16. $8\overline{)\$4.40}$	**17.** $8\overline{)\$6.48}$	**18.** $9\overline{)\$5.67}$
19. $9\overline{)568}$	**20.** $8\overline{)690}$	**21.** $7\overline{)611}$

Lesson 28

1. What time is shown on the clock?

2. What time was it 2 hours ago?

3. What time will it be in 2 hours?

4. What time was it half an hour ago?

5. What time will it be in a half hour?

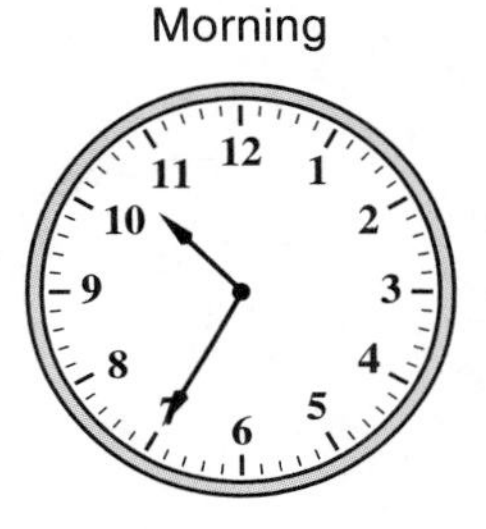

6. What time is shown on the clock?

7. What time will it be in 12 hours?

8. What time was it 2 hours ago?

9. What time was it half an hour ago?

10. How many minutes is it until 1:00 p.m.?

11. What time is shown on the clock?

12. What time will it be in 24 hours?

13. What time was it half an hour ago?

14. What time will it be in $1\frac{1}{2}$ hours?

15. How many minutes is it until noon?

16. What time is 10 minutes before noon?

17. What time is $1\frac{1}{2}$ hours after midnight?

18. What time is 5 minutes after two in the afternoon?

19. What time is 5 minutes before six in the morning?

20. What time is a quarter after three in the afternoon?

Lesson 29 Multiply:

1. 10 × 36 **2.** 47 × 30 **3.** 50 × 78

4. 34 × 70 **5.** 90 × 37 **6.** 45 × 10

7. 20 × 35 **8.** 73 × 40 **9.** 60 × 38

10. 74 × 80 **11.** 10 × 271 **12.** 932 × 30

13. 70 × 674 **14.** 465 × 20 **15.** 60 × 793

16. 81 × 100 **17.** 500 × 36 **18.** 64 × 900

19. 400 × 84 **20.** 96 × 800

Lesson 33 Round each number to the nearest ten:

1. 46 **2.** 37 **3.** 61 **4.** 58

5. 43 **6.** 79 **7.** 85 **8.** 96

Round each number to the nearest hundred:

9. 375 **10.** 216 **11.** 850 **12.** 781

13. 460 **14.** 329 **15.** 198 **16.** 748

Round each number to the nearest ten:

17. 121 **18.** 127 **19.** 358 **20.** 341

21. 769 **22.** 532 **23.** 477 **24.** 265

Lesson 34 Divide:

1. $3\overline{)31}$ **2.** $4\overline{)83}$ **3.** $2\overline{)61}$

4. $3\overline{)122}$ **5.** $4\overline{)243}$ **6.** $5\overline{)404}$

7. $6\overline{)365}$ **8.** $6\overline{)305}$ **9.** $8\overline{)407}$

10. $3\overline{)\$3.15}$ **11.** $4\overline{)\$8.24}$ **12.** $5\overline{)\$5.40}$

13. $2\overline{)415}$ **14.** $3\overline{)920}$ **15.** $4\overline{)433}$

16. $7\overline{)\$7.42}$ **17.** $3\overline{)\$6.06}$ **18.** $4\overline{)\$9.60}$

Lesson 37

1. Draw a square and shade $\frac{1}{2}$ of it.

2. Draw a square and shade $\frac{1}{2}$ of it another way.

3. Draw a square and shade $\frac{1}{2}$ of it another way.

4. Draw a square and shade $\frac{1}{4}$ of it.

5. Draw a circle and shade $\frac{1}{2}$ of it.

6. Draw a circle and shade $\frac{3}{4}$ of it.

7. Draw a circle and shade $\frac{1}{3}$ of it.

8. Draw a rectangle and shade $\frac{1}{2}$ of it.

9. Draw a rectangle and shade $\frac{1}{4}$ of it.

10. Draw a rectangle and shade $\frac{1}{3}$ of it.

11. Draw a rectangle and shade $\frac{1}{5}$ of it.

12. Draw a square and shade $\frac{3}{4}$ of it.

13. Draw a circle and shade $\frac{2}{3}$ of it.

14. Draw a rectangle and shade $\frac{2}{3}$ of it.

15. Draw a rectangle and shade $\frac{2}{5}$ of it.

16. Draw a circle and shade $\frac{1}{6}$ of it.

17. Draw a rectangle and shade $\frac{1}{6}$ of it.

18. Draw a rectangle and shade $\frac{3}{5}$ of it.

19. Draw a circle and shade $\frac{5}{6}$ of it.

20. Draw a rectangle and shade $\frac{5}{6}$ of it.

Lesson 38 Use a fraction or mixed number to name each point marked with an arrow on these number lines:

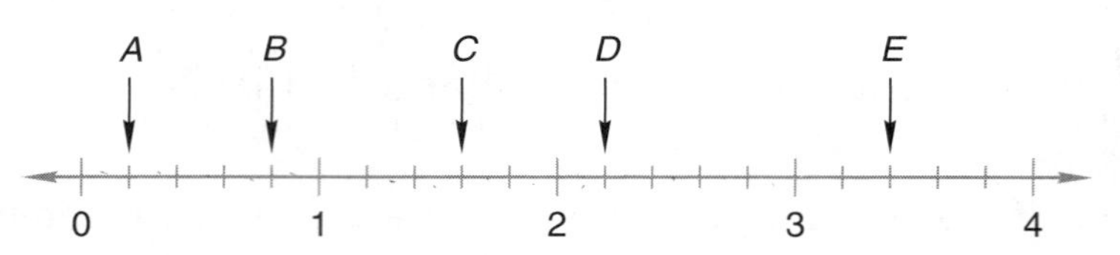

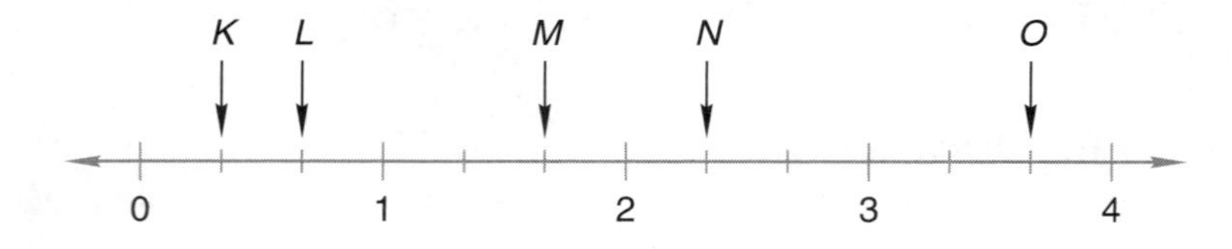

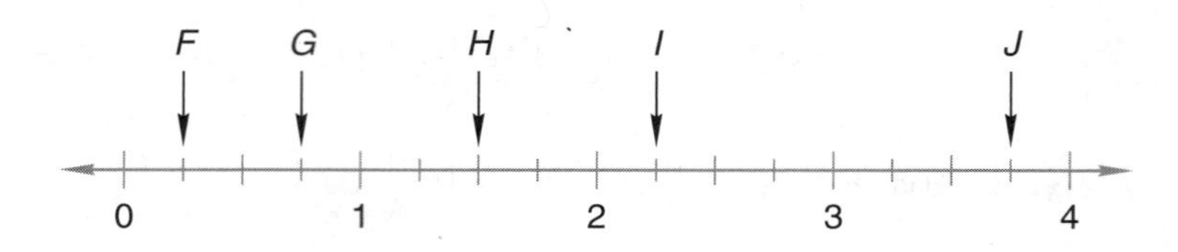

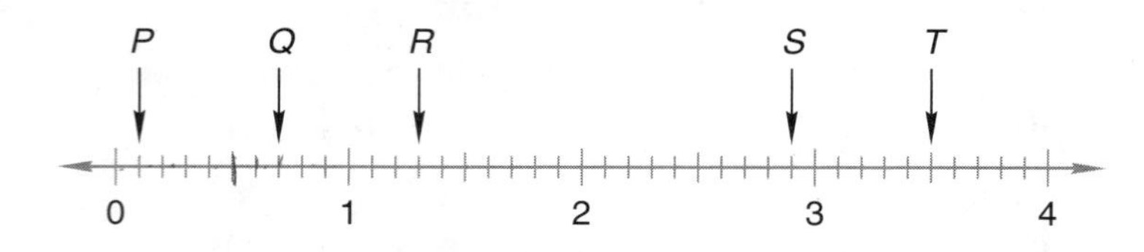

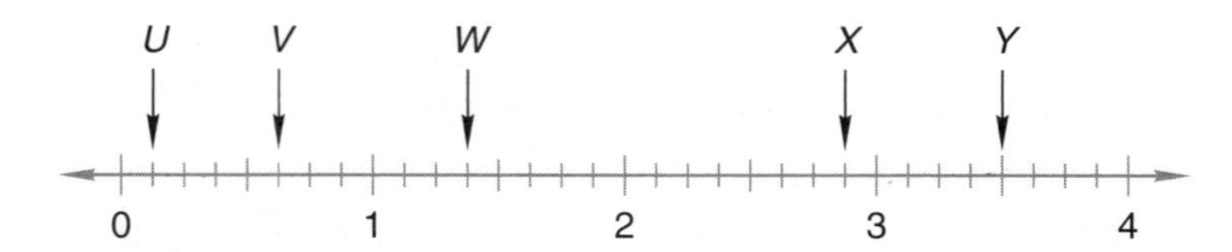

Lesson 43 Find each sum or difference:

1. $3\frac{1}{3} + 1$

2. $2\frac{1}{3} + 3\frac{1}{3}$

3. $5 + 1\frac{2}{5}$

4. $1 + \frac{3}{4}$

5. $6\frac{4}{8} + 1\frac{3}{8}$

6. $7\frac{2}{3} + 5$

7. $4 + \frac{3}{10}$

8. $3\frac{5}{10} + 1\frac{4}{10}$

9. $9 + 7\frac{1}{2}$

10. $\frac{5}{6} + 1$

11. $6\frac{2}{3} - 4$

12. $3\frac{3}{4} - 1\frac{2}{4}$

13. $7\frac{1}{2} - \frac{1}{2}$

14. $1\frac{5}{8} - 1$

15. $8\frac{3}{4} - 2\frac{3}{4}$

16. $3\frac{1}{2} - 3\frac{1}{2}$

17. $10\frac{7}{10} - 1\frac{4}{10}$

18. $4\frac{3}{5} - \frac{1}{5}$

19. $9\frac{4}{5} - 4$

20. $1\frac{7}{8} - \frac{7}{8}$

Lesson 48 Write each number in standard form:

1. $(5 \times 1000) + (2 \times 100) + (8 \times 10)$

2. $(6 \times 100) + (4 \times 10) + (2 \times 1)$

3. $(4 \times 10{,}000) + (5 \times 1000) + (6 \times 10) + (7 \times 1)$

4. $(5 \times 1000) + (4 \times 100) + (9 \times 10) + (2 \times 1)$

5. $(7 \times 10{,}000) + (1 \times 1000) + (4 \times 100)$

6. $(6 \times 1000) + (4 \times 100) + (3 \times 1)$

7. $(7 \times 1000) + (8 \times 10) + (9 \times 1)$

8. $(1 \times 10{,}000) + (4 \times 100) + (7 \times 1)$

9. $(6 \times 1000) + (1 \times 10)$

10. $(1 \times 10{,}000) + (6 \times 1000) + (5 \times 1)$

Write each number in expanded notation:

11. 65

12. 742

13. 320

14. 506

15. 7500

16. 2001

17. 1040

18. 1760

19. 1492

20. 25,000

Lesson 50

Find the average of each group of numbers:

1. 3, 3, 6

2. 4, 5, 7, 8

3. 5, 6, 8, 9

4. 15, 17, 19

5. 21, 19, 26

6. 1, 2, 3, 4, 5

7. 3, 5, 7, 9

8. 36, 44

9. 65, 47, 32

10. 6, 7, 8, 9, 10

11. 112, 124

12. 47, 52, 54

13. 6, 6, 6, 10

14. 11, 12, 13, 14, 15

15. 33, 34, 35

16. 30, 40, 50, 60

17. 22, 24, 26, 28

18. 163, 197

19. 97, 101, 111

20. 43, 62, 56, 63

Lesson 51 Multiply:

1. 38×49
2. 96×97
3. $\$0.78 \times 76$
4. $\$0.52 \times 47$
5. 63×85
6. 69×81
7. $\$0.58 \times 59$
8. $\$0.16 \times 74$
9. 96×36
10. 27×73
11. $\$0.85 \times 96$
12. $\$0.47 \times 72$
13. 74×18
14. 36×83
15. $\$0.74 \times 58$
16. $\$0.67 \times 64$
17. 92×47
18. 63×49
19. $\$0.18 \times 85$
20. $\$0.46 \times 89$

Lesson 52 Write the value of the 1 in each number:

1. 315,275,486
2. 21,987,564
3. 128,675
4. 7,351,487
5. 125,386,794
6. 97,315,248

Name the value of the place held by the zero in each number:

7. 20,675,482
8. 123,450,683
9. 5,046,912
10. 17,954,068
11. 805,423,796
12. 8,907,485

Which digit is in the millions place in each number?

13. 654,297,801
14. 37,591,846

Which digit is in the ten-millions place in each number?

15. 752,931,468
16. 246,801,357

Write the value of the 5 in each number:

17. 375,286,420
18. 17,576,284
19. 56,234,196
20. 123,456,786

Use digits to write each number:

21. one million, two hundred fifty thousand

22. five million, three hundred twelve thousand

23. ten million, one hundred twenty-five thousand, two hundred

24. thirteen million, two hundred ten thousand, five hundred

25. twenty-five million, one hundred ninety-six thousand, one hundred

26. three hundred twenty-seven million

27. six hundred forty-five million, six hundred thousand, two hundred

28. seven hundred sixteen million, nine hundred eleven thousand

29. one hundred twenty million, six hundred fifteen thousand

30. nine hundred eighty-four million, two hundred thousand

Use words to name each number:

31. 1,500,000

32. 10,200,000

33. 15,352,000

34. 25,740,000

35. 42,164,000

36. 78,345,200

37. 120,000,000

38. 253,000,000

39. 412,520,000

40. 635,154,000

Lesson 54

Divide:

1. $20\overline{)420}$	2. $30\overline{)450}$	3. $40\overline{)\$4.80}$
4. $50\overline{)700}$	5. $60\overline{)800}$	6. $70\overline{)\$7.00}$
7. $80\overline{)900}$	8. $20\overline{)560}$	9. $30\overline{)\$5.70}$
10. $40\overline{)650}$	11. $50\overline{)850}$	12. $60\overline{)\$9.00}$
13. $70\overline{)800}$	14. $20\overline{)614}$	15. $30\overline{)\$7.80}$
16. $40\overline{)876}$	17. $50\overline{)987}$	18. $60\overline{)\$9.60}$

Lesson 56

Multiply:

1. $\begin{array}{r} 135 \\ \times\ 246 \\ \hline \end{array}$	2. $\begin{array}{r} 650 \\ \times\ 473 \\ \hline \end{array}$	3. $\begin{array}{r} \$4.08 \\ \times\ 592 \\ \hline \end{array}$	4. $\begin{array}{r} \$3.54 \\ \times\ 260 \\ \hline \end{array}$
5. $\begin{array}{r} 625 \\ \times\ 403 \\ \hline \end{array}$	6. $\begin{array}{r} 754 \\ \times\ 365 \\ \hline \end{array}$	7. $\begin{array}{r} \$3.47 \\ \times\ 198 \\ \hline \end{array}$	8. $\begin{array}{r} \$6.80 \\ \times\ 743 \\ \hline \end{array}$
9. $\begin{array}{r} 503 \\ \times\ 936 \\ \hline \end{array}$	10. $\begin{array}{r} 418 \\ \times\ 650 \\ \hline \end{array}$	11. $\begin{array}{r} \$9.73 \\ \times\ 409 \\ \hline \end{array}$	12. $\begin{array}{r} \$3.49 \\ \times\ 156 \\ \hline \end{array}$
13. $\begin{array}{r} 760 \\ \times\ 394 \\ \hline \end{array}$	14. $\begin{array}{r} 507 \\ \times\ 938 \\ \hline \end{array}$	15. $\begin{array}{r} \$2.43 \\ \times\ 671 \\ \hline \end{array}$	16. $\begin{array}{r} \$9.53 \\ \times\ 870 \\ \hline \end{array}$
17. $\begin{array}{r} 740 \\ \times\ 698 \\ \hline \end{array}$	18. $\begin{array}{r} 486 \\ \times\ 203 \\ \hline \end{array}$	19. $\begin{array}{r} \$7.05 \\ \times\ 258 \\ \hline \end{array}$	20. $\begin{array}{r} \$5.78 \\ \times\ 369 \\ \hline \end{array}$

Lesson 58

Divide. Write each quotient as a mixed number.

1. $6\overline{)19}$	2. $5\overline{)36}$	3. $4\overline{)27}$
4. $16 \div 7$	5. $25 \div 8$	6. $56 \div 9$
7. $\frac{10}{3}$	8. $\frac{50}{7}$	9. $\frac{81}{10}$
10. $2\overline{)45}$	11. $3\overline{)46}$	12. $4\overline{)47}$
13. $56 \div 5$	14. $79 \div 6$	15. $61 \div 10$
16. $\frac{33}{8}$	17. $\frac{125}{3}$	18. $\frac{95}{6}$
19. $100 \div 7$	20. $100 \div 9$	21. $100 \div 3$

Lesson 62 Estimate each answer by rounding before doing the arithmetic. Round numbers less than 100 to the nearest ten. Round numbers more than 100 to the nearest hundred.

1. $36 + 43$ **2.** $38 + 49$ **3.** $73 - 31$

4. $59 - 31$ **5.** 51×39 **6.** 78×42

7. $88 \div 29$ **8.** $81 \div 19$ **9.** $397 + 214$

10. $688 + 291$ **11.** $687 - 304$ **12.** $915 - 588$

13. 503×491 **14.** 687×298 **15.** $395 \div 21$

16. $589 \div 29$ **17.** $87 + 93$ **18.** $786 + 495$

19. $893 - 514$ **20.** $980 - 217$

Lesson 63 Subtract:

1. $1 - \frac{1}{3}$ **2.** $2 - \frac{2}{3}$ **3.** $3 - \frac{1}{4}$

4. $4 - \frac{3}{4}$ **5.** $2 - 1\frac{1}{5}$ **6.** $3 - 1\frac{1}{6}$

7. $4 - 2\frac{5}{6}$ **8.** $5 - 3\frac{1}{8}$ **9.** $6 - 1\frac{3}{8}$

10. $8 - 5\frac{5}{8}$ **11.** $7 - 6\frac{7}{8}$ **12.** $10 - \frac{1}{2}$

13. $4 - 2\frac{1}{10}$ **14.** $6 - 3\frac{3}{10}$ **15.** $3 - 2\frac{1}{2}$

16. $5 - 1\frac{1}{12}$ **17.** $10 - \frac{1}{10}$ **18.** $8 - 4\frac{2}{5}$

19. $1 - \frac{11}{12}$ **20.** $3 - 2\frac{3}{5}$

Lesson 68 Use words to name each decimal number:

1. 3.4 **2.** 0.23

3. 12.9 **4.** 7.14

5. 20.5 **6.** 15.15

7. 10.1 **8.** 1.10

9. 120.8 **10.** 21.04

Use digits to write each decimal number:

11. twenty-three and four tenths

12. thirty-two hundredths

13. ten and five tenths

14. two and twenty-five hundredths

15. fifty-two and one tenth

16. five hundredths

17. one hundred thirty-five and nine tenths

18. seventy-six and twelve hundredths

19. one and six hundredths

20. ninety-six and five tenths

Lesson 75 Simplify:

1. $\frac{8}{3}$ **2.** $\frac{7}{2}$ **3.** $\frac{12}{4}$ **4.** $\frac{7}{4}$

5. $\frac{10}{5}$ **6.** $\frac{100}{100}$ **7.** $5\frac{3}{2}$ **8.** $6\frac{6}{3}$

9. $4\frac{8}{5}$ **10.** $9\frac{4}{4}$ **11.** $3\frac{11}{8}$ **12.** $4\frac{9}{4}$

13. $5\frac{8}{3}$ **14.** $7\frac{9}{5}$ **15.** $8\frac{7}{3}$

Add. Simplify each answer.

16. $\frac{2}{3} + \frac{2}{3}$ **17.** $\frac{3}{4} + \frac{3}{4} + \frac{3}{4}$ **18.** $\frac{2}{3} + \frac{2}{3} + \frac{2}{3}$

19. $1\frac{1}{2} + 1\frac{1}{2}$ **20.** $3\frac{2}{3} + 1\frac{2}{3}$ **21.** $\frac{5}{3} + \frac{4}{3}$

Lesson 76 Multiply:

1. $\frac{1}{2} \times \frac{1}{2}$ **2.** $\frac{1}{2} \times \frac{3}{4}$ **3.** $\frac{2}{3} \times \frac{2}{3}$

4. $\frac{1}{3} \times \frac{1}{3}$ **5.** $\frac{5}{6} \times \frac{1}{2}$ **6.** $\frac{3}{4} \times \frac{1}{2}$

7. $\frac{1}{2} \times \frac{1}{3}$ **8.** $\frac{1}{5} \times \frac{2}{3}$ **9.** $\frac{3}{7} \times \frac{2}{5}$

10. $\frac{1}{4} \times \frac{1}{4}$ **11.** $\frac{2}{3} \times \frac{1}{3}$ **12.** $\frac{1}{2} \times \frac{1}{5}$

13. $\frac{1}{2} \times \frac{1}{4}$ **14.** $\frac{3}{4} \times \frac{3}{4}$ **15.** $\frac{5}{8} \times \frac{1}{2}$

16. $\frac{1}{3} \times \frac{1}{4}$ **17.** $\frac{3}{4} \times \frac{1}{4}$ **18.** $\frac{3}{4} \times \frac{3}{5}$

19. $\frac{1}{10} \times \frac{1}{10}$ **20.** $\frac{5}{8} \times \frac{3}{4}$

Lesson 79 Find the fraction name for 1 used to make each equivalent fraction:

1. $\frac{1}{2} \times \frac{?}{?} = \frac{2}{4}$ **2.** $\frac{1}{2} \times \frac{?}{?} = \frac{6}{12}$

3. $\frac{2}{3} \times \frac{?}{?} = \frac{4}{6}$ **4.** $\frac{2}{3} \times \frac{?}{?} = \frac{8}{12}$

5. $\frac{3}{4} \times \frac{?}{?} = \frac{6}{8}$

6. $\frac{3}{4} \times \frac{?}{?} = \frac{9}{12}$

7. $\frac{1}{2} \times \frac{?}{?} = \frac{5}{10}$

8. $\frac{5}{6} \times \frac{?}{?} = \frac{10}{12}$

Find the numerator that completes each equivalent fraction:

9. $\frac{2}{5} = \frac{?}{10}$

10. $\frac{1}{4} = \frac{?}{12}$

11. $\frac{4}{5} = \frac{?}{15}$

12. $\frac{3}{8} = \frac{?}{16}$

13. $\frac{2}{3} = \frac{?}{15}$

14. $\frac{1}{6} = \frac{?}{12}$

15. $\frac{1}{3} = \frac{?}{18}$

16. $\frac{1}{2} = \frac{?}{20}$

17. $\frac{3}{10} = \frac{?}{20}$

18. $\frac{3}{4} = \frac{?}{20}$

19. $\frac{4}{5} = \frac{?}{20}$

20. $\frac{1}{10} = \frac{?}{100}$

Lesson 82 Find the greatest common factor (GCF) of each pair of numbers:

1. 4 and 6
2. 4 and 8
3. 6 and 8
4. 6 and 9
5. 6 and 10
6. 6 and 12
7. 8 and 12
8. 9 and 12
9. 10 and 12
10. 5 and 10
11. 3 and 5
12. 8 and 16

Reduce each fraction by dividing the terms of the fraction by the GCF of the terms:

13. $\frac{12}{16}$

14. $\frac{12}{18}$

15. $\frac{9}{15}$

16. $\frac{8}{16}$

17. $\frac{12}{20}$

18. $\frac{16}{24}$

Lesson 86 Multiply. Simplify answers when possible.

1. $\frac{1}{3} \times 2$ **2.** $\frac{1}{2} \times 3$ **3.** $\frac{2}{3} \times 2$

4. $\frac{2}{3} \times 3$ **5.** $\frac{1}{4} \times 5$ **6.** $\frac{3}{4} \times 3$

7. $2 \times \frac{4}{5}$ **8.** $3 \times \frac{3}{5}$ **9.** $4 \times \frac{2}{3}$

10. What is $\frac{1}{3}$ of 9? **11.** What is $\frac{2}{3}$ of 9?

12. What is $\frac{1}{4}$ of 8? **13.** What is $\frac{3}{4}$ of 8?

14. What is $\frac{1}{5}$ of 10? **15.** What is $\frac{3}{5}$ of 10?

16. What is $\frac{1}{6}$ of 12? **17.** What is $\frac{5}{6}$ of 12?

18. What is $\frac{1}{7}$ of 21? **19.** What is $\frac{4}{7}$ of 21?

20. What is $\frac{1}{8}$ of 16? **21.** What is $\frac{5}{8}$ of 16?

Lesson 90 Reduce each fraction or mixed number to lowest terms:

1. $\frac{2}{8}$ **2.** $\frac{3}{9}$ **3.** $\frac{4}{6}$ **4.** $\frac{4}{10}$

5. $\frac{5}{10}$ **6.** $\frac{3}{12}$ **7.** $1\frac{2}{4}$ **8.** $3\frac{6}{8}$

9. $2\frac{3}{6}$ **10.** $4\frac{8}{10}$ **11.** $1\frac{2}{6}$ **12.** $5\frac{6}{9}$

13. $\frac{4}{8}$ **14.** $\frac{6}{12}$ **15.** $\frac{8}{12}$ **16.** $\frac{10}{20}$

17. $\frac{4}{20}$ **18.** $\frac{8}{16}$ **19.** $\frac{12}{18}$ **20.** $\frac{10}{100}$

21. $\frac{18}{24}$ **22.** $\frac{50}{100}$ **23.** $\frac{16}{20}$ **24.** $\frac{60}{100}$

Find each sum, difference, or product. Reduce your answers to lowest terms.

25. $\frac{3}{8} + \frac{3}{8}$ **26.** $\frac{9}{10} - \frac{3}{10}$ **27.** $\frac{2}{3} \times \frac{1}{4}$

28. $\frac{1}{6} \times 3$ **29.** $1\frac{1}{4} + 2\frac{1}{4}$ **30.** $3\frac{5}{6} - 1\frac{1}{6}$

31. $5\frac{5}{9} + 1\frac{1}{9}$ **32.** $\frac{5}{12} + \frac{5}{12}$ **33.** $\frac{7}{12} + \frac{1}{12}$

34. $\frac{15}{16} - \frac{3}{16}$ **35.** $\frac{3}{10} \times \frac{2}{3}$ **36.** $\frac{2}{3} \times \frac{3}{8}$

37. $\frac{9}{24} + \frac{7}{24}$ **38.** $\frac{17}{18} - \frac{11}{18}$ **39.** $\frac{6}{10} \times \frac{5}{10}$

40. $\frac{1}{12} \times 6$ **41.** $\frac{7}{20} + \frac{1}{20}$

Write each percent as a reduced fraction:

42. 25% **43.** 10% **44.** 2% **45.** 60%

46. 80% **47.** 90% **48.** 30% **49.** 1%

50. 50% **51.** 20% **52.** 5% **53.** 70%

54. 99% **55.** 4% **56.** 40% **57.** 75%

Lesson 91

Simplify each fraction or mixed number:

1. $\frac{8}{6}$ **2.** $\frac{9}{6}$ **3.** $\frac{10}{6}$ **4.** $\frac{10}{8}$

5. $\frac{12}{8}$ **6.** $\frac{12}{10}$ **7.** $\frac{14}{4}$ **8.** $\frac{27}{6}$

9. $\frac{20}{8}$ **10.** $\frac{15}{6}$ **11.** $\frac{15}{10}$ **12.** $\frac{14}{8}$

13. $3\frac{6}{4}$ **14.** $4\frac{16}{10}$ **15.** $5\frac{10}{4}$

Find each sum or product. Simplify your answers.

16. $\frac{8}{9} + \frac{8}{9} + \frac{8}{9}$

17. $3\frac{7}{8} + 4\frac{7}{8}$

18. $\frac{3}{4} \times 10$

19. $\frac{6}{5} \times \frac{9}{2}$

Lesson 94 Divide:

1. $12\overline{)432}$
2. $24\overline{)432}$
3. $18\overline{)432}$
4. $27\overline{)432}$
5. $13\overline{)235}$
6. $29\overline{)401}$
7. $32\overline{)516}$
8. $19\overline{)399}$
9. $23\overline{)490}$
10. $14\overline{)500}$
11. $25\overline{)700}$
12. $33\overline{)1000}$
13. $41\overline{)464}$
14. $39\overline{)800}$
15. $17\overline{)422}$
16. $22\overline{)657}$
17. $15\overline{)218}$
18. $31\overline{)943}$

Lesson 96 Divide. Remember to simplify your answers.

1. $\frac{2}{3} \div \frac{1}{2}$
2. $\frac{1}{2} \div \frac{2}{3}$
3. $\frac{1}{3} \div \frac{3}{4}$
4. $\frac{3}{4} \div \frac{1}{3}$
5. $\frac{3}{4} \div \frac{1}{4}$
6. $\frac{1}{4} \div \frac{3}{4}$
7. $2 \div \frac{1}{2}$
8. $\frac{1}{2} \div 2$
9. $2 \div \frac{1}{3}$
10. $\frac{1}{3} \div 2$
11. $\frac{1}{6} \div \frac{1}{3}$
12. $\frac{1}{3} \div \frac{1}{6}$
13. $\frac{3}{4} \div \frac{1}{2}$
14. $\frac{1}{2} \div \frac{3}{4}$
15. $3 \div \frac{2}{3}$
16. $\frac{2}{3} \div 3$
17. $3 \div \frac{3}{4}$
18. $\frac{3}{4} \div 3$

Lesson 99 Find each sum or difference:

1. 3.47 + 6.4
2. 23.51 − 17
3. 25.3 + 0.421
4. 6.57 − 0.8
5. 3.842 + 1.6
6. 20.45 − 12
7. 4.2 + 4 + 0.1
8. 5.423 − 1.4
9. 4.28 + 0.6 + 3
10. 1.00 − 0.84

11. $7.45 + 12.383$

12. $1.000 - 0.625$

13. $3 + 4.6 + 0.27$

14. $36.27 - 12$

15. $14.2 + 6.4 + 5$

16. $3.427 - 1$

17. $5.2 + 3 + 0.47$

18. $32.47 - 5.8$

19. $5.36 + 12$

20. $16.25 - 15$

Lesson 102

Subtract:

1. $0.4 - 0.15$

2. $0.3 - 0.23$

3. $3.5 - 0.35$

4. $4.2 - 1.25$

5. $0.2 - 0.12$

6. $8.6 - 4.31$

7. $5.0 - 1.4$

8. $0.75 - 0.375$

9. $0.8 - 0.75$

10. $4.3 - 0.125$

11. $0.6 - 0.599$

12. $1.25 - 0.625$

13. $4.0 - 1.25$

14. $4.1 - 0.14$

15. $0.25 - 0.125$

16. $7.0 - 1.6$

17. $0.5 - 0.425$

18. $4.8 - 3.29$

19. $6.0 - 0.6$

20. $0.34 - 0.291$

21. $3 - 2.1$

22. $4 - 3.21$

23. $1 - 0.2$

24. $3.45 - 1$

25. $6 - 4.7$

26. $1 - 0.01$

27. $3.4 - 2$

28. $1 - 0.23$

29. $12 - 6.4$

30. $15 - 1.5$

31. $4.3 - 1$

32. $8 - 7.9$

33. $1 - 0.9$

34. $4 - 3.99$

35. $25 - 12.5$

36. $16.7 - 8$

37. $14 - 5.6$

38. $8 - 1.35$

39. $4 - 2.77$

40. $1 - 0.211$

Lesson 104 Round each number to the nearest whole number:

1. $7\frac{1}{8}$
2. 3.8
3. 4.18
4. $5\frac{5}{6}$
5. 5.2
6. 4.93
7. $12\frac{1}{3}$
8. 16.9
9. 14.23
10. $3\frac{2}{3}$
11. 6.7
12. 5.41
13. $16\frac{1}{5}$
14. 24.4
15. 12.75
16. $9\frac{9}{10}$
17. 9.6
18. 9.87

Lesson 109 Multiply:

1. 0.3×5
2. 4×0.6
3. 0.7×0.8
4. 0.6×6
5. 0.4×0.4
6. 0.8×9
7. 0.25×3
8. 2.5×5
9. 2.5×0.7
10. 0.12×6
11. 1.2×0.8
12. 0.15×5
13. 0.18×3
14. 4.7×0.5
15. 0.3×0.8
16. 1.23×0.7
17. 6.25×8
18. 0.15×1.5
19. 0.45×0.3
20. 0.06×8

Lesson 110 Multiply:

1. 0.3×0.3
2. 0.2×0.4
3. 0.12×0.3
4. 0.05×0.07
5. 0.08×0.7
6. 0.12×0.12
7. 0.12×0.08
8. 0.42×0.2
9. 0.25×0.3
10. 0.23×0.4
11. 0.03×0.07
12. 1.23×0.04
13. 0.4×0.2
14. 0.25×0.1

15. 0.025 × 0.7

16. 6.5 × 0.01

17. 0.03 × 0.03

18. 0.01 × 0.1

19. 0.24 × 0.3

20. 0.12 × 0.06

Lesson 112 Find the least common multiple (LCM) of each pair of numbers:

1. 3 and 4 **2.** 4 and 5 **3.** 4 and 6

4. 3 and 6 **5.** 4 and 8 **6.** 6 and 8

7. 6 and 9 **8.** 6 and 10 **9.** 6 and 12

10. 8 and 10 **11.** 8 and 12 **12.** 8 and 16

13. 10 and 15 **14.** 5 and 15 **15.** 5 and 10

16. 5 and 6 **17.** 10 and 20 **18.** 10 and 25

19. 20 and 30 **20.** 20 and 40

Lesson 113 Name the number of shaded circles as a mixed number and as an improper fraction:

1.  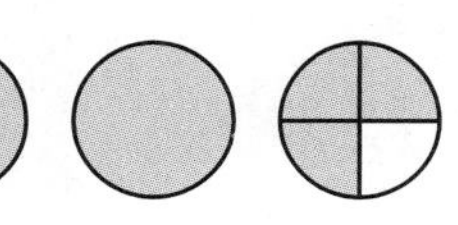

2.  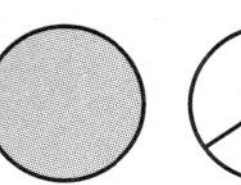 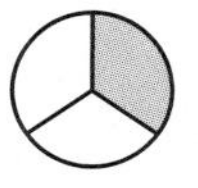

3. 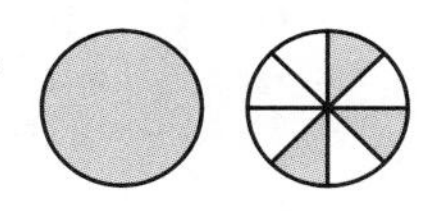

4. 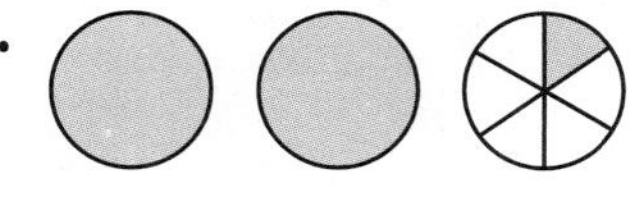

Change each mixed number to an improper fraction:

5. $3\frac{1}{2}$ **6.** $2\frac{1}{3}$ **7.** $3\frac{2}{3}$

8. $4\frac{1}{2}$ **9.** $1\frac{1}{8}$ **10.** $2\frac{1}{5}$

11. $5\frac{1}{2}$ **12.** $4\frac{1}{3}$ **13.** $3\frac{1}{4}$

14. $7\frac{1}{2}$ **15.** $3\frac{1}{3}$ **16.** $4\frac{1}{5}$

Lesson 116 Find each sum or difference:

1. $\frac{1}{2} + \frac{1}{4}$
2. $\frac{3}{4} - \frac{1}{2}$
3. $\frac{1}{2} + \frac{3}{8}$
4. $\frac{5}{8} - \frac{1}{2}$
5. $\frac{1}{4} + \frac{1}{8}$
6. $\frac{7}{8} - \frac{1}{4}$
7. $\frac{3}{4} + \frac{1}{8}$
8. $\frac{1}{3} - \frac{1}{9}$
9. $\frac{1}{2} + \frac{1}{10}$
10. $\frac{8}{9} - \frac{2}{3}$
11. $\frac{1}{5} + \frac{1}{10}$
12. $\frac{9}{10} - \frac{1}{2}$
13. $\frac{2}{5} + \frac{3}{10}$
14. $\frac{3}{10} - \frac{1}{5}$
15. $\frac{1}{6} + \frac{7}{12}$
16. $3\frac{1}{2} + 1\frac{1}{4}$
17. $3\frac{3}{4} + 1\frac{1}{8}$
18. $5\frac{3}{8} + 1\frac{1}{2}$
19. $5\frac{1}{6} + 1\frac{1}{3}$
20. $4\frac{1}{2} + 1\frac{1}{6}$
21. $3\frac{2}{3} + 1\frac{1}{6}$
22. $4\frac{3}{8} + 1\frac{1}{4}$
23. $6\frac{3}{10} + 1\frac{1}{2}$
24. $3\frac{3}{10} + 2\frac{3}{5}$
25. $5\frac{1}{2} + 1\frac{5}{12}$
26. $4\frac{5}{12} + 1\frac{1}{3}$
27. $6\frac{3}{4} + 1\frac{1}{12}$
28. $4\frac{7}{8} - 1\frac{1}{2}$
29. $4\frac{3}{4} - 2\frac{3}{8}$
30. $6\frac{1}{4} + 1\frac{5}{12}$
31. $5\frac{7}{10} - 1\frac{1}{2}$
32. $8\frac{2}{3} - 1\frac{1}{6}$
33. $4\frac{5}{6} - 1\frac{1}{2}$
34. $6\frac{7}{8} - 1\frac{3}{4}$
35. $7\frac{7}{12} - 3\frac{1}{2}$
36. $\frac{1}{2} + \frac{1}{3}$
37. $\frac{1}{2} - \frac{1}{3}$
38. $\frac{1}{3} + \frac{1}{4}$
39. $\frac{1}{3} - \frac{1}{4}$
40. $\frac{1}{2} + \frac{1}{5}$
41. $\frac{1}{2} - \frac{1}{5}$
42. $\frac{1}{4} + \frac{1}{5}$
43. $\frac{1}{4} - \frac{1}{5}$
44. $\frac{2}{3} + \frac{1}{4}$
45. $\frac{2}{3} - \frac{1}{4}$
46. $\frac{3}{4} + \frac{1}{3}$
47. $\frac{3}{4} - \frac{1}{3}$

48. $\frac{1}{4} + \frac{1}{6}$ **49.** $\frac{1}{4} - \frac{1}{6}$ **50.** $\frac{5}{6} + \frac{3}{4}$

51. $\frac{5}{6} - \frac{3}{4}$ **52.** $\frac{3}{4} + \frac{2}{3}$ **53.** $\frac{3}{4} - \frac{2}{3}$

54. $3\frac{1}{3} + 1\frac{1}{4}$ **55.** $5\frac{2}{5} + 2\frac{1}{2}$ **56.** $4\frac{1}{6} + 3\frac{3}{4}$ **57.** $4\frac{3}{4} - 1\frac{1}{2}$

58. $5\frac{5}{6} - 1\frac{1}{4}$ **59.** $4\frac{7}{8} - 1\frac{3}{4}$ **60.** $9\frac{1}{3} + 3\frac{2}{5}$ **61.** $4\frac{3}{5} + 1\frac{1}{4}$

62. $6\frac{1}{2} + 1\frac{1}{3}$ **63.** $4\frac{5}{6} - 1\frac{1}{2}$ **64.** $8\frac{3}{4} - 1\frac{2}{3}$ **65.** $7\frac{5}{8} - 4\frac{1}{3}$

66. $3\frac{3}{5} + 1\frac{3}{10}$ **67.** $5\frac{1}{2} - 1\frac{1}{3}$

68. $7\frac{2}{3} + 1\frac{1}{6}$ **69.** $7\frac{2}{3} - 1\frac{3}{5}$

70. $4\frac{3}{5} + 3\frac{1}{4}$ **71.** $6\frac{1}{4} - 6\frac{1}{6}$

72. $3\frac{1}{8} + 2\frac{3}{4}$ **73.** $9\frac{3}{4} - 7\frac{3}{5}$

Lesson 117 Divide:

1. $3\overline{)3.42}$ **2.** $4\overline{)5.2}$ **3.** $5\overline{)0.85}$

4. $6\overline{)4.2}$ **5.** $7\overline{)0.84}$ **6.** $8\overline{)9.6}$

7. $2\overline{)0.36}$ **8.** $4\overline{)7.2}$ **9.** $5\overline{)7.5}$

10. $6\overline{)1.32}$ **11.** $7\overline{)12.6}$ **12.** $8\overline{)3.44}$

13. $6.4 \div 4$ **14.** $0.64 \div 2$

15. $6.5 \div 5$ **16.** $0.63 \div 3$

17. $3.24 \div 6$ **18.** $12.8 \div 8$

19. $1.44 \div 9$ **20.** $23.8 \div 7$

21. $3\overline{)0.15}$ **22.** $4\overline{)0.28}$ **23.** $5\overline{)1.35}$

24. $6\overline{)0.144}$ **25.** $7\overline{)0.63}$ **26.** $8\overline{)0.144}$

27. $9\overline{)0.45}$ **28.** $3\overline{)0.012}$ **29.** $2\overline{)0.054}$

30. $4\overline{)0.36}$ **31.** $5\overline{)0.30}$ **32.** $6\overline{)0.138}$

33. $0.18 \div 3$ **34.** $1.54 \div 7$

35. $0.36 \div 9$ **36.** $0.144 \div 6$

37. $0.08 \div 2$ **38.** $0.095 \div 5$

39. $0.64 \div 8$ **40.** $0.036 \div 4$

Lesson 118

Divide:

1. $4\overline{)3.4}$ **2.** $5\overline{)0.12}$ **3.** $6\overline{)2.7}$

4. $8\overline{)0.52}$ **5.** $2\overline{)3.1}$ **6.** $4\overline{)0.54}$

7. $5\overline{)0.7}$ **8.** $6\overline{)1.5}$ **9.** $8\overline{)3.6}$

10. $2\overline{)0.5}$ **11.** $4\overline{)1.5}$ **12.** $5\overline{)0.12}$

13. $0.5 \div 4$ **14.** $0.6 \div 5$

15. $1.2 \div 8$ **16.** $3.3 \div 6$

17. $0.9 \div 2$ **18.** $0.9 \div 5$

19. $0.18 \div 4$ **20.** $0.18 \div 8$

Lesson 119

Divide:

1. $0.3\overline{)0.15}$ **2.** $0.4\overline{)2.4}$ **3.** $0.5\overline{)0.15}$

4. $0.2\overline{)0.32}$ **5.** $0.3\overline{)1.23}$ **6.** $0.4\overline{)0.56}$

7. $0.6\overline{)0.72}$ **8.** $0.7\overline{)0.98}$ **9.** $0.8\overline{)1.52}$

10. $0.5\overline{)6.5}$ **11.** $0.4\overline{)0.132}$ **12.** $0.6\overline{)1.26}$

13. $4.6 \div 0.2$

14. $0.64 \div 0.4$

15. $4.5 \div 0.3$

16. $0.45 \div 0.5$

17. $3.21 \div 0.3$

18. $1.23 \div 0.3$

19. $0.95 \div 0.5$

20. $1.74 \div 0.6$

Lesson 120

Multiply:

1. $1\frac{1}{2} \times \frac{2}{3}$

2. $\frac{3}{4} \times 1\frac{1}{4}$

3. $2\frac{1}{2} \times 3$

4. $4 \times 2\frac{1}{2}$

5. $1\frac{1}{3} \times 1\frac{1}{3}$

6. $1\frac{1}{2} \times 1\frac{1}{4}$

7. $\frac{1}{2} \times 1\frac{2}{3}$

8. $2\frac{1}{3} \times \frac{1}{2}$

9. $2 \times 3\frac{1}{2}$

10. $3\frac{1}{3} \times 3$

11. $1\frac{2}{3} \times 2\frac{1}{2}$

12. $3\frac{1}{2} \times 1\frac{3}{4}$

13. $\frac{1}{3} \times 2\frac{2}{3}$

14. $2\frac{3}{4} \times \frac{1}{2}$

15. $4\frac{1}{2} \times 4$

16. $3 \times 1\frac{2}{3}$

17. $2\frac{1}{4} \times 1\frac{1}{2}$

18. $1\frac{3}{4} \times 1\frac{2}{3}$

acute angle An angle whose measure is more than 0° and less than 90°.

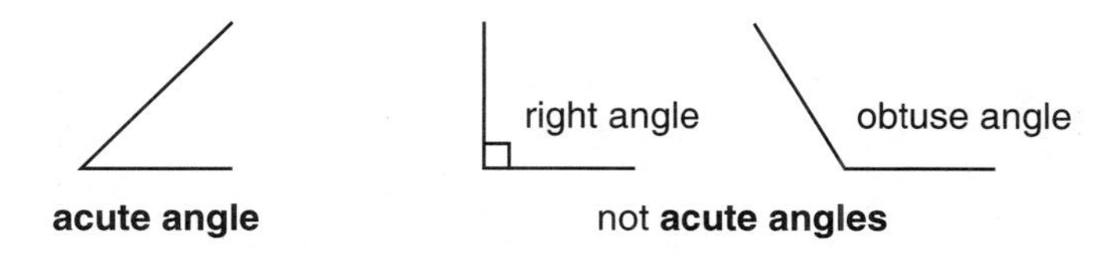

*An **acute angle** is smaller than both a right angle and an obtuse angle.*

acute triangle A triangle whose largest angle measures less than 90°.

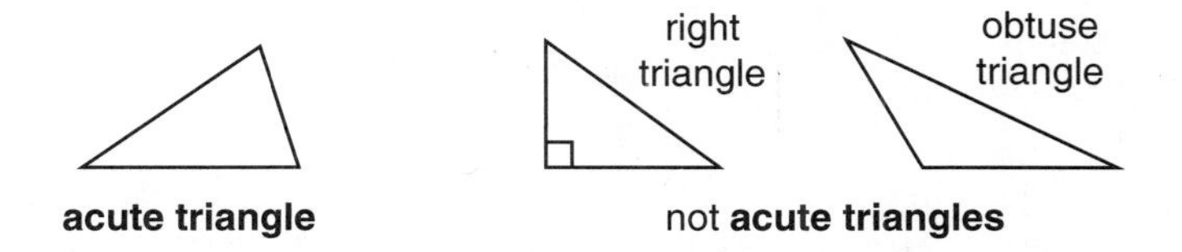

addend Any one of the numbers added in an addition problem.

7 + 3 = 10 *The **addends** in this problem are 7 and 3.*

algorithm Any process for solving a mathematical problem.

*In the addition **algorithm** we add the ones first, then the tens, and then the hundreds.*

a.m. The period of time from midnight to just before noon.

*I get up at 7 **a.m.**, which is 7 o'clock in the morning.*

angle The opening that is formed when two lines, line segments, or rays intersect.

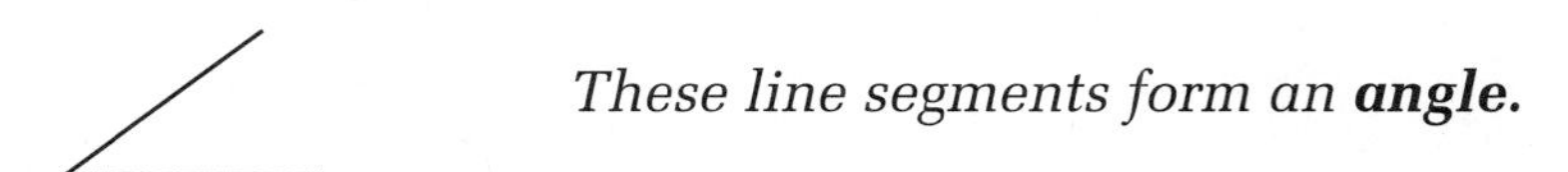

*These line segments form an **angle.***

area The number of square units needed to cover a surface.

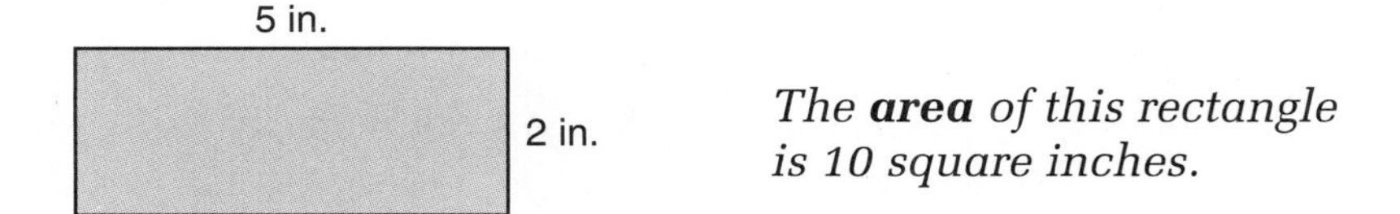

*The **area** of this rectangle is 10 square inches.*

arithmetic sequence A sequence in which each term is found by adding a fixed amount to the previous term.

$$3, \overset{+3}{\frown} 6, \overset{+3}{\frown} 9, \overset{+3}{\frown} 12, \overset{+3}{\frown} 15, \ldots$$

*This **arithmetic sequence** counts up by 3's.*

array A rectangular arrangement of numbers or symbols in columns and rows.

X X X
X X X
X X X
X X X

*This is a 3-by-4 **array** of X's. It has 3 columns and 4 rows.*

associative property of addition The grouping of addends does not affect their sum. In symbolic form, $a + (b + c) = (a + b) + c$. Unlike addition, subtraction is not associative.

$(8 + 4) + 2 = 8 + (4 + 2)$ — *Addition is **associative.***

$(8 - 4) - 2 \neq 8 - (4 - 2)$ — *Subtraction is not **associative.***

associative property of multiplication The grouping of factors does not affect their product. In symbolic form, $a \times (b \times c) = (a \times b) \times c$. Unlike multiplication, division is not associative.

$(8 \times 4) \times 2 = 8 \times (4 \times 2)$ — *Multiplication is **associative.***

$(8 \div 4) \div 2 \neq 8 \div (4 \div 2)$ — *Division is not **associative.***

average The number found when the sum of two or more numbers is divided by the number of addends in the sum; also called *mean.*

*To find the **average** of the numbers 5, 6, and 10, first add.*

$$5 + 6 + 10 = 21$$

Then, since there were three addends, divide the sum by 3.

$$21 \div 3 = 7$$

*The **average** of 5, 6, and 10 is 7.*

bar graph A graph that uses rectangles (bars) to show numbers or measurements.

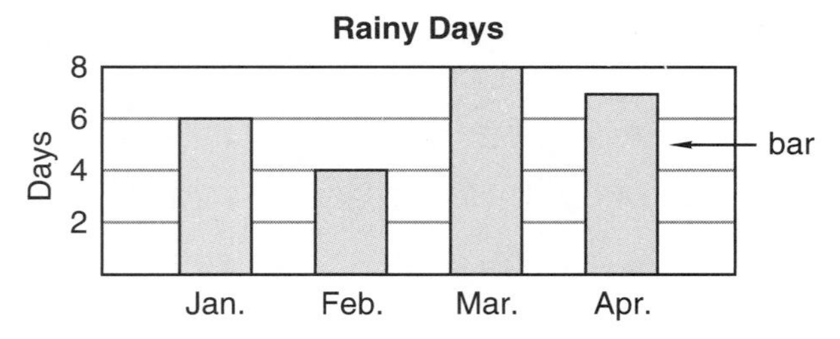

*This **bar graph** shows how many rainy days there were in each of these four months.*

base (1) The lower number in an exponential expression.

base → 5^3 ← *exponent*

5^3 means 5 × 5 × 5 and its value is 125.

(2) A designated side or face of a geometric figure.

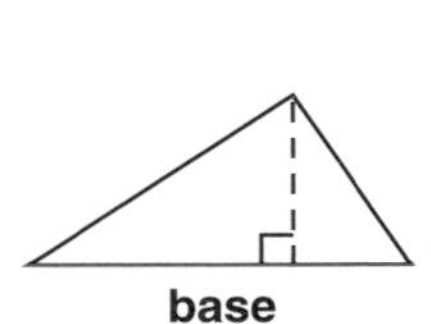

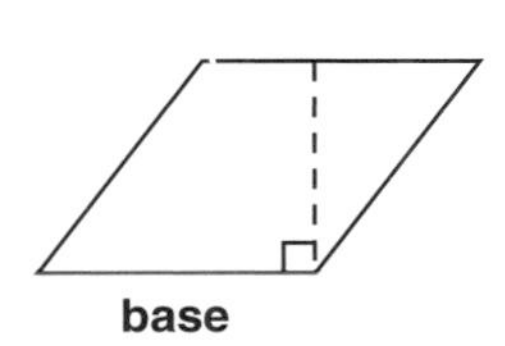

capacity The amount of liquid a container can hold.

*Cups, gallons, and liters are units of **capacity.***

Celsius A scale used on some thermometers to measure temperature.

*On the **Celsius** scale, water freezes at 0°C and boils at 100°C.*

center The point inside a circle from which all points on the circle are equally distant.

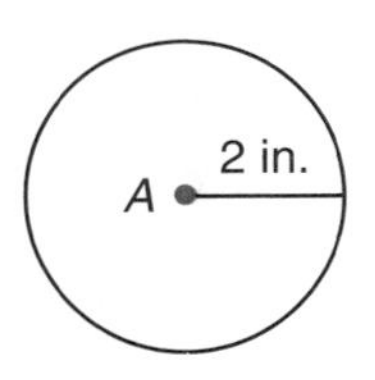

*The **center** of circle A is 2 inches from every point on the circle.*

century A period of one hundred years.

*The years 2001–2100 make up one **century.***

chance A way of expressing the likelihood of an event; the probability of an event expressed as a percentage.

*The **chance** of snow is 10%. It is not likely to snow.*

*There is an 80% **chance** of rain. It is likely to rain.*

circle A closed, curved shape in which all points on the shape are the same distance from its center.

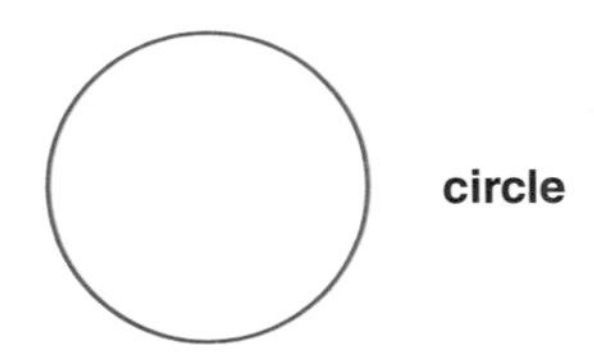

circle graph A graph made of a circle divided into sectors. Also called *pie chart* or *pie graph.*

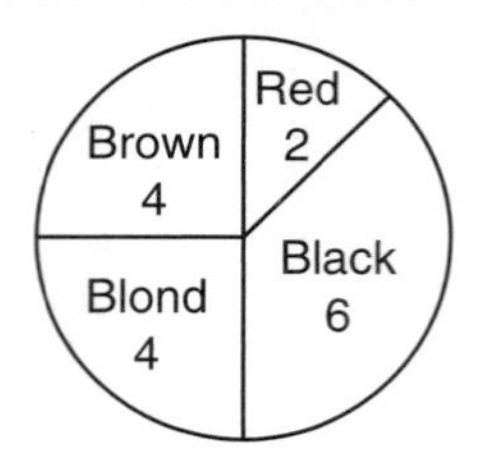

*This **circle graph** displays data on students' hair color.*

circumference The distance around a circle; the perimeter of a circle.

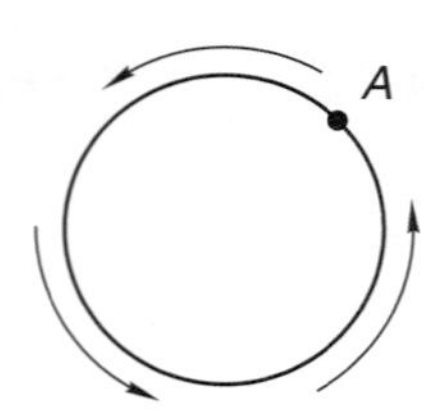

*If the distance from point A around to point A is 3 inches, then the **circumference** of the circle is 3 inches.*

cluster A group of data points that are very close together.

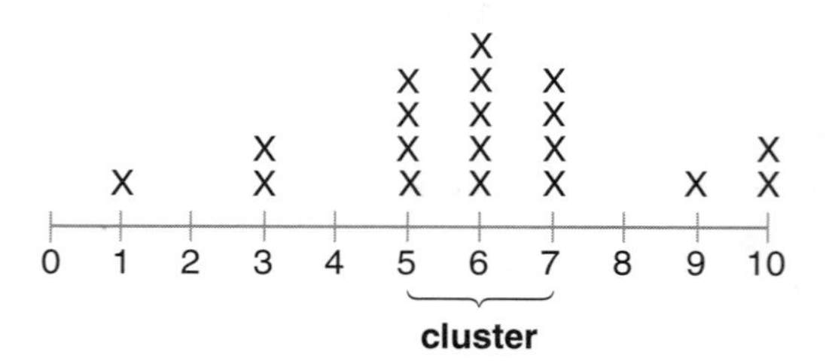

common denominators Denominators that are the same.

*The fractions $\frac{2}{5}$ and $\frac{3}{5}$ have **common denominators.***

common fraction A fraction with whole-number terms.

$\frac{1}{2}$ $\frac{5}{7}$ $\frac{4}{3}$

common fractions

$\frac{1.2}{2.4}$ $\frac{3}{4.5}$ $\frac{\pi}{2}$

not **common fractions**

common year A year with 365 days; not a leap year.

*The year 2000 is a leap year, but 2001 is a **common year.** In a **common year** February has 28 days. In a leap year it has 29 days.*

commutative property of addition Changing the order of addends does not change their sum. In symbolic form, $a + b = b + a$. Unlike addition, subtraction is not commutative.

$8 + 2 = 2 + 8$ — *Addition is **commutative.***

$8 - 2 \neq 2 - 8$ — *Subtraction is not **commutative.***

commutative property of multiplication Changing the order of factors does not change their product. In symbolic form, $a \times b = b \times a$. Unlike multiplication, division is not commutative.

$8 \times 2 = 2 \times 8$ — *Multiplication is **commutative.***

$8 \div 2 \neq 2 \div 8$ — *Division is not **commutative.***

comparative bar graph A method of displaying data, usually used to compare two or more related sets of data.

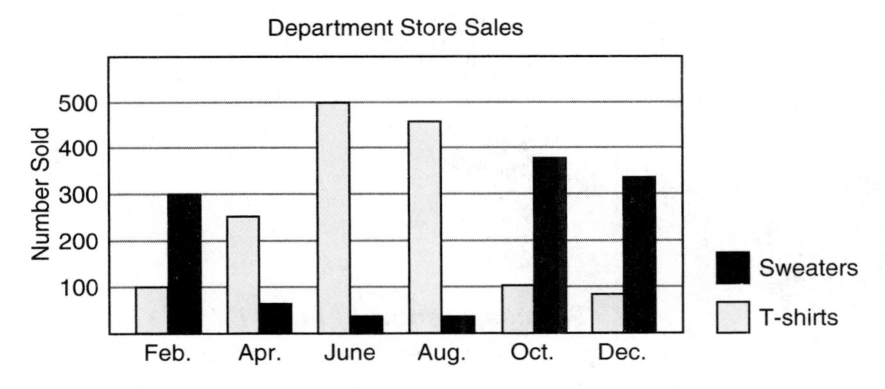

*This **comparative bar graph** compares how many sweaters were sold with how many t-shirts were sold in each of these six months.*

composite number A counting number greater than 1 that is divisible by a number other than itself and 1. Every composite number has three or more factors. Every composite number can be expressed as a product of two or more prime numbers.

*9 is divisible by 1, 3, and 9. It is **composite.***

*11 is divisible by 1 and 11. It is not **composite.***

cone A three-dimensional solid with a circular base and a single vertex.

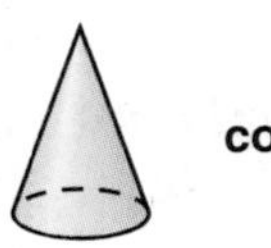

congruent Having the same size and shape.

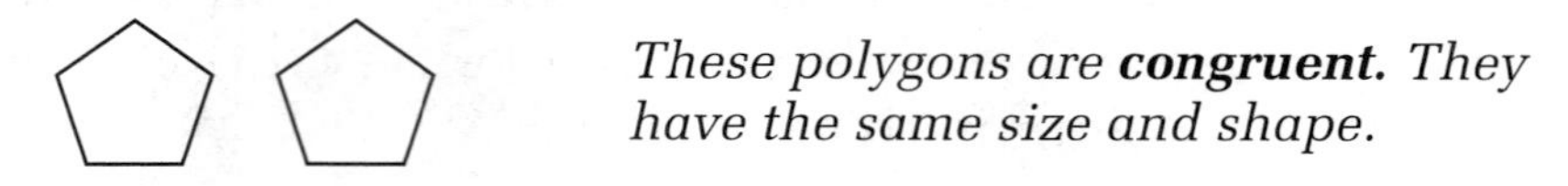

These polygons are ***congruent.*** *They have the same size and shape.*

coordinate(s) (1) A number used to locate a point on a number line.

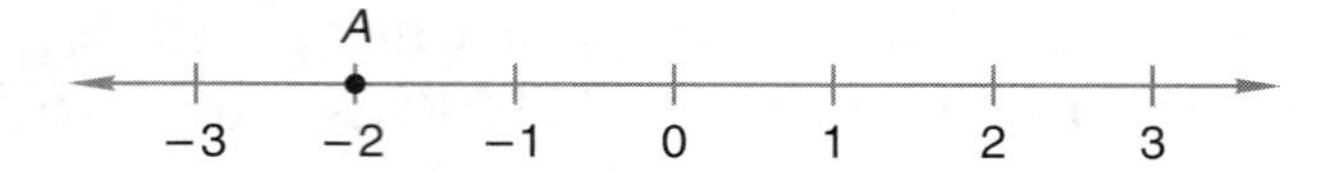

The ***coordinate*** *of point A is −2.*

(2) A pair of numbers used to locate a point on a coordinate plane.

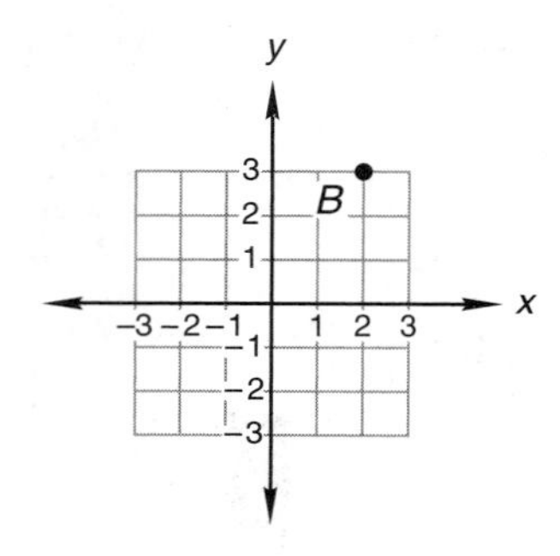

The ***coordinates*** *of point B are (2, 3). The x-coordinate is listed first, the y-coordinate second.*

coordinate plane A grid on which any point can be identified by its distances from the x- and y-axes.

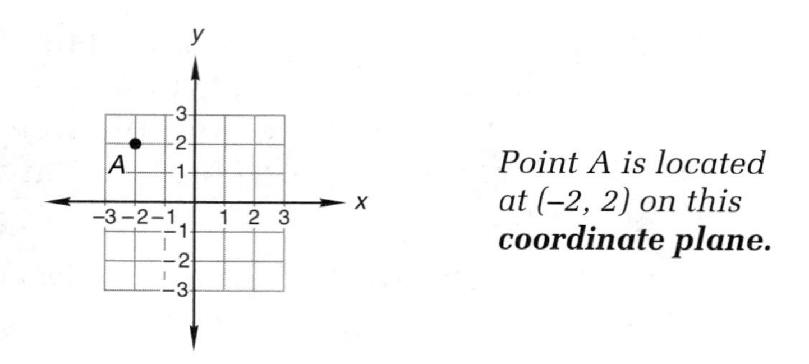

Point A is located at (−2, 2) on this ***coordinate plane.***

counting numbers The numbers used to count; the numbers in this sequence: 1, 2, 3, 4, 5, 6, 7, 8, 9, ...

The numbers 12 and 37 are ***counting numbers,*** *but 0.98 and $\frac{1}{2}$ are not.*

cube A three-dimensional solid with six square faces. Adjacent faces are perpendicular and opposite faces are parallel.

cubic unit A cube with edges of designated length. Cubic units are used to measure volume.

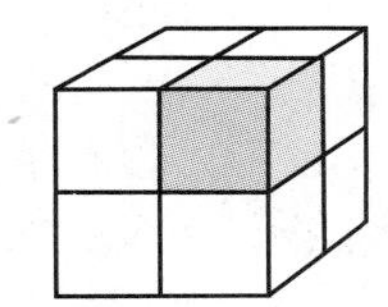

The shaded part is 1 ***cubic unit.*** *The volume of the large cube is 8* ***cubic units.***

cylinder A three-dimensional solid with two circular bases that are opposite and parallel to each other.

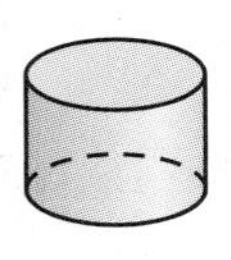

data (Singular: datum) Information gathered from observations or calculations.

82, 76, 95, 62, 98, 97, 93

These ***data*** *are Skylar's first 7 test scores.*

decade A period of ten years.

The years 2001–2010 make up one ***decade.***

decimal number A numeral that contains a decimal point.

23.94 is a ***decimal number*** *because it contains a decimal point.*

decimal places Places to the right of the decimal point.

5.47 has two ***decimal places.***

6.3 has one ***decimal place.***

8 has no ***decimal places.***

decimal point A symbol used to separate the ones place from the tenths place in decimal numbers.

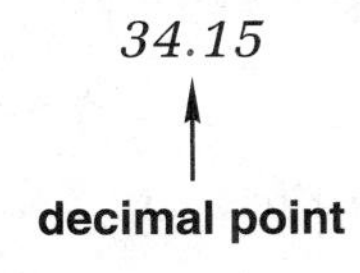

degree (°) (1) A unit for measuring angles.

There are 90 ***degrees*** *(90°) in a right angle.*

There are 360 ***degrees*** *(360°) in a circle.*

(2) A unit for measuring temperature.

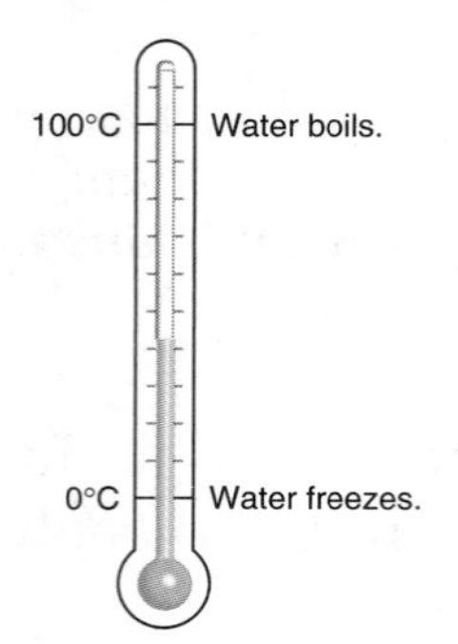

There are 100 ***degrees*** *between the freezing and boiling points of water on the Celsius scale.*

denominator The bottom number of a fraction; the number that tells how many parts are in a whole.

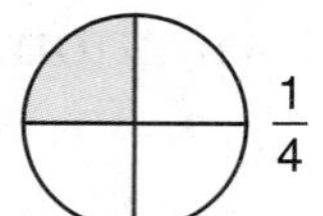

The ***denominator*** *of the fraction is 4. There are 4 parts in the whole circle.*

diameter The distance across a circle through its center.

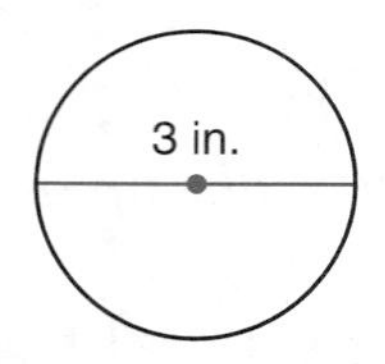

The ***diameter*** *of this circle is 3 inches.*

difference The result of subtraction.

$12 - 8 = 4$ *The* ***difference*** *in this problem is 4.*

digit Any of the symbols used to write numbers: 0, 1, 2, 3, 4, 5, 6, 7, 8, 9.

The last ***digit*** *in the number 7862 is 2.*

distributive property A number times the sum of two addends is equal to the sum of that same number times each individual addend: $a \times (b + c) = (a \times b) + (a \times c)$.

$$8 \times (2 + 3) = (8 \times 2) + (8 \times 3)$$

Multiplication is ***distributive*** *over addition.*

dividend A number that is divided.

$12 \div 3 = 4$ $\quad 3\overline{)12}$ with quotient 4 $\quad \frac{12}{3} = 4$

*The **dividend** is 12 in each of these problems.*

divisible Able to be divided by a whole number without a remainder.

$4\overline{)20}$ with quotient 5 — *The number 20 is **divisible** by 4, since $20 \div 4$ has no remainder.*

$3\overline{)20}$ with quotient 6 R 2 — *The number 20 is not **divisible** by 3, since $20 \div 3$ has a remainder.*

division An operation that separates a number into a given number of equal parts or into a number of parts of a given size.

$21 \div 3 = 7$ *We use **division** to separate 21 into 3 groups of 7.*

divisor A number by which another number is divided.

$12 \div 3 = 4$ $\quad 3\overline{)12}$ with quotient 4 $\quad \frac{12}{3} = 4$

*The **divisor** is 3 in each of these problems.*

double-line graph A method of displaying a set of data, often used to compare two performances over time.

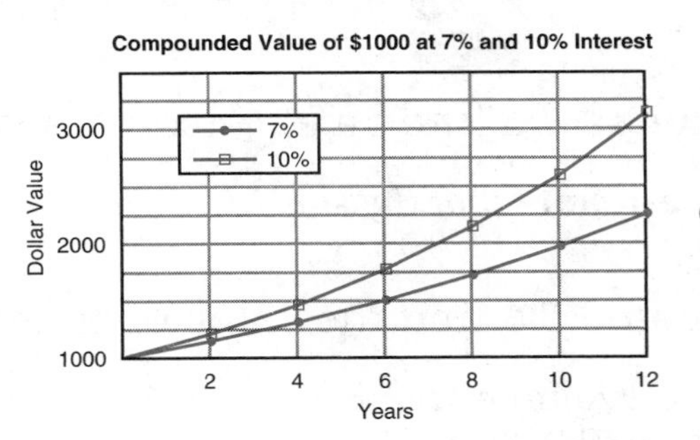

edge A line segment formed where two faces of a solid intersect.

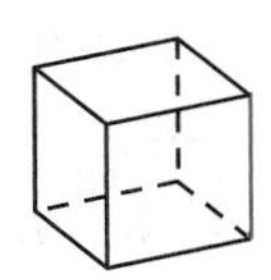

*One **edge** of this cube is colored blue. A cube has 12 **edges.***

elapsed time The difference between a starting time and an ending time.

*The race started at 6:30 p.m. and finished at 9:12 p.m. The **elapsed time** of the race was 2 hours 42 minutes.*

endpoint A point at which a line segment ends.

A ———— B

*Points A and B are the **endpoints** of line segment AB.*

equation A number sentence that uses the symbol "=" to show that two quantities are equal.

$x = 3$	$3 + 7 = 10$	$4 + 1$	$x < 7$
equations		not **equations**	

equilateral triangle A triangle in which all sides are the same length.

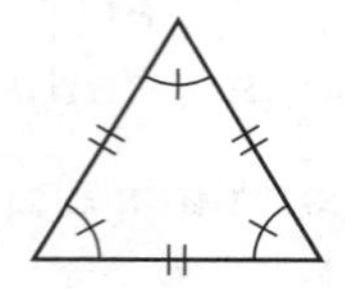

*This is an **equilateral triangle.***
All of its sides are the same length.

equivalent fractions Different fractions that name the same amount.

$\frac{1}{2}$ = $\frac{2}{4}$

$\frac{1}{2}$ *and* $\frac{2}{4}$ *are **equivalent fractions.***

estimate To find an approximate value.

*I **estimate** that the sum of 199 and 205 is about 400.*

evaluate To find the value of an expression.

*To **evaluate** $a + b$ for $a = 7$ and $b = 13$, we replace a with 7 and b with 13:*

$$7 + 13 = 20$$

even numbers Numbers that can be divided by 2 without a remainder; the numbers in this sequence: 0, 2, 4, 6, 8, 10, ...

***Even numbers** have 0, 2, 4, 6, or 8 in the ones place.*

event An outcome or group of outcomes in an experiment involving probability.

*The **event** of rolling a 4 with one roll of a standard number cube has a probability of $\frac{1}{6}$.*

exact number A number that has not been rounded.

*Corissa estimated that about 50 tickets were sold, but the **exact number** of tickets sold was 49.*

expanded form A way of writing a number that shows the value of each digit.

*The **expanded form** of 234 is 200 + 30 + 4.*

expanded notation A way of writing a number as the sum of the products of the digits and the place values of the digits.

*In **expanded notation** 6753 is written*

$$(6 \times 1000) + (7 \times 100) + (5 \times 10) + (3 \times 1).$$

experiment A test to find or illustrate a rule.

*Flipping a coin and selecting an object from a collection of objects are two **experiments** that involve probability.*

exponent The upper number in an exponential expression; it shows how many times the base is to be used as a factor.

base → 5^3 ← ***exponent***

5^3 means $5 \times 5 \times 5$ and its value is 125.

exponential expression An expression that indicates that the base is to be used as a factor the number of times shown by the exponent.

$$4^3 = 4 \times 4 \times 4 = 64$$

*The **exponential expression** 4^3 uses 4 as a factor 3 times. Its value is 64.*

face A flat surface of a geometric solid.

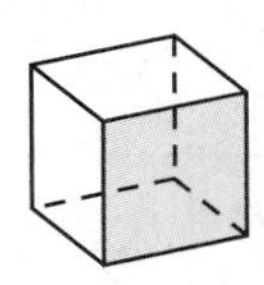

*One **face** of the cube is shaded.*
*A cube has six **faces**.*

fact family A group of three numbers related by addition and subtraction or by multiplication and division.

*The numbers 3, 4, and 7 are a **fact family**. They make these four facts:*

$3 + 4 = 7$ $4 + 3 = 7$ $7 - 3 = 4$ $7 - 4 = 3$

factor (1) Noun: Any one of the numbers multiplied in a multiplication problem.

2 × 3 = 6 *The **factors** in this problem are 2 and 3.*

(2) Noun: A whole number that divides another whole number without a remainder.

*The numbers 2 and 3 are **factors** of 6.*

(3) Verb: To write as a product of factors.

*We can **factor** the number 6 by writing it as 2 × 3.*

Fahrenheit A scale used on some thermometers to measure temperature.

*On the **Fahrenheit** scale, water freezes at 32°F and boils at 212°F.*

fraction A number that names part of a whole.

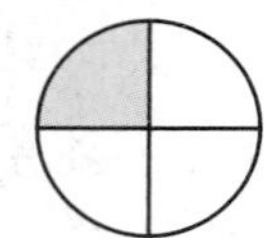

$\frac{1}{4}$ of the circle is shaded.

*$\frac{1}{4}$ is a **fraction**.*

frequency The number of times an event or outcome occurs.

Quiz Results

Number Correct	Tally	Frequency
0		0
1	|	1
2	||||	4
3	~~||||~~ ||	7
4	~~||||~~ ~~||||~~	10
5	|||	3

*This table shows the **frequency** of recent quiz scores.*

frequency table A table that is used to tally and display the number of times an event or outcome occurs.

Quiz Results

Number Correct	Tally	Frequency
0		0
1	|	1
2	||||	4
3	~~||||~~ ||	7
4	~~||||~~ ~~||||~~	10
5	|||	3

*This **frequency table** summarizes the class's performance on the most recent quiz.*

function A rule for changing an "in-number" to an "out-number."

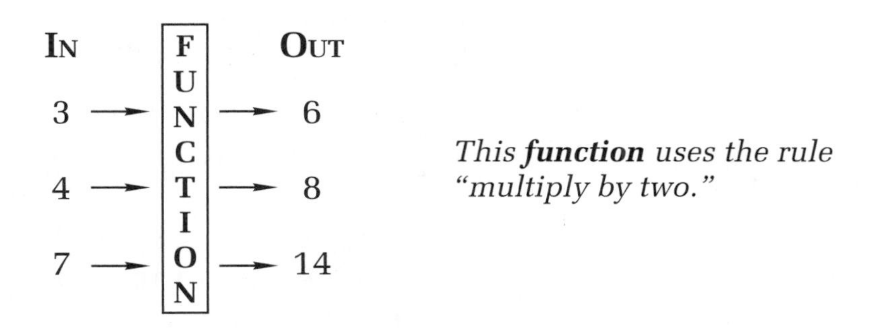

*This **function** uses the rule "multiply by two."*

geometric sequence A sequence in which each term is found by multiplying the previous term by a fixed amount.

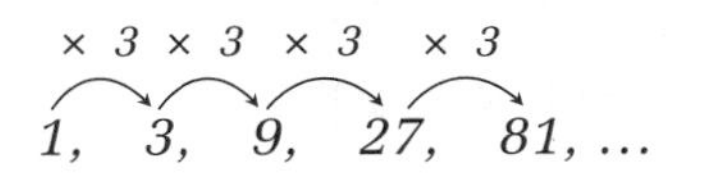

*We multiply a term by 3 to find the term that follows it in this **geometric sequence.***

geometric solid A shape that takes up space.

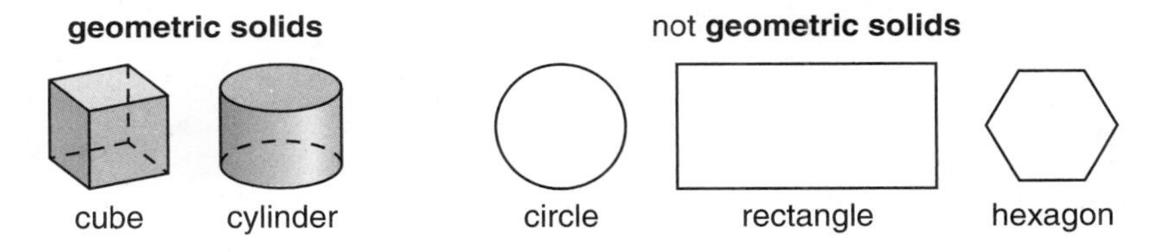

geometry A major branch of mathematics that deals with shapes, sizes, and other properties of figures.

*Some of the figures we study in **geometry** are angles, circles, and polygons.*

graph (1) Noun: A diagram that shows data in an organized way. *See also* **bar graph, circle graph, line graph,** *and* **pictograph.**

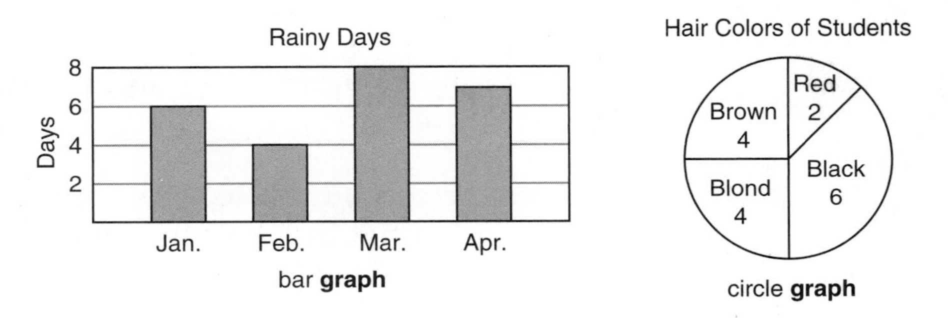

(2) Verb: To draw a point, line, or curve on a coordinate plane.

greatest common factor (GCF) The largest whole number that is a factor of two or more given numbers.

The factors of 20 are 1, 2, 4, 5, 10, and 20.
The factors of 30 are 1, 2, 3, 5, 6, 10, 15, and 30.
The common factors of 20 and 30 are 1, 2, 5, and 10.
*The **greatest common factor** of 20 and 30 is 10.*

histogram A method of displaying a range of data. A histogram is a special type of bar graph that displays data in intervals of equal size with no space between bars.

horizontal Side to side; perpendicular to vertical.

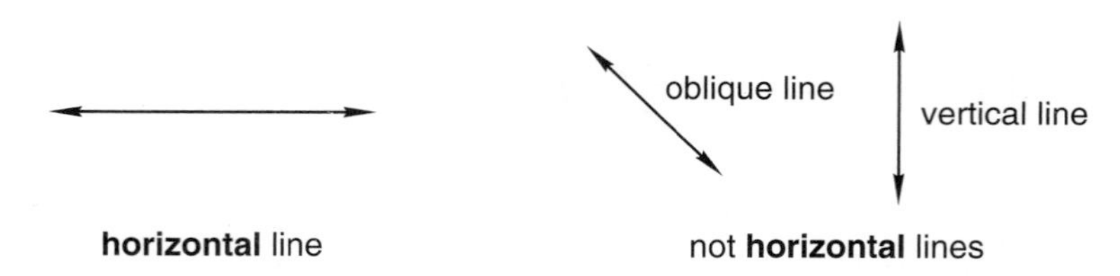

icon A symbol used in a pictograph to represent data.

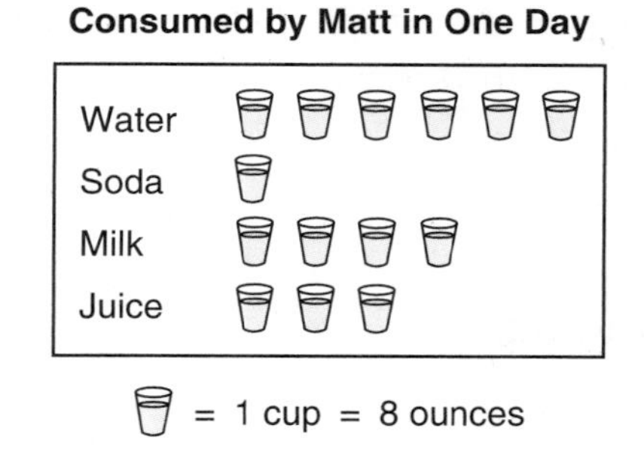

*Each **icon** in the pictograph represents 1 cup of liquid that Matt consumed.*

identity property of addition The sum of any number and 0 is equal to the initial number. In symbolic form, $a + 0 = a$. The number 0 is referred to as the *additive identity.*

*The **identity property of addition** is shown by this statement:*

$$13 + 0 = 13$$

identity property of multiplication The product of any number and 1 is equal to the initial number. In symbolic form, $a \times 1 = a$. The number 1 is referred to as the *multiplicative identity.*

*The **identity property of multiplication** is shown by this statement:*

$$94 \times 1 = 94$$

improper fraction A fraction with a numerator greater than or equal to the denominator.

$\frac{4}{3}$ $\frac{2}{2}$ *These fractions are **improper fractions.***

integers The set of counting numbers, their opposites, and zero; the members of the set {..., –2, –1, 0, 1, 2, ...}.

*–57 and 4 are **integers.** $\frac{15}{8}$ and –0.98 are not **integers.***

International System of Units *See* **metric system.**

intersect To share a point or points.

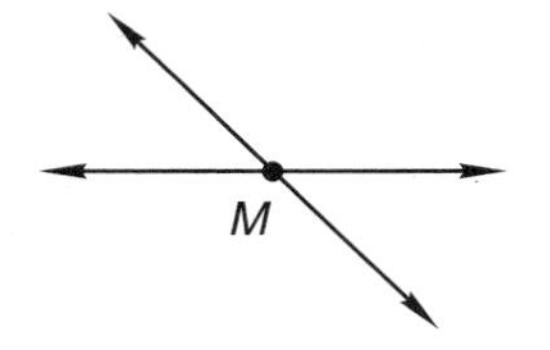

*These two lines **intersect.***
They share the common point M.

intersecting lines Lines that cross.

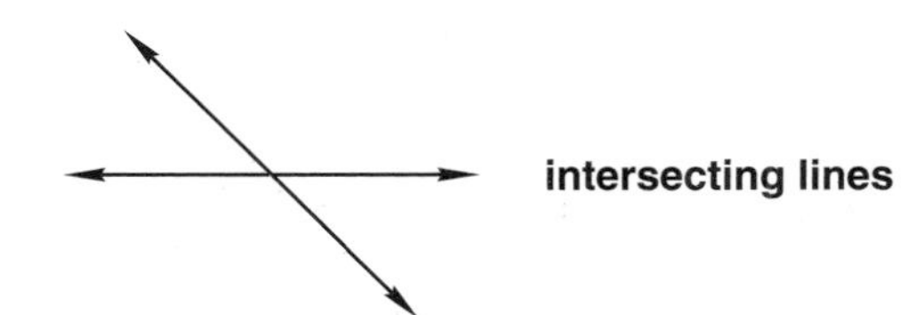

inverse operations Operations that "undo" one another.

$a + b - b = a$	*Addition and subtraction are*
$a - b + b = a$	***inverse operations.***
$a \times b \div b = a \quad (b \neq 0)$	*Multiplication and division are*
$a \div b \times b = a \quad (b \neq 0)$	***inverse operations.***
$\sqrt{a^2} = a \quad (a \geq 0)$	*Squaring and finding square*
$(\sqrt{a})^2 = a \quad (a \geq 0)$	*roots are **inverse operations.***

invert To switch the numerator and denominator of a fraction.

*If we **invert** the fraction $\frac{3}{4}$, we get $\frac{4}{3}$.*

isosceles triangle A triangle with at least two sides of equal length.

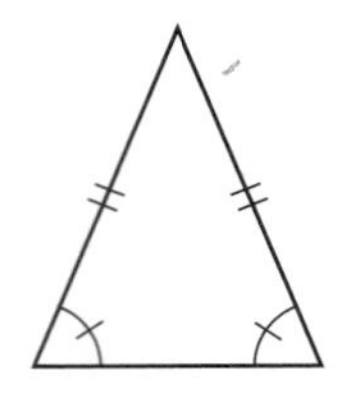

*Two of the sides of this **isosceles triangle** have equal lengths.*

leap year A year with 366 days; not a common year.

*In a **leap year** February has 29 days.*

least common multiple (LCM) The smallest whole number that is a multiple of two or more given numbers.

The multiples of 4 are 4, 8, 12, 16, 20, ...

The multiples of 6 are 6, 12, 18, 24, 30, ...

*The **least common multiple** of 4 and 6 is 12.*

legend A notation on a map, graph, or diagram that describes the meaning of the symbols and/or the scale used.

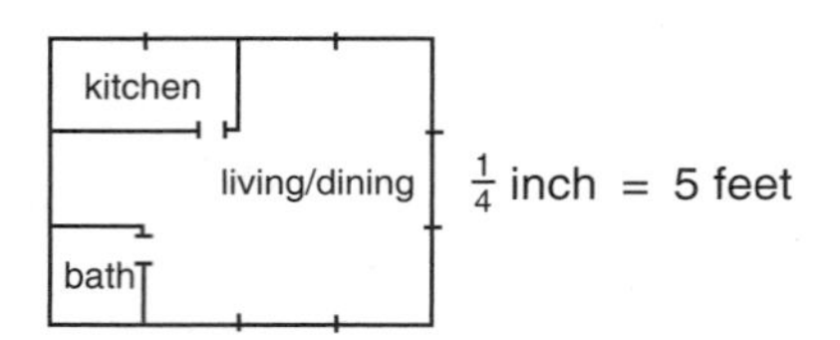

*The **legend** of this scale drawing shows that $\frac{1}{4}$ inch represents 5 feet.*

length A measure of the distance between any two points.

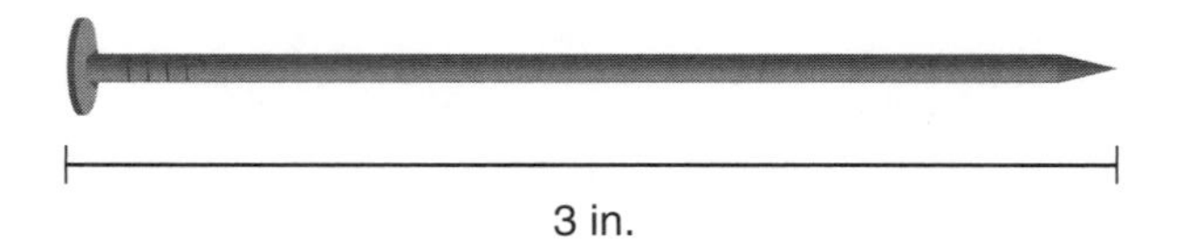

*The **length** of this nail is 3 inches.*

line A straight collection of points extending in opposite directions without end.

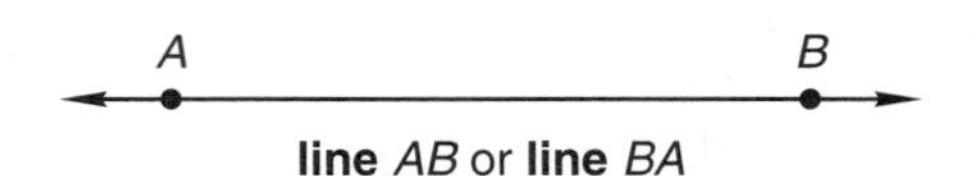

line graph A graph that connects points to show how information changes over time.

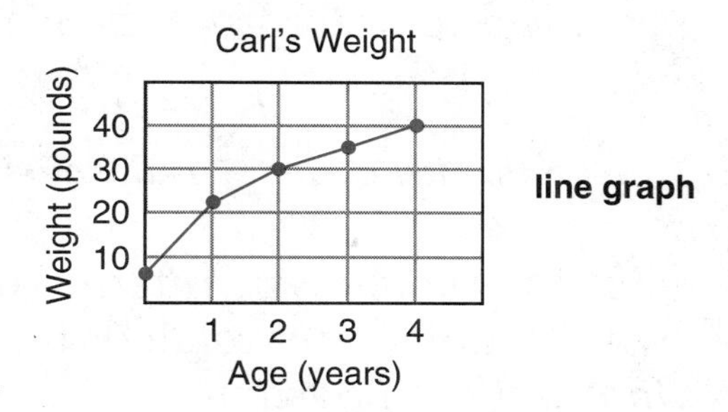

line of symmetry A line that divides a figure into two halves that are mirror images of each other.

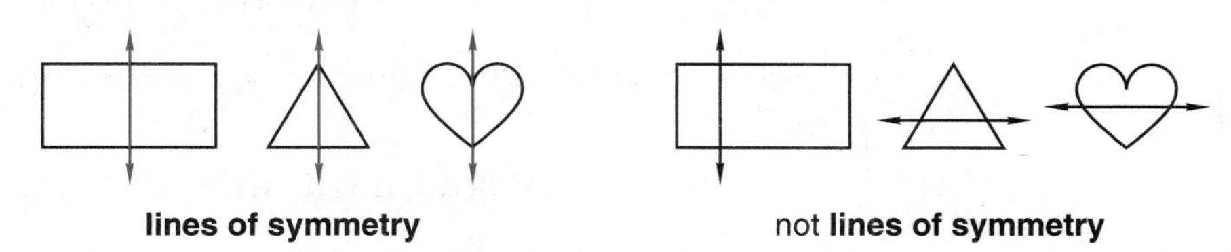

line plot A method of plotting a set of numbers by placing a mark above a number on a number line each time it occurs in the set.

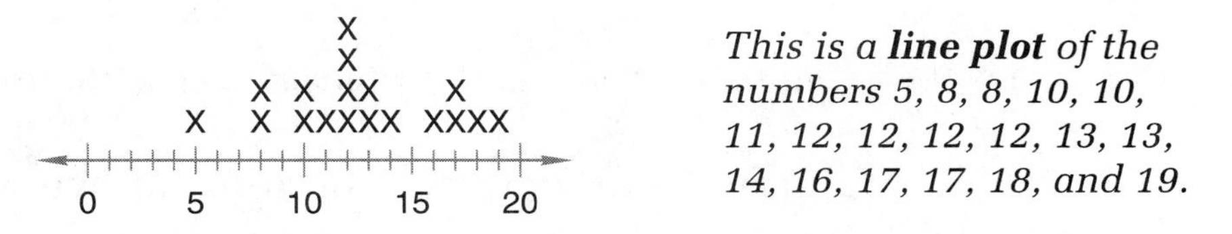

*This is a **line plot** of the numbers 5, 8, 8, 10, 10, 11, 12, 12, 12, 12, 13, 13, 14, 16, 17, 17, 18, and 19.*

line segment A part of a line with two distinct endpoints.

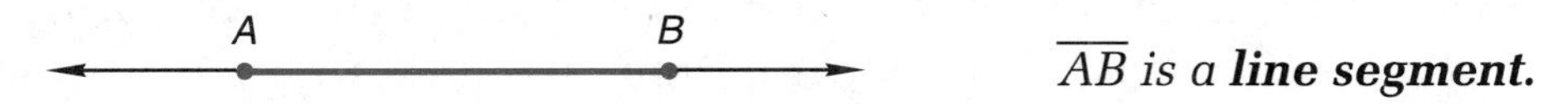

$\overline{AB}$ *is a **line segment.***

lowest terms A fraction is in *lowest terms* if it cannot be reduced.

*In **lowest terms,** the fraction* $\frac{8}{20}$ *is* $\frac{2}{5}$.

mean *See* **average.**

measure of central tendency A value that describes a property of a list of data, such as the middle number of the list or the number that appears in the list most often. *See also* **mean, median,** *and* **mode.**

1, 3, 5, 6, 8, 9, 13

*The median of this set is 6. The median of a set is one **measure of central tendency.***

median The middle number (or the average of the two central numbers) of a list of data when the numbers are arranged in order from the least to the greatest.

1, 1, 2, 4, 5, 7, 9, 15, 24, 36, 44

In this list of data 7 is the ***median.***

metric system An international system of measurement in which units are related by a power of ten. Also called *International System.*

Centimeters and kilograms are units in the ***metric system.***

millennium A period of one thousand years.

The years 2001–3000 make up one ***millennium.***

mixed number A whole number and a fraction together.

The ***mixed number*** $2\frac{1}{3}$ *means "two and one third."*

mode The number or numbers that appear most often in a list of data.

5, 12, 32, 5, 16, 5, 7, 12

In this list of data the number 5 is the ***mode.***

multiple A product of a counting number and another number.

The ***multiples*** *of 3 include 3, 6, 9, and 12.*

mutually exclusive Categories are *mutually exclusive* if each data point can be placed in one and only one of the categories.

When flipping one coin, the categories are "landing heads-up" and "landing tails-up." One coin cannot land both heads-up and tails-up on the same toss. Thus, the categories "landing heads-up" and "landing tails-up" are ***mutually exclusive.***

negative numbers Numbers less than zero.

–15 and –2.86 are ***negative numbers.***

19 and 0.74 are not ***negative numbers.***

number line A line for representing and graphing numbers. Each point on the line corresponds to a number.

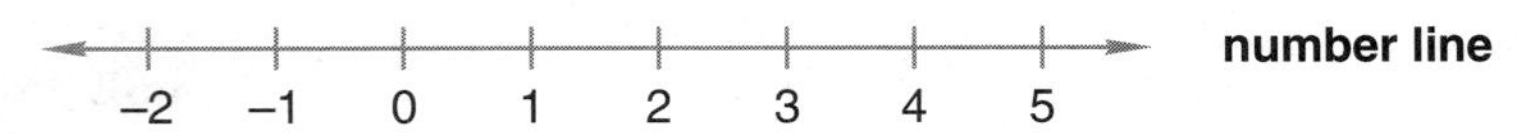

numeral A symbol or group of symbols that represents a number.

*4, 72, and $\frac{1}{2}$ are examples of **numerals.***
*"Four," "seventy-two," and "one half" are words that name numbers but are not **numerals.***

numerator The top number of a fraction; the number that tells how many parts of a whole are counted.

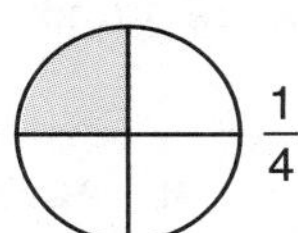

*The **numerator** of the fraction is 1. One part of the whole circle is shaded.*

oblique (1) Slanted or sloping; not horizontal or vertical.

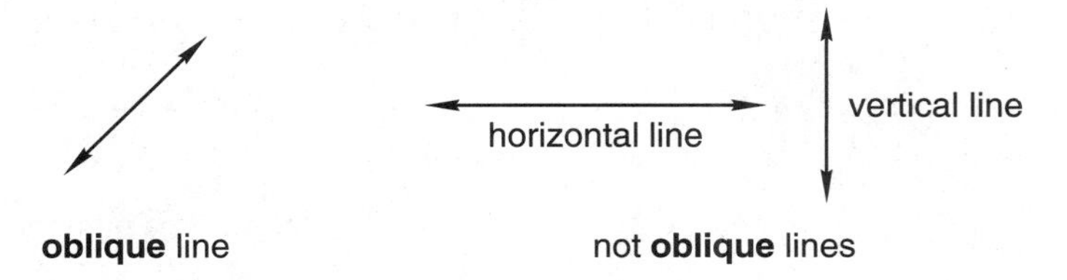

(2) Lines in the same plane that are neither parallel nor perpendicular.

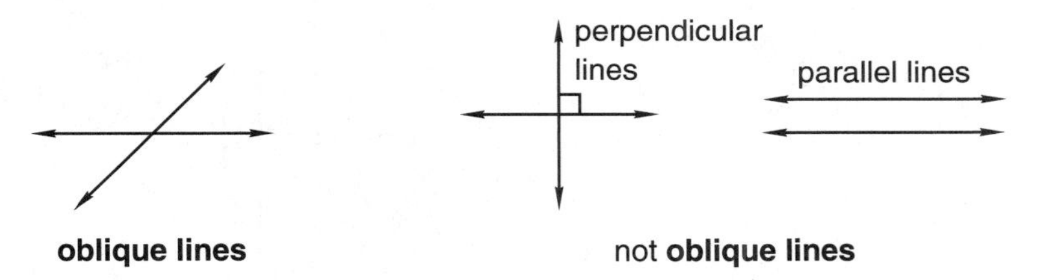

obtuse angle An angle whose measure is more than 90° and less than 180°.

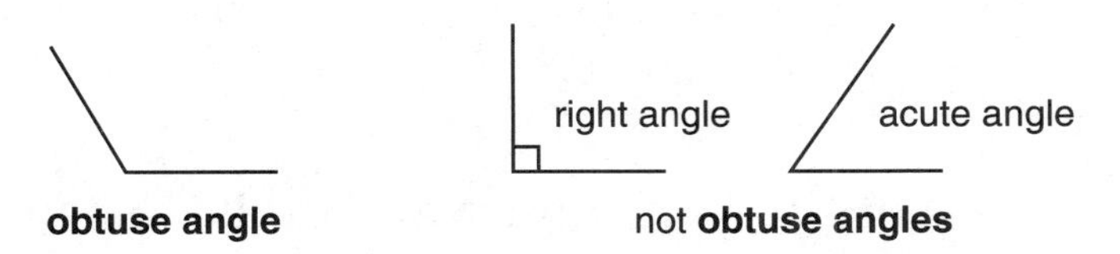

*An **obtuse angle** is larger than both a right angle and an acute angle.*

obtuse triangle A triangle whose largest angle measures more than 90° and less than 180°.

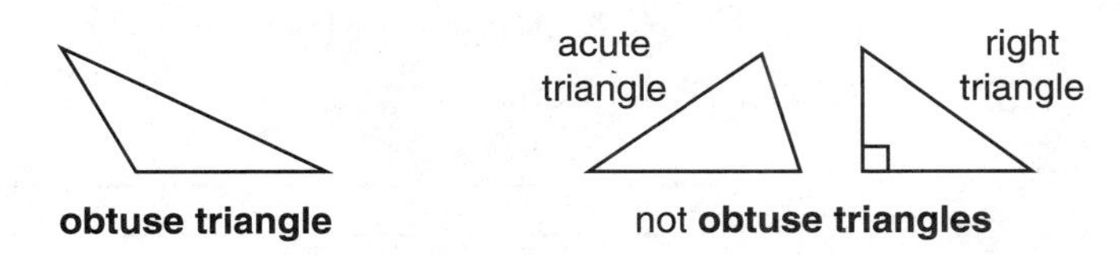

odd numbers Numbers that have a remainder of 1 when divided by 2; the numbers in this sequence: 1, 3, 5, 7, 9, 11, ...

__Odd numbers__ have 1, 3, 5, 7, or 9 in the ones place.

operations of arithmetic The four basic mathematical operations: addition, subtraction, multiplication, and division.

1 + 9 *21 − 8* *6 × 22* *3 ÷ 1*

the **operations of arithmetic**

ordinal numbers Numbers that describe position or order.

"First," "second," and "third" are __ordinal numbers.__

origin (1) The location of the number 0 on a number line.

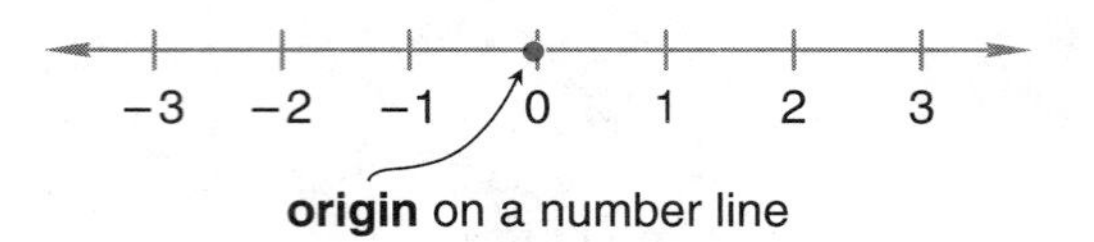

(2) The point (0,0) on a coordinate plane.

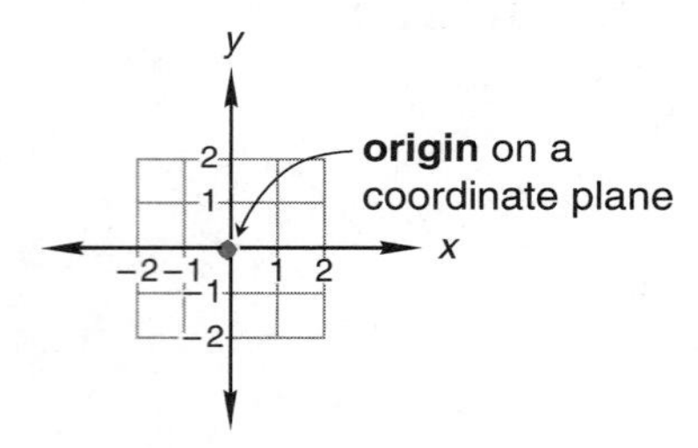

outcome Any possible result of an experiment.

When rolling a number cube, the possible __outcomes__ are 1, 2, 3, 4, 5, and 6.

outlier A number that is distant from most of the other numbers in a list of data.

In the data at right the number 28 is an __outlier,__ because it is distant from the other numbers in the list.

1, 5, 4, 3, 6, 28, 7, 2

parallel lines Lines that stay the same distance apart; lines that do not cross.

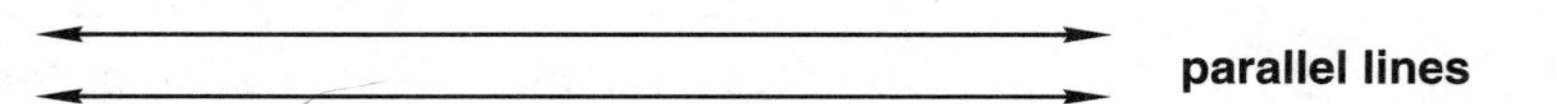

parallelogram A quadrilateral that has two pairs of parallel sides.

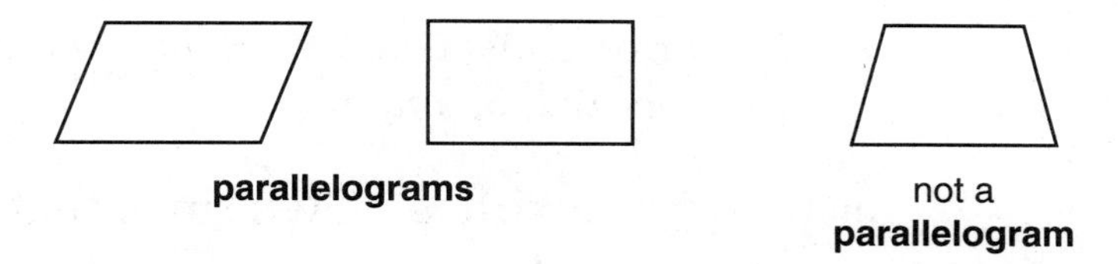

parentheses A pair of symbols used to set apart parts of an expression so that those parts may be evaluated first: ( ).

$$15 - (12 - 4)$$

In the expression 15 – (12 – 4), the ***parentheses*** *indicate that 12 – 4 should be calculated before subtracting the result from 15.*

partial product When multiplying using pencil and paper, a product resulting from multiplying one factor by one digit of the other factor. The final product is the sum of the shifted partial products.

```
   53
 × 26
  318  ← partial products
 106   ←
 1378
```

percent A fraction whose denominator of 100 is expressed as a percent sign (%).

$\frac{99}{100}$ = 99% = 99 ***percent***

perfect square The product when a whole number is multiplied by itself.

The number 9 is a ***perfect square*** *because 3 × 3 = 9.*

perimeter The distance around a closed, flat shape.

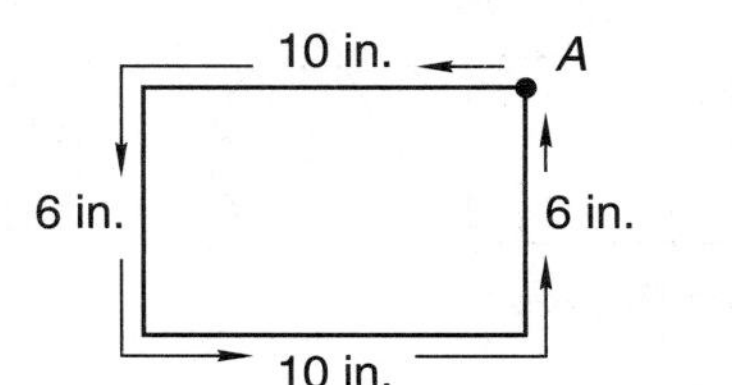

The ***perimeter*** *of this rectangle (from point A around to point A) is 32 inches.*

period The number of terms in a repeating unit of a sequence.

4, 1, 3, 5, 4, 1, 3, 5, 4, ...

The repeating unit of this sequence is "4, 1, 3, 5." There are four terms in the repeating unit. The ***period*** *of this sequence is four.*

permutation One possible arrangement of a set of objects.

2 4 3 1

The arrangement above is one possible ***permutation*** *of the numbers 1, 2, 3, and 4.*

perpendicular lines Two lines that intersect at right angles.

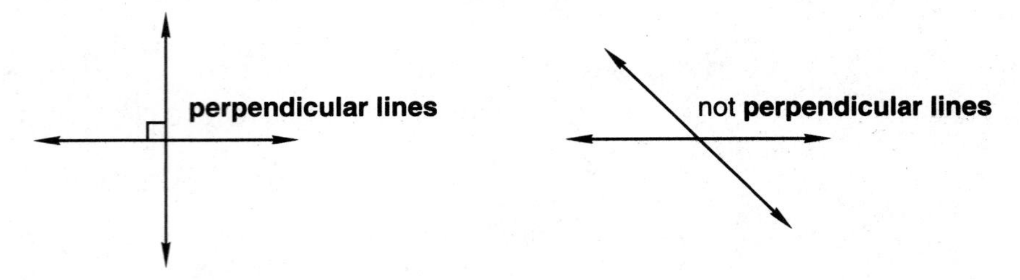

pictograph A graph that uses symbols to represent data.

Stars We Saw

Name	Stars
Tom	☆☆☆☆☆
Bob	☆☆
Sue	☆☆☆☆
Ming	☆☆☆☆☆
Juan	☆☆☆☆☆☆

This is a ***pictograph.*** *It shows how many stars each person saw.*

pie graph *See* **circle graph.**

place value The value of a digit based on its position within a number.

$$\begin{array}{r} 341 \\ 23 \\ +\ \ 7 \\ \hline 371 \end{array}$$

Place value *tells us that 4 in 341 is worth "4 tens." In addition problems, we align digits with the same* ***place value.***

plane A flat surface that has no boundaries.

The flat surface of a desk is part of a ***plane.***

plane figure A flat shape.

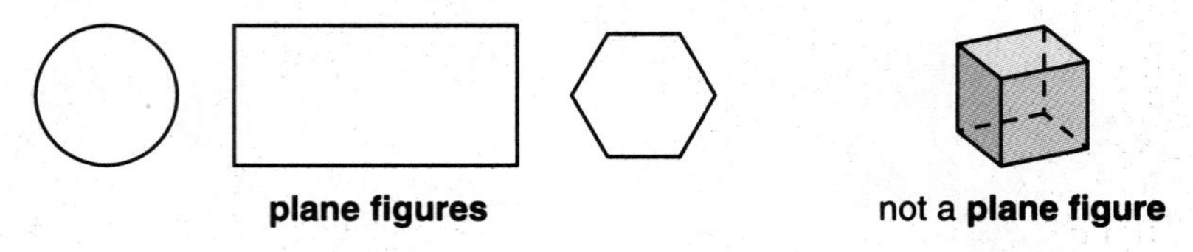

p.m. The period of time from noon to just before midnight.

I go to bed at 9 ***p.m.,*** *which is 9 o'clock at night.*

point An exact position.

•A *This dot represents* ***point*** *A.*

polygon A closed, flat shape with straight sides.

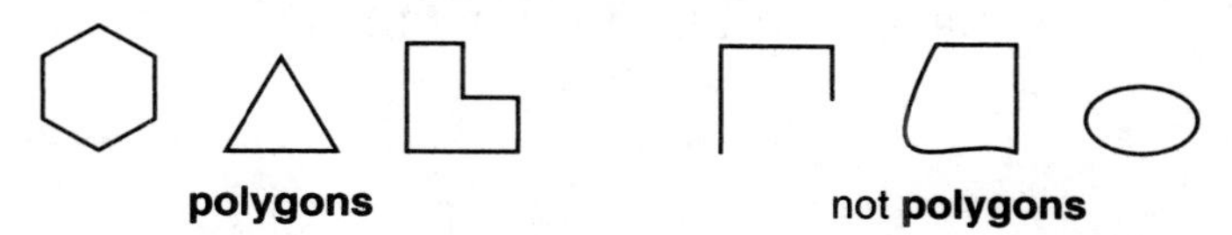

positive numbers Numbers greater than zero.

*0.25 and 157 are **positive numbers.***
*−40 and 0 are not **positive numbers.***

power (1) The value of an exponential expression.

*16 is the fourth **power** of 2 because $2^4 = 16$.*

(2) An exponent.

*The expression 2^4 is read "two to the fourth **power.**"*

prime number A counting number greater than 1 whose only two factors are the number 1 and itself.

*7 is a **prime number.** Its only factors are 1 and 7.*
*10 is not a **prime number.** Its factors are 1, 2, 5, and 10.*

probability A way of describing the likelihood of an event; the ratio of favorable outcomes to all possible outcomes.

*The **probability** of rolling a 3 with a standard number cube is $\frac{1}{6}$.*

product The result of multiplication.

$5 \times 4 = 20$ *The **product** of 5 and 4 is 20.*

proper fraction A fraction whose denominator is greater than its numerator.

*$\frac{3}{4}$ is a **proper fraction.***
$\frac{4}{3}$ is an improper fraction.

property of zero for multiplication Zero times any number is zero. In symbolic form, $0 \times a = 0$.

*The **property of zero for multiplication** tells us that $89 \times 0 = 0$.*

protractor A tool used to measure and draw angles.

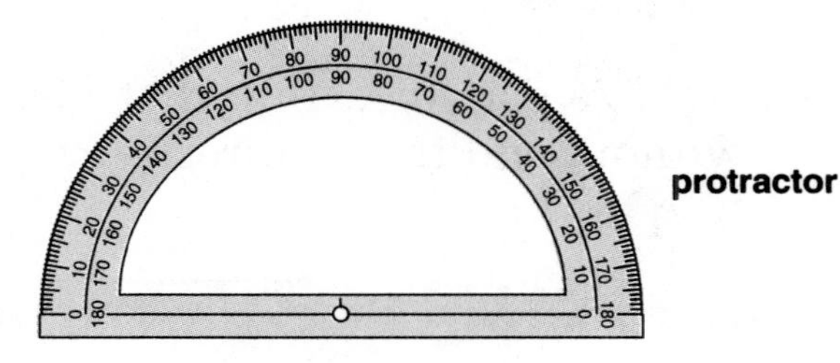

pyramid A three-dimensional solid with a polygon as its base and triangular faces that meet at a vertex.

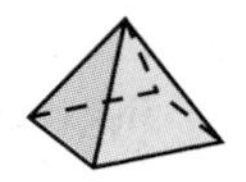

quadrilateral Any four-sided polygon.

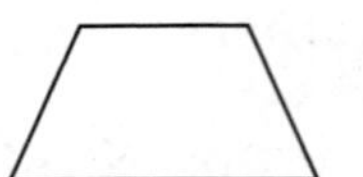 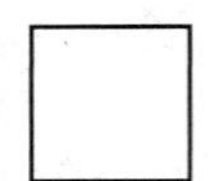 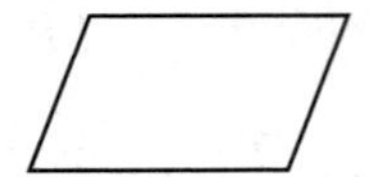  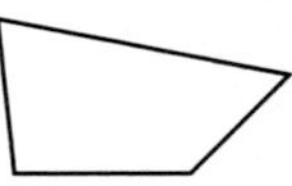

Each of these polygons has 4 sides. They are all ***quadrilaterals.***

quotient The result of division.

$12 \div 3 = 4$ $\quad 3\overline{)12}$ with quotient 4 $\quad \frac{12}{3} = 4$

The ***quotient*** *is 4 in each of these problems.*

radius (Plural: *radii*) The distance from the center of a circle to a point on the circle.

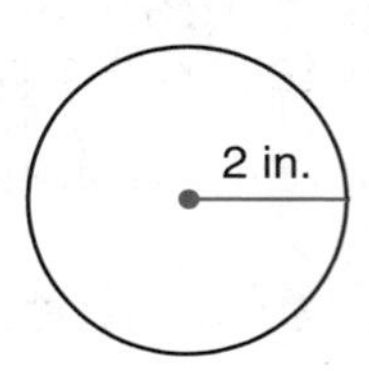

The ***radius*** *of this circle is 2 inches.*

range The difference between the largest number and the smallest number in a list.

5, 17, 12, 34, 29, 13

To calculate the ***range*** *of this list, we subtract the smallest number from the largest number. The* ***range*** *of this list is 29.*

ratio A comparison of two numbers by division.

There are 3 triangles and 5 stars. The ***ratio*** *of triangles to stars is "three to five," or* $\frac{3}{5}$.

ray A part of a line that begins at a point and continues without end in one direction.

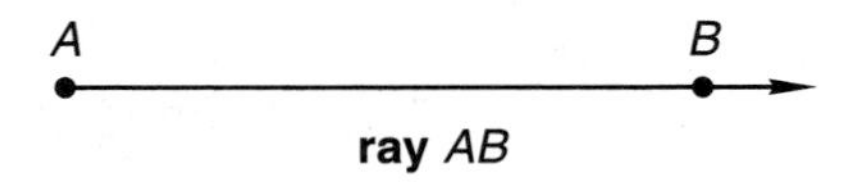

ray *AB*

reciprocals Two numbers whose product is 1.

$$\frac{3}{4} \times \frac{4}{3} = \frac{12}{12} = 1$$

Thus, the fractions $\frac{3}{4}$ and $\frac{4}{3}$ are ***reciprocals.*** *The* ***reciprocal*** *of $\frac{3}{4}$ is $\frac{4}{3}$.*

rectangle A quadrilateral that has four right angles.

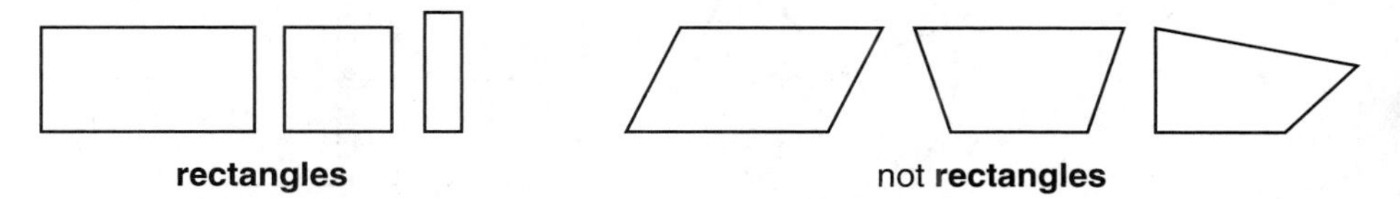

rectangles not **rectangles**

rectangular solid A three-dimensional solid having six rectangular faces. Adjacent faces are perpendicular and opposite faces are parallel.

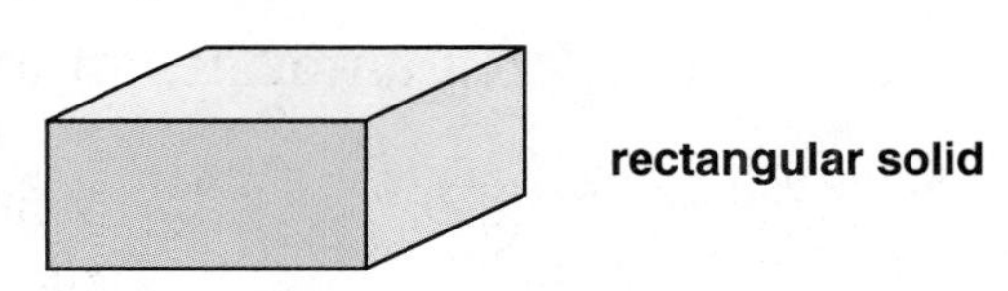

rectangular solid

reduce To rewrite a fraction in lowest terms.

If we ***reduce*** *the fraction $\frac{9}{12}$, we get $\frac{3}{4}$.*

reflection Flipping a figure to produce a mirror image.

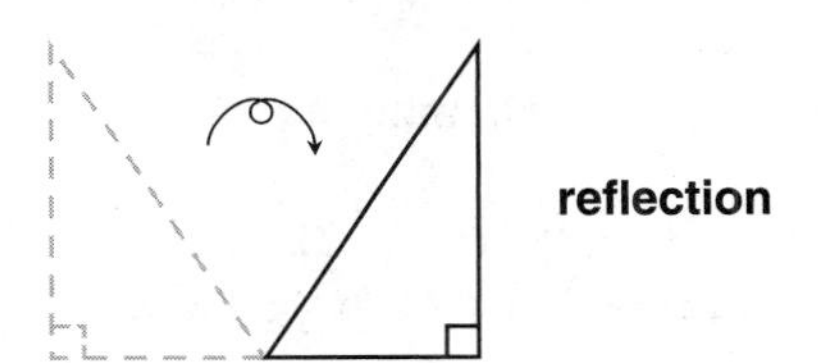

reflection

reflective symmetry A figure has *reflective symmetry* if it can be divided into two halves that are mirror images of each other. *See also* **line of symmetry.**

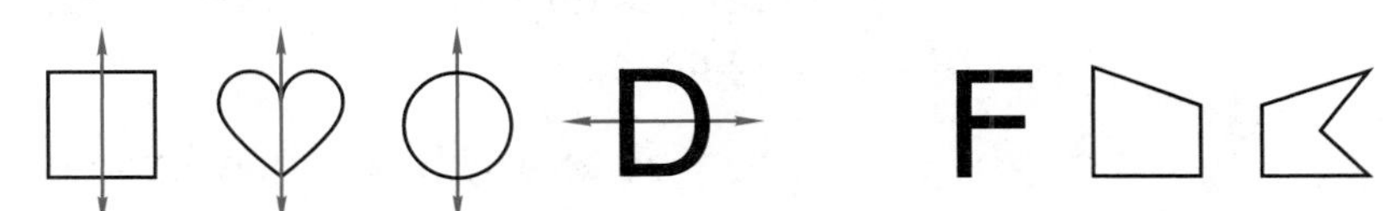

These figures have **reflective symmetry.** These figures do not have **reflective symmetry.**

regular polygon A polygon in which all sides have equal lengths and all angles have equal measures.

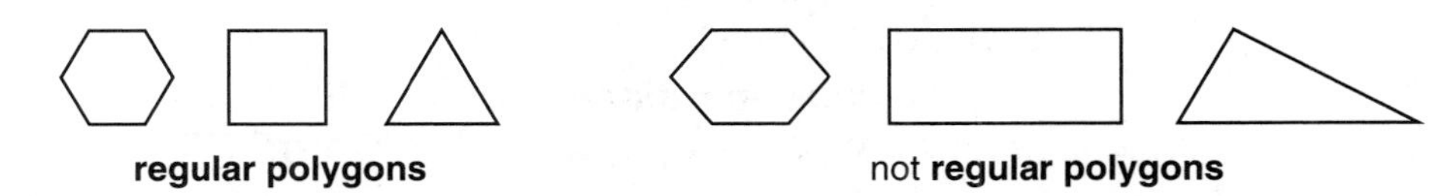

regular polygons not **regular polygons**

relative frequency table A frequency table in which the frequencies for all categories are displayed as the numerator of a fraction with the total number of outcomes as the denominator.

Outcome	Tally	Relative Frequency
1	𝍸 𝍸 𝍸 ||	$\frac{17}{50}$
2	𝍸 𝍸 𝍸 𝍸 𝍸 |||	$\frac{28}{50}$
3	𝍸	$\frac{5}{50}$

*This **relative frequency table** shows data obtained by spinning the spinner at left 50 times.*

remainder An amount left after division.

$$\begin{array}{r} 7\ R\ 1 \\ 2\overline{)15} \\ \underline{14} \\ 1 \end{array}$$

*When 15 is divided by 2, there is a **remainder** of 1.*

rhombus A parallelogram with all four sides of equal length.

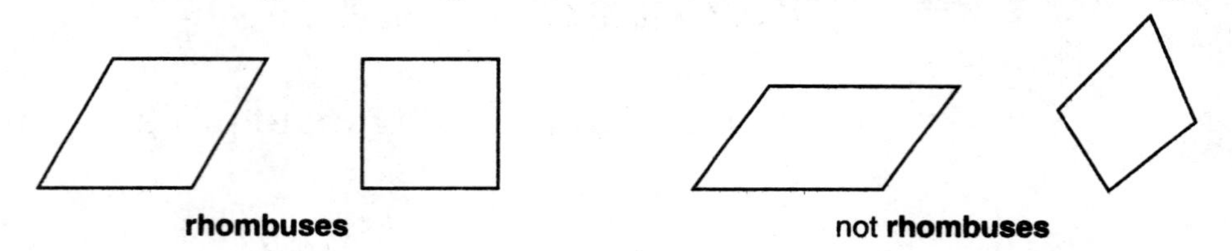

right angle An angle that forms a square corner and measures 90°. It is often marked with a small square.

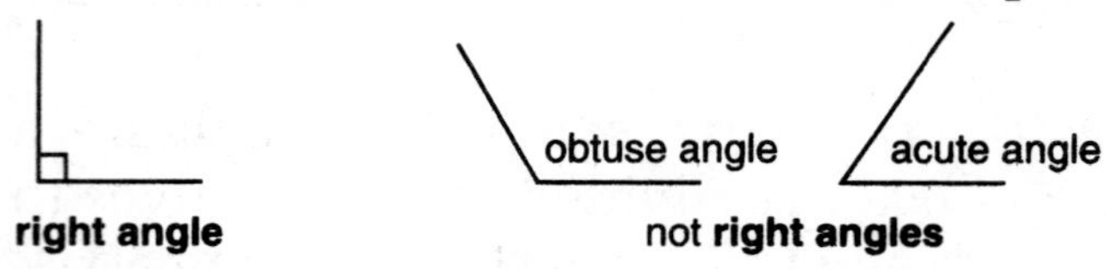

*A **right angle** is larger than an acute angle and smaller than an obtuse angle.*

right triangle A triangle whose largest angle measures 90°.

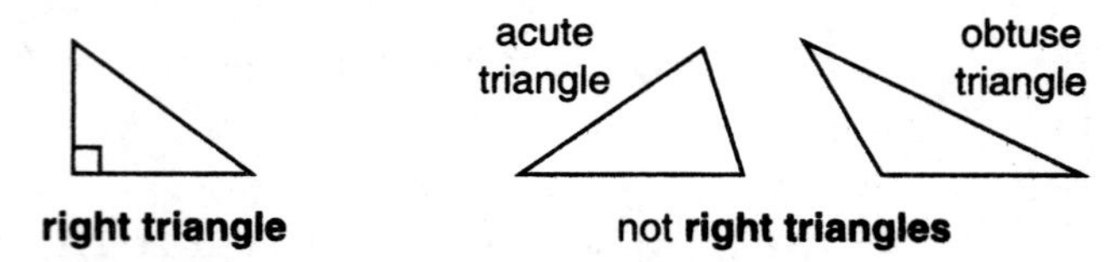

Roman numerals Symbols used by the ancient Romans to write numbers.

*The **Roman numeral** for 3 is III.*

*The **Roman numeral** for 13 is XIII.*

rotation Turning a figure about a specified point called the *center of rotation.*

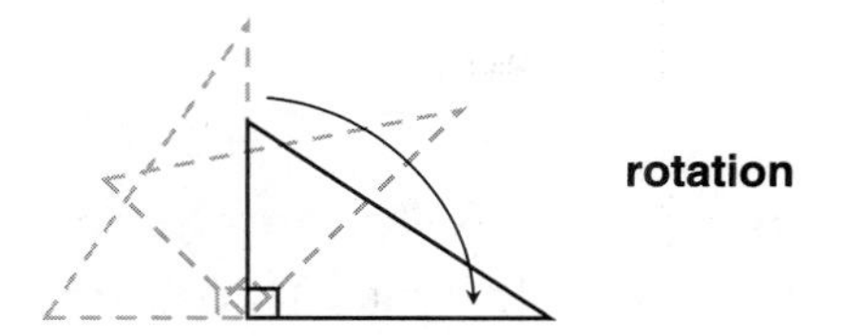

rotational symmetry A figure has *rotational symmetry* if it can be rotated less than a full turn and appear in its original orientation.

These figures have **rotational symmetry.** These figures do not have **rotational symmetry.**

round To express a calculation or measure to a specific degree of accuracy.

To find about how many hundred feet make a mile, we ***round*** *5280 feet to 5300 feet.*

round number A whole number that ends with one or more zeroes.

100, 20, 570, and 6380 are ***round numbers.***

101, 30.2, 573, and 6384 are not ***round numbers.***

scale (1) A type of number line used for measuring.

cm 1 2 3 4 5 6 7

The distance between each mark on this ruler's ***scale*** *is 1 centimeter.*

(2) A ratio that shows the relationship between a scale model and the actual object.

If a model airplane is $\frac{1}{24}$ the size of the actual airplane, the ***scale*** *of the model is 1 to 24.*

scale drawing A two-dimensional representation of a larger or smaller object.

Blueprints and maps are examples of ***scale drawings.***

scale model A three-dimensional rendering of a larger or smaller object.

Globes and model airplanes are examples of ***scale models.***

scalene triangle A triangle with three sides of different lengths.

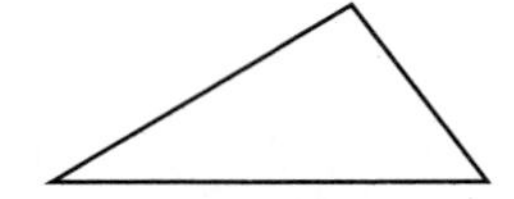

*All three sides of this **scalene triangle** have different lengths.*

schedule A list of events organized by the times at which they are planned to occur.

Sarah's Class **Schedule**	
8:15 a.m.	Homeroom
9:00 a.m.	Science
10:15 a.m.	Reading
11:30 a.m.	Lunch and Recess
12:15 p.m.	Math
1:30 p.m.	English
2:45 p.m.	Art and Music
3:30 p.m.	End of School

sector A region bordered by part of a circle and two radii.

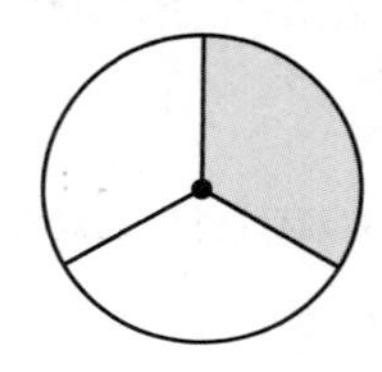

*This circle is divided into 3 **sectors.***

segment *See* **line segment.**

sequence A list of numbers arranged according to a certain rule.

*The numbers 2, 4, 6, 8, ... form a **sequence.** The rule is "count up by twos."*

side A line segment that is part of a polygon.

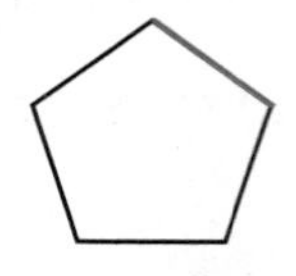

*This pentagon has 5 **sides.***

similar Having the same shape but not necessarily the same size. Similar figures are proportional.

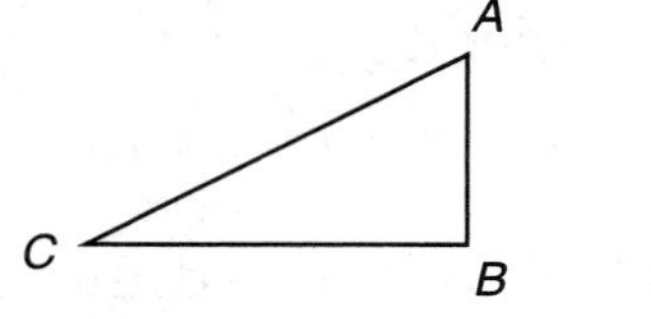

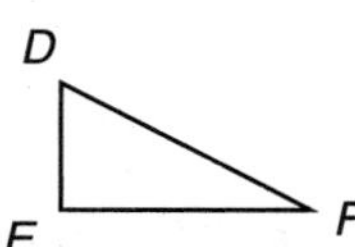

*$\triangle ABC$ and $\triangle DEF$ are **similar.** They have the same shape, but not the same size.*

solid *See* **geometric solid.**

sphere A round geometric solid having every point on its surface at an equal distance from its center.

sphere

spread A value that describes how the data in a set are distributed. *See also* **range.**

5, 12, 3, 20, 15

The range of this set is 17. Range, which is the difference between the greatest and least numbers, is one measure of the ***spread*** *of data.*

square (1) A rectangle with all four sides of equal length.

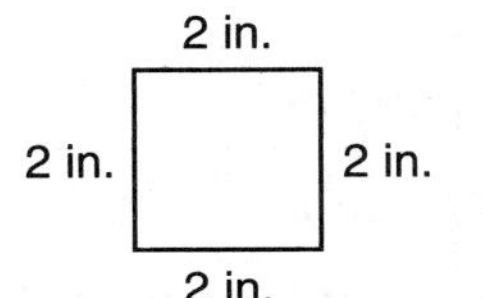

All four sides of this ***square*** *are 2 inches long.*

(2) The product of a number and itself.

The ***square*** *of 4 is 16.*

square root One of two equal factors of a number. The symbol for the principal, or positive, square root of a number is $\sqrt{\ }$.

A ***square root*** *of 49 is 7 because* $7 \times 7 = 49$.

$$\sqrt{49} = 7$$

statistics A branch of mathematics that deals with the collection, analysis, organization, and display of numerical data.

Some activities performed in ***statistics*** *are taking surveys and organizing data.*

stem-and-leaf plot A method of graphing a collection of numbers by placing the "stem" digits (or initial digits) in one column and the "leaf" digits (or remaining digits) out to the right.

Stem	Leaf
2	1 3 5 6 6 8
3	0 0 2 2 4 5 6 6 8 9
4	0 0 1 1 1 2 3 3 5 7 7 8
5	0 1 1 2 3 5 8

In this ***stem-and-leaf plot,*** *3|2 represents 32.*

straight angle An angle that measures 180° and thus forms a straight line.

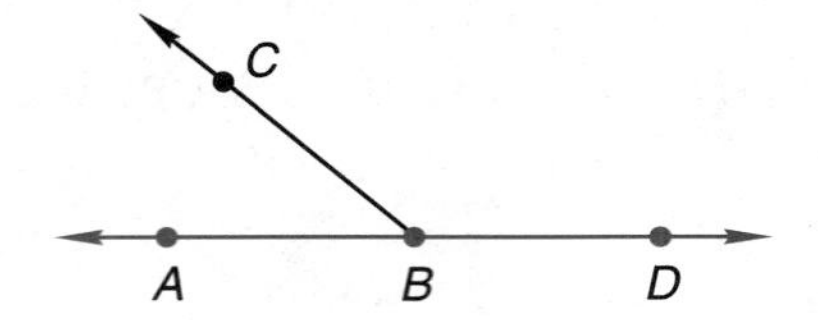

*Angle ABD is a **straight angle.** Angles ABC and CBD are not **straight angles.***

sum The result of addition.

7 + 6 = 13 *The **sum** of 7 and 6 is 13.*

tally mark A small mark used to help keep track of a count.

*I used **tally marks** to count the goals scored. I counted seven goals.*

term (1) A number that serves as a numerator or denominator of a fraction.

$\frac{5}{6}$ terms

(2) A number in a sequence.

1, 3, 5, 7, 9, 11, ...

*Each number in this sequence is a **term.***

tessellation The repeated use of shapes to fill a flat surface without gaps or overlaps.

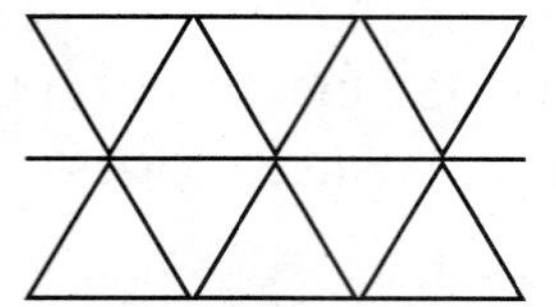

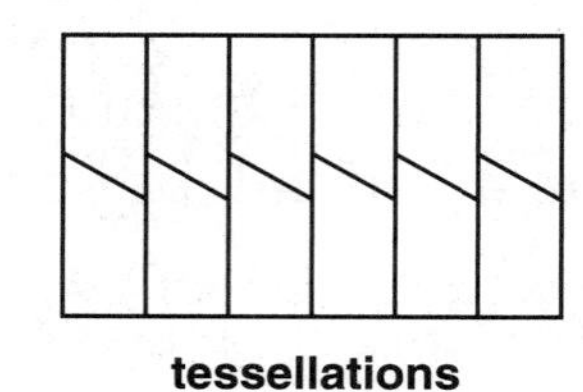

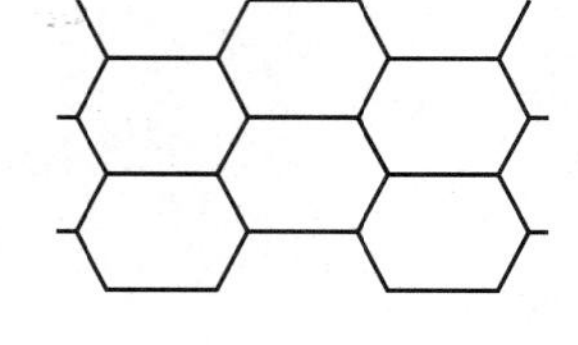

tessellations

transformation The changing of a figure's position through rotation, reflection, or translation.

Transformations

Movement	Name
flip	reflection
slide	translation
turn	rotation

translation Sliding a figure from one position to another without turning or flipping the figure.

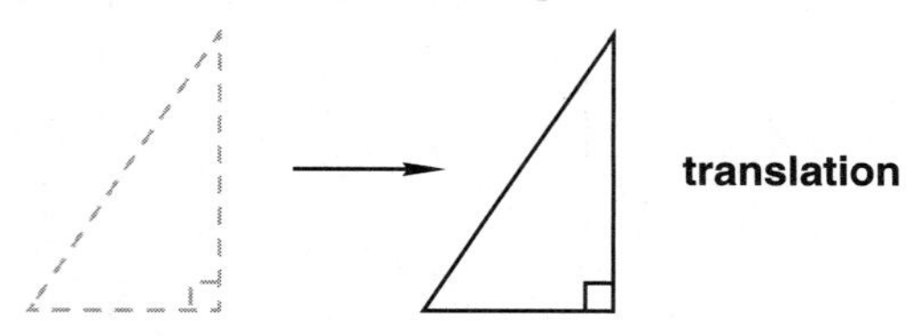

trapezium A quadrilateral with no parallel sides.

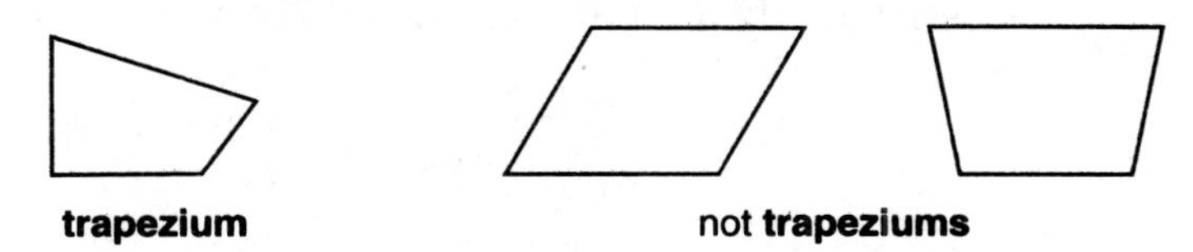

trapezoid A quadrilateral with exactly one pair of parallel sides.

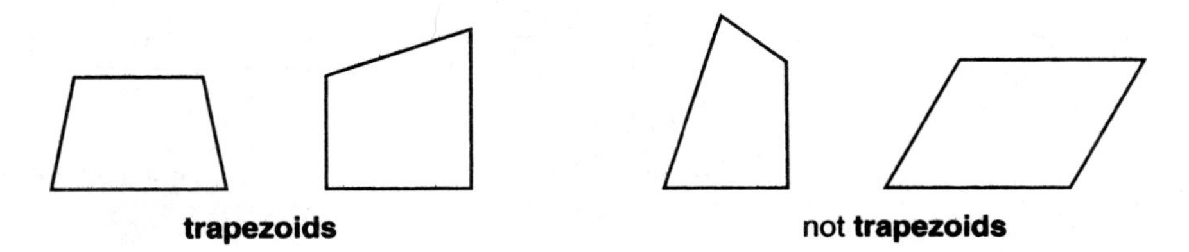

triangle A polygon with three sides and three angles.

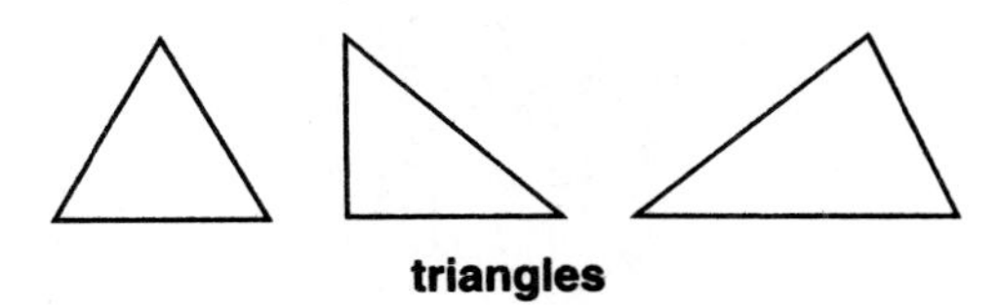

triangular numbers Numbers that can be represented by objects arranged in a triangular pattern.

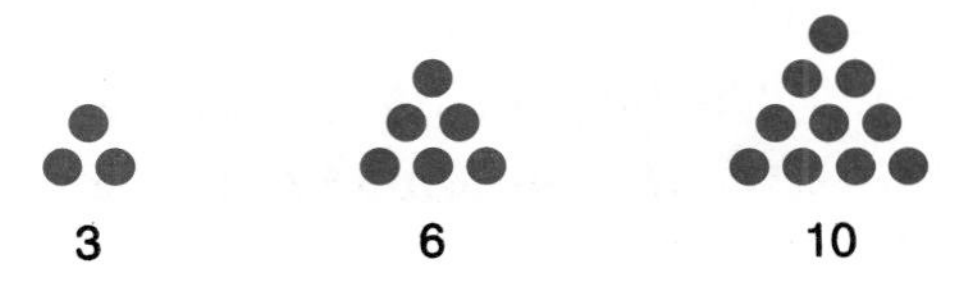

__Triangular numbers__ include all the numbers in this sequence:

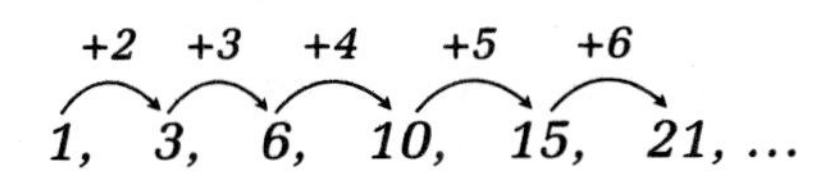

U.S. Customary System A system of measurement commonly used in the United States.

Pounds, quarts, and feet are units in the __U.S. Customary System.__

Venn diagram A diagram made of overlapping circles used to display data.

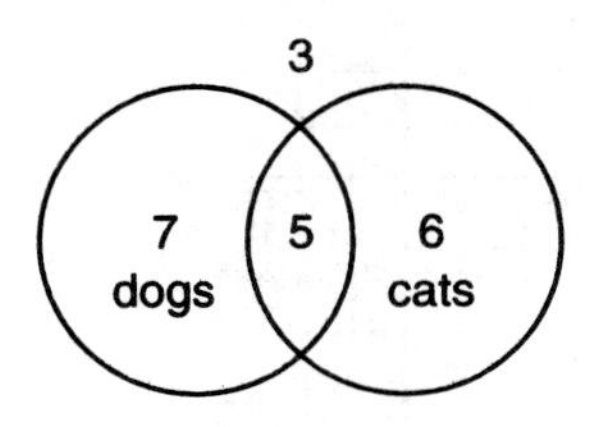

This __Venn diagram__ shows data on the students' pets. Three students do not have a cat or a dog. Seven students have a dog, but not a cat. Six students have a cat, but not a dog. And five students have both a dog and a cat.

vertex (Plural: *vertices*) A point of an angle, polygon, or solid where two or more lines or line segments meet.

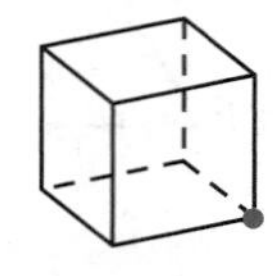

*One **vertex** of this cube is colored. A cube has eight **vertices.***

vertical Upright; perpendicular to horizontal.

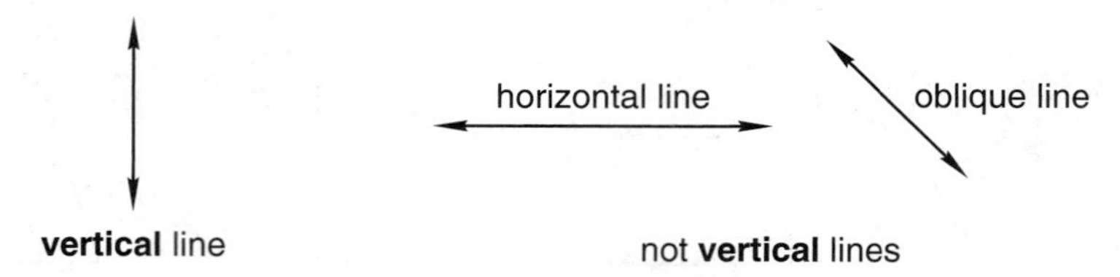

volume The amount of space a geometric solid occupies. Volume is measured in cubic units.

*This rectangular prism is 3 units wide, 3 units high, and 4 units deep. Its **volume** is 3 · 3 · 4 = 36 cubic units.*

whole numbers All the numbers in this sequence: 0, 1, 2, 3, 4, 5, 6, 7, 8, 9, ...

*The number 35 is a **whole number,** but $35\frac{1}{2}$ and 4.2 are not. **Whole numbers** are the counting numbers and zero.*

x-axis The horizontal number line of a coordinate plane.

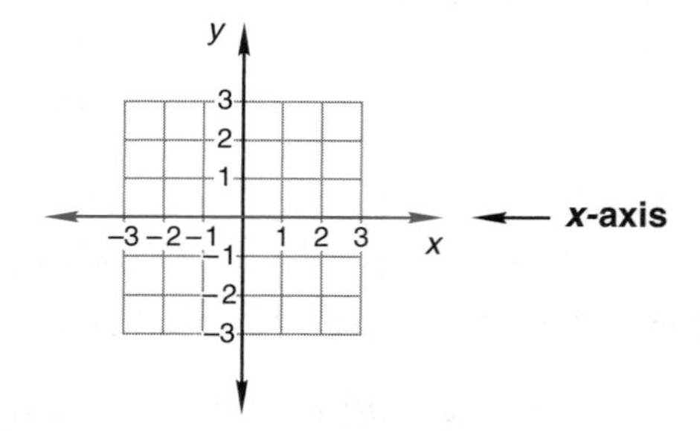

y-axis The vertical number line of a coordinate plane.

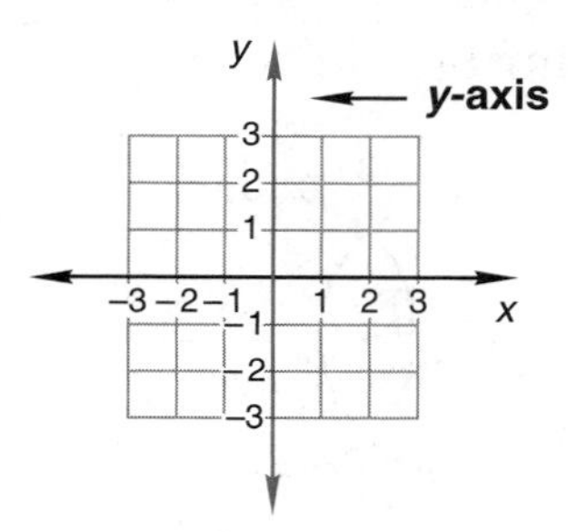

INDEX

Note: Asterisks (*) indicate that the cited topic is covered in a lesson Warm-up.
Page locators followed by the letter "n" are references to footnotes on the indicated pages.

A

D

G

K

L

M

N

O

U

V

W

X

Y

Z